This book provides the first coherent and comprehensive treatment of the thermodynamics and gas dynamics of the practical Stirling cycle. Invented in 1816, the Stirling engine is the subject of world-wide research and development on account of unique qualities – silence, indifference to heat source, low level of emissions when burning conventional fuels, and an ability to function in reverse as heat pump or refrigerator.

The working cycle embraces all aspects of engineering thermodynamics, heat transfer (conduction and convection) and unsteady, compressible fluid flow. A wide-ranging problem in thermodynamic design is reduced to manageable proportions through application of the key principle of *dynamic similarity*. This draws together and unifies a vast literature on the central issue of the thermal regenerator, permits a unified approach to computer simulation of the gas processes and allows reduction of a substantial problem in gas circuit design to the use of simple charts. Performance optimization is also treated. Comprehensive appendices include conversion factors and tables of the properties of gases and metals required for the design of practical gas circuits.

The student of engineering will discover an instructive and illuminating case study revealing the interactions of basic disciplines. The researcher will find the groundwork prepared for any type of computer simulation that is contemplated. Those involved in the use and teaching of solution methods for unsteady gas dynamics problems will find a comprehensive treatment of nonlinear (Method of Characteristics) and linear wave approaches, for the Stirling machine provides an elegant example of the application of each. The book will be indispensable for all those involved in researching, designing or manufacturing Stirling prime movers, coolers and related regenerative thermal machines.

# THERMODYNAMICS AND GAS DYNAMICS OF THE STIRLING CYCLE MACHINE

# THERMODYNAMICS AND GAS DYNAMICS OF THE STIRLING CYCLE MACHINE

ALLAN J. ORGAN

*Engineering Department, University of Cambridge*

CAMBRIDGE
UNIVERSITY PRESS

CAMBRIDGE UNIVERSITY PRESS
Cambridge, New York, Melbourne, Madrid, Cape Town, Singapore,
São Paulo, Delhi, Dubai, Tokyo

Cambridge University Press
The Edinburgh Building, Cambridge CB2 8RU, UK

Published in the United States of America by Cambridge University Press, New York

www.cambridge.org
Information on this title: www.cambridge.org/9780521131797



First published 1992
This digitally printed version 2010

*A catalogue record for this publication is available from the British Library*

*Library of Congress Cataloguing in Publication data*
Organ, Allan J.
Thermodynamics and gas dynamics of the Stirling cycle machine/Allan J. Organ.
p. cm.
Includes bibliographical references.
ISBN 0-521-41363-X
1. Stirling engines—Design and construction. 2. Thermodynamics.
I. Title.
TJ765.074 1992
621.4′2—dc20 91-29146 CIP

ISBN 978-0-521-41363-3 Hardback
ISBN 978-0-521-13179-7 Paperback

This book is dedicated to the memory of my parents,
Margaret and Albert Allan

# Contents

# Foreword

by Theodore Finkelstein

Much progress has been made in the technology and understanding of Stirling cycle machines in the 40 or so years during which I have been active in Stirling machine research and development. Although series production is currently limited to special purpose machines, there is hope that the type may soon be in common use, as witnessed by the large-scale efforts by research and development groups in the USA, Japan and various European countries. As a consequence of these efforts, sponsored both by government and private industry, there has in recent years also been a surge of interest worldwide in theoretical aspects.

Stirling cycle machines present a curious paradox in that they are at the same time well known and unknown. They are well known because they have existed as engines for longer than most other engines in common use, and as heat pumps and refrigerators for about as long as current systems. Furthermore, there is no lack of literature references to the general – as well as to the detailed – aspects of Stirling technology. Yet the general knowledge and understanding of Stirling engines is still at such a low level that, even among experts, a wide divergence of opinion can be found, not only as to their basic applicability or desirable constructional features, but even as to the analytical approach appropriate for their design and optimization. This may be verified by anyone who attends one of the specialized meetings on the subject.

This dichotomy is the reason that, over the years, unrealistic applications have been proposed for Stirling machines, and prototypes have been constructed whose performance frequently fell short of theoretical predictions. Thus, the absence of commercial success in spite of evident advantages may possibly be correlated to the lack of adequate modelling techniques and of sound theoretical predictions of what these machines can actually accomplish.

An up-to-date, authoritative and comprehensive treatment of the thermodynamic design of Stirling machines has therefore been overdue for a long time, and now it has finally arrived. The author of this specialized text-book is one eminently qualified for the task: Dr Organ is one of the leading theoreticians worldwide in this field, and his thorough and extensive treatise has been eagerly awaited by his co-workers in the subject. I have been privileged to know and exchange ideas with my friend and co-worker for a long time. Our relationship spans a quarter of a century, or longer if one counts correspondence conducted between us when he was working on Stirling machines in São Paulo, Brazil. Since then Dr Organ has returned to England and has become known as one of the few scientists worldwide who have devoted most of their working lives to the study of this area of scientific engineering. I am proud to say that since then he and I have maintained a two-man mutual help society engendered by a commonality of purpose, and agreement with each other's analytical, theoretical and practical approaches to Stirling cycle machines.

Dr Organ is one of the foremost innovators in this speciality, having originated several advanced concepts, including the analysis of the fluid circuit in terms not of *Eulerian* coordinates (grid boundaries fixed in the space domain) but of *Lagrange* coordinates (grid boundaries movable in the space domain). It was he who first pointed out that the finite difference methods applied to fixed boundary elements in Stirling machine analysis are unable to deal with propagating discontinuities. He has been a pioneer in the application of the Method of Characteristics and in the use of linear wave analysis. His unorthodox techniques have not been widely understood when first published. His contributions to the field have been so advanced that comparatively few co-workers have adopted his techniques at this time, and the publication of this book will undoubtedly make use of his advanced analytical approaches more widespread.

Since the work is based on advanced principles of analysis and fluid dynamics, one might easily miss the central message: a very significant question to ask is 'what should be the relative importance accorded in the first place to accurate modelling of a thermodynamic cycle, or in the second place to making an interpretation of the results of such modelling work so that it may be usable in machine design?' The development of Stirling cycle machines must necessarily rest upon design procedures that can be presented in a simple and easily digested form. Thus, the researcher should have available to him such design tools as Z-charts, multi-variate graphs and optimization diagrams which cover a wide range of design parameters and which can easily be applied to a candidate design. These distilled design data must all be presented in normalized form involving non-dimensional groupings of parameters. An examination of the text will clearly show that its ultimate aim is the achievement of design data in this form.

The reader will readily appreciate that, in order to compare different mechanizations adequately one has to define a valid normalized criterion, referred to in this text as *specific cycle work*, applicable universally to elementary, Schmidt-type analyses, through advanced treatments to computerized 'number-crunching' solutions of the appropriate partial differential equations. This abstraction is essential for the purpose of making the simulation results accessible to the design engineer.

Dr Organ's investigational philosophy is best summed up in his own words: In a private communication to me he has stated (sic):

> . . . my view of cycle analysis and simulation is, and always has been, that it is a tool to be developed, tested, placed at the service of the designer and then forgotten so that we may get on with the problems of layout, thermal stress, wear and the production of specimen engines . . . the top priority is simplification of the approach to thermodynamic and flow design – the ultimate being elimination of the requirement for the computer in the process of designing a specific machine . . .

The gist of this approach is that the ultimate aim is not an elaborate computer code, however elegant and sophisticated it may be in terms of physical modelling techniques or 'structured' programming, but the end result – concise and reliable, simple design charts. These tools must be based on individually computed points, which may obscure the fact that a comprehensive 'number-crunching' capacity was utilized for individual data points. Electronic computer studies can be made to range over any number of parametric variations, and it is ultimately not the computer study but the resulting design chart which is the usable end product. The argument is eloquently made in this textbook and is supported by sound theoretical deductions and examples.

This overall strategy is by no means in conflict with the fact that the author has delved as deeply into the black art of computer programming for Stirling cycle machines as any other person known to me. The painstaking quality of Dr Organ's work is evident from even a cursory review of any section of the book. The accuracy of the material presented, combined with the diversity of carefully planned and always graceful illustrations make this book a pleasure to look through and a valuable addition to any reference library. It is refreshing to note that, in spite of the weighty subject the book is neither pedantic nor pompous, but that the wording has originality and even lightheartedness in places and is a pleasure to read.

This text may be used at two levels: A first use is for the experienced practitioner in the field who may concentrate on one particular topic, and thus study the section devoted to it without reference to the other parts, since one of the outstanding merits of this book is that most parts segregated by subject matter are written in a stand-alone fashion. A second use will be for the student wishing to acquire an understanding, and a working knowledge of most of the practical approaches to the thermodynamic analysis of Stirling cycle machines. Thus this book should be suitable for reference as well as for a text of an advanced academic course. It is confidently expected to become a standard textbook and reference work on this topic.

The thermodynamic analysis of Stirling cycle machines is treated on the assumption that the reader has adequate background in the mathematical methods required, depending on the specific chapter. In some cases this includes such advanced topics as partial differential equations or matrix operations. Also presupposed is the fact that most serious applications of the theories presented require the use of up-to-date computers.

The text places into perspective and comprehensively deals with the main topic implied by the title. To this purpose it is structured so that first the reader is exposed to certain basics, then to progressively more advanced concepts until, towards the end of the book, he is thoroughly familiar with the fundamentals and methods deemed to be practical for the scientific analysis. Dr Organ does not attempt to make the book an all-inclusive reiteration of all that has been published before, unless it contributes to the current 'state of the art'. Thus the main text is not cluttered with lengthy algebraic procedures that are archaic. Instead, the reader is referred to the original sources in the literature. Similarly, the author consigns related, but not directly applicable, portions of the text to appendices, where the smooth flow of the textual presentation might otherwise be impaired. This includes the closed-form solutions first derived by Schmidt, and amplified considerably by other investigators, and which are the basis required for the theoretical reference case only.

A prominent feature of this book is the organized manner in which the theoretical cycle analysis is presented, leading up from the author's basic perception of the available avenues of approach to a rational analysis (Chapter 1) through a historical overview of the development of analytical tools in Chapter 2. The latter is not merely a recitation of scientific papers in the order in which they appeared, but a critical evaluation of the scientific foundation upon which they were based with reference to the conservation laws which were obeyed or violated in certain instances by previous investigators.

I am glad to benefit now from Dr Organ's in-depth research and shall improve my own analysis by utilizing his formulations. It is my sincere hope that the publication of this book will also make it easier for other workers in this field to optimize Stirling cycle machine proportions based on the improved analytical techniques here presented. This should ultimately be reflected in the emergence of a new generation of advanced and improved types of Stirling cycle machine.

Los Angeles *Theodore Finkelstein*

# Preface

An internationally-recognized authority on Stirling cycle machines† is on record as affirming that

It is clear that for these engines the principal impediment to success is not theoretical analysis, but practical design and application . . .

Against this claim, it is common experience that the simple analyses (Schmidt, isothermal, adiabatic, basic finite cell) are useless for *ab initio* thermodynamic design, that the acquisition of design know-how of proven worth in the form of licences is prohibitively costly, and that the work of developing suitable analytical and computational tools from scratch is monumental.

If Stirling cycle machines are to become a commercial reality there is clearly an urgent need for a raising of the threshold from which the work of thermodynamic design may start. It is accordingly a purpose of this text to provide the design methods which have so far been lacking. There is first of all (Chap. 2) a review – and, more importantly, a classification – of existing analytical methods. This will serve to introduce the terminology and some essential concepts. It will also highlight a number of anomalies between the way certain principles of thermodynamics and gas dynamics are used in the context of Stirling machines and elsewhere.

Chap. 3 will establish a proper ideal reference cycle in place of the thermodynamically inconsistent reference standard commonly referred to. It will demonstrate that the margin between the efficiency of the ideal cycle and that of the practical counterpart is inevitable rather than the result of inappropriate design.

Chap. 4 demonstrates the benefits of applying to Stirling cycle analysis and design the principles of dynamic similarity. (While these principles are

† Naso V Chairman's introduction, Conference prospectus, 3rd. International Stirling Engine Conference, Rome, June 23–6 1986

regarded as indispensable in most branches of engineering – including the crucially pertinent areas of heat transfer and fluid flow – the benefits of working in terms of characteristic dimensionless groups have clearly not been apparent to the majority of Stirling analysts.) The chapter introduces for the first time the notion of *thermodynamic similarity* for use at the preliminary stages of design.

Crucial to the performance of Stirling machines is the regenerator, so it is ironic that detailed investigation into the inner workings of this component has tended to be a less attractive option than the 'black box' approach. Chap. 6 directs attention to the vast body of work dealing with flow through the wire gauze, very little of which has found its way into the science of Stirling machine design.

Chap. 5 develops an approach to cycle modelling having special relevance to the task of thermodynamic design: formulation in terms of the principles of Chap. 4 permits a single computational run to map the entire performance envelope of a whole class of thermodynamically similar machines. The result is equivalent to tens of thousands of simulations of individual machines operating at conditions specified in terms of absolute variables (rpm, $p_{\text{ref}}$, etc.). Performance is mapped graphically, and the result is a first step towards the thermodynamic design of Stirling machines by sole reference to charts. Chap. 6 extends the treatment to take into account the intricate transient conduction phenomenon internal to individual wires of the regenerator gauze.

A facility essential to any area of design is *scaling*: if it is not possible to create a new machine of arbitrarily increased (or decreased) size from a prototype of known performance, then the work of design and development becomes an inordinately costly and unattractive proposition. The key to scaling lies with dynamic similarity: Chap. 9 identifies the dimensionless combinations of variables which determine similarity of thermodynamic performance between geometrically identical machines of different sizes, and illustrates the rôle of the performance map in the scaling process.

Chap. 10 shows that the lumped-parameter approach to modelling reformulates directly in terms of the same combinations of dimensionless variables. It deals with the problem of specifying realistic coefficients of convective heat transfer for the variable-volume spaces.

Chap. 2 expressed concern at the fact that much computational work in the area of Stirling cycle machines makes use of numerical techniques which are considered unsuitable for solution of the same equations applied to other engineering problems. Chap. 11, with the aid of Appendix XI, introduces the Method of Characteristics, universally recognized as yielding correct numerical solutions to the equations of one-dimensional, unsteady, compressible flow. The chapter applies the method to a searching test – a temperature discontinuity propagating in a heat exchanger passage. Comparison with the analytical solution reveals none of the arbitrary 'numerical diffusion' introduced by *ad hoc* fixed-grid schemes. Examples are given of simulations based on the Method of Characteristics, and the prospects discussed for the routine modelling of Stirling machines by this means.

The argument in favour of rigorous solution of the one-dimensional defining equations is weakened by recognition of the fact that there are respects in which these equations themselves fail to represent the gas processes of interest. The shortcoming is most marked in respect of flow through the regenerator where the assumption of one-dimensional flow cannot be justified on any grounds (Chap. 5). It would be unrealistic to consider looking into individual, local flow patterns with either fixed-grid or Characteristics methods, since several thousand subdivisions of the flow field would be called for, with no means of checking experimentally the realism of assumptions applied within individual computational cells. There is, on the other hand, a simplification which permits at least some features of the flow through the individual gauze to be recognized, and which allows each and every gauze (in a stack of some thousands) to contribute its individual effect to the overall computation. The assumption is that of linearity of pressure wave effects. Chap. 12 looks at the Stirling cycle gas circuit as a linear wave phenomenon and demonstrates the capability of linear methods to deal individually and collectively with an arbitrary number of regenerator gauzes.

In the search for a relatively straightforward integration scheme free of the vexed problem of artificial diffusion, Chap. 13 examines solutions based in Lagrange coordinates. There emerges immediately a gas process description so simple that it is virtually a 'one-line' cycle model. The Lagrange approach is amplified into a comprehensive cycle simulation in the form of a novel 'building bloc' architecture. One of several advantages of this form is the unprecedented scope it affords for checks of correct functioning. The simulation is applied to the Philips MP1002CA machine and comprehensive output obtained and discussed.

The ultimate use of cycle analysis is design of the *optimum* thermodynamic circuit. Given the problems of formulating a straightforward cycle analysis, the difficulties and pitfalls of extending to optimization promise to be numerous. Chap. 14 nevertheless makes a start on this most challenging of all aspects of thermodynamic modelling. Chap. 15 presents an analysis of the deceptively simple hot-air engine. Chap. 16 concludes.

It is pertinent to record that I have previously drafted two complete texts on the thermodynamics of Stirling cycle machines, discarding them wholesale on coming to the realization that they were 'writing for the sake of writing' – mere compendia of thought current at the time. The present treatment differs in a number of respects, most importantly in uniting all the material around the principle of dynamic similarity (and its more flexible derivative, thermodynamic similarity). If unity had been the sole gain, a further attempt at a text would not have been justified. However, there has resulted in addition substantial simplification: that which previously appeared a monumental task of analytical and numerical modelling is now seen to lie within manageable bounds, the limits of which are set in terms of a modest number of characteristic dimensionless groups.

If the would-be designer will study the chapters which follow, he should at least be in a position to start work somewhat ahead of where the author began 28 years ago. In particular, if he takes on board the dynamic and thermodynamic similarity approaches, and calculates values of the characteristic dimensionless parameters for some specimen machine of known and acceptable performance, he may scale to a new design in the confidence that his new machine *has no option* but to deliver the predicted specific performance, since it is thermodynamically identical to its prototype. He must, of course, ensure that the new design does not introduce thermal shorting or seal leakage out of proportion to those of the prototype, and must bear in mind that when a machine is scaled, the quantity whose numerical value remains constant is *specific cycle work*: if this desired value of specific cycle work is attained in the new machine at very low rpm, (as determined by the scaling process) then power output is likely to be unacceptable; if at very high rpm, then inertia loads may be a problem. Guidance on dealing with these and other aspects of the fascinating problem of Stirling machine design is offered in the pages which follow.

My essential acknowledgement must be to my late father, Albert Allan: without the background of his enthusiasm for the toy hot-air engines he cherished as a child I should never have selected as my final year research project at the University of Birmingham a topic entitled 'An Appraisal of the Hot Air Engine'. Supervisors and heads of department who have encouraged and made possible my subsequent work in this area include Professor B. N. Cole, (at that time Reader in Plasticity at the University of Birmingham), Professor F. C. Hooper at the University of Toronto, Professor S. A. Tobias at the University of Birmingham, Professor W. B. Gosney of King's College, London, and Professors W. A. Mair, A. Shercliffe and J. Heyman of my present Department at Cambridge. Those familiar with the work of Dr T. Finkelstein will recognize his imprint on the aims, approach and style of every aspect of this text. My experimental and theoretical work has been generously supported over the years by the Science and Engineering Research Council and by the Procurement Executive of the Ministry of Defence.

The literature review of Chap. 2 and the linear wave treatment of Chap. 12 were originally published in the *Proceedings of the Institution of Mechanical Engineers*, Part C, in 1987 and 1989 respectively, and are reproduced here with the permisson of the Council of the Institution. The Council also gives its approval for the use of all seven figures of Chap. 15.

The linear wave treatment was made possible through considerable assistance from Dr Ann Dowling of Cambridge University Engineering Department.

Reverting briefly to the opening theme, a statement by Glegg† is pertinent:

> Theory is practice. Every device you can invent behaves exactly as the theory of its operation in the context of its surroundings

With allowance for the terse wording, the claim identifies the indispensable rôle of analysis in design: to focus attention on the interaction between the natural laws underlying the behaviour of the physical system. When practical experiment identifies an unexpected feature of performance (frequently a shortfall) it is through manipulation of the symbolic model that the incomplete insight is most readily adjusted and a course of remedial action identified. Any who doubt may care to study developments in one of the most demanding of all areas of design – that of human powered flight. The recent sequence of emotive successes awaited Larrabee's elegant definition of the optimum low-speed propellor in terms of classical airscrew theory.‡

The imperatives of engineering design are to be reckoned with. Expensive disappointment is the price of imagining otherwise.

Cambridge University Engineering Department *AJO*

† Glegg G *A design engineer's pocket book*. Macmillan, London, 1983

‡ Burke J D The Gossamer Condor and Albatross: a case study in aircraft design. Report No. AV-R-80/549, AeroVironment Inc., Pasadena, California 91107, USA, June 1980 (published in the AIAA Professional Study Series)

# Notation

Where possible, lower-case English letter characters represent variables having dimensions. A Greek character tends to denote a dimensionless quantity. Where a suitable Greek character is not available for the purpose, an English character in **bold** may be used.

### English letter symbols

| | | |
|---|---|---|
| $a$ | isentropic acoustic speed = $\sqrt{(\gamma RT)}$ | m/s |
| | constant in equation for van der Waals' gas | $m^6$ Pa/$kg^2$ |
| $a, b$ | elements of 2 × 2 matrix, $a_{11}$, $a_{12}$, $b_{11}$, $b_{12}$ etc. | |
| | constants for general use, as required | |
| $b$ | constant in equation for van der Waals' gas | $m^3$/kg |
| $A_w$ | wetted or exposed area | $m^2$ |
| $A_x$ | free-flow or cross-sectional area | $m^2$ |
| $B$ | Beale number, $W_{brake}/(p_{ref} V_{ref})$ | |
| $c$ | specific heat (of solid) | J/kg K |
| $c_v$, $c_p$ | specific heats at constant volume and pressure respectively | J/kg K |
| $C_d$ | coefficient of discharge | |
| $C_f$ | friction factor, $\tau_{wall}/\frac{1}{2}\rho u^2$ | |
| $d$ | diameter | m |
| $d_w$ | diameter of individual wire of regenerator matrix | m |
| D | substantial derivative | |
| $D$ | diameter, nominal diameter of displacer | m |
| $D^*$ | effective diameter | m |
| $E$ | Euler number, 0.577215 | |

| | | |
|---|---|---|
| $f(\,), g(\,)$ | functions defining left- and right-running waves | Pa |
| $f_{ve}, f_{vc}$ | variable volume – $f_{ve}(\phi), f_{vc}(\phi)$ made dimensionless by dividing through by respective amplitude, $V_E$, $V_C$, so that values lie in range 0.0–1.0 | |
| $F$ | force | N |
| | friction parameter $(\frac{1}{2}C_f u^2/r_h)u/\lvert u\rvert$ | $m/s^2$ |
| $F$, $G$ | complex constant with units of pressure | Pa |
| $\mathbf{F}_e$ | dimensionless friction factor defined in connection with lumped-parameter formulation | |
| $g$ | acceleration due to gravity | $m/s^2$ |
| $G$ | mean mass velocity, $\rho u$ | $kg/m^2$ s |
| $h$ | coefficient of convective heat transfer | $W/m^2$ K |
| $h$ | specific enthalpy | J/kg K |
| $H$ | enthalpy | J/K |
| $\mathbf{H}$ | $N_{ST} \cdot d\phi/2\pi$, e.g., $\mathbf{H}_{he} = N_{ST_{he}}\, d\phi/2\pi$ (expansion exchanger) | |
| $i$ | square root of negative unity | |
| $k$ | (dimensional) discharge coefficient defined by Finkelstein | 1/m s |
| | thermal conductivity of gas | J/s m K |
| | permeability coefficient defined by Darcy's law | $m^2$ |
| $k_r$ | thermal conductivity of individual, solid element of regenerator | J/s m K |
| $l$ | length | m |
| $L$ | a fixed length – nominal distance between opposed faces of expansion and compression pistons in equivalent 'one-dimensional' Stirling cycle machine | m |
| $L_j$ | length of $j$th flow passage element | m |
| $L_d$ | nominal length of displacer | m |
| $m$ | mass | kg |
| $m'$ | mass flow rate | kg/s |
| $M$ | mass of working fluid taking part in cycle | kg |
| $N$ | revolutions per minute | 1/min |
| $N_c$ | Courant number, $u\Delta t/\Delta x$ | |
| $N_f$ | Fourier modulus, $\alpha\Delta t/\Delta x^2$ | |

| | | |
|---|---|---|
| $N_{nu}$ | local, instantaneous Nusselt number, $hd/k$ | |
| $N_{ma}$ | local, instantaneous Mach number, $u/a$ | |
| $N_{pr}$ | local, instantaneous Prandtl number, $\mu c_p/k$ | |
| $N_{re}$ | local, instantaneous Reynolds number, $4\rho u r_h/\mu$ | |
| $N_{re}{}^*$ | a Reynolds number defined in Chapter 5 | |
| $N_{red}$ | local, instantaneous Reynolds number based on diameter, $\rho u d/\mu$ | |
| $N_s$ | revolutions per second | 1/s |
| $N_{st}$ | Stanton number, $h/c_p G$ | |
| $N_C$ | characteristic Courant number, $\Delta\phi/\Delta\lambda$ | |
| $N_F$ | characteristic Fourier number, $\alpha/L_{ref}{}^2 N_s$ | |
| $N_H$ | heat flux number $= p_{ref} L_{ref} N_s R_H/K_r T_{ref}$ | |
| $N_{MA}$ | characteristic pressure coefficient or Mach number, $N_s L_{ref}/\sqrt{(RT_{ref})}$ | |

*NB* In a review paper reported extensively in Chapter 2 the symbol $N_F$ was used for $N_{MA}$, and the definition included constant terms, viz, $N_F = 2\pi N_s L_{ref}/\sqrt{(2RT_{ref})}$. The definition of Reynolds number was also different in the paper in question, *viz*: $N_{RE} = 8\pi p_{ref} L^2{}_{ref} N_s/\mu_{ref} R T_{ref}$. Thus, $N_F \approx 2\pi N_{MA}/\sqrt{2}$ and $N_{RE}$(Chap. 2) $\approx 8\pi N_{RE}$

| | | |
|---|---|---|
| $N_{RE}$ | characteristic Reynolds number, $p_{ref} L^2{}_{ref} N_s/\mu_{ref} R T_{ref}$ | |
| $N_{SG}$ | Stirling number, $p_{ref}/N_s\mu = N_{RE}/N_{MA}{}^2$ | |
| $N_{ST}$ | a reference Stanton number, $hA_{ref}/c_p M N_s$ | |
| $N_{TC}$ | thermal capacity ratio $= T_{ref}\rho_r C_r/p_{ref}$ | |
| $N_\alpha$ | diffusivity number $= N_s L_{ref}{}^2 \rho_r C_r/k_r$ | |
| $N_\tau$ | rationalized temperature ratio, $T_e/T_c$ | |
| $n$ | polytropic index | |
| | dimensionless number | |
| | number of (items) | |
| $p$ | absolute pressure | Pa, N/m$^2$ |
| $P$ | wetted perimeter | m |
| $q^*$ | rate of heat exchange per unit mass | J/kg s |
| $q'$ | rate of heat exchanged | J/s or W |
| $q_L{}'$ | rate of heat exchange per unit length of duct | J/m s |
| $Q$ | heat exchanged | J |
| $r$ | crank pin offset | m |
| $r_h$ | hydraulic radius, $A_x/P$ | m |
| $R_H$ | characteristic hydraulic radius, $V_{ref}/A_{ref}$ | m |

| | | |
|---|---|---|
| $\mathbf{r}$ | compression ratio, $V_{min}/V_{max}$ (=inverse of expansion ratio, ε) | |
| $\mathbf{r}^*$ | a ratio of volumes in the ideal cycle in which piston motion is discontinuous. = (vol. of comp. space at which const. vol. transfer begins)/$V_C$ = $\mathbf{r}(\Sigma\mu_d + 1) - \Sigma\mu_d$ | |
| $R$ | specific gas constant | J/kg K |
| | thermal resistance, $1/hA$ | K/W |
| $\underline{R}$ | universal gas constant | J/kmol K |
| $s$ | specific entropy | J/kg K |
| | stroke (of piston or displacer) | m |
| $s_w$ | length of curve (e.g., of centre line of curved wire element) | m |
| $S$ | entropy | J/K |
| $t$ | independent variable, time | s |
| $t_w$ | radial thickness of wall of cylindrical heat exchanger tube | m |
| $T$ | absolute temperature | K |
| $T_{su}$ | Sutherland temperature | K |
| $u$, $U$ | velocity, 'superficial' velocity in porous materials (Chap. 5) | m/s |
| $\boldsymbol{u}$ | velocity normalized by reference speed $N_s L_{ref}$ or $a_o$) | |
| $U$ | internal energy | J |
| $U(\,)$ | step function defined in text | |
| $v$ | specific volume | $m^3$/kg |
| $V$ | volume | $m^3$ |
| $V_E$ | volume displaced during stroke of expansion piston | $m^3$ |
| $V_C$ | volume displaced during stroke of compression piston | $m^3$ |
| $W$ | work or work/cycle | J |
| $W'$ | work rate | J/s, W |
| $x$, $y$, $z$ | independent variables, length | m |
| $X_{VE}$ | $V_E/r^3$ | |
| $Y$ | regenerator gap width in hot-air engine | m |
| $Y_d$, $Y_p$, $Y_{cl}$ | fixed lengths defined in Appendix V | m |
| $\mathbf{Z}$ | matrix 2 × 2 | |

## Greek characters

| | | |
|---|---|---|
| $\alpha$ | phase angle (rad) between volume variations. Equal to piston phase angle in opposed-piston machine | |
| | thermal diffusivity – $k/\rho c$ | $m^2/s$ |
| $\alpha_c$ | angle between axes of cylinders in V-configuration machine | |
| $\alpha_{¶}$ | ratio of free flow area to frontal area | |
| $\varepsilon$ | expansion ratio, $V_{max}/V_{min}$ $(=1/\mathbf{r})$ | |
| $\varepsilon^*$ | corresponds to $\mathbf{r}^*$. Ratio $V_C$/(vol. of comp. space at which constant volume transfer begins) $= 1/\mathbf{r}^*$ | |
| $\phi$ | crank angle $(=\omega t)$ | |
| $\Phi$ | dissipation function | $s^{-2}$ |
| $\Phi^*$ | dimensionless dissipation function, $\Phi/\omega^2$ | |
| $\gamma$ | ratio of specific heats, $c_p/c_v$ | |
| $\lambda$ | dimensionless length, $x/L$ or $x/L_{ref}$ | |
| | (in coaxial machine) ratio piston swept volume/displacer swept volume | |
| $\lambda_{bs}$ | ratio of bore to stroke | |
| $\lambda_h$ | dimensionless hydraulic radius, $r_h/L_{ref}$ or $r_h/R_H$ | |
| $\eta$ | efficiency (thermal or mechanical depending on subscript) | |
| $\theta$ | dimensionless discharge coefficient | |
| | dimensionless time in Mach plane, $t/(L/a_o)$ | |
| $\kappa$ | volume ratio $V_C/V_E = r_c/r_e$ for machine with equal bores | |
| $\mu$ | ratio of a volume to a reference volume, $V/V_{ref}$ | |
| | coefficient of dynamic viscosity | Pa s |
| $\mu_D$ | normalized dead space, $V_D/V_{ref}$, e.g., $V_D/V_E$ | |
| $\mu_D^{+}$ | 'additional' dead space – amount by which the normalized compression-end dead space of a displacer-type (gamma) machine exceeds that of the otherwise identical coaxial (beta) machine. (Defined in text.) | |
| $\mu_v$ | ratio $V_E/V_{ref}$ $(=1$ if $V_{ref} = V_E)$ | |

| | | |
|---|---|---|
| $v$, $v(N_\tau)$ | reduced dead space $= \sum V_{di}T_{ref}/V_{ref}T_{di}$ | |
| $v^*$, $v^*(N_\tau)$ | approximate expression for $v$, $v(N_\tau)$ employed by some authors $= 2\sum V_D/(V_E(1 + N_\tau))$ | |
| $\pi$ | a constant, 3.141592654 | |
| $\Pi$ | pressure ratio of ideal cycle $(1 + \xi)/(1 - \xi)$ | |
| $\rho$ | density | kg/m³ |
| | $\sigma$, $\sigma'$ dimensionless mass, $m/M$, mass flow rate, $d(m/M)/d\phi$ | |
| $\tau$ | normalized temperature, $T/T_e$ | |
| $\tau$ | characteristic temperature ratio $T_c/T_{ref}$. *Subsequent to the review chapter (Chap. 2) a new characteristic temperature ratio is defined as* $N_\tau = T_e/T_c$. *This new variable is thus the inverse of the characteristic quantity used elsewhere in the literature.* (For reasons, see Chap. 2.) | |
| $\xi$ | $\sqrt{[1/N_\tau^2 + \kappa^2 + (2/N_\tau)\kappa \cos \alpha]}/[1/N_\tau + \kappa + 2v(N_\tau)]$ | |
| $\zeta$ | specific cycle work – $W/p_{ref}V_{ref}$ or $W/MRT_{ref}$ etc. | |
| $\omega$ | angular speed | 1/s |

## Other characters

| | |
|---|---|
| ¶$_v$ | volumetric porosity (e.g. of regenerator matrix) |
| $\wp$ | dimensionless density, $\rho V_{ref}/M$ |

## Subscripts

| | |
|---|---|
| brake | relates to work or power measured at brake |
| c, C | compression space |
| cv | control volume |
| d | dead (unswept) volume |
| ds | dynamic similarity |
| e, E | expansion space |
| gc | gas circuit – equivalent to wfc (below) |
| gen | generated (as in entropy generated) |
| h | relating to heat exchanger – he, expansion exchanger, hc compression |

| | |
|---|---|
| k | kinematics |
| mech | mechanical |
| $n$ | $n$th item of |
| nom | nominal |
| o | optimum value of – e.g., $\kappa_o$, $\alpha_o$ |
| p | relating to pressure |
| r (or reg) | relating to regenerator |
| ref | a reference quantity |
| Sch | relating to Schmidt analysis |
| Su | Sutherland |
| sw | swept |
| th | thermal |
| ts | thermodynamic similarity case |
| w (or wetted) | wetted (as in heat exchanger wetted area) |
| | wall (as in $T_w$, $t_w$) |
| | wire (as in $d_w$) |
| wfc | working fluid circuit |
| $x$, $y$, $z$ | in $x$, $y$, $z$ coordinate direction |
| underline | mean value |

# 1

# The Stirling cycle machine

## 1.1 Origin

The Stirling cycle has its origins in 1816, being first described in a patent[1] granted in that year to a Scottish minister, the Revd Robert Stirling. The history of the patent specification has for a long time held an element of engaging mystery, dispelled only recently in the authoritative account by Hargreaves[2] of the evolution of the modern Stirling cycle machine (which took place largely in the laboratories of NV Philips Gloeilampenfabrieken, Eindhoven, Netherlands).

There has never been doubt about the substance of Stirling's specification which discloses, for the first time in any formal way, the concept of thermal regeneration. At the time of the invention, the caloric theory of heat was current, and against that background the novelty of Stirling's insight is remarkable enough. Even more noteworthy is that the entirely novel engine described to illustrate application of the invention was evidently considered by Stirling to be a logical development of the regenerator principle – a fact sadly neglected[3] by the majority of later innovators. The patent description will be referred to again, and a transcript is included as Appendix I.

It is claimed (see e.g. Ref. 4) that a 2 hp. air engine embodying the principle drove a water pump at an Ayrshire quarry during 1818. Figure 1.1 is Finkelstein's[5] impression of how the engine might have appeared, although it is to be emphasized that specific technical and graphical records are not available.

Finkelstein's scholarly account[3] of the history of the Stirling engine to recent times cannot be improved upon. The interim will therefore not receive treatment here, except to remark that through the historical literature dealing with machines operating on the various regenerative cycles (principally the Stirling and Ericsson) may be traced the evolution of modern engineering thermodynamics from its beginnings in the caloric theory.

Stirling's engine survived into the early part of the twentieth century, mainly

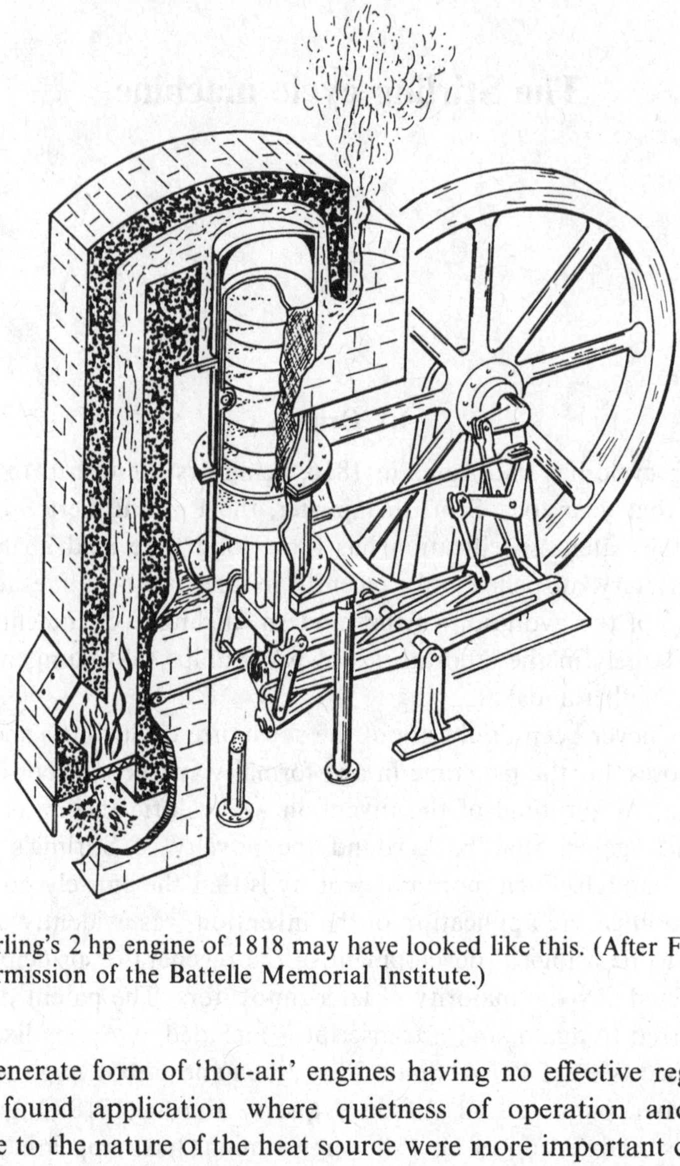

Fig. 1.1 Stirling's 2 hp engine of 1818 may have looked like this. (After Finkelstein[5] with the permission of the Battelle Memorial Institute.)

in the degenerate form of 'hot-air' engines having no effective regenerator. The type found application where quietness of operation and relative indifference to the nature of the heat source were more important considerations than unattractive thermal efficiency and ratio of power-to-weight.

The great disparity between the measured thermodynamic performance of available air engines and that indicated by theory was the motivation for a programme of analytical and development work begun at the laboratories of N.V Philips Gloeilampenfabrieken in 1937.[6] Philips were seeking a quiet prime mover to power electric generators for radio equipment, and preferably one which would not cause interference. By the end of the war, the company had improved power-to-weight ratios by a factor of 50 and power per unit swept volume by about 125.[7] The improvements were attributed to the

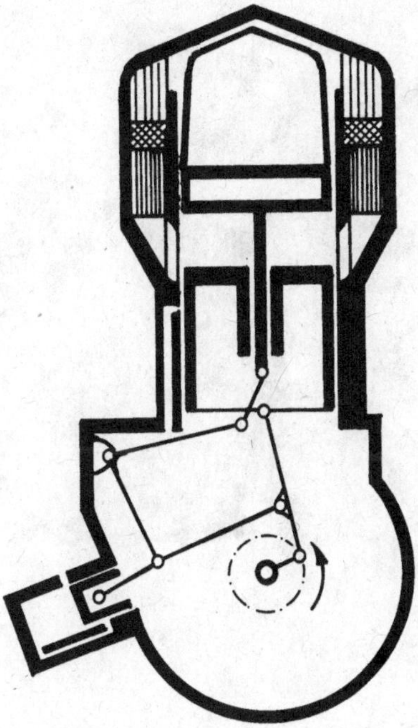

Fig. 1.2 Schematic of air engine of about 1946. (Courtesy NV Philips Gloeilampenfabrieken.)

application of modern materials and of contemporary knowledge of thermodynamics, heat transfer and fluid flow, and, above all, to a recognition of the fundamental importance of an efficient regenerator. Figure 1.2 is a schematic cross-section of Philips' single-acting air engine of 1946, illustrating striking points of similarity with Stirling's original (Fig. 1.1).

## 1.2 The Stirling cycle machine in more recent times

For all the claims for the contribution of contemporary theoretical knowledge, it was some years before analytical material published in support of development work went any further than a cycle analysis which had been described by Schmidt[8] in 1871. When this work was eventually extended (with analytical finesse by Finkelstein[9] and with the aid of numerical techniques by Walker[10] and Kirkley[11]) it was in an attempt to define values for piston phase angle, $\alpha$, and for swept volume ratio, $\kappa$, giving the maximum numerical value of the Schmidt formula for specific, indicated cycle work. Any such optimum is, of course, only as reliable as the expression being optimized. The Schmidt treatment embodies drastic simplifications, and it is thus only by

Fig. 1.3 200 W Stirling powered portable generator set put into batch production by Philips in the 1940s.

chance that a configuration determined to be favourable by experiment bears any relation to that predicted by this early analytical work.

Philips pursued their development of a silent electrical generator to the stage of batch production of the impressively engineered unit illustrated in Fig. 1.3 and rated at 200 W. Launch of the unit coincided with the appearance of the transistor and a consequent dramatic reduction in the power requirements of radio equipment. The company accordingly switched its efforts to the development of larger units for mechanical propulsion, and to reversed cycle machines (refrigerators). Further improvements in specific power and efficiency were rapidly achieved by a change of working fluid from air to hydrogen and helium. The unit shown in Fig. 1.4 is typical of this next generation of engines, and would have had a reported thermal efficiency (not qualified) of some 38% and a specific power of about 100 hp/l.[12] The values corresponded to heater temperatures of 700 K and charge pressures in excess of 100 atm.

The descriptive literature made increasing reference to computer code for thermodynamic design (e.g. Ref. 13). Philips' analytical treatment was never disclosed in any detail, although it is thought to have been a conceptually simple approach embodying empirical factors resulting from extensive practical

Fig. 1.4 Philips 1-98 engine of about 7.5 kW. The rhombic drive crank mechanism was capable of complete dynamic balance. (Courtesy NV Philips Gloeilampenfabrieken.)

experience. Independent analytical work had become increasingly intricate (see classified account in Chap. 2), with emphasis on analysis rather than on design. Indeed, with the exception of claims by Philips which they chose not to substantiate in the open literature, there is to this point no evidence of an explicit method of thermodynamic design for Stirling cycle machines.

The sale by Philips of licences to such organizations as United Stirling (Sweden) and the consortium MAN/MWM in Germany resulted in a new generation of Stirling engines typified by the V-160 unit shown in Fig. 1.5 and the four-cylinder experimental propulsion engine of Fig. 1.6. The

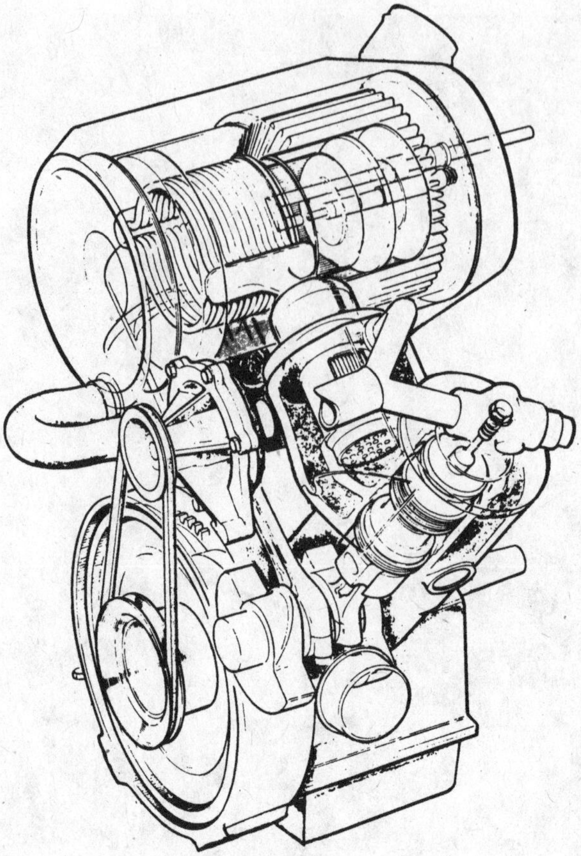

Fig. 1.5 United Stirling V-160 engine: rpm range 1000–3000. Approximately 12 kW maximum continuous output.

independent theoretical literature also burgeoned, with increasingly sophisticated algebra and computer code exemplified by that of Urieli.[14] With the exception of an admirable contribution by Creswick,[15] however, emphasis remained on analysis rather than design. Moreover, theoretical formulations had taken on a structure which discouraged any attempt at the sort of manipulation required by the processes of formal optimization.

The relatively recent entry of Japanese organizations into Stirling engine development[16–18] has resulted in somewhat improved access to the theoretical and experimental methods of organizations with substantial programmes of work and large resources, but in no evidence of an explicit method of thermodynamic design. Current work by the independent researchers is, however, just beginning to show recognition of the need for such a resource (see the classified account in Chap. 2). Despite this, we are as distant as ever

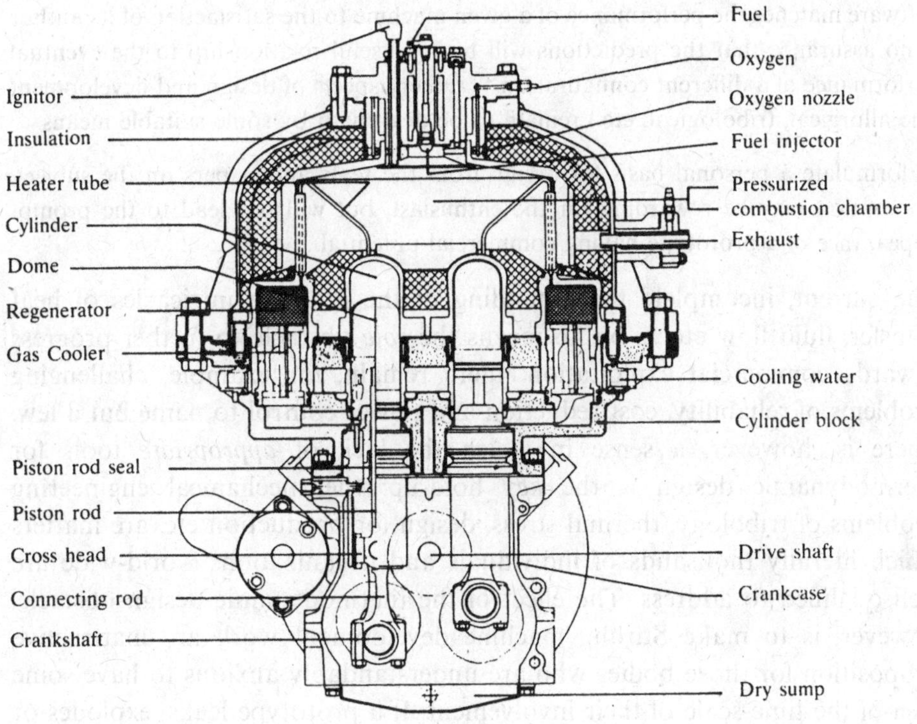


Fig. 1.6 United Stirling 4-95 engine.

from a situation whereby one might sit down and design from scratch a Stirling machine with any assurance that performance will be as intended. What does this mean?

## 1.3 Thermodynamic design – the *status quo*

The individual or organization wishing to embark for the first time on the development of a Stirling cycle machine (prime-mover or cooler) has a choice of starting points:

to take a licence from an established manufacturer. The option is known to be sufficiently costly to deter all but substantial corporate bodies. If he negotiates successfully, the licensee acquires designs and know-how of proven worth, but is then constrained to yesterday's technology until he has made the considerable investment in educating himself to the point where he can innovate.

to acquire proprietary computer analysis software from one of several sources now available. The outlay is orders of magnitude less than that for a licence (there is software which sells for around \$500.00[19]). However, the fact that a given piece of software matches the performance of a given machine to the satisfaction of its author is no assurance that the predictions will bear a useful relationship to the eventual performance of a different configuration.[20] Other aspects of design and development (metallurgical, tribological, etc.) remain to be dealt with by some suitable means.

to formulate a personal basis for design from the texts and papers on the subject. The option may be rewarding for the enthusiast, but will not lead to the prompt appearance of a prototype having commercial potential.

The current, incomplete understanding of the ultimate intricacies of heat transfer, fluid flow etc. is by no means the *sole* obstacle to further progress towards commercial exploitation: there remain, for example, challenging problems of reliability, cost reduction and power control to name but a few. There is, however, a sense in which the lack of *appropriate* tools for thermodynamic design is the *key* hold-up: the mechanical engineering problems of tribology, thermal stress, design for production etc. are matters which literally thousands of individuals and organizations world-wide are well qualified to address. The effect of the thermodynamic design obstacle, however, is to make Stirling machine development work an unattractive proposition for those bodies who are understandably anxious to have some idea of the time-scale of their involvement: If a prototype leaks, explodes or wears out, one knows what to do. If it runs without fault for 10 000 hours, but at a fraction of required specific power or efficiency, the search for a reason and then a remedy can be of indeterminate duration.

The chapters which follow attempt to provide the methods of thermodynamic design which have so far been lacking.

# 2

# Review of available theoretical treatments

## 2.1 Background

The theoretical literature which has accumulated in parallel with technical progress includes individual contributions of inspired quality. By comparison with other theoretical branches of engineering, however, it lacks any structure or unifying influence which might make for an agreed method of thermodynamic design. The lack will be especially evident to those to whom it has fallen to present an academic course on the subject, and is arguably a factor in the Stirling engine's slow progress towards commercial exploitation.

It also makes difficult the task of critical review: Noteworthy attempts are those due to Walker,[1] Urieli,[2,3] Martini,[4] Chen and Griffin,[5] and West.[6] The survey of Chen and Griffin is pertinent in being an account of work under contract specifically to rank the various theoretical treatments in order of usefulness for thermodynamic design. The authors declared themselves unable to reach any firm conclusion, recommending instead the programming of candidate treatments for a common computer and the running of each in turn with the same set of data corresponding to a machine of known performance.

The difficulty encountered by Chen and Griffin is typical. A remedy which will be explored here is first of all to propose a framework within which the analysis of the practical Stirling cycle *might* have evolved. By superimposing this framework retrospectively upon salient, existing treatments, these will be found to fall into place as components of an analytical picture of some generality. A means will be described whereby the numerical predictions of certain cycle descriptions may be compared objectively. Optimization of the thermodynamic cycle is considered.

The discussion infers prime movers, but applies equally to coolers. It is assumed that volume variations are kinematically controlled, but all concepts dealt with extend to free-piston variants.

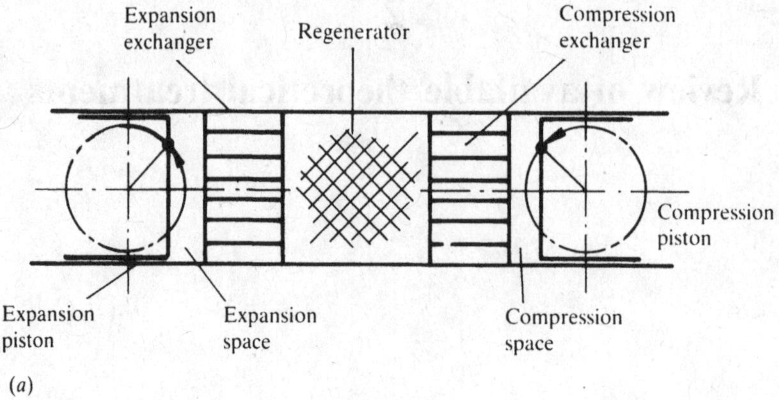


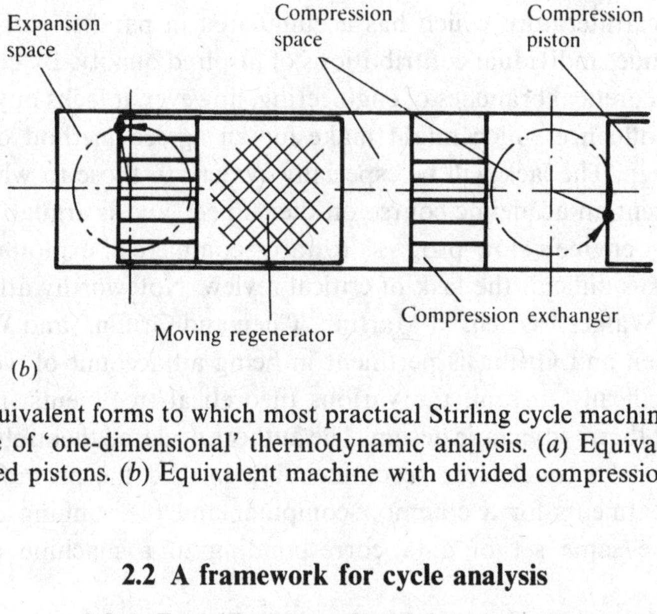


Fig. 2.1 Equivalent forms to which most practical Stirling cycle machines reduce for the purpose of 'one-dimensional' thermodynamic analysis. (*a*) Equivalent machine with opposed pistons. (*b*) Equivalent machine with divided compression space.

## 2.2 A framework for cycle analysis

### *2.2.1 Equivalent, one-dimensional machine*

Figure 2.1 shows the two schematic forms to which the majority of Stirling cycle machines reduce for the purposes of analysis. Treatment of the configuration with split compression space (Fig. 2.1(*b*)) is a simple extension of that for the opposed-piston type, (Fig. 2.1(*a*)), so for present purposes only the latter will be pursued.

The working fluid circuit is made up from three basic elements:

(1) A finely-divided matrix or regenerator affording tortuous flow passages. (Regenerators are used elsewhere, and there is an extensive specialist literature which cannot be included in the present review. Instead we note that, corresponding to the wire sizes common in Stirling cycle machines instantaneous accelerations of

up to $10^5\ g$ are indicated (Chap. 5). The value undermines the frequently-drawn distinction between steady and unsteady flow in the matrix, and suggests a flow phenomenon distinct from that in other working spaces of the machine.)

(2) Heat exchanger tubes or slots whose length is very much greater than the characteristic cross-sectional dimension and in which flow is commonly supposed to be similar to simple pipe flow.

(3) Cylinders of variable volume in which flow is undoubtedly three-dimensional.

It is clear that any analytical description of the thermodynamic and flow processes occurring in such a system can be at best a crude approximation to reality.

### 2.2.2 *Problem formulation – finite cell* vs *finite difference*

The conservation laws for mass, momentum and energy exist in at least two symbolic forms. These might be called the 'engineering thermodynamics' and the 'flow-field' forms, and are illustrated schematically in Fig. 2.2. The corresponding analytical statements of the conservation laws are compared in Table 2.1.

Ultimately the two descriptions must be reconcilable. In the modelling of the gas processes in the Stirling cycle machine, however, they have come to form the bases of two, different approaches. The engineering thermodynamics form has led to the so-called 'lumped-parameter', 'finite-cell' and 'nodal' models, as explored by Finkelstein,[7–10] Kirkley,[11] Schock,[12] Heames, Uherka, Zabel and Daley,[13] Tew, Jeffries and Miao,[14] and Urieli[2] among others. The flow-field counterpart leads to 'gas dynamics' formulations of the types described by Organ,[15,16] Sirett,[17] and Rispoli.[18]

At the numerical analysis stage both approaches use finite difference relationships to approximate infinitesimals. The fundamental distinction lies in the fact that the finite-cell formulations treat as total differentials the increment terms of Eqns (2.1)–(2.3). By contrast, the gas dynamics formulations transform the partial differential relationships mathematically *before* discretization. The transformation is to new coordinates along which changes in fluid properties are justifiably in total differential form.

The level of analytical debate within the field of Stirling cycle modelling was elevated at a stroke by the appearance of a dissertation by Urieli,[2] who took on the daunting task of defining a one-dimensional finite-cell discretization scheme equivalent to the differential form of the one-dimensional compressible flow-field. For the first time, momentum effects were accorded their due importance. An objection to Urieli's approach is that it makes use

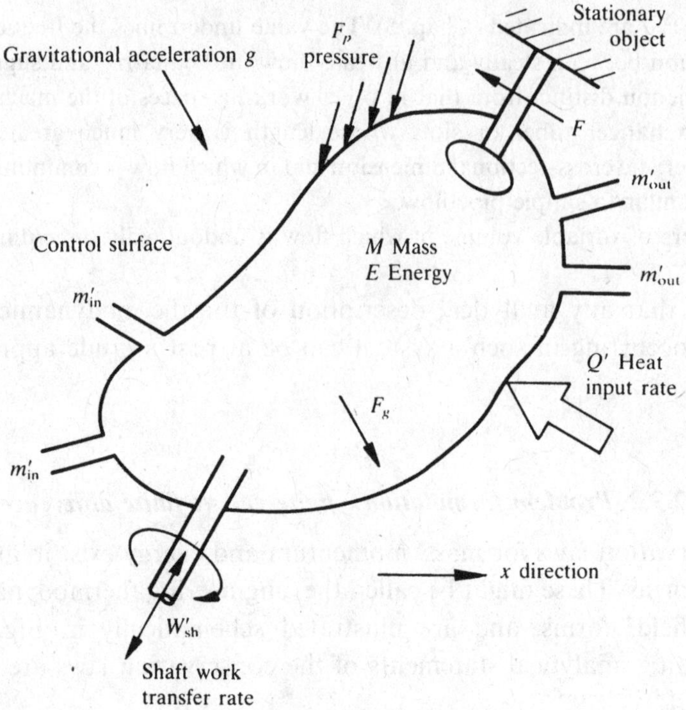


(a)

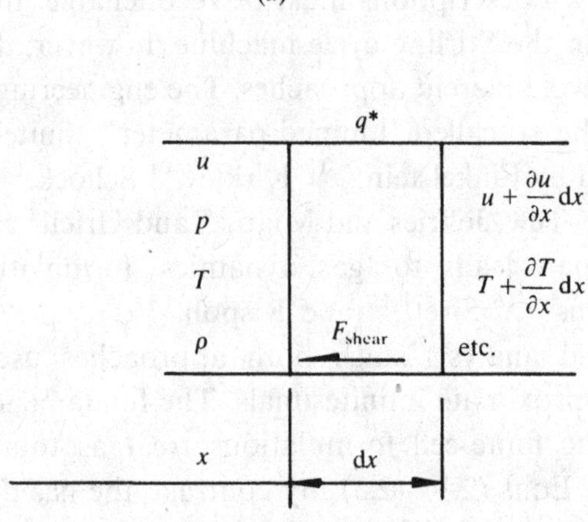


(b)

Fig. 2.2 Comparison between engineering thermodynamics and flow field interpretations of the working fluid system. (a) The system in engineering thermodynamics interpretation. (b) Flow-field interpretation of the system (one-dimensional case).

Table 2.1. *Comparison between engineering thermodynamics and flow field interpretations of the conservation laws (one-dimensional case).*

Conservation of mass

$$\sum_{\text{in}} m' - \sum_{\text{out}} m' = \frac{\mathrm{d}m}{\mathrm{d}t} \qquad (2.1a)$$

$$\frac{\partial \rho}{\partial t} + \frac{\partial}{\partial x}(\rho u) = 0 \qquad (2.1b)$$

Conservation of momentum

$$\sum_{x} F + \sum_{\text{in}} (m'u_x) - \sum_{\text{out}} (m'u_x) = \frac{\partial}{\partial t}(Mu_x)_{\text{cv}} \qquad (2.2a)$$

$$\frac{\partial u}{\partial t} + u\frac{\partial u}{\partial x} + \frac{1}{\rho}\frac{\partial p}{\partial x} + F = 0 \qquad (2.2b)$$

Conservation of energy

$$\sum_{\text{in}} m'(h + \tfrac{1}{2}u^2 + gz) - \sum_{\text{out}} m'(h + \tfrac{1}{2}u^2 + gz) + Q' - W_{\text{sh}}' = \frac{\partial E}{\partial t} \qquad (2.3a)$$

$$\frac{\partial}{\partial t}[\rho(c_v T + \tfrac{1}{2}u^2)] - q^*\rho = -\frac{\partial}{\partial x}[\rho u(c_v T + p/\rho + \tfrac{1}{2}u^2)] \qquad (2.3b)$$

of Gauss' theorem which requires the derivatives of the dependent variables to be continuous in the integration interval.[19] In fact, the one-dimensional, compressible flow-field contains an infinity of inherent discontinuities in those derivatives.[20] Urieli's presentation nevertheless stands as a model of exposition, and sets a standard of analytical virtuosity extremely difficult to follow.

Finite-cell schemes embody the implicit assumption that the incremental changes of mass, momentum and energy occur in a time interval determined only by the fact that it is less than the integration time step. In the balance of that time step, complete mixing of the contents within the control surface is supposed to take place, resulting in uniform properties (or in properties which vary linearly with distance within the cell).

Different ratios of $\Delta t$ and $\Delta x$ (i.e. different Courant numbers, $N_C$) thus signify different rates of diffusion and different rates of pressure information propagation.[21] The numerical solution of such formulations is thus a function of $N_C$. Increasing the number of flow-field subdivisions may thus be expected to lead to a *different* solution, but not necessarily to a more *accurate* one as is sometimes implied (for example. as in Ref. 22).

In numerical integration schemes appropriate to the gas dynamics formulation (for example, the Method of Characteristics), time and distance increments, $\Delta t$ and $\Delta x$, are the 'natural' coordinates[23] of the integration step. Their magnitudes are interdependent and related to acoustic wave propagation speed. Within limits, therefore, the solution is an accurate reflection of the defining equations (though not of the actual flow problem) regardless of integration step size.

Recognizing that rigorous description of the flow-field and energy transfer processes is impossible, the most that can be asked of an analytical treatment is twofold:

(*a*) that the form of the solution should have the greatest possible generality in the sense that it should represent more than a single operating point in the performance envelope of a specific machine; and
(*b*) that the solution should be a correct solution to the equations forming the flow-field description. This requires that the formulation should be such as to give notice of any violation of the underlying physical laws.

Where the foregoing have been achieved in other, related areas (aerodynamics, hydrodynamics etc.) the rôle of dimensional analysis has been prominent. The value of this approach appears to have been perceived at an early stage by Rinia and du Pré[24] and even more convincingly by Finkelstein[7,25] but then to have been overlooked until the recent appearance of papers by Organ[26,27] and Hutchinson.[28] A case for the reintroduction of the dimensionless formulation is essentially a restatement of the arguments of Annand[29] who so influenced thinking on the mathematical modelling of the internal combustion engine cycle.

### 2.2.3 Dimensional analysis as a basis for problem formulation

However intricate the defining equations, the eventual solution is a function only of the quantities independently specifying the flow-field and its boundaries, i.e., various lengths and characteristic values of pressure, temperature, specific heat, coefficient of dynamic viscosity etc. With length, mass, time and temperature as independent, basic variables, application of Buckingham's 'pi' theorem allows the variables to be made dimensionless and to be reduced in number by four.[29] A statement of normalized work per cycle, $\zeta$, thus becomes:

$$\zeta = \frac{W/\text{cycle}}{p_{\text{ref}} V_{\text{ref}}} = \zeta(\lambda_1, \lambda_2, \lambda_3, \ldots, \lambda_n, \tau, N_{\text{MA}}, N_{\text{SG}}, N_{\text{PR}} \text{ etc.}) \qquad (2.4)$$

The definitions of the groups are as under Notation. They may be varied by combination, but the total number of groups resulting from a given formulation remains the same after such combination.

An independent approach confirming Eqn (2.4) is to normalize all defining relationships (including the conservation laws) with respect to the reference variables as analysis proceeds, whereupon it is found that the dimensionless groups (or their combinations) emerge as multipliers. The discretization process has no effect on the multipliers, except to yield an additional group – a Courant number – based on $\Delta t$ and $\Delta x$ if a fixed-grid (for example, finite-cell) discretization scheme is applied. If different regions of the machine are subdivided differently then there are several different Courant numbers, or, alternatively, one Courant number and extra $\lambda_n$ for ratios of subdivision lengths. Inspection of the relationships thus normalized will reveal that the groups alone require specification as data in order to make numerical solution possible.

If Eqn (2.4) is multiplied by mechanical efficiency, $\eta_{\text{mech}}$, it represents normalized brake output per cycle – a quantity which has become known as the Beale number, $B$. It has been claimed (as in Ref. 1) that, for a fully-developed Stirling engine,

$$B = \eta_{\text{mech}}\zeta(\tau \text{ only}) \tag{2.5}$$
$$\approx 0.15 \text{ for } \tau = 0.33$$

Equation (2.4) confirms that such a notion is misleading, but is consistent with the interpretation by West[6] that the value of $B$ realized in a given engine may be uncharacteristically low at the developmental stage (i.e., while the $\eta_{\text{mech}}$, $N_{\text{MA}}$ etc. are being explored which maximize the function $\zeta$).

Not all cycle analyses are represented by the same function $\zeta$. However, the number and combination of dimensionless groups which will appear is predetermined at problem formulation stage, when the decision is taken as to which of the conservation laws are to be embodied. The combination is of fundamental importance in deciding which method of numerical solution will be appropriate (i.e., which method will minimize the unwanted influence of Courant number(s), $N_{\text{C}}$). It is also the basis of a useful scheme for classifying cycle descriptions.

### 2.2.4 *Classification of cycle models – the CL-n classification*

The most comprehensive description possible of the flow processes is that which includes all the conservation laws – mass, momentum and energy –

together with an equation of state and provision for energy conservation for the solid wall.

Any description of the cycle, or of a thermodynamic subprocess, must include at least a statement of geometry, the mass conservation law and an equation of state. Such a minimum treatment will be labelled CL-1 to denote the single conservation law. An example of the CL-1 type is the Schmidt analysis which, as is well known, yields an algebraic expression for work per cycle of the form:

$$\zeta_{\mathrm{Sch}} = \frac{W_{\mathrm{Sch}}}{p_{\mathrm{ref}} V_{\mathrm{ref}}} = \zeta_{\mathrm{Sch}}[\kappa, \alpha, \tau, \nu(\tau)] \tag{2.6}$$

The $\kappa$ and $\alpha$ are simple geometric ratios and $\nu(\tau)$ is a product of geometric ratios with temperature ratio, $\tau$. Equation (2.6) is thus of the form:

$$\zeta_{\mathrm{Sch}} = \zeta_{\mathrm{Sch}}[\lambda_1, \lambda_2, \lambda_3, \ldots, \lambda_n, \tau] \tag{2.6a}$$

The formulation may be made a more complete description of the flow-field by including further conservation laws. Table 2.2 contains a selection of possible combinations and suggests a label of the general form CL-$n$( ) for each. Where an incomplete form of the conservation law is employed (e.g., the energy equation without the term for kinetic energy, or the momentum equation without the fluid acceleration term) a system of self-explanatory super- or subscripts is easily invoked.

Not all of the possible combinations in Table 2.2 make for useful cycle descriptions. Later, however, when the 'building bloc' approach to computer modelling is introduced, it will be seen that the more unlikely combinations can provide a valuable check of correct functioning of the numerical solution.

Table 2.3 relates the choice of conservation laws to the functional form of the solution which the combination represents, i.e., to the resulting combination of dimensionless groups in terms of which the solution may be expressed.

### 2.2.5 Perturbation

A variable fluid property, say, $p$, may be expressed as the sum of an approximate value, $p_i$, and an error (or perturbation) component, $p'$:

$$p = p_i + p' \tag{2.7}$$

The approximate value, $p_i$, might be that derived from an analytical solution

Table 2.2. *Relationship between type of analytical model and numerical integration scheme. Criteria for stability and convergence are a separate (but related) matter*

| Analysis type | Essential features | Integration scheme |
|---|---|---|
| CL-1 | Solution for $p$, $u$, $T$ not affected by sound wave propagation or temperature gradient discontinuities. Integration intervals independent and open to free choice | Fixed grid method (cf. Ref. 15) – $\Delta x$, $\Delta t$ chosen for computational convenience |
| CL-2$MF^-$ (i.e., without term in $\partial(\rho u)/\partial t$ | Governing equations suggest diffusion-type flow | $\Delta x$, $\Delta t$ chosen to allow appropriate diffusion rate |
| CL-2$MF$ | Pressure information in numerical model must propagate at more or less the same speed as in the real gas circuit | Method of Characteristics (cf. Ref. 15) or fixed grid scheme – $\Delta x$, $\Delta t$ set by ref. to local sound speed |
| CL-3$ME$ | Heat transfer and friction act on individual fluid particles giving rise[20] to fluid property gradient discontinuities | Lagrange coordinates |
| CL-3$MFE$ | Discontinuities[20] in gradients of all properties. Full gas dynamics treatment | Method of Characteristics |

such as the Schmidt. Taking the momentum equation as an example (Eqn (2.2b)), substituting Eqn (2.7) and collecting terms gives

$$\frac{\partial u'}{\partial t} + u_i \frac{\partial u'}{\partial x} + u' \frac{\partial u_i}{\partial x} + \frac{1}{\rho_i}\frac{\partial p'}{\partial x} + F' = -\frac{\partial u_i{}'}{\partial t} - u_i \frac{\partial u_i}{\partial x} - u_i{}' \frac{\partial u_i}{\partial x} - F_i$$

$$+ \text{ products of terms of perturbation magnitude – which include } \partial p/\partial x.) \tag{2.8}$$

From the definition of $F$ (but neglecting the flow direction term, $u/|u|$),

$$F_i = \frac{1}{2r_h} u_i{}^2 C_f(N_{re_i})$$

Table 2.3. *Relationship between conservation laws embodied in cycle model and corresponding functional form of solution*

| $\zeta = \zeta($ | $\lambda_1, \lambda_2, \ldots, \lambda_n$ | $\tau$ | $\gamma$ | $N_{RE}$ | $N_{SG}$ | $N_{PR}$ | $\ldots$ | $N_C)$ |
|---|---|---|---|---|---|---|---|---|
| CL-1 | * | * | | | | | | |
| CL-2$MF^-$ (diffusion – fixed grid) | * | * | | * | * | | | * |
| CL-2$MF$ (Characteristics) | * | * | | * | * | | | |
| CL-2$MF$ (fixed grid) | * | * | | * | * | | | * |
| CL-2$ME$ | * | * | * | * | * | * | | * |
| CL-3 (fixed grid) | * | * | * | * | * | * | | * |
| CL-3 (Characteristics) | * | * | * | * | * | * | | |

$$F' = \frac{1}{2r_h}\left[u_i^2 \frac{\partial C_f}{\partial N_{re}} N_{re}' + 2u_i u' C_f(N_{re_i})\right]$$

An equation which was non-linear is now linear in the unknowns $p'$, $u'$ etc. If all the defining equations (including the gas law) are so perturbed and an appropriate method used to solve for these new unknowns, then the values of the original unknowns, $p$, $u$, etc. may be found by substituting back into equations of the form of Eqn (2.7). In suitable cases, the solution may be made identical to that for the original (unlinearized) equations by feeding successively improved values of the perturbation quantities into the right-hand sides as the solution iterates within the integration interval. (This process *must* be applied at least to the mass conservation equation to prevent drifting of the solution due to artificial loss or gain of mass).

Solutions based on the principle have been presented by Organ[30] and Rix.[31] However, it has since been appreciated that unless such perturbation permits substantial simplification (e.g., by allowing at least part of the solution to proceed analytically) then the linearization is superfluous, since discretization for numerical solution is, itself, a linearization process.

In the theoretical study of Stirling cycle machines there exist several variants of formal perturbation.

### 2.2.6 *Offset linearization – 'de-coupling'*

The mass flow rates predicted by a simple (e.g. CL-1) analysis are used to estimate local, instantaneous pressures and flow rates, and hence friction factors and heat-transfer coefficients. Thus, for pressure gradients, for example,

$$\frac{\partial p}{\partial x} \approx \tfrac{1}{2}\rho_i u_i^2 C_f(N_{re_i})/r_h \tag{2.9}$$

Integrating pumping work corresponding to Eqn (2.9) throughout the machine and around a complete cycle gives a work quantity which may be subtracted (along with other losses computed in similar fashion) from ideal cycle work to give an estimate of net output. Comparison of Eqn (2.9) with Eqn (2.8) reveals how many terms have been arbitrarily dropped from the momentum equation.

No estimate of accuracy has even been made. However, the effect of perturbing by reference to a limited number of unknowns is to offset the computed thermodynamic properties from the values which would have resulted from a solution of the perturbed equations in full form (cf. Eqn (2.8)). A form of perturbation known as linear harmonic analysis (or LHA) acknowledges this problem, and some variants of LHA attempt a remedy.

### 2.2.7 *Linear harmonic analysis (LHA)*

The pioneers appear to have been Rios and Smith,[32] and the work has since been taken up by Rauch[33] and by Chen, Griffin and West[34] among others. The confidential work carried out at the Philips Research Laboratories may have been essentially of LHA type.[35]

In a typical formulation the cyclic variations of local mass, $m$ and temperature, $T$, are expressed as simple harmonic functions of crank angle, $\omega t$,

$$m_e(\omega t) = \underline{m}_e + c_1 \sin(\omega t) + c_2 \cos(\omega t) \tag{2.10}$$

where the underline denotes an average. Substitution into the conservation equations and neglect of non-linear products converts the former into algebraic relationships in the (unknown) coefficients, $c_i$. Matrix methods have been used[6] to obtain a solution for the coefficients and hence to the periodic variations of the required thermodynamic quantities.

In the version of LHA described by Qvale and Smith[36] and by Rios and Smith[32] pressure and mass flow rates were assumed to be simple harmonic functions of crank angle, $\omega t$. To satisfy mass conservation it was necessary to solve for compression space volume as an unknown. The volume variation was thus 'offset' from its specified sinusoidal form by a variable amount of perturbation magnitude. The treatment provided for retrospective correction of this effect.

Claims for LHA include high computational speed and mathematical accuracy.[6] The latter would seem a little optimistic even if the assumption of near-sinusoidal variations were a good one: there is practically no experimental information on working space temperature variations, but values computed from theoretical considerations (see, e.g., Ref. 7) have anything but simple harmonic form.

Chen, Griffin and West[34] have claimed that an LHA taking account of three conservation laws is necessarily of type CL-3. On the other hand, the solution from an LHA will not, in general, be the same as that from the 'parent' CL-3 treatment, and the difference requires explanation. The fact is that the LHA equation system conserves not mass, momentum and energy, but rather the simple harmonic form of the specified variables. To this extent, the LHA appears to combine the worst of all numerical worlds.

A method has been proposed recently by the author[27] which, while admittedly leading to numerical offset, at least avoids the embarrassment of arbitrarily discarding the awkward unsteady terms from the perturbed form of the mass and energy conservation laws.

### 2.2.8 *Perturbation and availability theory*

The Second Law is used to express the rate of entropy generation per unit volume for unsteady flow in a duct with friction and heat transfer. The formulation of Bejan[37] is followed. After substitution of the unsteady form of the mass conservation equation:

$$S_{\mathrm{gen}_v}{}' = \frac{1}{T}\nabla q - \frac{1}{T^2} q + \rho \frac{\mathrm{D}s}{\mathrm{d}t}$$

Invoking the First Law of thermodynamics and considering the one-dimensional case:

$$S_{\mathrm{gen}}{}' = \left( q' \frac{\Delta T}{T^2} + \frac{\rho u F A}{T} \right) \mathrm{d}x \qquad (2.11)$$

The simplicity of this expression belies the fact that it derives from the differential equations of energy and mass conservation for compressible flow taking account of unsteady effects. The latter effects are not inherently irreversible, and the absence from the expressions of differentials with respect to time might have been anticipated.

In the use of Eqn (2.11) to describe conditions in the Stirling cycle machine the steady flow terms are approximated by corresponding local, instantaneous values from a reference cycle. The treatment described by Organ uses Finkelstein's generalized thermodynamic analysis[7] as the reference cycle. The product $T_o\, dS_{gen}$ is local, instantaneous lost work which is summed throughout the machine and over a complete cycle. The result is subtracted from ideal cycle work to give an estimate of net value.

## 2.3 Experimental corroboration

A critical assessment of the worth of a theoretical model would ideally involve comparison of predicted variations in pressure, temperature and particle speed with corresponding values measured in the machine being modelled. Although pressure has been recorded as a function of time,[11,13] the other quantities are relatively inaccessible.

One practice which has thus evolved is to judge the worth of a theoretical treatment by the extent to which predicted indicated thermal efficiency and indicated work per cycle agree with values inferred from brake measurements on the machine in question.

The approach has been called 'validation'[3,5,38] and appears to cover the arbitrary adjustment of friction and heat transfer correlations, discharge coefficients etc. for the purpose of closing the gap between measured and predicted output quantities. Validation having this meaning is referred to here on the strength of its wide currency. However, if a cycle model based on one-dimensional flow gives a specific result which is *exactly* that measured from a machine in which heat and fluid flows are three-dimensional, then it is safe to conclude that there is something wrong with the cycle model or with the data supplied to it.

A more reasonable approach to experimental corroboration involves examining *trends* rather than point values. Two approaches offer potential which has barely yet been tapped:

(1) If Eqn (2.4) has any validity, then the cycle of thermodynamic processes in the real machine will be little changed by alterations to operating conditions which leave the values of the defining dimensionless groups unchanged. Accordingly, in an experiment devised by Organ and reported by Rix[31] pressure–time traces were acquired from the working spaces of the SM1 laboratory engine over a range of crankshaft speeds, charge pressures and working fluids which kept the dominant dimensionless groups, $\tau$, $N_{MA}$ and $N_{RE}$ constant. (The $\lambda$ remain constant by definition in tests on a machine of fixed geometry.) Figure 2.3 is a summary of results which speak clearly for themselves.

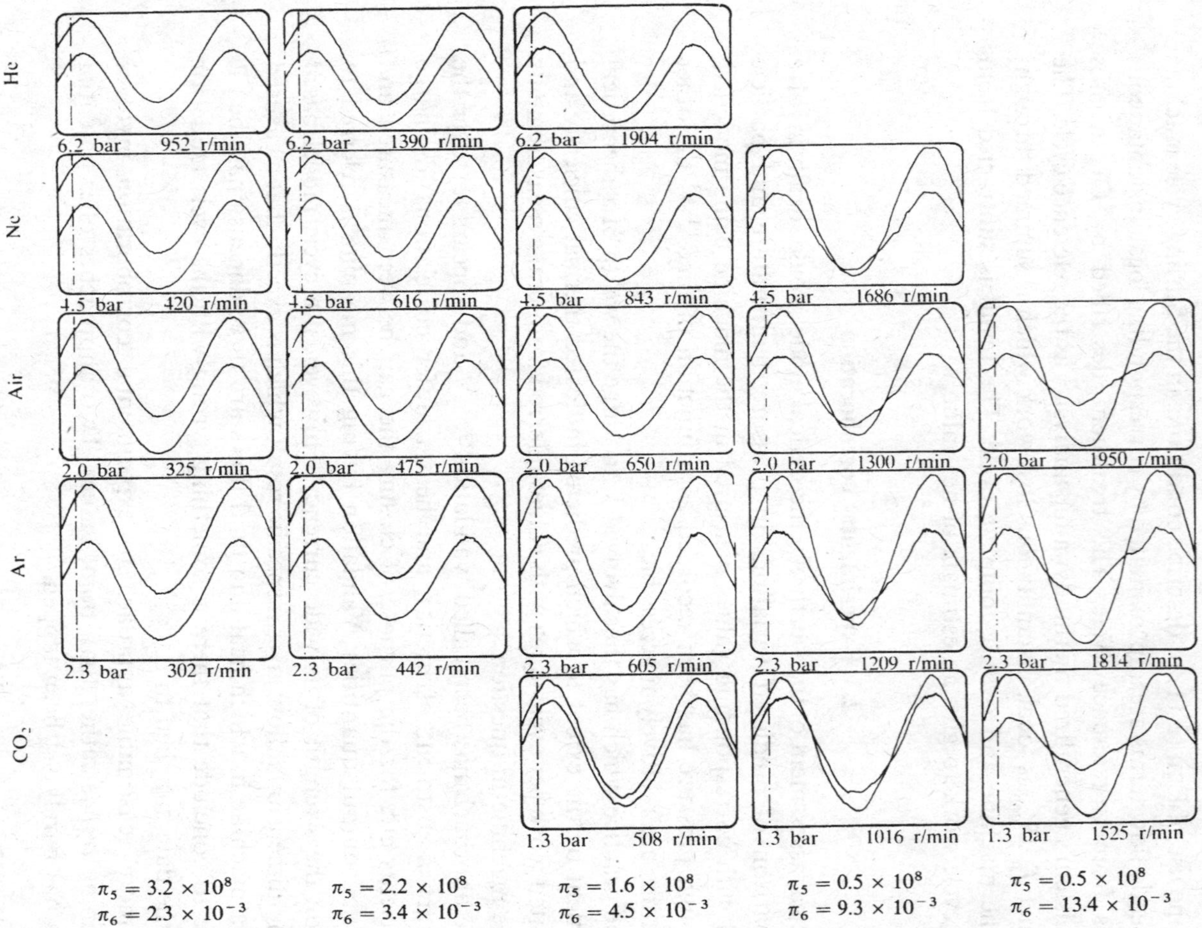


Fig. 2.3 Dynamic similarity demonstrated by traces of working space pressures vs time when $N_{MA}$ and $N_{RE}$ are held constant (after Rix[31]). $\pi_6$ is essentially $N_{MA}$ while $\pi_5 = N_{RE}/\pi_6{}^2$. The interchangeability has been discussed in the main text.

(2) For a given machine operating at a given temperature ratio, computed performance may be plotted as 'maps' of $\zeta$ as a function of $N_{MA}$ and $N_{RE}$.[27] (Specific examples will be given later.) Equivalent maps – or partial maps – may be plotted for the actual machine. Quantitative correspondence is unlikely, but it is probable that the general form of the computed map will resemble the experimental counterpart if the underlying analysis is well founded. The trends implied in the computed maps are likely to represent more valuable information than the point values from analyses 'validated' in conventional fashion.

## 2.4 Established analytical approaches

### *2.4.1 Basic machine operation*

Figure 2.4(a) indicates an idealized machine in which $M$ kg of compressible working fluid is contained between the faces of two movable pistons. Expansion and compression space contents remain uniformly at temperatures $T_e$ and $T_c$ respectively regardless of volume and pressure. There is a linear gradient of temperature across the regenerator. Piston motion is such that there is an overall reduction in volume when expansion space mass is near to minimum, and an overall volume increase when compression space mass is near to minimum. Over a cycle there is thus a net rejection of heat at temperature $T_c$ from the compression space and a net addition at $T_e$ to the expansion space. The discontinuous piston motion produces the indicator diagram of Fig. 2.4(*b*). The diagram area is positive for $T_e > T_c$ and negative when $T_e < T_c$.

This well-worn description of an ideal, regenerative machine is repeated here to draw attention to the universally ignored fact that lines 1–2 and 3–4 are isothermals ($pV = MRT$) ***only*** *for the special case in which* $T_e = T_c$, *and in which these lines therefore coincide.* Only for this case can a $T$–$s$ diagram be constructed, and Fig. 2.4(*b*) is therefore not an ideal thermodynamic cycle diagram in the normal sense.[39]

Real Stirling machines rely on continuous piston motion, and further discussion will assume such motion. For convenience it will be assumed that the working fluid behaves as an ideal gas.

### *2.4.2 Analyses of type CL-1*

The first credible analysis is due to Schmidt.[40] The treatment was made accessible in English in a reformulation by Rinia and du Pré.[24] Schmidt neglected pressure drops (i.e., variable pressure was considered to be uniform throughout the system at any instant) and assumed unrestricted heat transfer.

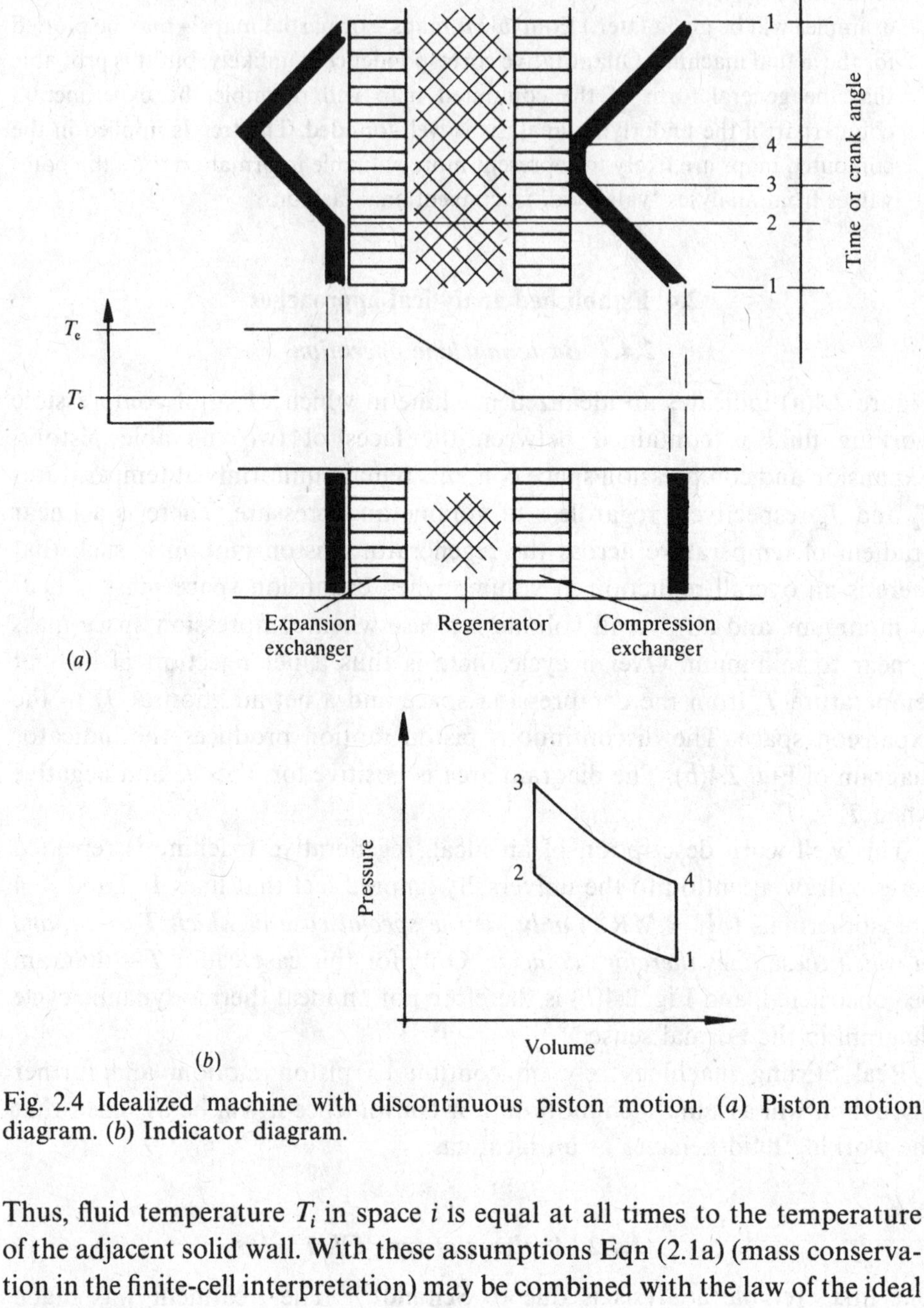


Fig. 2.4 Idealized machine with discontinuous piston motion. (*a*) Piston motion diagram. (*b*) Indicator diagram.

Thus, fluid temperature $T_i$ in space $i$ is equal at all times to the temperature of the adjacent solid wall. With these assumptions Eqn (2.1a) (mass conservation in the finite-cell interpretation) may be combined with the law of the ideal gas to give:

$$M = \sum m_i = \frac{p}{R} \sum \frac{V_i}{T_i}$$

With working space volume variations known for all crank angles, $\phi$, and with either total working fluid mass $M$, or a reference pressure, $p_{ref}$, decided upon, the expression may be inverted algebraically to give pressure $p$ as a function of $\phi$:

$$\frac{p(\phi)}{p_{ref}} = \frac{\sum V_i(\phi_{ref})/T_i}{\sum V_i(\phi)/T_i} \tag{2.12}$$

When $p(\phi)$ is multiplied by the simple harmonic volume increment the resulting expression for incremental work is a standard integral. Ideal, normalized work per cycle, $\zeta$, then follows directly as an explicit algebraic expression equivalent to Eqn (2.6).

Two Stirling machines of greatly different proportions can have the same numerical value of the dead-space parameter, $\nu(\tau)$, which does not therefore determine geometric similarity. Without geometric similarity there cannot be dynamic similarity.[41] This fact helps to explain why the Schmidt expressions with parameters corresponding to a satisfactorily functioning machine have an (optimistic) finite value, while a machine designed by reference only to the Schmidt equation can have zero measured output.

When the idealization of simple harmonic piston motion is dispensed with the single conservation law treatment becomes the slightly less restricted 'isothermal' analysis. Analytical integration to find cycle work is no longer straightforward, and numerical means are generally used to compute $\zeta$ corresponding to given piston motion. The solution has the form of the more general Eqn (2.6a).

An alternative CL-1 analysis formulated by Organ[15] is based on the mass conservation law in differential form (Eqn (2.1b)). Because the flow-field with uniform pressure and predetermined temperature contains no moving discontinuities, numerical integration leads to values of $\zeta$ identical for a given machine to the predictions of the isothermal analysis above. The functional equivalent of the solution is again Eqn (2.6a).

The CL-1 formulation was for many years the backbone of Stirling cycle analysis. Thus, Finkelstein[42] ingeniously reduced Eqn (2.6) to a simple nomogram relating $\zeta$ to $\tau$ and $\nu(\tau)$ (Fig. 2.5). The nomogram is exactly equivalent to the Schmidt formula for the symmetrical machine ($\kappa = 1$, $\alpha = \pi/2$), and is a close approximation for $\kappa$ and $\alpha$ not greatly different from those values. This anticipates by some years the proposal that $\zeta = B(\tau\ (\text{only}))$.

Walker[1,43] has used the Schmidt cycle to gain an impression of the motion of the working fluid within the heat exchanger system. The results drew attention for the first time to the fact that given working fluid particles do not, in general, traverse the entire heat exchanger system.

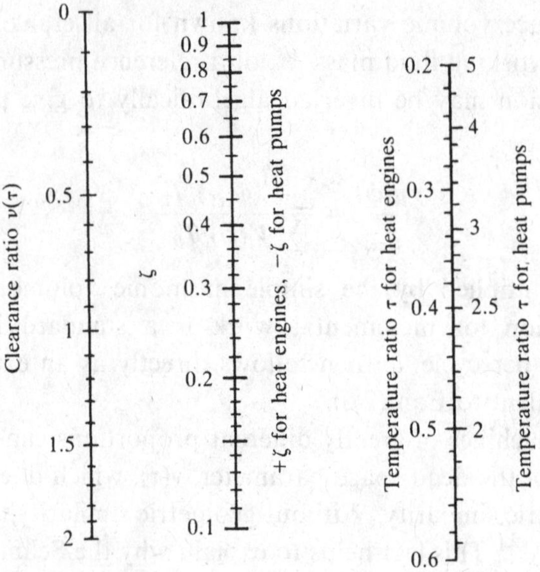


Fig. 2.5 Nomogram relating $\zeta$ (=mean effective pressure, $\psi$, in Finkelstein's terminology) to temperature ratio, $\tau$, and clearance ratio, $\nu(\tau)$, for a symmetrical machine ($\kappa = 1$, $\alpha = 90°$). (After Finkelstein.[42])

With an estimate available for pressure as a function of crank angle, instantaneous, local heat exchange per radian between the solid surface at temperature $T_w$ (Fig. 2.6) and fluid flowing at the same temperature is readily expressed:

$$q' = \left( -A\frac{\mathrm{d}p}{\mathrm{d}t} + \frac{R\gamma}{\gamma - 1}m'\frac{\mathrm{d}T}{\mathrm{d}x}\right)\mathrm{d}x \tag{2.13}$$

Creswick[44] may have been first to use expressions of this type to plot ideal heat exchange as a function of crank angle, $\phi$, for heater, cooler and regenerator of the Stirling machine. His results confirm prior claims by workers at Philips Laboratories[45] that regenerator heat flux rates considerably exceed those in the heat exchangers.

Organ[39] combined the investigations of Walker and Creswick to examine the variations in specific volume and specific enthalpy of selected working fluid particles. This permitted definition of true cycles of thermodynamic properties for an ideal Stirling cycle based plausibly on the real machine.

The grossness of the assumption that conditions in the variable-volume spaces are isothermal had been noted by Rinia and du Pré,[24] who imply that they had investigated an alternative rationalization, namely that of adiabatic conditions. A tantalizingly restricted insight into Philips' confidential work

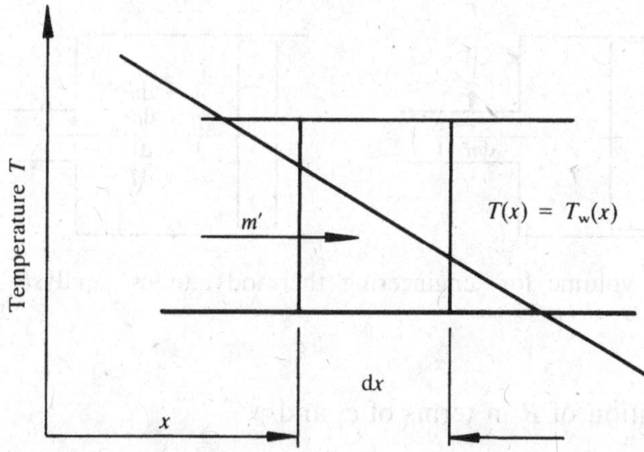


Fig. 2.6 Exchange of heat without temperature difference between flowing fluid and wall with linear temperature gradient.

was to be provided shortly by Cuttil, Malik and Decker[46] in a report which appears not to have been widely circulated. As far as the open literature is concerned, however, it was Finkelstein[7] who first proposed and solved a cycle description based on more than one conservation law. The novelty of his analysis lay in the treatment of the variable-volume spaces. His approach has since found its way into virtually all subsequent analyses of lumped-parameter or finite-cell type, and therefore merits presentation in some detail.

### 2.4.3 *More than one conservation law – lumped-parameter analyses*

Figure 2.7 indicates a space whose volume is controlled by a moving piston. Mass may enter or leave via a port. Wall temperature, $T_w$, is assumed constant and uniform. When a mass element enters it does so at $T_w$. An element of mass leaving does so at the instantaneous temperature, $T$, of the cylinder contents. The thermodynamic properties of the latter are assumed uniform at any instant, and the instantaneous mixing which this implies typifies the 'engineering thermodynamics' approach.

The First Law states:

$$dQ + dH - p\,dV = dU$$

Assuming a known coefficient of convective heat transfer, $h$, the inflow case is defined by

$$hA(T_w - T)\,dt + dmc_pT_w - p\,dV = c_v(T\,dm + m\,dT)$$

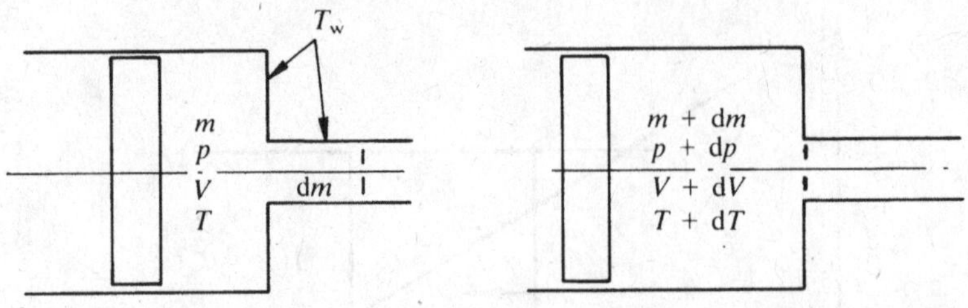


Fig. 2.7 Control volume for 'engineering thermodynamics' analysis of variable-volume space.

From the definition of $R$ in terms of $c_v$ and $\gamma$

$$\frac{(\gamma - 1)}{\gamma} hA(T_w - T)\,dt + RT_w\,dm = \frac{V}{\gamma}\,dp + p\,dV \quad \text{(inflow)} \qquad (2.14a)$$

$$\frac{(\gamma - 1)}{\gamma} hA(T_w - T)\,dt + RT\,dm = \frac{V}{\gamma}\,dp + p\,dV \quad \text{(outflow)} \qquad (2.14b)$$

Equations (2.14) differ only in the second term of the left-hand sides and may be combined if a step function is defined:

$$U(dm) = \begin{cases} 1 & \text{for } dm \geqslant 0 \quad \text{(inflow)} \\ 0 & \text{otherwise} \quad \text{(outflow)} \end{cases}$$

Then

$$\frac{(\gamma - 1)}{\gamma} hA(T_w - T)\,dt + R\,dm[T(1 - U(dm)) + U(dm)T_w] = \frac{V}{\gamma}\,dp + p\,dV \qquad (2.15)$$

For $h = 0$ Eqn (2.15) reduces to the 'adiabatic' case:

$$R\,dm[T(1 - U(dm)) + U(dm)T_w] = \frac{V}{\gamma}\,dp + p\,dV \qquad (2.15a)$$

Replacing $\gamma$ by unity and setting $T_w = T$ gives the well-known 'isothermal' case:

$$RT\,dm = V\,dp + p\,dV \qquad (2.15b)$$

Finkelstein's original treatment applies Eqn (2.15) to the two variable-volume spaces. The assumption that pressure at any instant is uniform throughout the heat exchangers and that fluid temperature in the heat exchangers is equal to that of the adjacent solid wall permits a mass balance to be written. With subscript e to indicate the expansion space, c to indicate the compression space, and d to denote dead (or unswept) volume, total working space mass, $M$, may be expressed as

$$M = m_e + m_c + m_d$$
$$= m_e + m_c + \frac{p}{R}\sum \frac{V_j}{T_j} \tag{2.16}$$

Equation (2.16) permits $p$ and d$p$ to be expressed in terms of the variables $m_e$, $m_c$, $m_e'$ and $m_c'$; $T$ may be expressed via the ideal gas equation in terms of $p$ and $V$ and hence may also be expressed in terms of $m_e$ and $m_c$. Equations (2.15), one for each of the two working spaces, thus reduce to a pair of ordinary differential equations in $m_e'$ and $m_c'$, integrable by straightforward numerical means. With $m_e$ and $m_c$ known as functions of $\phi$, $p(\phi)$ and hence work per cycle follow.

An essential aspect of Finkelstein's approach not so far highlighted is that he works in terms of *normalized variables.* Thus, Eqns (2.15) are divided before substitution by $MRT_{ref}$ and Eqn (2.16) by $M$. His final result is equivalent to a statement for normalized work (and for corresponding heat quantities) of the form:

$$\zeta = \zeta(\lambda_1, \lambda_2, \lambda_3 \ldots \lambda_n, \tau, \gamma, \eta_e, \eta_c) \tag{2.17}$$

If piston motion is sinusoidal the geometric ratios in Eqn (2.17) are again $\kappa$, $\alpha$ and $\nu(\tau)$. Then $\eta_e$ and $\eta_c$ are Finkelstein's dimensionless heat transfer coefficients:

$$\eta = \frac{hA}{\omega c_p M}$$

It is self-apparent that $\eta$ has the form of a Stanton number, $N_{ST}$, defined in turn as

$$N_{ST} = \frac{N_{NU}}{N_{RE} N_{PR}}$$

A correlation between local, instantaneous $N_{nu}$ and corresponding $N_{re}$ and $N_{pr}$ for the Stirling machine cylinder has not yet been proposed (but see Ref. 2 in Chapter 10). Finkelstein assumed constant values of $N_{st}$ (*viz*, $N_{ST}$) for his examples. Nevertheless, there is for the first time a formulation which may, in principle, be written as

$$\zeta = \zeta(\lambda_1, \lambda_2, \lambda_3 \ldots \lambda_n, \tau, \gamma, N_{RE}, N_{PR}) \tag{2.18}$$

Figure 2.8 is from Finkelstein's original paper and shows the computed variation with crank angle, $\phi$, of working space temperature. The marked

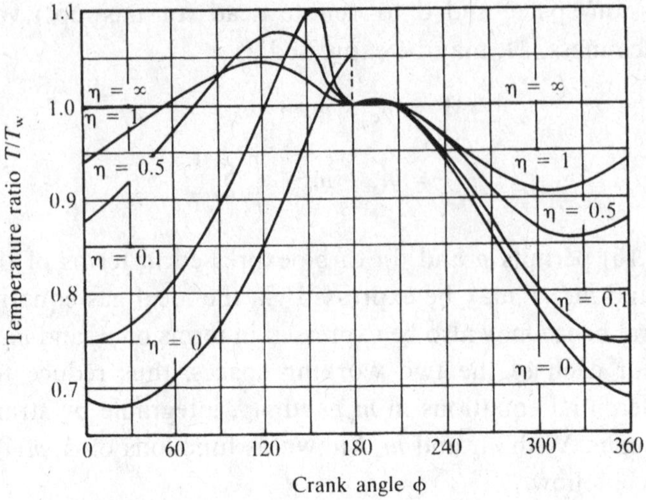


Fig. 2.8 Temperature ratio, $T/T_w$ for expansion space as a function of crank angle with dimensionless heat transfer coefficient as parameter. (After Finkelstein[7] and reproduced with permission of the Society of Automotive Engineers, Inc.)

deviation from the sinusoidal form assumed by West[6] is due in part to the fact that Finkelstein's example allows dead space at top dead centre to take the value zero.

In a report dated 1962 on work completed in August 1960 the primary concern of Cuttil *et al.*[46] was not engine design but development of a power control system for an existing 3 kW engine intended for the generation of electrical power in space. For this purpose they required an electronic model of the response of the engine to changes in load and control signals (charge pressure, heat input). The cycle description which they implemented was nevertheless the most comprehensive to date, and apparently superseded the contemporary work of Philips, of which the authors wrote:

> The . . . Philips analysis assumed that the pressure was identical throughout the system, and that the mass transfer within the system was dependent only on the displacement of the pistons and the temperature changes within the cylinders.

The assertion suggests Philips' cycle description to that point was similar to Finkelstein's.

Cuttil *et al.* considered the engine to be divided into five control volumes corresponding to each of the working spaces, the two heat exchangers and the regenerator. The engineering thermodynamics form of the laws of mass and energy conservation were written (without a kinetic energy term) for the control volumes. A degenerate form of the momentum equation, *viz*,

$$\frac{dm}{dt} = k(p_i - p_{i+1})$$

was applied at the four interfaces between the control volumes $i$ and $i + 1$. Volume variations were derived exactly from the kinematics of the rhombic drive. Loads due to pressure on piston and displacer were equated to piston friction, bearing friction, inertia and output load. Indicator diagrams were constructed and brake power and a plausible brake thermal efficiency calculated. On the other hand, heat exchanger thermal flux was not scrutinized as a function of crank angle, otherwise it would not have been reported as being unidirectional. This and the attempts to find a polytropic index, $n$, to fit a curve of type $pV^n = C$ to the indicator diagram ($V$ is not a property of the open system) detract from the authority of an account which in other respects represents a milestone in Stirling cycle modelling.

Possibly unaware of the work reported by Cuttil *et al.*, Finkelstein built on his generalized thermodynamic analysis in a rapid succession of papers. In the first[8] he dispensed with the assumption of uniform pressure in favour of an approximation to flow losses. This was achieved by defining a variable pressure, $p_r$, at the regenerator mid-plane, and discharge coefficients, $f$ (having dimensions), for the two halves of the heat exchanger system. The instantaneous mass flow rate, $m'$, and corresponding pressure drop between cylinder port and regenerator mid-plane were then related by expressions of the form

$$p_r(p_r - p_e) = f_e m'$$

The treatment retained the assumption of a linear temperature gradient through the regenerator, but fluid temperature distribution within the exchangers had the steady state (exponential) form. At any instant the steady-flow solution corresponded to that for the flow computed at the adjacent cylinder port.

The assumption implies instantaneous switching of the temperature profile at the instant of change of sign of cylinder flow. Although this may be regarded as arbitrary, it has never been shown to be less representative of reality than results from the more intricate finite-cell model of an unsteady heat exchanger with its equally arbitrary numerical diffusion.

Eleven simultaneous equations resulted and were solved by coding in a form acceptable to a program for a digital computer which, in turn, emulated a differential analyser. Finkelstein's solution for the machine with sinusoidal volume variations required values of 18 dimensionless groups as input, and thus had the symbolic form:

$$\zeta = \zeta(\lambda_1, \lambda_2, \lambda_3, \ldots, \lambda_n, \tau, \gamma, N_{ST1}, N_{ST2}, \ldots, \theta_1, \theta_2, \ldots) \tag{2.19}$$

In Eqn (2.19) the $\theta_n$ are dimensionless discharge coefficients. Extracting the common factors, $N_{ST}$, $\theta$, from the various $N_{STn}$, $\theta_n$ leaves only length ratios $\lambda_n$, which are independent geometric variables. The $N_{ST}$, $\theta$ may be re-expressed in terms of a single $N_{RE}$ and $N_{PR}$, while $\kappa$ and $\alpha$ are merely particular values of $\lambda$. The functional representation of Finkelstein's CL-3($FE$) analysis is thus:

$$\zeta = \zeta(\lambda_1, \lambda_2, \lambda_3, \ldots, \lambda_n, \tau, \gamma, N_{RE}, N_{PR}) \tag{2.4a}$$

and appears to be the last to have been formulated in terms of dimensionless parameters.

Finkelstein may have been the first to perceive the usefulness of the Sankey diagram to represent the internal energy conversion processes for a typical cycle, and the paper includes a revealing example. This form of presentation has since been extended by Urieli and Berchowitz.[47]

In the five or so years which followed, the lumped-parameter approach was extended further by Finkelstein[9,48] and was taken up (apparently independently) by Kirkley.[11] Kirkley's algebraic description of the problem was far less compact than that of Finkelstein, partly because he used the full form of the inflow and outflow equations for the working spaces (instead of combining these with the aid of a step function). Moreover, being in terms of absolute (rather than normalized) $p$, $T$, etc., the treatment lacked generality. Nevertheless, it remains one of few accounts offering a direct comparison between computed and measured thermodynamic quantities – in this case, pressure in expansion and compression spaces.

By 1965 Finkelstein's lumped-parameter cycle description had grown to 28 separate defining equations, including a balance between instantaneous force due to gas pressure and corresponding acceleration of the mechanical parts. Solution was by analogue computer and yielded the variations with crank angle of pressure, temperature and mass content (and hence mass flow rate) at five locations within the machine. The formulation was thus similar in content and scope to that of Cuttil *et al.* Differences were that piston motion was assumed to be sinusoidal and that seal leakage was accounted for.

Possibly in response to diminishing marginal returns for escalating computational work, cycle modelling in terms of a separate description for each subsystem of the machine was not taken further by Finkelstein or, as far as is known, by others. Instead, there followed rationalization along two lines.

Firstly, it was noted that, in a machine operating without excessive losses, cyclic variations in pressure, temperature and flow rate might be sufficiently close to those of a corresponding reference cycle for discrepancies to add linearly. The pioneer work appears to have been done by Qvale and Smith[36] who used the principle to de-couple the terms describing flow friction in the heat exchangers and regenerator. These terms were then integrated separately to give net pumping loss per cycle and hence net indicated cycle work. The treatment presupposed sinusoidal variation in pressure and mass flow rate. In this respect it is the precursor of linear harmonic analysis.

Secondly, it was eventually recognized that, if the conservation laws may be written for a specific control volume within the machine, then they may be written for an arbitrary but typical control volume. A description of the gas processes throughout the machine as a whole then follows from repeated application of the same, basic set of equations to all subdivisions of the flow system. The potential economy of algebra and computer coding by comparison with the original, lumped-parameter concept is self-evident, and the result has been the so-called nodal or finite-cell analyses.

In a description published in 1970 Finkelstein, Walker and Joshi[10] anticipated the eventual, definitive form of the genre while retaining a feature of the lumped-parameter description. Figure 2.9 shows the mass and energy flow paths in nodal representation. The thermal conduction path is self-explanatory. CF (cathode-follower) nodes represent the choice between inflow or outflow forms of the energy equation (cf. Eqn (2.15a)). Local, instantaneous values of the heat transfer coefficient are used, and properties such as thermal conductivity can be functions of temperature. Kinetic energy was not considered.

Given the need for the numerical solution scheme to scan all nodes systematically, it is not clear why a general statement of momentum conservation was not applied between each. Instead, three significant pressures, $p_e$, $p_c$ and $p_r$ were defined as for the earlier, lumped-parameter scheme. Momentum conservation was then expressed in terms of a discharge coefficient and mass flow rate, and the calculation performed once per sweep of the node system.

Nodal representation of the processes of mass and energy conservation owes much to the methods of analogue computing originally employed for solution. It does not give the most revealing picture of the physical processes, which now tend to be represented by a volume cell. The pictorial representation, however, is the only essential difference between treatments respectively labelled nodal or finite-cell.

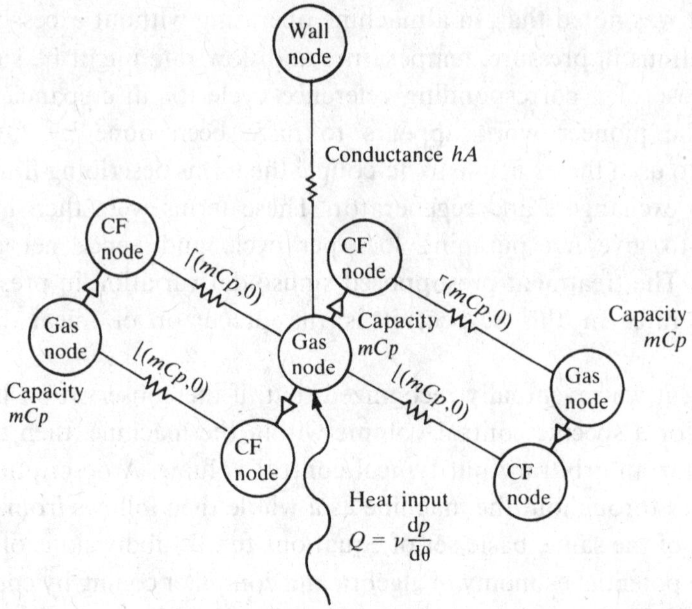


Fig. 2.9 Finkelstein's nodal representation of energy exchange. (After Finkelstein *et al.*[10])

### 2.4.4 *Finite-cell* vs *gas dynamics formulation*

To date there had been no questioning of the suitability of finite difference methods in general for obtaining numerical solutions representative of the underlying flow problem. Using the simplification of incompressibility, Organ[49] demonstrated, however, that one-dimensional, cyclically reversing flow in a duct is inevitably accompanied by discontinuities in the lengthwise distribution of temperature. Figure 2.10 is a relief of temperature as a function of time and distance for one of the cases solved. The phenomenon may have been observed several years previously by Cuttil *et al.*,[46] but their account is perfunctory, and in any case they suppressed the phenomenon in question, which they found 'unacceptable'. The problem underlying potential discontinuities has since been recognized by Fokker and van Eckelen[21] and by Gedeon.[50]

If the matter had any bearing on the modelling of flow phenomena in Stirling cycle machines, its original revelation came too late or was too obscure to influence the substantial number of finite-cell analyses which have since been developed. (An exception was an untitled, undated report from about 1976 apparently produced within the American Dynatech Corporation. This described a computer simulation formulated in Lagrange coordinates in

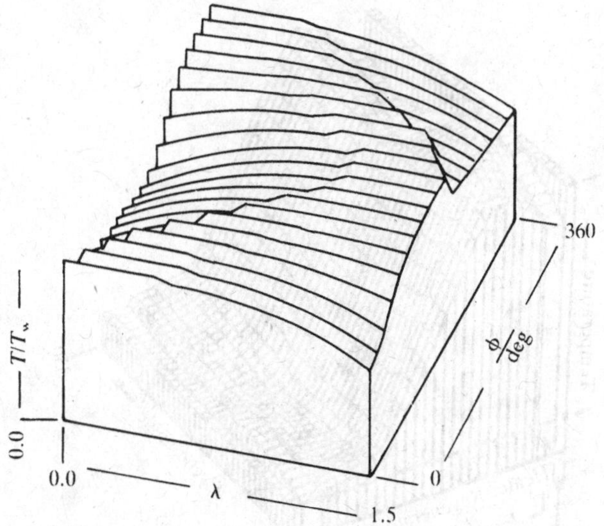


Fig. 2.10 Relief of gas temperature ratio, $T/T_w$ as a function of crank angle and normalized distance within isothermal heat exchanger. (After Organ.[49] © IEEE 1975.)

direct response to Organ's findings. Dynatech were concerned with the thermal design of the wire mesh regenerator. Coefficients of convective heat transfer tend to be much higher in this component than in the heat exchanger tubes, and they found the discontinuities to be too small to be of significance.) Among the conventional (Eulerian) formulations are those of Schock,[12,51] Heames *et al.*,[13] Tew *et al.*,[14] and, most influential of all, Urieli[2] and his collaborators.[3,22] The treatments all have this in common: the variable-volume spaces are described by an adiabatic model (cf. Eqn (2.15a)). The flow passages are subdivided into a number of cells. For a typical cell the engineering form of the conservation equations is written in terms of finite differences. A given treatment may or may not include all the terms in the momentum and energy equations, and an energy balance for the wall may or may not be part of the formulation. The majority of numerical solutions have been of explicit form, although Gedeon[52] and Organ[27] (see later section) have shown that implicit (matrix) methods can have advantages. Formulations tend to be in terms of absolute variables, but are functions by implication of one or more Courant numbers.

An attraction of nodal and finite-cell methods (as opposed to the lumped-parameter types) is that computation proceeds in sufficient detail to generate a picture of gas and metal temperature distributions as a function of location and time. Figure 2.11 is taken from Schock's account and shows computed variation of regenerator temperature distribution after attainment of steady

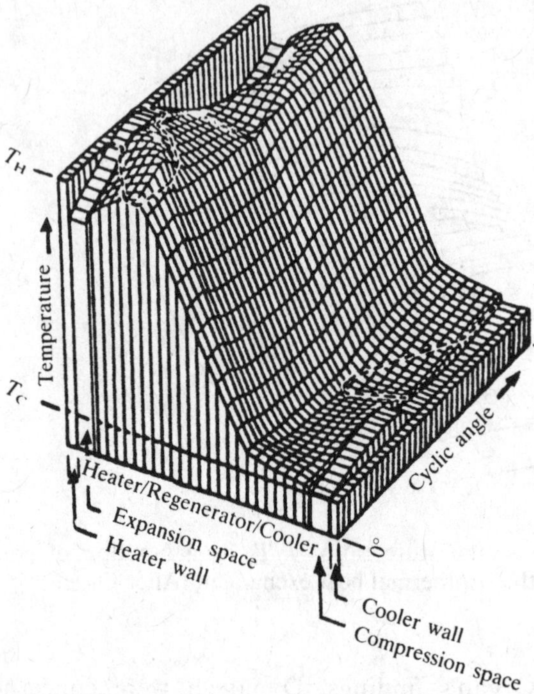


Fig. 2.11 Relief of working fluid temperature variation. (After Schock[51] and reproduced with permission of the Society of Automotive Engineers Inc.)

state. The graphical detail gives the impression of rigour, and so it should be borne in mind that the heat transfer process represented by the finite-cell method does not correspond to the classic analytical models of the regenerator as explored by Hausen, Nussel and others.

Leach and Fryer[53] appear to have been first to note the link between the flow processes in the Stirling cycle machine and the one-dimensional, unsteady compressible flow-field of classical gas dynamics. Accordingly they proposed the Method of Characteristics as the appropriate solution method, but judged that contemporary computing power (in 1968) was not adequate to permit implementation.

Concern that finite-cell schemes had incorrectly come to be regarded as the 'rigorous'[5] numerical method led Organ[16] to program a solution based on the Method of Characteristics. The approach was simplified to the extent that fluid temperature was assumed equal at all times to that of the immediately adjacent solid wall. This allowed computation to proceed along Mach lines without the need for simultaneous solution along path lines. Figure 2.12 shows a typical Mach plane, which is believed to be the first characteristics mesh generated for the Stirling cycle machine. The figure

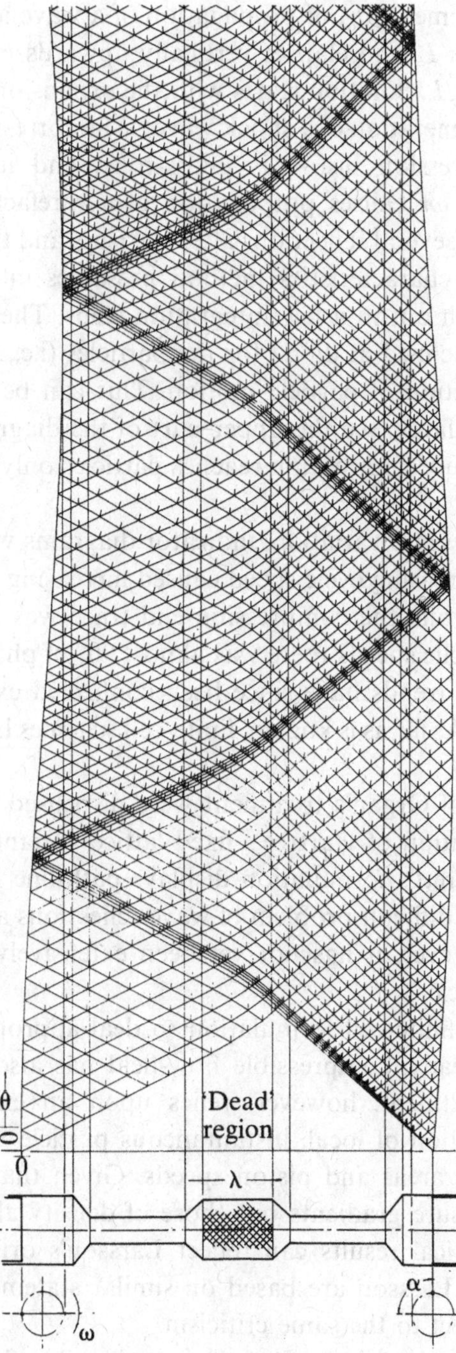


Fig. 2.12 Mach line net corresponding to operation of the machine described by Organ[16] at 4000 rpm.

employs nomenclature and format standard in gas dynamics work: $\theta$ is the ratio of time, $t$, to the time, $L/a_0$, of propagation of a wave at acoustic speed $a_0$ over reference length $L$, and elapses vertically upwards in the diagram. $\lambda$ is the reduced length, $x/L$. Solution starts with the expansion piston at outer dead centre (to avoid generation of a shock wave). The corresponding starting position of the compression piston is mid-stroke, and its instantaneous attainment of velocity $\omega r$ causes propagation of a rarefaction wave. That wave is still in evidence several reflections after start-up, and the fact evidences the way in which the characteristics method preserves information which finite-cell methods wash out in a few integration steps. The 'flags' at Mach line intersections are inclined at $d\theta/d\lambda$ for the particles (i.e., at $1/u$) and thus indicate the local direction of particle motion. This can be opposite to the motion of both piston faces (see the upper part of the diagram), supporting the principle that pressure information reaches particles only along the Mach lines.

From piston face pressure variation, indicator diagrams were constructed. These revealed that increasing cycle speed caused increasing distortion of the expansion-end trace, while the compression-end trace was significantly less affected. At extreme cyclic speeds there was almost 90° of phase shift between the two diagrams. The trends have since been confirmed experimentally by Rix (see Section 2.2.5). As far as is known, finite-cell schemes have not revealed anything comparable.

The simplifications of Organ's treatment were dispensed with by Sirett[17] in an ingenious adaptation of Hartree's fixed-lattice variant of the Method of Characteristics.[54] Sirett's formulation displays path-line information and successfully tackles the difficult problem of wave reflections at discontinuities in free-flow area. However, it has not yet been extensively used to model performance over a cycle.

The inability of the finite-cell formulation to deal appropriately with the one-dimensional, unsteady, compressible flow-field has also been noted by Larsson,[55] whose treatment, however, relies upon an empirical, explicit expression for distribution of local, instantaneous particle velocity in terms only of cross-sectional areas and piston speeds. Given that the expression neglects not only pressure gradients but those of density also, it is difficult to see how the numerical results can reflect Larsson's original intentions. Subsequent papers by Larsson are based on similar statements for velocity and are therefore subject to the same criticism.

Other attempts to apply solutions of the compressible flow equations to the Stirling machine include those of Rispoli[18] and Taylor.[56] Rispoli chose the so-called 'lamba' fixed-grid integration scheme (Fig. 2.13). The method

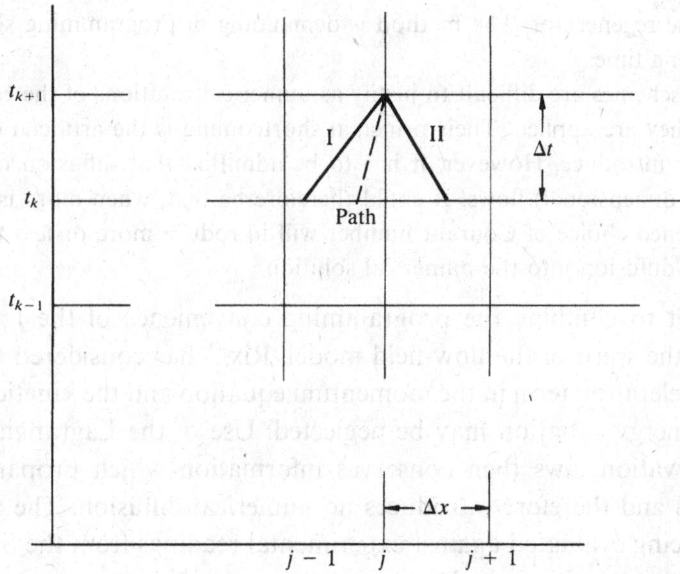


Fig. 2.13 Stencil of an integration step in the 'lambda' variant of the Method of Characteristics.

gives an initial impression of daunting complexity, but is in reality the Hartree method with provision for examination of the slopes of the I and II characteristics as a routine part of each integration step or 'unit process'. In this way an appropriate choice may be made between points along the line $t = t_k$ from which information could or could not reach the point at $j$, $k + 1$. However, the option is an unnecessary luxury in the absence of shocks, and the lambda method offers no advantages over the Hartree scheme in this case.

Rispoli's adaptation involves an analytical transformation which converts the moving faces of the opposed pistons to fixed boundaries. No explanation is offered as to the effect which the transformation has on the flow area discontinuities internal to the machine which are fixed to the original framework. Taylor's account goes no further than stating the characteristics form of the defining equations and acknowledging the insight promised into Stirling cycle thermodynamics.

In summary, the gas dynamics and engineering thermodynamics methods compare as follows:

(1) The Method of Characteristics is an analytically justifiable method of integrating numerically the equations of one-dimensional, unsteady, compressible flow. On the other hand, these equations themselves are a poor representation of conditions inside the variable-volume spaces, and a hopelessly inadequate model of flow

through the regenerator. The method is demanding of programming skills and of computing time.

(2) Finite-cell schemes are difficult to justify as numerical solutions of the equations to which they are applied. Their principal shortcoming is the artificial diffusion which they introduce. However, it has to be admitted that diffusion occurs in real (three-dimensional) flows. It could therefore be that, when more is known, an enlightened choice of Courant number will introduce more or less the right amount of diffusion into the numerical solution.

In an attempt to combine the programming convenience of the fixed-grid scheme with the spirit of the flow-field model, Rix[31] has considered the case where the acceleration term in the momentum equation and the kinetic energy term in the energy equation may be neglected. Use of the Lagrangian form of the conservation laws then conserves information which propagates at particle speed and therefore introduces no numerical diffusion. The analysis is currently being evaluated against experimental readings from the SM1 test machine with encouraging results.

A paper by Gedeon presented to a Stirling engine session of the 1984 IECEC[50] dealt with what he called 'numerical advection errors' in cycle modelling. This appears to mark the taking on board by Stirling cycle analysts of the fact that they cannot indefinitely regard their flow problem as being outside the laws which apply throughout the rest of engineering. If so, there will no doubt arise interesting debate as to the appropriate course of action for future work. The choice appears to be between:

(1) A much more critical look at actual flow processes and at the analytical models best suited to describing these. Corresponding numerical analysis and computer code promise to be more substantial and more intricate even than those based on the Method of Characteristics.

(2) Acknowledgement that appropriately linearized solutions are probably at least as accurate as current implementation of the non-linear conservation equations. With relatively little streamlining, such linearized solutions can be called many times by a suitable optimization algorithm (see Section 2.6). A recent paper by Gedeon[52] shows that his simulation work is evolving along these lines.

Even if it is assumed that lumped-parameter, nodal and finite difference schemes are obsolescent, the problem which faced Chen and Griffin, namely that of deciding which combination of defining equations gives the best representation under given operating conditions, is still of considerable interest. Indeed, a means is now available for carrying out the evaluation. It has emerged from a search for a numerical solution scheme which would provide a method of self-checking, so that a numerical solution could at least be claimed to be free of coding errors.

## 2.5 Anatomy of a flexible 'building bloc' approach to cycle simulation

In an earlier section a cycle analysis was thought of as invariably comprising the same elements regardless of complexity, i.e., of always embodying a geometric and kinematic statement, a gas law and four conservation laws or corresponding 'dummy' laws.

If attention is restricted to those solution methods which do not rely on analytical combination of the conservation laws (i.e., to the fixed-grid schemes such as the lumped-parameter or finite-cell types) it is possible to propose a common solution structure which embodies all possible analytical models and which thus permits an objective comparison of the predictions of each. The method relies on choosing the implicit discretization scheme for numerical solution. This immediately gives a common skeleton of the numerical processes in the form of a matrix (invariably sparse) of the coefficients of the unknowns ($p$, $u$, $T$ etc.) in the integration step.

The matrix is set up in five blocs – one each for the gas law and for the conservation laws (Fig. 2.14). Individual blocs are the same size for a conservation law and for the corresponding dummy, because if the flow field is so divided up as to give, say, ten values of pressure, then ten pressures are

| $i$ \ $j$ | $a_{i,j}$ | | | | | $b_i$ |
|---|---|---|---|---|---|---|
| | $\Delta\psi$ | $\Delta\tau$ | $\Delta\rho$ | $\Delta u$ | $\Delta\tau_w$ | |
| CL-1 Mass | $a_{11}a_{12}\ldots$ $a_{21}$ $a_{31}$ : | | | | | $b_1$ . |
| CL-2 Momentum | | | | | | |
| CL-3 Energy | | | | | | |
| CL-4 Energy (wall) | | | | | | |
| State Gas equation | | | | | | |

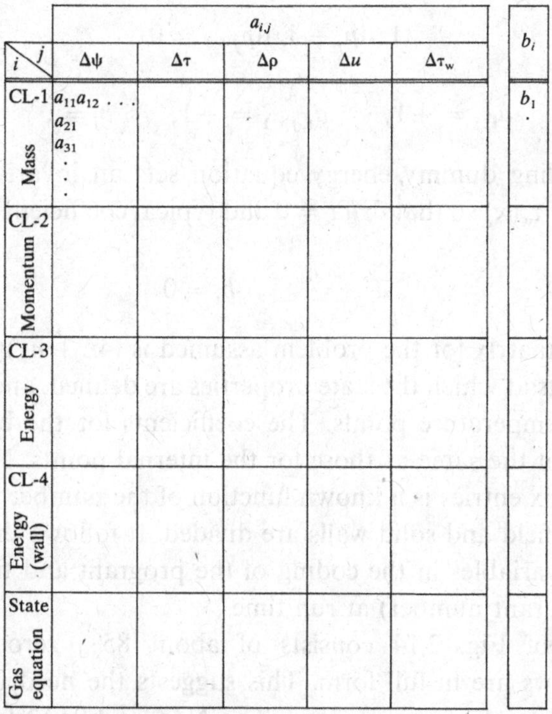

Fig. 2.14 Stencil of matrix of coefficients for 'building bloc' implicit scheme.

solved for even if they are all equal. The approach has been applied to formulations in Lagrangian as well as Eulerian coordinates.

The conservation laws and gas law are put into difference form individually (i.e., with no attempt at combination; this permits interchange of the ideal gas law with, say, the van der Waals form). Coefficients of unknowns are grouped in vertical bands, one for each dependent variable ($p$, $u$, $T$, $T_w$), as indicated in Fig. 2.14. By normalizing the conservation laws the defining dimensionless groups are factored out, thereby confirming that the solution is of functional form.

The corresponding computer code is written using one, separately compiled subroutine per bloc. Within that subroutine 'switches' capable of being set to +1 or 0 multiply the coefficients of 'optional' terms (those corresponding to acceleration in the momentum equation or to kinetic energy in the energy equation). Alternative 'dummy' subroutines are prepared corresponding to momentum, energy and wall thermal balances. These have the same names as the 'real' subroutines, but set coefficients corresponding to a 'dummy' conservation law. For example, the dummy momentum bloc sets coefficients equivalent to uniform (dimensionless) pressure, i.e., it sets $p_j = p_{j+1}$ through the discretized relationship:

$$1 \cdot \mathrm{d}p_j - 1 \cdot \mathrm{d}p_{j+1} = 0$$

or

$$a_{i,j} = +1, \qquad a_{i,j+1} = -1, \qquad b_i = 0$$

The corresponding dummy energy equation sets an invariant temperature distribution $\tau = \tau_w(x)$ so that $\partial\tau/\partial t = 0$ and typical coefficients for normalized $\tau$ are

$$a_{i,j} = 1, \qquad b_i = 0$$

The size of the matrix for the problem assumed is $(4n + n_w)^2$, where $n$ is the number of points at which the state properties are defined, and $n_w$ the number of metal wall temperature points. The coefficients for the boundary points are generally not the same as those for the internal points. Nevertheless, the number of matrix entries is a known function of the number of intervals into which the flow-field and solid walls are divided. It follows that the $n$ and $n_w$ may be left as variables in the coding of the program and fixed in terms of input data (Courant number) at run time.

The matrix of Fig. 2.14 consists of about 85% zeros when all the conservation laws are in full form. This suggests the need for an inversion routine tailored to the sparse matrix. The NAG[57] F04AXF in conjunction with F01BRF (decomposition) requires only the values of the non-zero elements (and a vector of integers indicating their location in the matrix) to be stored.

If the subprogram which supplies the coefficients for a conservation law has the same high-level programming name (e.g., CL2( ) in FORTRAN) as the subroutine which supplies the coefficients for the corresponding dummy case, then the calling program will accept whichever of the alternative versions is loaded at run time. It therefore becomes possible to set up combinations of conservation laws and dummies. This not only permits a comparison between computed results from the different combinations, but, more importantly, provides a check on the correct functioning of the program as a whole: it is self-apparent that the combination of building blocs (CL1 + DUM2 + DUM3 + DUM4 + ideal gas law) is equivalent to the CL-1 analysis. With piston motion sinusoidal and phase angle constant it is equivalent to the Schmidt analysis, for which pressure variation, mass flow rates etc. are readily obtained from closed-form algebraic expressions. Accordingly a searching check is available for correct functioning of the calling program, dummy subroutines and CL-1 matrix bloc.

In fact, the inclusion of *any* dummy bloc in isolation provides a useful check on the correctness of the discretization algebra and of programming, since the matrix is actually 'solved' for the conditions assumed in the dummy routines. For example, the dummy momentum equation (uniform pressure case) may be run in conjunction with the full form of the other conservation laws. Since pressure distribution is solved for as an unknown, an error in any bloc will show up as a non-uniform distribution of pressure inconsistent with the assumptions of the dummy equation.

The number and range of results required of the building bloc formulation to compare the various possible combinations of conservation laws is considerable and reporting of the results is left to separate discussion. It suffices to say here that the usefulness of the approach does not end with such comparison: for example, the van der Waals' gas equation (which is extremely cumbersome to substitute) is readily put into difference form for use instead of the ideal gas law. This step is virtually essential if the Stirling cooler is to be modelled.

## 2.6 Optimization

The ultimate use for the theoretical cycle model is optimization – the systematic determination of geometry and operating conditions giving best performance within specified constraints. A simple example of the process is the maximization of Schmidt cycle work for given reference pressure and reference volume. In the present discussion $V_{ref}$ has consistently been chosen to be the net volume displaced per piston/displacer set, and so the optimization process amounts to maximizing Eqn (2.6) for $\zeta$.

The relatively simple form of Eqn (2.6) invites a search for the values of $\alpha$ and $\kappa$ (i.e., of the quantities which the designer is free to vary) which give optimimum $\zeta$ for given values of the 'fixed' parameters $\tau$ and $\nu(\tau)$. The first successful attempt in this direction is due to Finkelstein.[25] His treatment is an algebraic *tour de force* which leads to almost-explicit expressions for the $\alpha$ and $\kappa$ giving best cycle work per unit mass of working fluid. The alternative problem of finding the $\alpha$ and $\kappa$ giving best work per characteristic cycle pressure and net swept volume was tackled by Walker[58] and Kirkley[59] – apparently simultaneously. These authors used numerical rather than analytical means, and presented their results in the now familiar form of the 'isothermal design chart'. There is considerable inconsistency between sets of results presented by the two authors. Finkelstein's approach has since been extended by Organ[60] to the cases covered numerically by Walker and Kirkley, and the results agree more closely with those of the former than with those of the latter. Figure 2.15 is a specimen isothermal design chart.

Takase[61] has used numerical means to find as a function of $\tau$ and $\nu(\tau)$ the proportions of the rhombic drive mechanism giving the maximum $\zeta$ predicted by the CL-1 (isothermal) analysis. Since the CL-1 treatments all indicate an efficiency equal to that of the Carnot cycle between the given values of $\tau$, the condition for efficiency to be a maximum is obvious.

It is now agreed that the results of optimization based on CL-1 treatments are of little practical value at the design stage. A starting point which intuition suggests should be somewhat better[47] is the 'adiabatic' formulation. It will be recalled that for simple harmonic volume variations this has the form:

$$\zeta_{adi} = \zeta_{adi}[\kappa, \alpha, \tau, \nu(\tau), \gamma], \quad \eta_{adi} = \eta_{adi}[\kappa, \alpha, \tau, \nu(\tau), \gamma] \qquad (2.20)$$

Walker and Khan[62] went some way towards using Eqns (2.20) to optimize $\zeta_{adi}$ and $\eta_{adi}$ by computing $\zeta_{adi}$ over a range of values of one parameter with the others held constant. Had they recalled that the only useful values of $\gamma$ are 1.4 and 1.66 (the value used by Walker and Khan is not stated) they might have been led to see that all possible values of $\alpha_{opt}$ and $\kappa_{opt}$ may be plotted into two charts, one for $\gamma = 1.4$ and one for $\gamma = 1.66$.

There are no accounts of elegant optimization work based on the more comprehensive analyses: a daunting computational effort is required for a systematic search for the best values of even a small number of variables affecting a complex simulation. In any case, if the underlying theoretical model is not a very good representation of reality when fixed data are

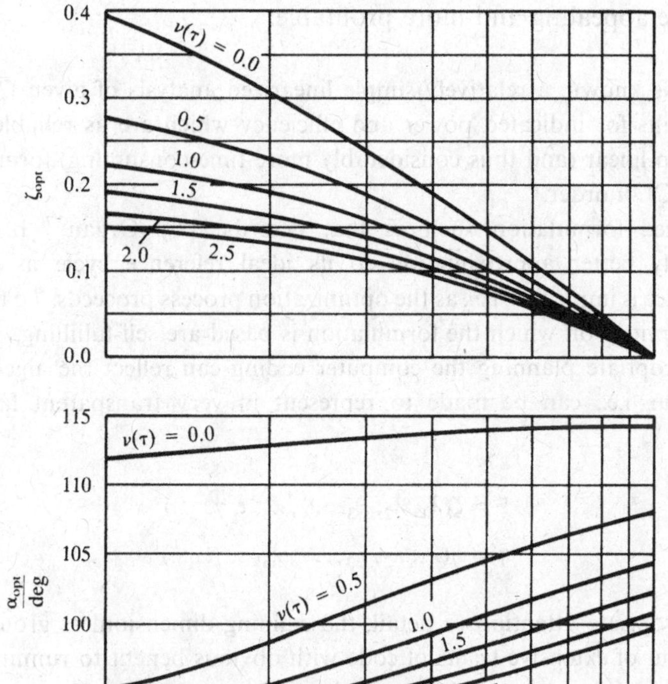

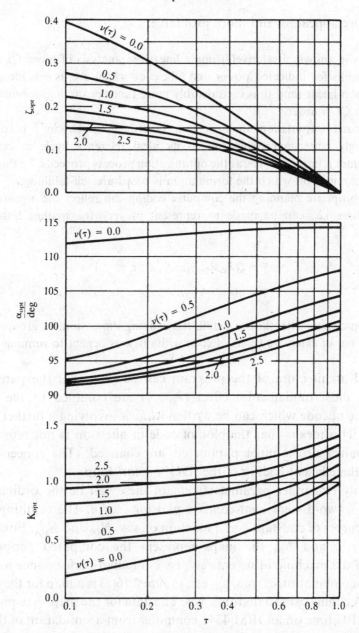

Fig. 2.15 'Isothermal' design chart for the equivalent opposed-piston machine.

supplied, then any 'optima' found by systematic variation of such data will, except by chance, be misleading. Heames and Daley[63] and Gedeon[38] are nevertheless pursuing this approach.

There is an alternative which, if the contentions below are accepted, may be both more appealing and more profitable:

(1) As far as is known, a relatively simple linearized analysis of given CL-$n$ order yields results for indicated power and efficiency which are as reliable as those from a non-linear (and thus considerably more time-consuming) formulation of the same CL-$n$ order.
(2) A linearized formulation such as that described by Organ[27] becomes an increasingly better approximation to its ideal reference cycle as computed performance is improved, i.e., as the optimization process proceeds. To this extent, the assumptions on which the formulation is based are self-fulfilling.
(3) With appropriate planning the computer coding can reflect the algebra of the formulation, i.e., can be made to represent in very transparent fashion the functions

$$\zeta = \zeta(\lambda_1, \lambda_2, \lambda_3, \ldots, \lambda_n, \tau, \gamma, \ldots)$$
$$\eta = \eta(\lambda_1, \lambda_2, \lambda_3, \ldots, \lambda_n, \tau, \gamma, \ldots)$$

(4) With appropriate attention to detail, the defining dimensionless groups can be factored out of extensive tracts of code with obvious benefit to running time.

The overall architecture of the program can be such that the parameters having the most fundamental effect ($\lambda_n$, $\tau$, $\gamma$) are confined to the largest possible bloc of code which can be written *without* involving a further 'inner' parameter. This means that the bloc of code in question is not reprocessed when the values of the inner parameters are changed. (The concept has a parallel in the efficient coding of the 'DO' or 'for' loop.)

When only two inner parameters remain these can be the ordinate and abscissa of a two-dimensional contour plot for $\zeta$ or $\eta$. The resulting 'map' comprises curves of constant $\zeta$ as a function of, say, $N_{RE}$ and $N_{MA}$. For chosen geometry, $\tau$, $\gamma$ and $N_{PR}$ the maps represent the computed performance envelope of the machine in its entirety, i.e. computed performance for *every* practicable combination of rpm, $p_{ref}$ etc. Figure 2.16(a) is a map for the Philips MP1002CA engine using a diatomic gas. The data for the map were produced in 13 s of CPU time on an IBM 4341 computer from a simulation of the type described by Organ.[27] CPU time on the same machine to produce the same quantity of information using a simulation formulated conventionally in terms of absolute variables would be in terms of hours.

For the rpm and charge pressure cited the $N_{RE}$ and $N_{MA}$ for the Philips engine are 1.64E + 04 and 0.30E − 02 respectively, giving $\zeta \approx 0.2$ at the intercept indicated. Corresponding indicated power is readily calculated to be 636 W. Assuming a mechanical efficiency of 75%, $\zeta_{brake}$ = Beale number,

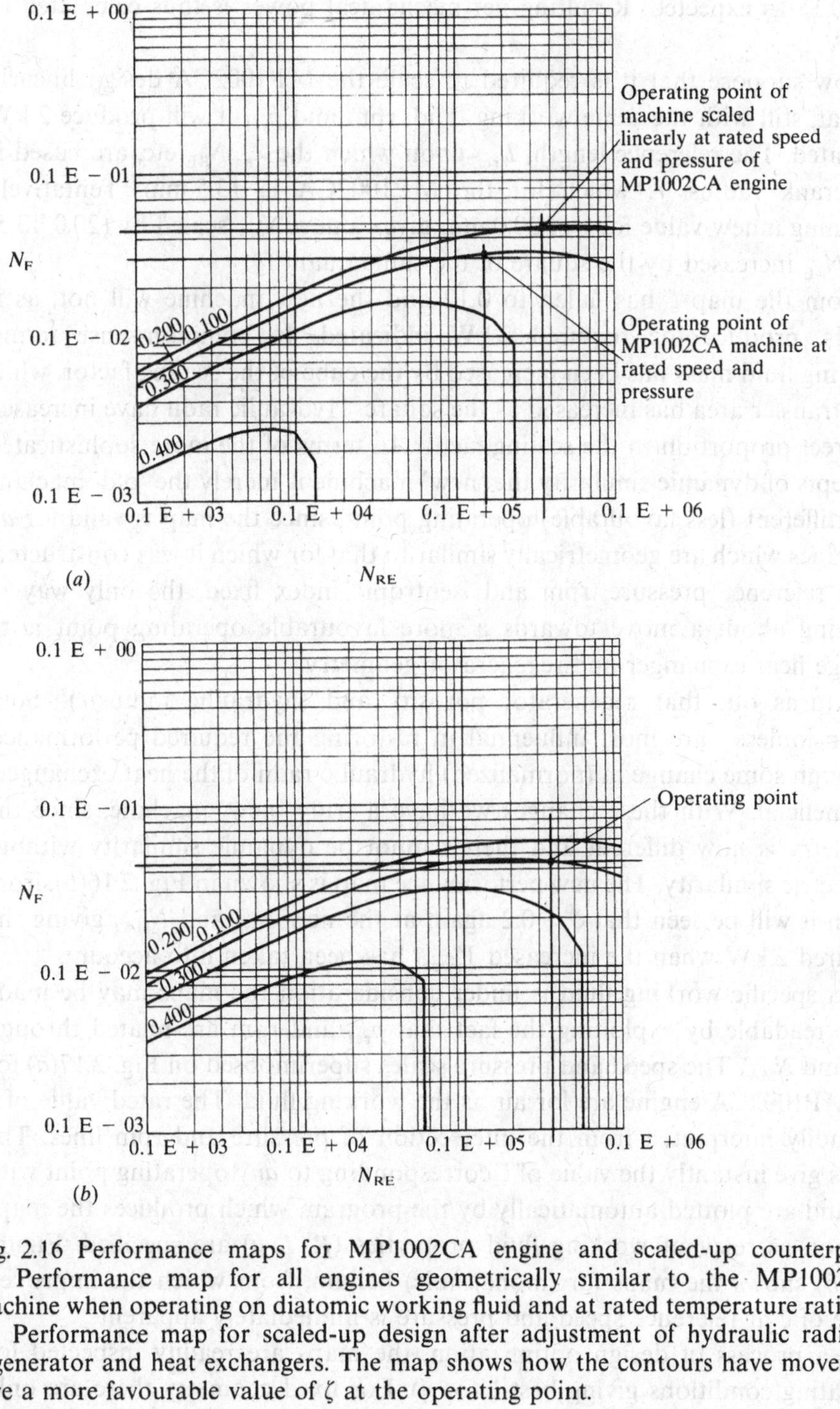


Fig. 2.16 Performance maps for MP1002CA engine and scaled-up counterpart. (*a*) Performance map for all engines geometrically similar to the MP1002CA machine when operating on diatomic working fluid and at rated temperature ratio, $\tau$. (*b*) Performance map for scaled-up design after adjustment of hydraulic radii of regenerator and heat exchangers. The map shows how the contours have moved to give a more favourable value of $\zeta$ at the operating point.

$B = 0.15$ as expected. Resulting net mechanical power is thus calculated at 477 W.

Now suppose that it is required to scale the MP1002CA design linearly so that, still with the same working fluid, rpm and $p_{ref}$ it will produce 2 kW indicated. The reference length, $L_{ref}$ upon which the $\lambda_n$, $N_{RE}$ etc. are based is the crank radius, $r$, which for the MP1002CA is 13.5 mm. Tentatively assuming a new value for $r$ of 20.0 mm gives a new $N_{MA}$ scaled by (20.0/13.5) and $N_{RE}$ increased by the square of the same quantity.

From the map $\zeta$ has fallen to 0.12 and the new machine will not, as it stands, produce the required 2 kW indicated. In common-sense terms, working fluid mass has been increased by the cube of the scaling factor, while heat transfer area has increased as the square. Hydraulic radii have increased in direct proportion to the scaling factor. In terms of the more sophisticated concepts of dynamic similarity the 'new' machine is merely the 'old' machine at a different (less favourable) operating point, since the map is valid for *all* machines which are geometrically similar to that for which it was constructed. With reference pressure, rpm and isentropic index fixed, the only way of bringing about a move towards a more favourable operating point is to change heat exchanger and regenerator geometry.

It turns out that regenerator porosity and (hydraulic radius/$r$) – both dimensionless – are most influential in restoring the required performance, although some change in (normalized) hydraulic radii of the heat exchangers is beneficial. With these changes we have a truly 'new' machine, since the geometry is now different and there cannot be dynamic similarity without geometric similarity. The new performance map is shown in Fig. 2.16(*b*), from which it will be seen that $\zeta \approx 0.2$ again at the new $N_{RE}$ and $N_{MA}$, giving the required 2 kW when the increased $V_{swept}$ has been taken into account.

If a specific working fluid is under consideration the maps may be made more readable by exploiting the fact that $p_{ref}$ and rpm are related through $N_{RE}$ and $N_{MA}$. The speed and pressure scales superimposed on Fig. 2.17(*a*) for the MP1002CA engine are for air as the working fluid. The rated value of $\zeta$ is readily interpreted from the intersection of pressure and rpm lines. The scales give instantly the value of $\zeta$ corresponding to *any* operating point with air, and are plotted automatically by the program which produces the maps as soon as relevant working fluid properties ($R$, $T$, $\mu$) are specified. Figure 2.17(*b*) shows the maps for (monatomic) helium, from which the improved value of $\zeta$ at reference speed and pressure is immediately apparent.

In a process of design optimization, the maps are readily inspected for operating conditions giving best $\zeta$ (or $\eta$). For fixed geometry there are only two basic maps, and at the design evaluation stage $\tau$ is essentially fixed.

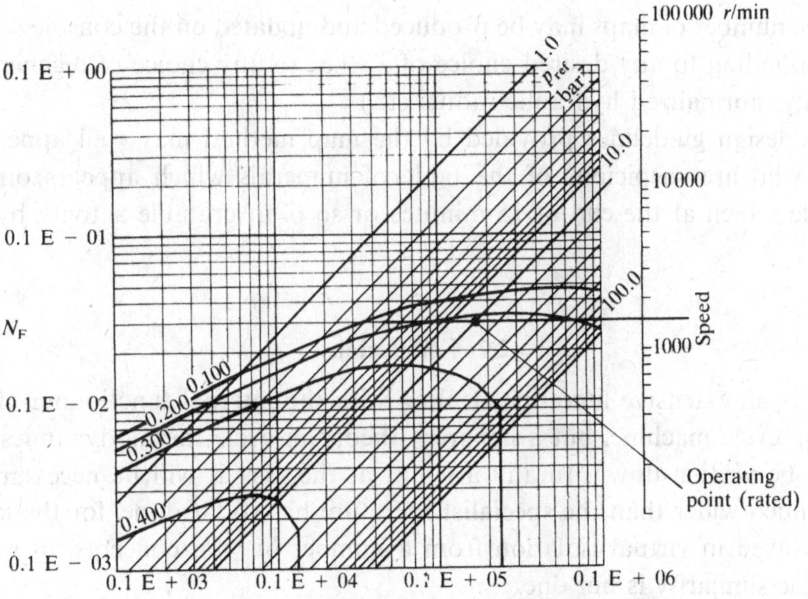


(a)

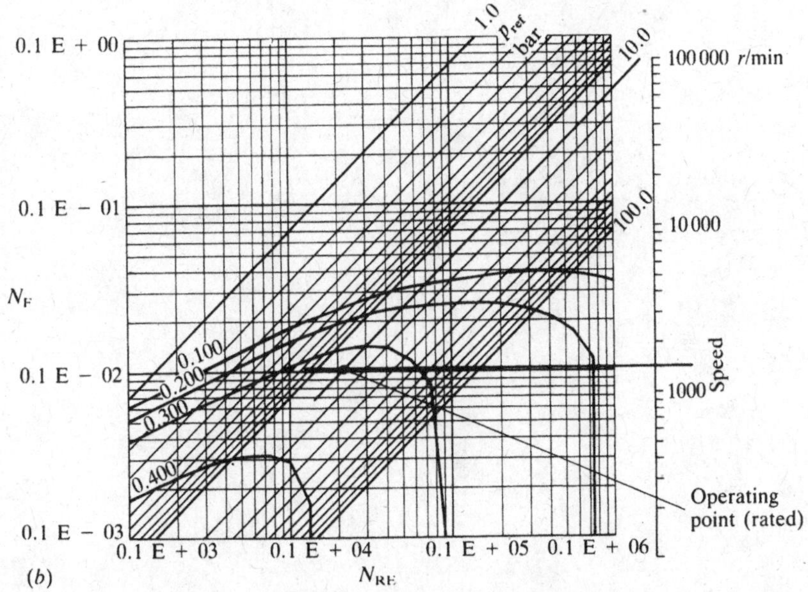


(b)

Fig. 2.17 Specimen performance maps showing how engine speed and pressure scales may be superimposed when working fluid is specified. (*a*) Performance map of MP1002CA engine at rated temperatures and with air as working fluid. (*b*) Performance map with monatomic gas helium (rated temperatures).

A large number of maps may be produced and updated on the console screen corresponding to any desired choice of $\lambda_n$ (i.e., to any choice of regenerator porosity, normalized hydraulic radius etc.).

The design guidelines provided by the map method may well appeal to those who are suspicious of the table of numerals which appears on the console screen at the end of 15 minutes or so of inscrutable activity by the CPU.

## 2.7 Conclusions

There is an extensive literature dealing with the internal functioning of the Stirling cycle machine, but a coherent theory of cycle thermodynamics has yet to be written down. In any attempt at the task it will be necessary to look much wider than the specialist literature has so far done, for the latter has evolved in virtual isolation from a number of vital concepts, of which dynamic similarity is but one.

The chapters which follow make inroads into this important undertaking.

# 3

# An ideal thermodynamic reference cycle

## 3.1 Ideal cycles

The true reason for the existence of ideal thermodynamic cycles is probably to be found in the insatiable academic urge to order and to classify. Appropriately interpreted, however, such cycles can serve the function of indicating limits to the thermodynamic potential of the machine to which they relate.

To be of value in this function it must be possible to conceive of conditions for which the thermodynamic processes of the prototype machine and those of the ideal cycle coincide. Such conditions may be in opposition to the demands of practicality (e.g., infinitely slow speed of operation) but are valid if they represent an *extrapolation* to the extreme of some condition which it would be possible, at least in principle, to vary experimentally. A practical example is the extrapolation to zero of the variable fuel/air ratio of the diesel engine during experimental determination of the Willans' line.

The standard 'text-book' ideal Stirling cycle does not meet the criterion. A text committed to the definition of a proper approach to thermodynamic design will therefore require some alternative ideal cycle definition which does, and the purpose of this chapter is thus identified. In order to explain the inadequacy of the text-book cycle it will be presented in full. Thereafter, usage of the term 'ideal cycle' will be reserved for a sequence of thermodynamic processes yet to be explained.

The matter of finite heat transfer rates will be considered. From this and from the treatment of the ideal cycle will emerge two insights which (*a*) are not revealed by the 'text-book' cycle and which (*b*) can become lost in more sophisticated analyses:

(1) *All other things being equal, a diatomic working fluid requires a significantly larger cyclic regenerative heat exchange than does a monatomic gas. Implications for performance and design are self-evident.*

(2) *Attainment of the ideal, limiting thermal efficiency (or COP) is as undesirable as it is impossible.*

For the purposes of the present chapter, and for the rest of this book, the definition of cycle temperature ratio is changed from that which has been in universal use so far (see Notation section). Thus the symbol $N_\tau$ will from now on stand for the ratio $T_e/T_c$ in recognition of the fact that in the case of both prime movers and coolers it is nominal expansion-end temperature, $T_e$, which is the design variable, and compression-end temperature, $T_c$, which is generally linked to ambient conditions and which is thus the appropriate choice of reference quantity. ($T_c$ will frequency coincide numerically with $T_0$, the reference temperature for expressions involving the concept of availability, and for this reason alone is the appropriate choice for the denominator.)

## 3.2 'Text-book' cycle

The cycle is commonly explained[1,2,3] by reference to a diagram such as that shown at Fig. 3.1

1–2: Isothermal compression of unit mass of working fluid at $T_c$ from specific volume $v_1$ to $v_2$ and specific entropy $s_1$ to $s_2$. Heat transferred/kg (negative) = work output, $W_c$/kg (negative) = $RT_c \ln(\mathbf{r}) = T_c(s_2 - s_1)$, where $\mathbf{r} = v_2/v_1$.

2–3: Reversible heat exchange at constant specific volume with regenerator having infinite thermal capacity. Temperature of entire mass changes uniformly from $T_c$ to $T_e$ and specific entropy from $s_2$ to $s_3$.

3–4: Isothermal expansion at $T_e$ with work output $W_e/\text{kg} = RT_e \ln(1/\mathbf{r})$. Magnitude of heat exchange $Q_e/\text{kg} = T_e(s_4 - s_3) = W_e/\text{kg}$.

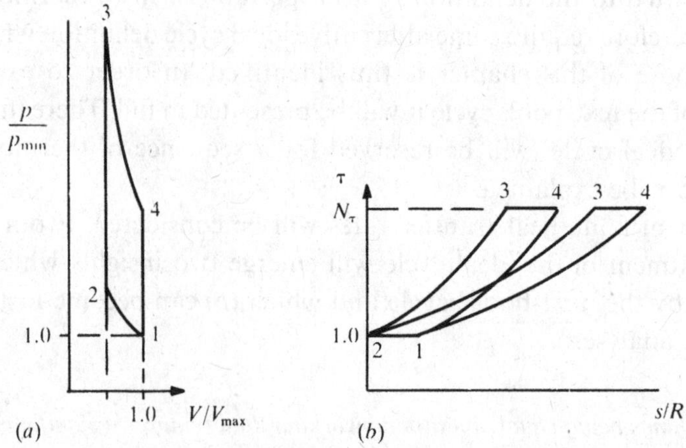


Fig. 3.1 (*a*) Indicator and (*b*) fluid property diagrams for the 'text-book' cycle.

4–1: Reversible exchange of heat with regenerator at constant specific volume. Change in specific entropy from $s_4$ to $s_1$.

For the text-book cycle:

(1) if mass $M$ of gas is involved, Fig. 3.1(*a*) becomes the indicator ($p$–$V$) diagram when scaled by $M$;
(2) plots of $p/p_{ref}$ vs $V/V_{ref}$, where $V_{ref}$ is the value of $V$ for which $p = p_{ref}$, are identical for all ideal gases;
(3) from the standard thermodynamic relationship $(s_b - s_a)/R = (\ln(T_b/T_a))/(\gamma - 1) + \ln(v_b/v_a)$ the entropy change between different temperature limits at constant specific volume is greater for a diatomic than for a monatomic gas. *All other things being equal, the diatomic gas imposes a heavier cyclic thermal load on the regenerator.*

With $W$ for work per cycle and $V_{sw} = M(v_1 - v_2)$ for swept volume:

$$\zeta = \frac{W}{MRT_c} = \frac{W}{p_{min}V_{sw}/(1 - 1/\varepsilon)} = (N_\tau - 1)\ln(\varepsilon) \tag{3.1}$$

$$\varepsilon = \text{expansion ratio} = 1/\mathbf{r}$$

Fig. 3.2 displays Eqn (3.1) in the form of a $Z$-chart for instant evaluation of $\zeta$ as a function of $N_\tau$ and $\varepsilon$. It shows that, although there is no optimum of either $N_\tau$ or $\varepsilon$ there are diminishing returns for increasing the latter.

There is no conceivable mechanical embodiment of the above text-book cycle, even allowing the traditional assumptions of infinitesimal temperature gradients, absence of viscous effects, zero unswept volume etc. Those texts (e.g., Refs 2 and 3) which explain the foregoing thermodynamic sequence by reference to pistons, cylinders etc. perpetuate a conceptual inconsistency. The text-book cycle should be set aside as a criterion of work-producing potential; it does not bear the same relationship to the realities of a practical Stirling machine as do the air standard Otto and Brayton cycles respectively to the realities of the petrol engine and gas turbine. Definition of a viable reference cycle begins with a conceivable physical configuration.

## 3.3 Configuration

In Fig. 3.3:

$M$ kg of elastic working fluid are contained between the faces of two movable pistons.

The working fluid everywhere adopts the temperature of the immediately adjacent containing surface.

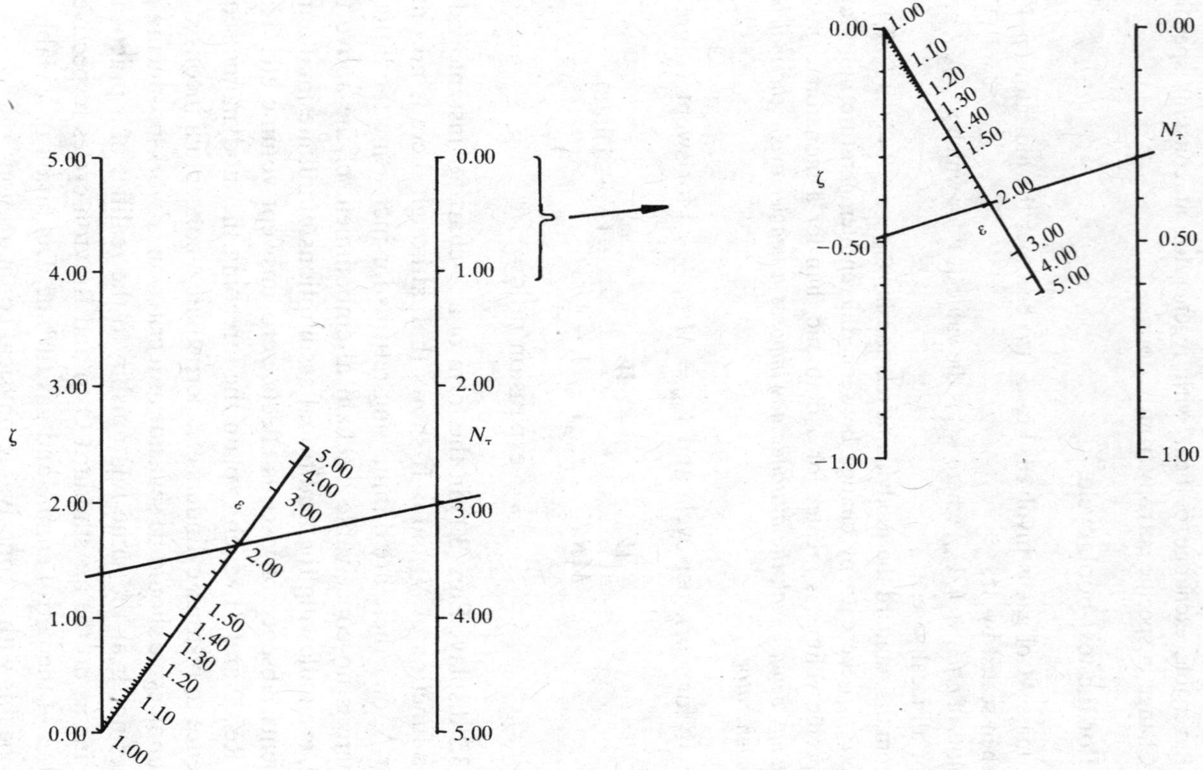


Fig. 3.2 *Z*-charts for graphical evaluation of Eqn (3.1).

The (variable) compression space and its adjacent heat exchanger are held at fixed temperature $T_c$. The variable-volume expansion space and its associated heat exchanger are at $T_e$.

There is a linear variation of temperature from $T_c$ to $T_e$ across the regenerator.

All spaces are in open communication with each other.

Effects which depend on time are not taken into account: instantaneous pressure is common to all working fluid elements.

## 3.4 Ideal reference cycle and indicator diagram

In Fig. 3.4 $T_c < T_e (N_\tau > 1)$ so the cycle of operations to be described is that of a prime mover. Analogous reasoning describes operation as cooler or heat pump.

1–2: Overall reduction in volume from $V_1$ to $V_2$ as compression space piston moves from right to left. In the indicator diagram (Fig. 3.4(*a*)) the relationship between pressure, $p$, and volume $V$ is *not* isothermal except for the trivial case of $N_\tau = 1$ and/or dead space = 0. Specific volume and specific entropy of particle $x$ reduce isothermally (fluid property diagrams – Figs. 3.4(*b*) and (*c*)) provided the particle does not reach the regenerator.

2–3: Shift of working fluid in direction of expansion space at constant overall volume. *Specific* volume and entropy of particle $x$ decrease isothermally with rise in pressure (2–2(*a*)) and then increase at variable temperature on the path through the regenerator (2(*a*)–3).

3–4: Increase in overall volume with movement of expansion piston. Again the pressure volume trace of the indicator diagram is *not* isothermal except for the special cases mentioned. Particle $x$ never emerges from the expansion end of the regenerator, but we can see what would happen if it did by focussing on particle $x'$ to its left. This continues through the regenerator (3–3(*a*)) with increase in $v$ and $s$ until it emerges into the expansion space. Thereafter (3(*a*)–4) $v$ and $s$ increase isothermally.

4–1: Both pistons move in such a way as to keep total enclosed volume constant and to move the working fluid in the direction of the compression space. The increasing fraction of $M$ now occupying the space at lower temperature causes overall pressure to fall and an accompanying increase in $v$ and $s$ of particle $x$ (4–4(*a*)) for as long as it remains in the expansion space at $T_e$. Thereafter, the fall in temperature during reverse passage through the regenerator (4(*a*)–(*b*)) causes $v$ and $s$ to decrease. There is a final, isothermal expansion of particle $x$ from 4(*b*)–1 with resultant increases in $v$ and $s$. The corresponding phase on the indicator diagram is an isochoric (constant volume).

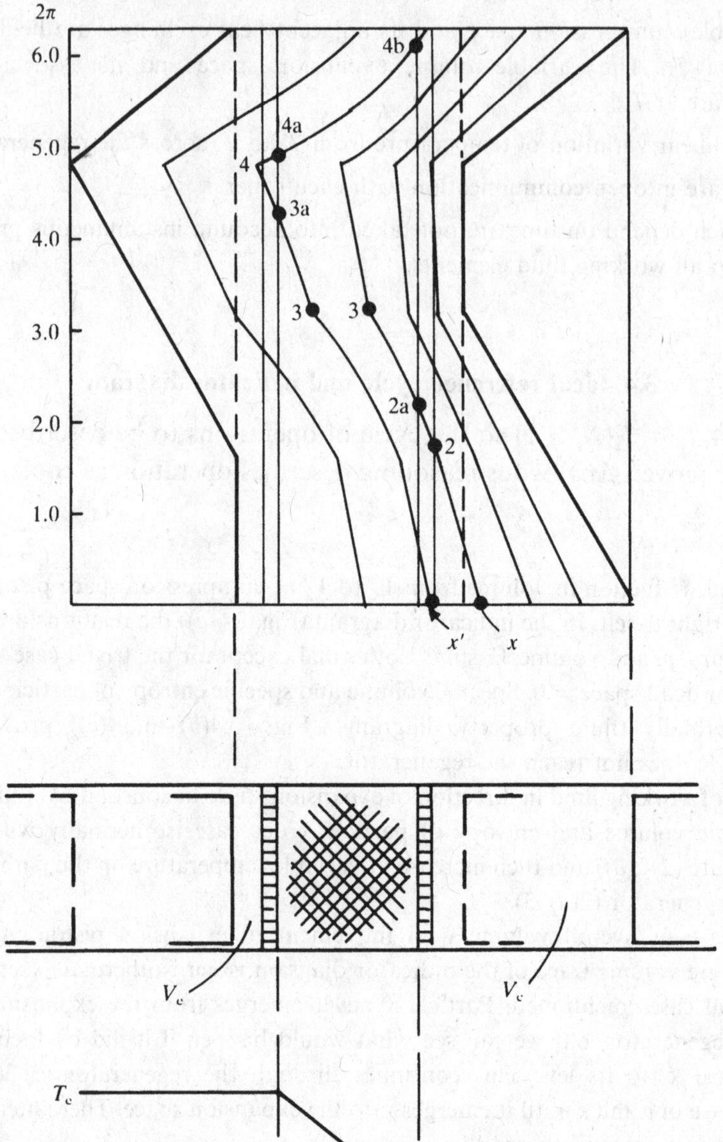


Fig. 3.3 Schematic of practical opposed-piston Stirling machine. The discontinuous piston motion is a function of 'crank' angle, $\phi$, which increases vertically upwards in the motion diagram from 0 to a cycle maximum of $2\pi$.

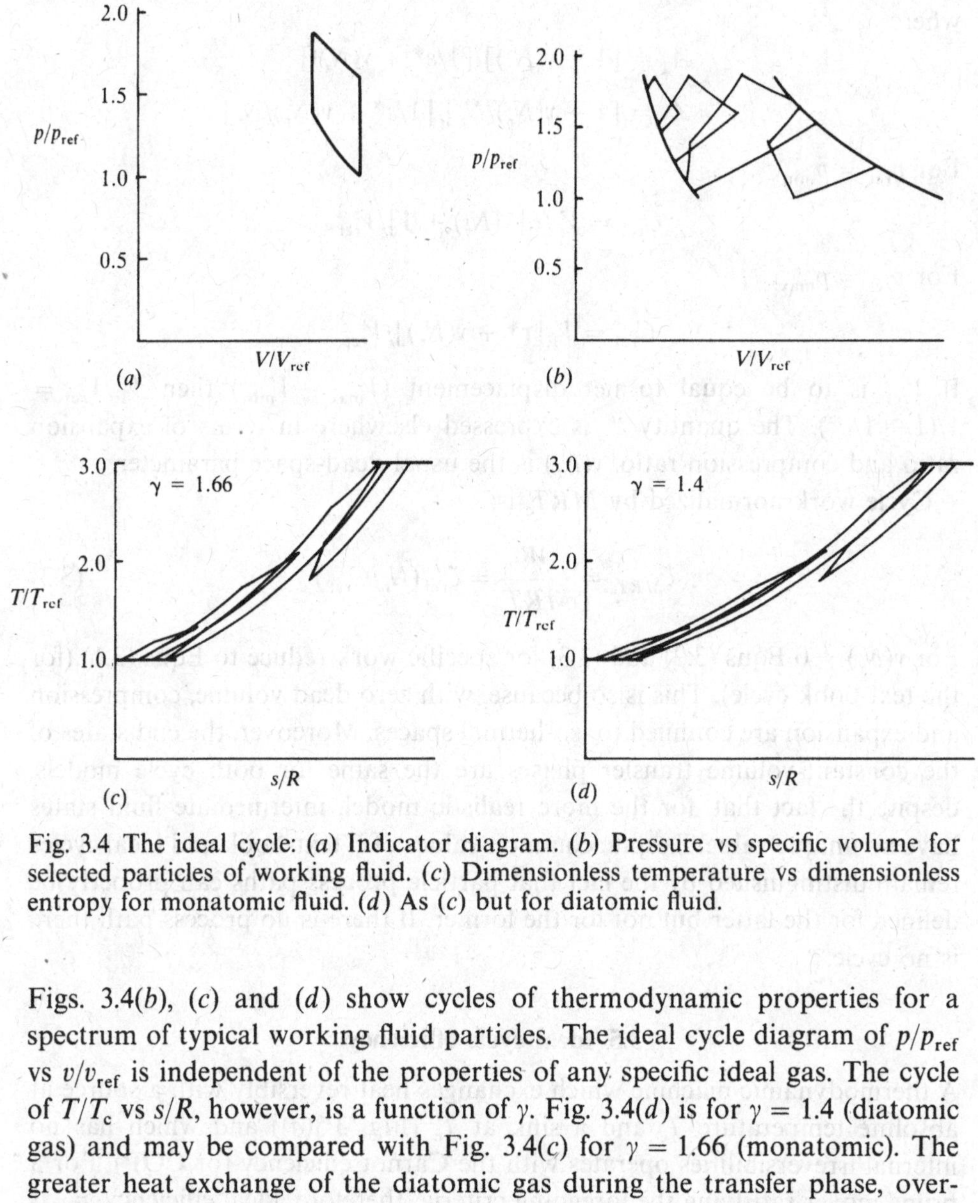


Fig. 3.4 The ideal cycle: (*a*) Indicator diagram. (*b*) Pressure vs specific volume for selected particles of working fluid. (*c*) Dimensionless temperature vs dimensionless entropy for monatomic fluid. (*d*) As (*c*) but for diatomic fluid.

Figs. 3.4(*b*), (*c*) and (*d*) show cycles of thermodynamic properties for a spectrum of typical working fluid particles. The ideal cycle diagram of $p/p_{ref}$ vs $v/v_{ref}$ is independent of the properties of any specific ideal gas. The cycle of $T/T_c$ vs $s/R$, however, is a function of $\gamma$. Fig. 3.4(*d*) is for $\gamma = 1.4$ (diatomic gas) and may be compared with Fig. 3.4(*c*) for $\gamma = 1.66$ (monatomic). The greater heat exchange of the diatomic gas during the transfer phase, overlooked by the text-book cycle, shows up in this more realistic ideal model. *It indicates that regenerator design for air as working fluid may have to be different from that for helium if the lower thermal conductivity of the former gas is not to give rise to a performance penalty.*

Thermal efficiency is equal to that of the text-book cycle. Specific work of the ideal cycle, $\zeta$, may be calculated (Appendix II) from:

$$\zeta_{ref} = W/p_{ref}V_{ref} = C_{ref}[\ln(A_1) - \ln(A_2)/N_\tau] \qquad (3.2)$$

where

$$A_1 = [1 + \nu(N_\tau)]/[1/\varepsilon^* + \nu(N_\tau)]$$

$$A_2 = [1 + \nu(N_\tau)/N_\tau]/[1/\varepsilon^* + \nu(N_\tau)/N_\tau]$$

For $p_{ref} = p_{min}$,

$$C_{ref} = V_E N_\tau[\nu(N_\tau) + 1]/V_{ref}.$$

For $p_{ref} = p_{max}$,

$$C_{ref} = V_E[\mathbf{r}^* + \nu(N_\tau)]/V_{ref}.$$

If $V_{ref}$ is to be equal to net displacement $(V_{max} - V_{min})$ then $V_E/V_{ref} = 1/(1 - 1/\varepsilon^*)$. The quantity $\varepsilon^*$ is expressed elsewhere in terms of expansion ratio and compression ratio. $\nu(N_\tau)$ is the usual dead-space parameter.

Cycle work normalized by $MRT_c$ is:

$$\zeta_{MRT_c} = \frac{W}{MRT_c} = \zeta_{ref}(N_\tau/C_{ref}) \tag{3.3}$$

For $\nu(N_\tau) = 0$ Eqns (3.2) and (3.3) for specific work reduce to Eqns (3.1) (for the text-book cycle). This is so because, with zero dead volume, compression and expansion are confined to isothermal spaces. Moreover, the end states of the constant-volume transfer phases are the same for both cycle models, despite the fact that, for the more realistic model, intermediate fluid states have no unique value. Despite common values of $\zeta$, text-book and ideal cycles remain distinguished by the fact that particle process paths can properly be defined for the latter but not for the former. If there is no process path there is no cycle.

## 3.5 Ideal cycle efficiency

A thermodynamic machine which exchanges heat reversibly with a source at absolute temperature $T_e$ and a sink at $T_c$ (Fig. 3.5(*a*)) and which has no internal irreversibilities operates with the Carnot efficiency (or COP). For a prime mover satisfying the foregoing criteria, therefore, ideal efficiency, $\eta_c$, is given by

$$\begin{aligned}\eta_c &= W/Q_e = (Q_e - Q_c)/Q_e = (T_e\,ds - T_c\,ds)/T_e\,ds \\ &= 1 - 1/N_\tau \end{aligned} \tag{3.4}$$

The Stirling cycle machine embodies a thermal regenerator, the purpose of which is to exchange with the working fluid that heat which would otherwise be exchanged with the environment at temperatures intermediate to $T_e$ and

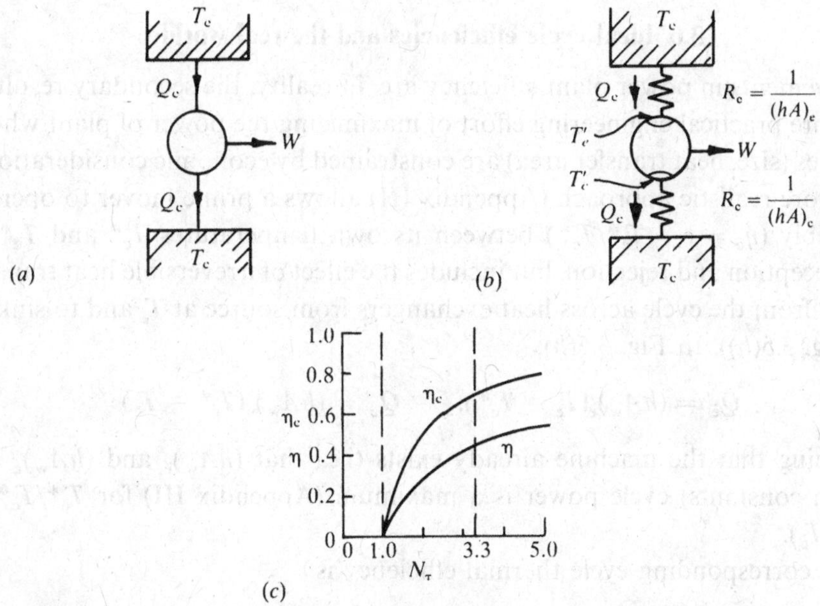


Fig. 3.5 Efficiency of the ideal cycle modified for the effects of heat exchange. (*a*) Ideal machine exchanging heat reversibly with source at $T_e$ and sink at $T_c$. (*b*) Thermal resistance between source and reversible machine and between reversible machine and sink. (*c*) $\eta_c$ and $\eta$ as a function of temperature ratio, $N_\tau$.

$T_c$. It thus has $\eta_c$ as the limiting value of its efficiency. This fact has been used in justification of research into and development of Stirling prime movers as potential replacements for conventional reciprocating plant, and there are reports (as, e.g., in Ref. 4) of Stirling engines returning thermal efficiencies of the order of 40% in the laboratory. The heat transfer processes of the ideal cycle suppose (*a*) a reservoir whose temperature does not change as heat is removed or added and (*b*) thermal equilibrium during heat transfer interactions.

The supposed equilibrium requires either an infinitely slow cycle or an infinitely large product of heat transfer area, $A_w$, with mean effective heat transfer coefficient, $h$. Incorporation of these assumptions reflects an outmoded view which deserves to be replaced by the new emphasis in fundamental engineering thermodynamics exemplified by the work of Bejan[5]. In the world of real engineering that transfer area and time are expensive commodities. Infinite heat transfer and infinite time are unaffordable. The thermal efficiency of the ideal cycle is thus not merely an unattainable goal, but a goal whose hypothetical attainment would have unacceptable consequences.

## 3.6 Ideal cycle efficiencies and the real world

Achievements in power plant efficiency are, in reality, the secondary result of the more practical engineering effort of maximizing the power of plant whose qualities (size, heat transfer area) are constrained by economic considerations. The more realistic approach (Appendix III) allows a prime mover to operate reversibly ($\eta_c = 1 - T_c^*/T_e^*$) between its own temperatures $T_e^*$ and $T_c^*$ of heat reception and rejection, but includes the effect of irreversible heat transfer to and from the cycle across heat exchangers from source at $T_e$ and to sink at $T_c$ (Fig. 3.5(*b*)). In Fig. 3.5(*b*):

$$Q_e' = (hA_w)_e(T_e - T_e^*), \qquad Q_c' = (hA_w)_c(T_c^* - T_c)$$

Assuming that the machine already exists (i.e., that $(hA_w)_e$ and $(hA_w)_c$ are known constants) cycle power is a maximum (Appendix III) for $T_e^*/T_c^* = \sqrt{(T_e/T_c)}$.

The corresponding cycle thermal efficiency is

$$\eta = 1 - 1/\sqrt{N_\tau} \tag{3.5}$$

Fig. 3.5(c) shows $\eta_c$ and $\eta$ as functions of $N_\tau$. For the typical prime mover case, $T_e = 1000$ K and $T_c = 300$ K, $N_\tau = 3.3$ and $\eta_c = 66.6\%$; the corresponding $\eta = 42.3\%$. The latter figure tallies with figures claimed for highly-developed Stirling prime movers. The foregoing picture suggests that higher efficiencies (between $\eta$ and $\eta_c$) could be achieved in practice only at unacceptable cost of excessive heat exchange area and low specific power.

The ultimate value of a method of thermodynamic design resides in the scope it offers for *scaling*, i.e., of predicting performance of a new, untried design from a different design of known performance. The ideal cycle is inadequate for the purpose, but affords an insight into the principles involved.

Specific cycle work at maximum efficiency allowed by thermal resistance at source and sink may be written (Appendix III):

$$\zeta_{max} = 30N_\tau[1 - \sqrt{(1/N_\tau)}]^2 \frac{hT_{ref}}{Np_{ref}r_{he}}\mu_{de} \tag{3.6}$$

Suppose it is proposed to extrapolate a design of proven performance at given temperature ratio, $N_\tau$, and given working fluid by a linear scaling of all dimensions. A factor of 2, for example, would give an increase in swept volume by a factor of 8, and all other things remaining equal, including pressure and rpm, would imply an increase in power by the same factor. Dimensionless expansion-end dead space will remain the same with the increased size, but hydraulic radius, $r_{he}$, will double. $h$ is not going to increase by a factor of 2 to compensate. $\zeta_{max}$ is unlikely to increase, and should preferably not decrease, so the change in $r_{he}$ must be offset by a corresponding decrease in the product of speed, $N$, with reference (charge) pressure, $p_{ref}$. Such a decrease is equivalent to an undesirable reduction in specific power. The required eightfold increase may yet be attainable, but only with a change in the ratio of exchanger surface area to mass of enclosed working fluid (effectively of $\mu_{de}$ etc.), i.e., only if geometric similarity is abandoned.

# 4
# Dynamic similarity

## 4.1 Towards a coherent analysis of the practical cycle

To the extent that a substantial number of Stirling cycle machines have now demonstrated viable values of specific power and of brake thermal efficiency, the problem of thermodynamic design is essentially that of having a means of *scaling* from one proven operating point to give satisfactory performance at another. The fact is acknowledged implicitly in Creswick's elegant and perceptive treatment,[1] and explicitly by Gedeon.[2] Creswick stops short of cataloguing the defining dimensionless parameters (otherwise there might have remained little for the rest of us to contribute!). Gedeon's treatment merges dimensional analysis with empirical correlation and with solution of simultaneous equations. In so doing it forfeits the simplicity essential for its declared task. Organ[3–5] has isolated the defining groups for a simplified cycle model which supposed invariant regenerator matrix temperature distribution. Work by Rix[6] has provided strong confirmation that the groups in question are part of the set which determine dynamic similarity.

The approach adopted here will be to define the similarity conditions (i.e., to identify the defining dimensionless groups) for a specified, one-dimensional model of the practical cycle. Application to cycle models of any additional degree of sophistication is a direct extension of the principles dealt with. It will be shown that an analytical treatment – together with any corresponding computer code – intended for use in performance optimization should be formulated in terms of these or equivalent dimensionless groups.

Where exploratory thermodynamic design is to be undertaken, a modified approach, to be called 'thermodynamic similarity', will be recommended. This relieves the designer of the need to specify intimate geometric details of the tentative design as a starting point: it leads to a specification permitting considerable leeway in heat exchanger and regenerator flow passage geometry and in crank kinematics while ensuring required, nominal performance.

## 4.2 A one-dimensional gas-side model

Fig. 4.1 is a schematic of a representative gas circuit. The following are assumed:

(1) The working fluid behaves as an ideal gas.
(2) Flow is one-dimensional, and described by the partial differential (flow-field form) of the laws of conservation of mass, momentum and energy.
(3) Piston motion, and hence working space volume variations, are capable of description in terms of some independent variable (time or crank angle) and the kinematics of a drive mechanism. (The approach is applicable to free-piston variants, but the fact is not demonstrated here.)
(4) Flow passage geometry is specified in terms of cross-sectional flow area, $A_x$, hydraulic radius, $r_h$ and length, $L$.
(5) The temperatures of the inside walls of heater and cooler do not vary with time (but see item (7) below).
(6) The structure of the regenerator may be described geometrically (e.g., in terms of mesh number, $m$ and wire diameter, $d_w$ of a woven wire screen).
(7) The temperature of individual solid elements of the regenerator varies with time and location. Variation within a given solid element in a plane perpendicular to flow direction is typical of that within all elements in the same plane. The treatment can be extended (see Chap. 9) to the solid walls of the heat exchangers, to allow simplification (5) above to be dispensed with.
(8) Published steady-flow correlations may be used for a first estimate of heat transfer coefficient and friction factor in terms of geometry and local, instantaneous Reynolds number. (In Chap. 5 it will be shown that standard correlations for the regenerator in particular are inappropriately formulated *even for the steady-flow case*. Alternative, steady-flow correlations will be proposed.)

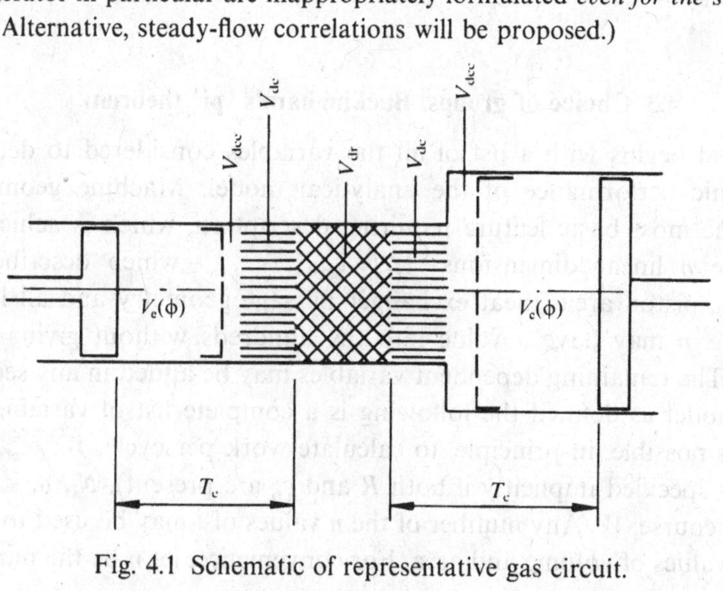


Fig. 4.1 Schematic of representative gas circuit.

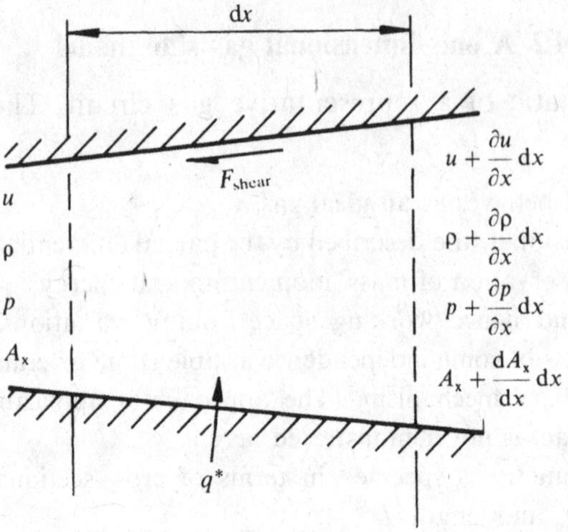


Fig. 4.2 Control volume for differential form of the equations of conservation of mass, momentum and energy.

In Fig. 4.2 a differential element of fluid exchanges heat and momentum with an element of solid, conducting wall. Heat flow within the wall is confined to a direction perpendicular to the solid surface. For convenience, a tubular wall element has been indicated, but this could equally be a conducting element typical of the regenerator matrix.

The formal approach to defining the conditions for dynamic similarity makes use of Buckingham's 'pi' theorem.

## 4.3 Choice of groups. Buckingham's 'pi' theorem

The method begins with a list of all the variables considered to determine the dynamic performance of the analytical model. Machine geometry is perhaps the most basic feature requiring description, which is achieved by stating the $n$ linear dimensions – $l_1, l_2, l_3, l_4, \ldots, l_n$ – which describe drive kinematics, piston areas, heat exchanger passage geometry and all internal dead space. $n$ may have a value into the hundreds without giving rise to difficulty. The remaining dependent variables may be added in any sequence. For the model as defined the following is a complete list of variables from which it is possible, in principle, to calculate work per cycle, $W$: $p_{ref}$, $T_e$, $T_c$, $R$, $c_p$, ($\gamma$ is specified implicitly if both $R$ and $c_p$ are present), $N_s$, $\mu$, $k$, $\rho_r$ and $C_r$ and, of course, $W$. Any number of the $n$ values of $l$ may be used to specify reference values of volume and area. For convenience, let $n$ be the number of free length values after specification of $A_{ref}$ and $V_{ref}$, and let the latter two variables be included with $p_{ref}$, $T_e$, etc. This gives a total of $n + 13$ variables.

The 'pi' theorem states that, with four basic dimensions, mass, M, length, L, time, T and temperature, K, a reduction of up to four may be hoped for in the number of variables. A saving of four when the number, $n$, of linear dimensions alone is likely to be substantial may not seem impressive. However, between geometrically similar machines the $n$ values of $l$ scale linearly, i.e., their respective values when divided by any chosen reference length, $L_{ref}$, remain constant. Geometric similarity is a first requirement of dynamic similarity.[7] The saving of four is therefore achieved in the interesting variables, i.e., those involving $p_{ref}$, $T_e$, etc. Where there are two quantities having the same dimensions (as in the case of the $T$s) it is necessary to make a choice of reference quantity. In this case, and for reasons already given, the reference temperature, $T_{ref}$ will be $T_c$.

It is therefore anticipated that the process of applying the 'pi' theorem will result in $n - 1$ length ratios of the type $\lambda_j = l_j/L_{ref}$, and $13 - 4 = 9$ dimensionless groups involving the thermodynamic quantities – a total of $n + 8$ groups. One of these will normalize $W$. There is a choice of algebraic techniques for determining a consistent set of groupings. Applying any preferred technique results in the following set of groups (or in an equivalent set achieved by combination):

| | |
|---|---|
| $\zeta = W/(p_{ref}V_{ref})$ | normalized cycle work (brake or indicated according to subscript (brake or ind) on $\zeta$) |
| $N_\tau = T_e/T_c$ | Second Law limitation |
| $\gamma = 1/(1 - R/c_p)$ | thermal load due to volume change |
| $N_{MA} = N_s L_{ref}/\surd(RT_{ref})$ | pressure coefficient or Mach number |
| $N_{pr} \approx N_{PR} = \mu c_p/k$ | Prandtl number – usually considered constant and close to unity |
| $N_{SG} = p_{ref}/N_s\mu$ | ratio of pressure effects/viscosity effects |
| $N_H = L_{ref}p_{ref}N_sR_H/k_rT_{ref}$ | compatibility of heat transfer processes either side of gas/metal interface |
| $N_F = \alpha/N_sL_{ref}^2$ | characteristic Fourier number |

A characteristic Reynolds number might have been anticipated. It will be shown shortly that an appropriate combination of $N_{MA}$ with $N_{SG}$ is equivalent to an $N_{RE}$. By a similar token, the absence of a characteristic group corresponding to $N_{st}$ is covered by the inclusion of $N_{PR}$ and ($N_{MA}$ with $N_{SG}$). Dimensionless parameters such as regenerator volumetric porosity, $\P_v$, and the ratio $R_H/L_{ref}$ are absorbed among the various $\lambda$.

If the groups represent a correct choice, then they may be expected to emerge naturally from normalization of the equations defining the thermodynamic phenomena of the gas circuit. It will be reassuring to have confirmation that this occurs.

## 4.4 Confirmation of groups by normalization – energy equation for the gas

With a friction parameter, $F$, defined as in the Notation, and with $q^*$ for rate of heat transfer per unit mass of working fluid, a form of the energy equation developed by Shapiro[8] is appropriate. This has been simplified from the basic form by substitution of both mass conservation and momentum conservation equations, and reads:

$$(\gamma - 1)\rho(q^* + uF) = \frac{\partial p}{\partial t} + u\frac{\partial p}{\partial x} - a^2\left(\frac{\partial \rho}{\partial t} + u\frac{\partial \rho}{\partial x}\right) \tag{4.1}$$

With $m'$ for instantaneous mass flow rate across free-flow area $A_x$, $u$ may be expressed:

$$u = \frac{m'}{\rho A_x} = \frac{m'RT}{pA_x} = \frac{\omega MRT}{pA_x}\sigma'$$

where $\sigma'$ is mass flow-rate normalized by total working fluid mass, $M$, and referred to crank angle, $\phi$, i.e., $\sigma' = \partial(m/M)/\partial\phi$. Normalizing pressure, $p$, by $p_{ref}$ to give $\psi$ and temperature, $T$, by $T_{ref}$ to give $\tau$:

$$u = \frac{2\pi N_s MRT_{ref}}{p_{ref}A_x}\frac{\tau}{\psi}\sigma' \tag{4.2}$$

Normalizing lengths by $L_{ref}$, noting that $\phi = \omega t = 2\pi N_s t$ and that $q^* = h(T_w - T)/\rho r_h$ leads to:

$$(\gamma - 1)N_{st}\left[\frac{\gamma - 1}{\gamma}\frac{R_H}{L_{ref}}\frac{1}{\alpha_x\lambda_h}(\tau_w - \tau) + \frac{C_f}{N_{st}}\frac{N_{MA}{}^2}{2\lambda_h}4\pi^2\frac{R_H{}^3}{L_{ref}{}^3\alpha_x{}^3}\frac{\tau^2}{\psi^2}\frac{\sigma'}{|\sigma'|}\right]$$

$$= \frac{\partial}{\partial\phi}[\psi(1 - 1/\tau)] + \frac{R_H}{L_{ref}}\frac{\tau}{\psi}\frac{\partial}{\partial\lambda}[\psi(1 - 1/\tau)] \tag{4.3}$$

In Eqn (4.3) $\alpha_x$ is the normalized cross-sectional area, $A_x/A_{ref}$, $\lambda_h$ is the normalized hydraulic radius, $r_h/L_{ref}$. A number of terms have disappeared by defining $V_{ref}$ such that $p_{ref}V_{ref} = MRT_{ref}$; $\lambda = x/L_{ref}$; $N_{st}$ is defined as $h/Gc_p$, where $G$ is mean mass velocity, $\rho u$, and $h$ is local, instantaneous coefficient of convective heat transfer. The temperature dependence of $c_p$ may be taken into account. If reference area $A_{ref}$ is chosen to be total heat exchanger wetted

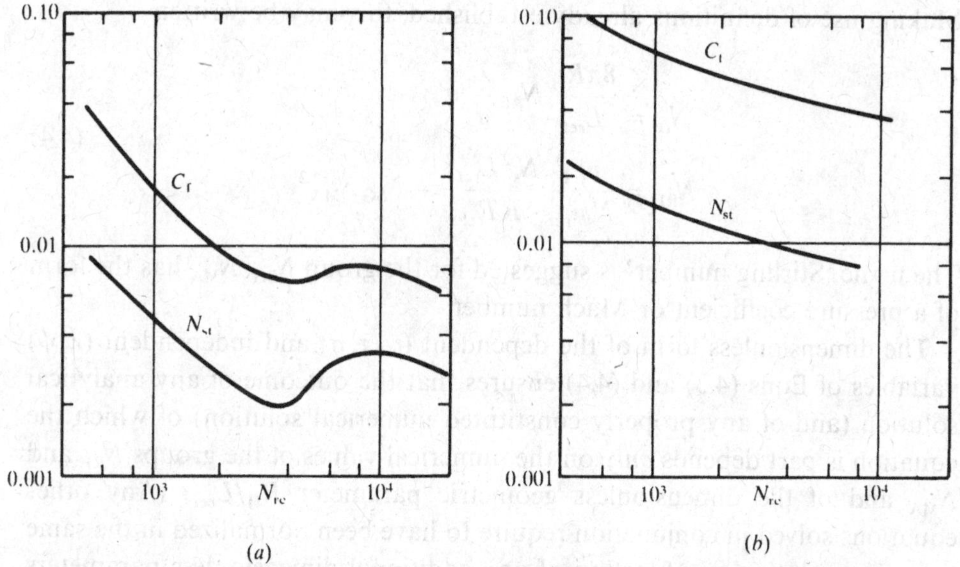


Fig. 4.3 Correlations of $C_f$ and $N_{st}$ with $N_{re}$ typical of smooth and interrupted heat exchanger surfaces. (*a*) Smooth flow passage. (*b*) Interrupted flow passage.

area, then a characteristic hydraulic radius may be defined for the machine, namely, $R_H = V_{ref}/A_{ref}$.

Given the near constancy of Prandtl number, $N_{pr}$, at a value close to 1.0, $N_{st}$ is essentially $N_{st}$(geometry, $N_{re}$). Pending reappraisal in Chap. 5 of flow friction correlations for Stirling machine work, friction factor, $C_f$, is effectively $C_f$ (geometry, $N_{re}$). Figs. 4.3(*a*) and (*b*) follow Creswick and typify correlations for the cases of smooth, parallel tube and interrupted passage respectively. Constant vertical separation on the log–log plot signifies a constant ratio of $C_f$ to $N_{st}$. Real correlations are not strictly parallel, but the trend is there, consistent with the fact that flow friction and heat transfer involve the same molecular phenomena, and that velocity and temperature profiles in the flowing fluid should be geometrically similar for $N_{pr} \approx 1.0$.

Eqn (4.3) yields information relevant to design, even in its unsolved form. It is not difficult to estimate the Reynolds number range within which the flow at a given exchanger location is concentrated. An exchanger geometry providing the minimum ratio of the quotient $C_f/N_{st}$ over this Reynolds number range is the tentative first choice pending investigation of the implications for other aspects of performance.

$$N_{re} = \frac{4\rho r_h u}{\mu}$$

Making use of definitions already established, this may be written

$$N_{re} = \frac{8\pi R_H}{L_{ref}} N_{RE} \frac{\lambda_h}{\alpha_x} \sigma'$$
$$N_{RE} = \frac{p_{ref}}{N_s \mu_{ref}} \frac{N_s^{\,2} L_{ref}^{\,2}}{R T_{ref}} = N_{SG} N_{MA}^{\,2} \tag{4.4}$$

The name 'Stirling number' is suggested for the group $N_{SG}$; $N_{MA}$ has the form of a pressure coefficient or Mach number.

The dimensionless form of the dependent ($\psi$, $\tau$, $\sigma$) and independent ($\lambda$, $\phi$) variables of Eqns (4.3) and (4.4) ensures that the outcome of any analytical solution (and of any properly constituted numerical solution) of which the equation is part depends *only* on the numerical values of the groups $N_{RE}$ and $N_{MA}$ and of the dimensionless geometric parameter $R_H/L_{ref}$. (Any other equations solved in conjunction require to have been normalized in the same way. The solution is a function of any additional dimensionless parameters introduced by the further normalization process.)

## 4.5 Heat flow at the gas–metal interface

Fig. 4.4 represents a cross-section through an element of solid wall showing the instantaneous temperature gradient through the solid and the fluid film. At any instant:

$$hP\,dx(T_w - T) = k_w P\,dx\,dT/dr$$

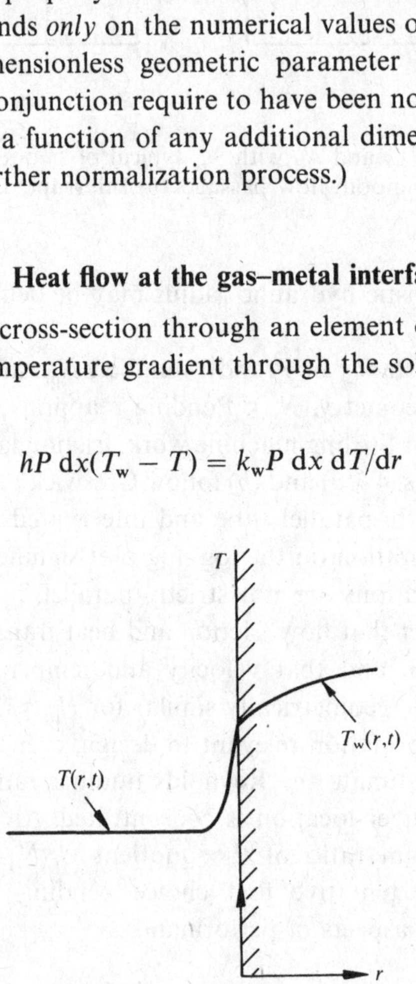


Fig. 4.4 Schematic of local, instantaneous temperature profile in fluid film and adjacent conducting wall.

Normalizing as before:

$$2\pi N_{\mathrm{H}} N_{\mathrm{st}} \frac{\gamma}{\gamma - 1} (\tau_{\mathrm{w}} - \tau)\sigma' = \frac{\mathrm{d}\tau}{\mathrm{d}\lambda} \tag{4.5}$$

$$N_{\mathrm{H}} = L_{\mathrm{ref}} p_{\mathrm{ref}} N_{\mathrm{s}} R_{\mathrm{H}} / k_{\mathrm{w}} T_{\mathrm{ref}}$$

Constancy of the dimensionless group $N_{\mathrm{H}}$ ensures dynamic similarity of the heat transfer processes immediately either side of the metal–gas interface.

## 4.6 Heat flow entirely within the solid

Transient, axi-symmetric thermal conduction is governed by the equation:

$$\frac{\partial^2 T}{\partial r^2} + \frac{1}{r}\frac{\partial T}{\partial r} = \frac{1}{\alpha}\frac{\partial T}{\partial t}$$

Normalizing

$$\frac{\partial^2 \tau}{\partial \lambda^2} + \frac{1}{\lambda}\frac{\partial \tau}{\partial \lambda} = \frac{2\pi}{N_{\mathrm{F}}}\frac{\partial \tau}{\partial \phi} \tag{4.6}$$

$$N_{\mathrm{F}} = \frac{k_{\mathrm{w}}}{N_{\mathrm{s}} L_{\mathrm{ref}}{}^2 C_{\mathrm{w}} \rho_{\mathrm{w}}} = \frac{\alpha_{\mathrm{w}}}{N_{\mathrm{s}} L_{\mathrm{ref}}{}^2}$$

$\alpha_{\mathrm{w}}$ is thermal diffusivity. With an appropriate choice of $\alpha_{\mathrm{w}}$, $N_{\mathrm{F}}$ is the characteristic Fourier number.

## 4.7 Compatibility of thermal capacity

Finally, in two machines operating under conditions of dynamic similarity the ratio of thermal capacity of the gas per pass to thermal capacity of the matrix must be the same in both cases:

$$\frac{\rho_{\mathrm{r}} C_{\mathrm{r}} (1 - \P_{\mathrm{v}}) V_{\mathrm{dr}}}{M c_p} = N_{\mathrm{TC}} \left[\frac{(\gamma - 1)}{\gamma}\right] (1 - \P_{\mathrm{v}}) \mu_{\mathrm{dr}} \tag{4.7}$$

$$N_{\mathrm{TC}} = \frac{T_{\mathrm{ref}} \rho_{\mathrm{r}} C_{\mathrm{r}}}{p_{\mathrm{ref}}}$$

$N_{\mathrm{TC}}$ might be called the thermal capacity ratio. Fixing its value alone fixes the required thermal capacity ratio, since the multipliers on the right-hand side of Eqn (4.7) are all dimensionless groups which also have common respective values between dynamically similar machines: $\P_{\mathrm{v}}$ is volumetric porosity and $\mu_{\mathrm{dr}}$ the ratio of unswept volume occupied by the regenerator structure $V_{dr}$, to reference volume, $V_{ref}$. Now, $N_{TC}N_H = N_s L_{ref} R_H \rho_r C_r/k_r = N_s L_{ref} R_H/\alpha_r = (R_H/L_{ref}) N_s L_{ref}^2/\alpha_r = (R_H/L_{ref})/N_F$, so one group (e.g., $N_{TC}$) is superfluous and one can work with $N_H$ and $N_F$.

The functional representation of indicated work per cycle for the one-dimensional Stirling model of Section 4.2 may now be written

$$\zeta = \zeta[\lambda_1, \lambda_2, \lambda_3, \ldots, \lambda_n, N_\tau, \gamma, N_{MA}, (N_{PR},) N_{SG}, N_H, N_F] \tag{4.8}$$

with a comparable expression for indicated thermal efficiency.

A way of approaching the task of detailed thermodynamic design is to propose a geometry (overall layout, drive kinematics, exchanger passage details, clearances, regenerator packing geometry etc.) and to investigate the effect of varying operating conditions. There are only two candidate values of $\gamma$, namely, 1.4 and 1.66 (but see Appendix XII, Section AXII.1) and $N_{PR}$ is essentially fixed. With, say, ten points to be computed per remaining variable, investigation of the performance envelope in dimensionless terms calls for $2 \times 10^5$ cycles of the code (two for $\gamma$, ten each for $N_\tau$, $N_{MA}$, $N_{SG}$, $N_H$, $N_F$). With four extra significant variables, the same task in terms of absolute quantities calls for some $2 \times 10^9$ cycles – a factor of $10^4$ in extra computing time, or in practical terms, one week rather than a single minute!

The comparison is, in reality, less unfavourable than the casual estimate suggests, since the range of candidate gases and of regenerator packing materials (and hence of properties such as $C_r$, $k_r$) is limited and values discrete. Moreover, computing time is not normally a consideration, since it costs about as much to operate a computer for 1 hour per day as for 24. The substantive attraction of the dimensionless approach is the opportunity it offers of eliminating use of the computer altogether from the routine work of Stirling machine design. In principle, this would be achieved by rendering the functional Eqn (4.8) in the form of charts covering suitable ranges of all variables. Even after normalization, the number of variables is still very great. A reduction of one variable may be achieved immediately by agreeing on the use of $R_H$ as reference length, $L_{ref}$, rather than the more usual crank radius.[4] The usage emphasizes the fact that the Stirling machine is first and foremost a thermodynamic device – a heat exchanger with provision for the transfer of work: a heat exchanger is characterized by its hydraulic radius.

Further, drastic simplification would follow if it were possible to relax the requirement of geometric similarity as a precondition of dynamic similarity. The section which follows will discuss the circumstances under which the precondition may indeed be set aside. These will lead to the novel concept of 'thermodynamic similarity'.

## 4.8 Thermodynamic similarity

Looking again at Eqns (4.2) and (4.3) it is seen that local, instantaneous flow is specified in terms of the quantity $\sigma'$ ($= \mathrm{d}(m/M)/\mathrm{d}\phi$); $\sigma'$ is the dimensionless mass rate term of simple cycle descriptions such as the single conservation law model of Schmidt, and Finkelstein's Generalized Thermodynamic Analysis.[5] Although these simple treatments are readily adapted to take into account the volume variations due to any desired piston drive mechanism, they most frequently assume simple harmonic piston motion and hence simple harmonic volume variations. For this special case, $\sigma'$ at any desired location in the machine (specified in terms of fractional dead space, $\mu_d$) is, like $\zeta$, *a function only of temperature ratio, $N_\tau$, volume ratio, $\kappa$, volume phase angle, $\alpha$, a dead space parameter, $\nu(N_\tau)$ and, in the case of the Finkelstein treatment, $\gamma$.*

Mass flow rate distribution in a real machine is not precisely that predicted by the simple analysis. On the other hand, the factors which cause flows in the real machine to differ from the ideal values are the irreversibilities due to flow and thermal resistance. In a well-designed machine operating at rated conditions, such irreversibilities are at a minimum. One is not normally interested in designing for poor performance, so for present purposes it will be assumed that $\sigma'$ from a simple, ideal cycle model will serve as an approximation to that in the real machine being designed. This is equivalent to assuming that *the simple analysis, applied to the real machine via values of $N_\tau$, $\kappa$, $\alpha$, and $\nu(N_\tau)$ determines, to the level of accuracy appropriate to the task, the fraction of total working fluid mass, $M$, involved in traversing the exchanger system and local rates of mass traverse as a function of crank angle.*

Specifying the cross-sectional area, $A_x$ (free-flow area) and the length of each exchanger element (heater, cooler, regenerator and any connecting ducts) effectively fixes $\nu(N_\tau)$, and, through $A_x$, determines conversion between the local, instantaneous mass flow rate and the corresponding velocity. On the other hand, there is not yet sufficient information to specify $C_f$ and $N_{st}$. Consultation of any suitable compendium of steady-flow heat transfer and flow friction data, however, will show (e.g., Ref. 9) how closely correlations of $C_f$ and $N_{ST}(N_{pr})^{2/3}$ with $N_{re}$ agree between different duct cross-sections when the $r_h$ is used as characteristic size.[10]

Indeed, an account by Miyabe *et al.*[11] gives the impression that for the wire screen regenerator a unique correlation is possible between $C_f$ and $N_{re}$ given a suitable definition of $r_h$. Any such uniqueness achieved in violation of geometric similarity is implausible, but Miyabe's proposition does at least support use, for present purposes, of single, specimen heat transfer and flow friction correlations for all candidate gauzes. With this accepted it remains only to specify characteristic size (i.e., hydraulic radius) for each heat exchanger element in order to complete the approximate thermodynamic definition of the entire machine. With $R_H$ as normalizing length, $\lambda_h$ for $r_h/R_H$, and $\alpha_x$ for $A_x/A_{ref}$ the functional form of the expression for $\zeta$ becomes:

$$\zeta = \zeta\left(N_\tau, \kappa, \alpha, \mu_v, \begin{matrix} \mu_{dce} & & \\ \mu_{de} & \alpha_{xe} & \lambda_{he} \\ \mu_{dr} & \alpha_{xr} & \lambda_{hr} \\ \mu_{dc} & \alpha_{xc} & \lambda_{hc} \\ \mu_{dcc} & & \end{matrix}, \gamma, N_{MA}, N_{SG}, N_H, N_F\right) \tag{4.9}$$

A total of 21 variables now completely define the thermodynamic performance of a Stirling cycle machine functioning according to the rationalizations of Section 4.2. Two such machines are thermodynamically similar if, in both cases, numerical values calculated for the quantities of Eqn (4.9) are the same. It should be noted that variables such as $N_{MA}$ are now defined in terms of $R_H$, *viz.*, $N_s R_H/\sqrt{(RT_c)}$. It has not been necessary to specify a quantity representing heat exchanger passage lengths explicitly, since these are fixed by respective ratios $\mu_d/\alpha_x$. The $\mu_{dce}$ and $\mu_{dcc}$ account for the volume unswept in expansion and compression cylinders respectively, as is necessary given the assumption of sinusoidal $V_e(\phi)$ and $V_c(\phi)$ and the corresponding minima of zero. $\mu_v$ is the ratio $V_E/V_{ref}$ and permits characteristic volume ratio, $\kappa$, to retain the traditional definition $V_C/V_E$.

## 4.9 Exploitation of the thermodynamic similarity condition

A full investigation of Eqn (4.9) calls for availability of an analysis rendered dimensionless in the manner suggested, embodying representative heat transfer and flow resistance correlations for the respective heat exchangers, and programmed for computer. Before embarking upon a treatment of such an analysis it will be profitable to look further into the relevance of steady-flow correlations to the reversing-flow case, and also into the validity of standard correlations for steady flow itself. In this connection the chapter which follows looks at flow through the wire gauze regenerator.

In the meantime, much can be inferred from the form of Eqn (4.9), noting first of all that $V_{ref}$ in the definition $(W/p_{ref}V_{ref})$ of $\zeta$ is the $V$ occupied by $M$ when $p = p_{ref}$. To convert to the standard definition of $\zeta$ it is necessary merely to multiply by the ratio $V_{ref}/V_{swept}$. The resulting numerical value of $\zeta$ is indicated Beale number, $B_i$, i.e., $B/\eta_{mech}$, so that

$$BV_{swept}/\eta_{mech}V_{ref} = \text{Eqn (4.9)}$$

The conventional view, on the other hand, is that

$$B = B(N_\tau \text{ only}) \qquad (4.9a)$$
$$= 0.15 \text{ for usual values of } N_\tau$$

The form of Eqn (4.9) shows how insensitive is the Beale number as a tool for use in thermodynamic design.

The simplest of all tenable cycle models is that due to Schmidt, and may be written

$$\zeta_{Sch} = \zeta(N_\tau, \kappa, \alpha, \mu_{de}, \mu_{dr}, \mu_{dc}, \mu_{dce}, \mu_{dcc}) \qquad (4.9b)$$

In work of numerical optimization based on the Schmidt expression, $N_\tau$ and the various $\mu_d$ are regarded as parameters, corresponding to which are to be found the $\kappa$ and $\alpha$ which optimize $\zeta$. The most favourable optima are determined for the case where the various $\mu_d = 0$, i.e., the case for which the real machine (and, indeed, the machine of Section 4.2) would return a $\zeta$ of zero! $\kappa$ and $\alpha$ are, in fact, the variables to whose adjustment the real machine is least sensitive. The effect of the factor on which performance depends heavily, namely, exchanger geometry, cannot be tested by such methods. It is safe to conclude that early optimization strategies based on the Schmidt cycle may not, in isolation, be relied upon as a guide to Stirling machine design.

An optimization process based on use of Eqn (4.9) would probably hold constant the terms $N_\tau$, $\kappa$, $\alpha$, and $\mu_v$, and would settle on a value (1.4 or 1.66) for $\gamma$; $\mu_{dce}$ and $\mu_{dcc}$ would be fixed at minimum practical values. There remain 14 quantities which may, in principle, be varied continuously within certain limits. For one or more values of each of these 14 there is a best value of $\zeta$. This number of variables is a realistic proposition for an exercise in numerical non-linear optimization. Such an exercise is undertaken in a later chapter, which will present results as part of a proposed recipe for thermodynamic design. In anticipation, it will pay to be aware of certain precautions in preparing Eqn (4.9) for the application of optimization techniques. These will concern the choice of suitable 'seed' values of the various dimensionless groups.

## 4.10 The meaning of an optimum value of Eqn (4.9)

It is important to bear in mind that the computation which leads to the value of $\zeta$, i.e., of specific cycle work, is carried out over a single cycle. To estimate power, a multiplication is required involving cycles/unit time, i.e., $N_s$. This

Table 4.1 *Numerical values of $R_H$ and of corresponding dimensionless groups calculated for two, specimen Stirling cycle engines. $V_{ref}$ is volume occupied by total working fluid mass, M, when cycle pressure is $p_{ref}$ (e.g., $p_{min}$). A complete set of numerical values for the dimensionless variables tabulated amounts to a comprehensive thermodynamic description of a Stirling cycle machine operating on the cycle defined in Section 4.2, Values for $N_H$, $N_F$ apply to the regenerator*

| Symbol | Definition | Calculated value | |
|---|---|---|---|
| | | MP1002CA | GPU-3 |
| rated working fluid | | air | He |
| $N_s$ | 1/s | 25 | 42 |
| $p_{ref}$ | Pa | 15.0E+05 | 42E+05 |
| $V_{ref} = V_{wfc}$ m$^3$ | | 0.12E−03 | 0.31E−03 |
| $A_{ref} = \sum A_w$ m$^2$ | | 0.55 | 1.9 |
| $\kappa$ – volume ratio $V_C/V_E$ | | 1.031 | 0.95 |
| $\alpha$ – equiv. phase angle, rad (deg) | | 2.1 (121°) | 2.1 (120°) |
| $R_H$ | $V_{wfc}/\sum A_w$ m | 0.216E−03 | 0.161E−03 |
| ¶$_v$ | vol. poros. regenerator | 0.80 | 0.7 |
| $\zeta_{ind}$ | $P_{ind}/N_s p_{ref} V_{ref})$ | 0.0667 | 0.0719 |
| $\mu_v$ | $V_E/V_{ref}$ | 0.514 | 0.3846 |
| $\mu_{dce}$ | $V_{dce}/V_{ref}$ | 0.0644 | 0.0401 |
| $\mu_{de}$ | $V_{de}/V_{ref}$ | 0.0473 | 0.2564 |
| $\mu_{dr}$ | $V_{dr}/V_{ref}$ | 0.1609 | 0.2080 |
| $\mu_{dc}$ | $V_{dc}/V_{ref}$ | 0.0473 | 0.0420 |
| $\mu_{dcc}$ | $V_{dcc}/V_{ref}$ | 0.0750 | 0.0673 |
| $\alpha_{xe}$ | $A_{xe}/A_{ref}$ | 0.27E−03 | 0.15E−03 |
| $\alpha_{xr}$ | $A_{xr}/A_{ref}$ | 1.302E−03 | 1.126E−03 |
| $\alpha_{xc}$ | $A_{xc}/A_{ref}$ | 0.27E−03 | 0.15E−03 |
| $\lambda_{he}$ | $r_{he}/L_{ref}$ | 0.7905 | 4.68 |
| $\lambda_{hr}$ | $r_{hr}/L_{ref}$ | 0.1847 | 0.174 |
| $\lambda_{hc}$ | $r_{hc}/L_{ref}$ | 0.7905 | 1.66 |
| $N_\tau$ | $T_e/T_c$ | 2.65 ($T_c$ = 60°C) | 3.39 |
| $\gamma$ | $c_p/c_v$ | 1.4 | 1.66 |
| $N_{MA}$ | $N_s R_H/\surd(RT_{ref})$ | 17.5E−06 | 8.6E−06 |
| ($N_{PR}$ | $\mu c_p/k$ | ≈1.0 | ≈1.0) |
| $N_{SG}$ | $p_{ref}/N_s\mu$ | 3.30E+09 | 5.5E+09 |
| $N_H$ | $p_{ref} N_s {R_H}^2/k_r T_{ref}$ | 3.33E−04 | 1.016E−03 |
| $N_F$ | $\alpha_r/N_s {R_H}^2$ | 4.26 | 5.58 |
| ($N_X = N_H/N_{SG}{N_{MA}}^2 = R\mu/k_r$ | | 3.6E−04 | 2.4E−03) |

latter variable does not have a definite value in the dimensionless approach. It is therefore necessary to take precautions against forming a value of power by multiplying by a numerical $N_s$ for which the computed $\zeta$ cannot be achieved.

One way is to decide upon an acceptable range of $N_s$, $p_{ref}$ and other variables over which the designer has control, and thus to establish corresponding numerical ranges for the dimensionless groups. The optimization algorithm must embody a means for confining to the respective declared ranges its search for the $N_{SG}$, $M_{MA}$ etc. which maximize $\zeta$.

In the case of certain groups the range calculation calls for a value of $R_H$ – a quantity which is not currently in common use and for which specimen values are therefore not available. Moreover, even those variables for which it is not meaningful to specify a range (e.g., the various $\mu$, $\alpha$, $\lambda_h$) must be given a 'seed' value at the start of the optimization process. Table 4.1 accordingly lists values of some constants and of the various groups calculated or estimated for two of the better-documented Stirling cycle engines – the Philips MP1002CA and the GPU-3. If the reader has access to suitable details of additional machines, he or she may care to fill in one or more of the extra, blank columns of the table.

# 5

# Flow within the regenerator

## 5.1 State of the art

There is something less than satisfactory about the way in which Stirling machine analysis handles flow within the regenerator:

(1) The flow case is treated by the method traditional for steady, two-dimensional (or axi-symmetric), incompressible viscous flow in pipes, i.e., in terms of a friction factor, $C_f$, correlated with geometry and Reynolds number, $N_{re}$. When analysis and computer simulation based on such correlations yield pressure distributions which do not tally with measurement from running machines, it is common practice to 'improve' matters by arbitrarily adjusting the correlations. The technique is part of a process which has become known as 'validation'. Exercises in validation have been reported[1–3] which called for $C_f$ at given $N_{re}$ to be multiplied by factors between 4 and 7.

(2) The discrepancy between experimental measurement and theoretical prediction has come to be attributed to the fact that steady-flow correlations do not take into account the unsteady effects which arise from the cyclic nature of the flow processes in the Stirling machine. An enquiry into the rôle of unsteadiness is certainly called for. At the same time, usage and interpretation of the steady-flow correlations has been parochial, having in most instances looked no further than the incompressible-flow cases documented by Kays and London.[4] There is, in fact, a vast literature on flow – compressible as well as incompressible – and a broad spectrum of porous media is treated. When this is taken into account the following points emerge:

Fig. 5.1 is adapted from the Kays and London compendium. It shows $C_f$ as a function of $N_{re}$ and the geometric parameter $\P_v$ (volumetric porosity). The correlation was obtained from experiments using wire screens and matrices of crossed rods simulating gauzes. The rods were 0.375 inch (9 mm) in diameter, and the working fluid was air. It is seen that the curves extend to $N_{re} = 10^5$. For the wire diameter quoted, the $\P_v = 0.832$ case gives $r_h \approx 10^{-2}$ m (or 0.39 in). Assuming ambient pressure, corresponding fluid speed from the definition of $N_{re}$ is some 35 m/s, or 0.1 of local acoustic speed. Converting the same $N_{re}$ of $10^5$

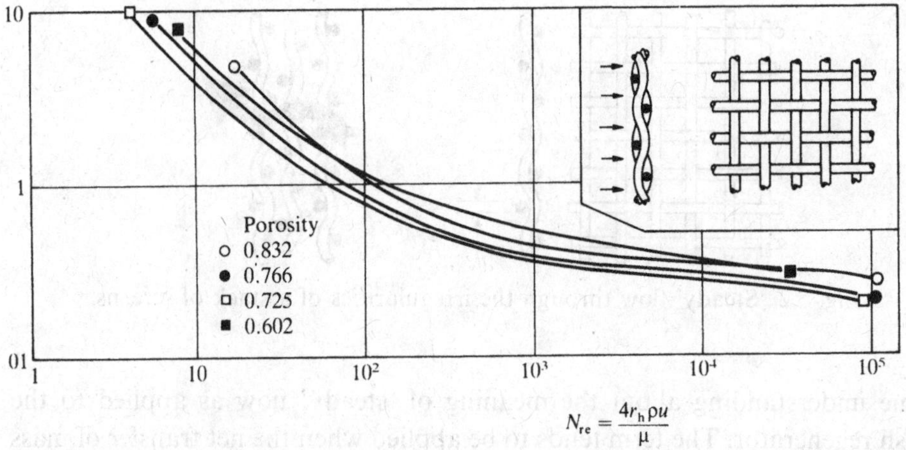


Fig. 5.1 Correlation by Kays and London[4] of data on flow through wire screens and stacks of cylindrical rods simulating wire screens. Perfect stacking (individual screens in contact) is assumed. The fluid was air – thought to be at more or less ambient temperature. (Reproduced with the permission of the McGraw-Hill Book Co.)

for the case of a gauze having $r_h$ typical of the Stirling regenerator (say $0.025 \times 10^{-3}$ m) gives a particle speed of $1.5 \times 10^4$, or about Mach 40! An $N_{re}$ two orders less (i.e., $10^3$) still gives a Mach number of 0.4 – more than high enough to cause choking problems. The parametric description of the data as $C_f = C_f(\P_v, N_{re})$ is thus incomplete as far as the requirements of Stirling machine design are concerned, *and to this extent, is not usable for Stirling regenerator design over its full published range.*

A reduction of experimental data (steady flow or otherwise) to a form suitable for the critical requirements of Stirling cycle modelling must, as will be demonstrated shortly, be carried out with the guidance of a suitable model of the physical processes involved. Measurement of a pressure drop and conversion to friction factor form by division wholesale by $\frac{1}{2}\rho u^2$ is inappropriate: the $\Delta p$ will, in the general case, be due to a combination of independent viscous ($\mu u/k$), inertial ($u\,du$) and compressibility ($u/a$) effects. The conservation equations used in a comprehensive Stirling cycle description embody separate terms for these effects. If the $C_f$ from the correlation has rolled the effects up together, then certain terms are accounted for twice, with indeterminate consequences.

Even the most comprehensive treatments of porous media flow assume either adiabatic or isothermal conditions. The primary function of the Stirling regenerator is to transfer heat, and it does so at high flux densities. Without enquiring about the likely effect of heat transfer on flow friction it is premature to blame cyclic unsteadiness for present difficulties.

Before looking at how matters might be rationalized it is worth coming to

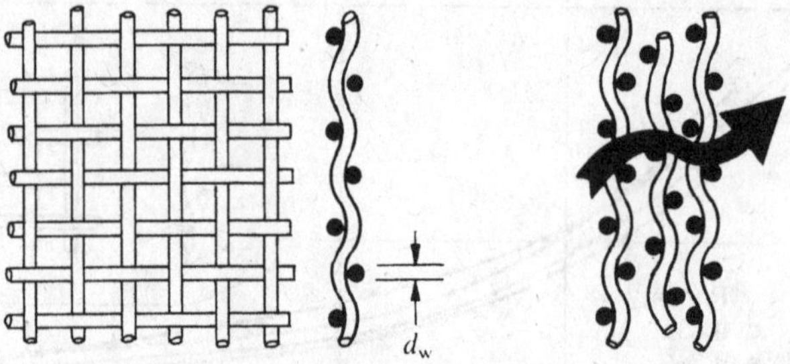


Fig. 5.2 'Steady' flow through the irregularities of a stack of screens.

some understanding about the meaning of 'steady' flow as applied to the mesh regenerator. The term tends to be applied when the net transfer of mass is unidirectional. Consider, however, Fig. 5.2, showing flow at a 'steady' speed $u\ (=m'/(\rho A))$ through a series of irregularities at pitch $2d_w$. The irregularities represent the successive interruptions to flow encountered in a stack of wire screens. Flow is subject to a disturbance on its passage left-to-right through the screens. The disturbance cannot be precisely quantified, but it is clear that it has a frequency of $u/2d_w$.

The disturbance might provisionally be considered to be simple harmonic. With an estimate for amplitude a calculation of peak acceleration, $\omega^2 r$, is possible. Supposing amplitude to be one half of wire diameter gives $r = 0.25d_w$. For this case:

$$\omega^2 r = (\pi u)^2/4d_w \tag{5.1}$$

For $d_w = 0.025 \times 10^{-3}$ (0.001 in) and a conservative 5 m/s for $u$:

$$\omega^2 r = 2 \times 10^6 \text{ m/s}$$
$$\approx 2 \times 10^5 g$$

No doubt there will be a tendency for flow to adjust locally to reduce this effect. Such adjustment results when slower-moving particles follow the tighter radii, the faster ones taking the straightest possible course. Even if the estimate is out by orders of magnitude, however, some substantial accelerations have to be accounted for.

It could well be that, when more is known, it will be possible to correlate regenerator pressure distribution with a parameter accounting for the unsteadiness due to the cyclic nature of the processes in the Stirling machine. In the meantime, there are several aspects of so-called 'steady' flow which would benefit from clarification.

## 5.2 Flow passage geometry

With the exception of the filament-wound arrangement used in the Philips MP1002CA air engine, regenerator construction tends to make use of materials developed for purposes other than heat storage and exchange. Stacks of discs punched from continuous wire gauze appear to have the most favourable thermodynamic performance, but are wasteful of sophisticated stock material. Such regenerators tend to be the most expensive single component of the Stirling machine. Other candidate materials are the sintered solids used for filtration, and proprietary flame-holder matrices such as Retimet™. Such materials are generally both cheaper and easier to form into regenerators than are wire gauzes.

It will be shown shortly that it is not essential to be able to specify precisely the geometry of a porous material in order to extract a viable correlation between $\Delta p$ and the parameters of flow. On the other hand, where geometry can be specified (or, rather, where a geometric parameter is available which determines geometric *similarity* between matrices of different absolute sizes) a special form of correlation is possible which affords certain advantages. The woven gauze offers an opportunity to define porosity geometrically.

Fig. 5.3 follows Squiers[5] and illustrates some of the weaves used in the production of wire screens used primarily for filtration. From the full selection there may possibly be a 'best' weave for Stirling regenerator use in the sense of a most favourable ratio of heat transfer performance to flow resistance. However, it is not yet possible to make such a choice, either from first principles or from available data. Only the performance of the square weave type has been documented in any detail, so this forms the appropriate example. Fig. 5.4(*a*) is an approximation to the geometry of the simple, square-woven mesh. Fig. 5.4(*b*) is the equivalent, rectilinear form on which Kays and London based their correlations.

Basic to all correlations involving flows through porous materials is a value of volumetric porosity, $\P_v$. Computation of the dead space parameter, $v(N_\tau)$ for use in application of ideal cycle models to real machines also calls for knowledge of $\P_v$ for the regenerator in use. $\P_v$ is defined as the ratio of void volume to enclosure volume, which is equivalent to (1 – volume of solid material/enclosure volume). Assuming homogeneous packing, an expression for $\P_v$ for the rectangular mesh may be based on the square element of Fig. 5.4. The elemental enclosure volume is $2d_w/m^2$. The corresponding solid volume is $\frac{1}{4}\pi d_w{}^2 2s_w$, where $s_w$ is the actual length of a strand of wire over the pitch distance $1/m$. On this basis:

$$\P_v = 1 - \frac{\pi m d_w}{4} m s_w \tag{5.2}$$

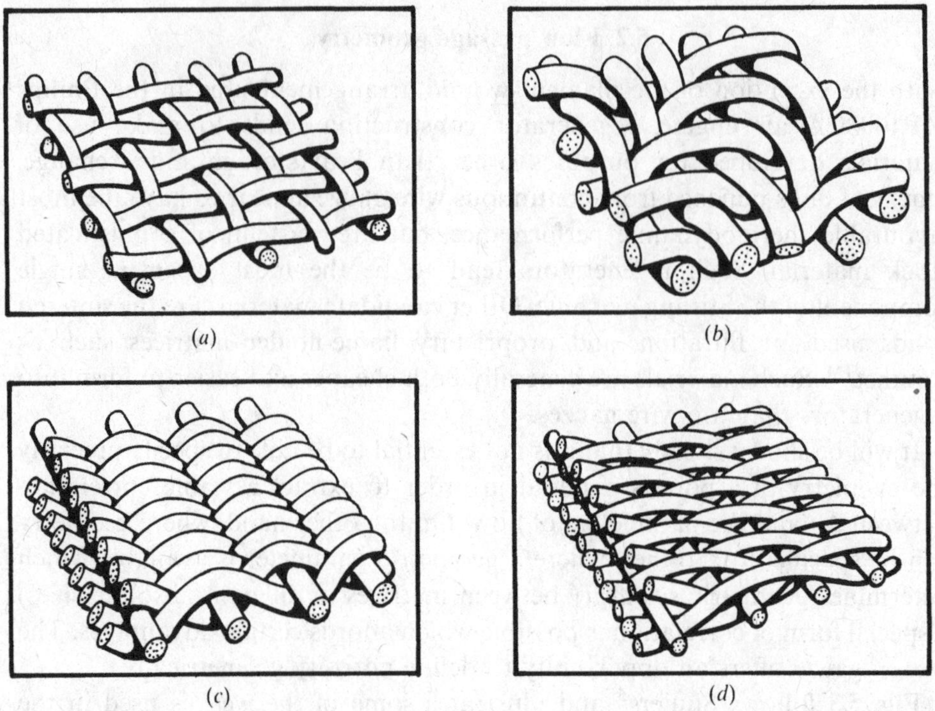


Fig. 5.3 Selection of weaves in which wire screens are available. (After Squiers.[5]) (*a*) Plain square weave. (*b*) Twilled square weave. (*c*) Plain Dutch weave. (*d*) Twilled Dutch weave.

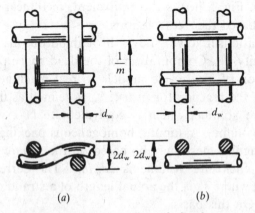

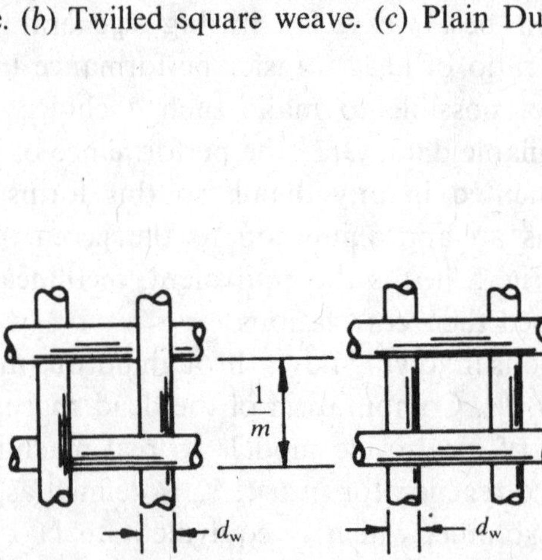


Fig. 5.4 Mesh geometry. (*a*) Representation of simple, square weave mesh. (*b*) A geometry having almost the same porosity and wetted area per unit volume as the actual mesh.

For the crossed-rod model of gauze geometry used by Kays and London[4] $ms_w = m/m = 1$, so that

$$\P_v = 1 - \frac{\pi m d_w}{4} \qquad \text{(crossed-rods)} \qquad (5.3)$$

It is as well to be aware of a physical limit on Eqn (5.3), namely, $md_w \leqslant 1$. It is intuitively evident that the geometry of Fig. 5.4(*b*) is unaltered by changes in $m$ and $d_w$ which maintain the ratio $md_w$ constant. $\P_v$ is a function of $md_w$ only, and thus determines geometric similarity.

Both Miyabe *et al.*[6] and, for different reasons, Pinker and Herbert,[7] found the need for an expression for $\P_v$ (and for $r_h$ and $\alpha_\P$ – see below) more sensitive to the true geometry of the matrix. The former authors envisage the woven gauze as per Fig. 5.5(*a*), and calculate total solid content from the volume of the inclined, straight-line segments. In present terminology the length of the segment is $ms_w = \sqrt{[1 + (md_w)^2]}$ (simple Pythagoras), so that

$$\P_v = 1 - \frac{\pi m d_w}{4}\sqrt{[1 + (md_w)^2]} \qquad \text{(inclined straight-line segments)} \quad (5.4)$$

which is seen to be Eqn (5.3) with the $md_w$ term corrected for the variable slight increase in wire length over the simpler case. $\P_v$ again determines geometric similarity. The physical limit in this instance is $md_w \leqslant 1/\sqrt{3}$ (Fig. 5.5(*d*)). The condition is artificial, since it cannot be approached without violating the picture of straight-line wire segments.

Arguably the most realistic picture is that of sinusoidal wire form (Fig. 5.5(*b*)). On the assumption that wire diameter is uniform at its nominal value, the quantity $ms_w$ may be obtained by integration along the sinusoidal spline of the wire. Carrying out the integration and substituting into Eqn (5.2):

$$\P_v = 1 - \tfrac{1}{4}\pi m d_w \int_0^{1/m} \sqrt{[1 + (\pi m d_w/2)^2 \sin^2(\pi x m)]}\, \mathrm{d}x$$

(sinusoidal wire form) (5.5)

The integral has a solution in terms of the error function, but may also be evaluated numerically. Strictly speaking, the true sinusoidal form is possible only for values of $md_w$ for which the minimum radius of curvature of the inner surface of the wire (Fig. 5.5(*c*)) exceeds $d_w/2$ (otherwise gauze thickness must exceed $2d_w$). On this basis Eqn (5.5) is valid for

$$\pi m d_w \leqslant \sqrt{2}$$

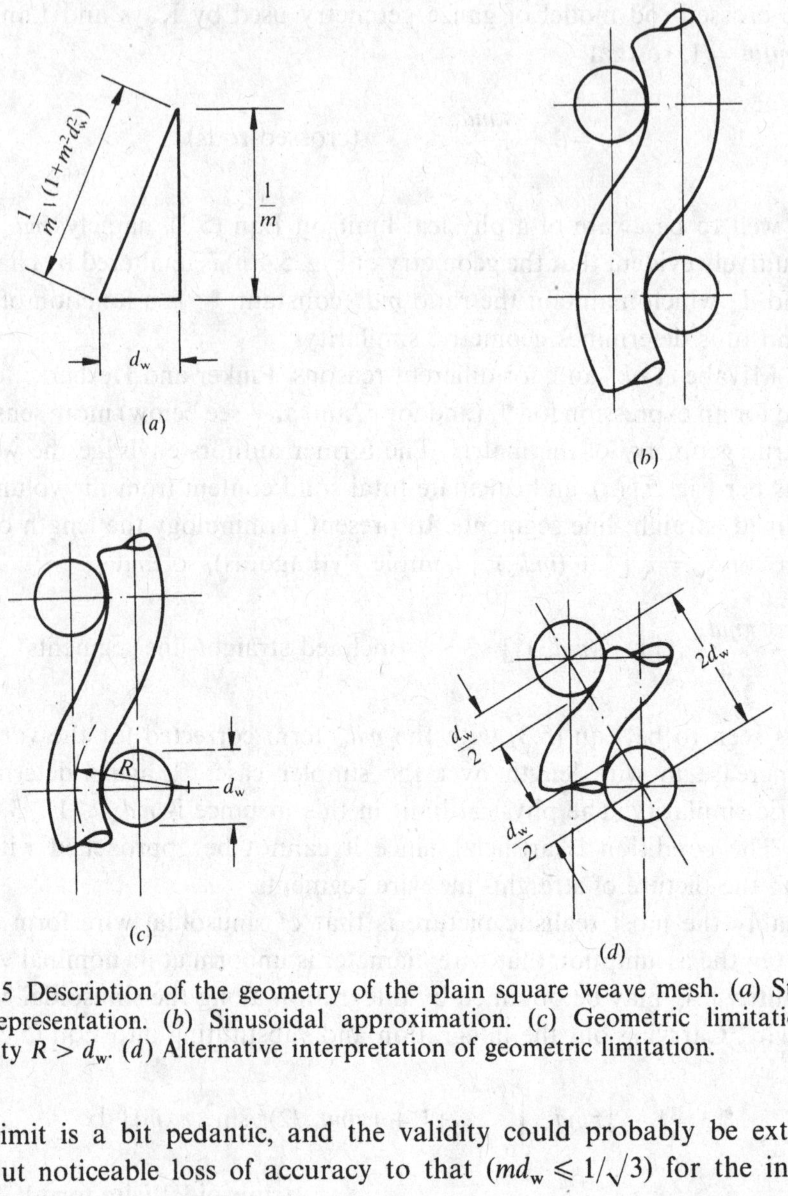


Fig. 5.5 Description of the geometry of the plain square weave mesh. (*a*) Straight line representation. (*b*) Sinusoidal approximation. (*c*) Geometric limitation on porosity $R > d_w$. (*d*) Alternative interpretation of geometric limitation.

The limit is a bit pedantic, and the validity could probably be extended without noticeable loss of accuracy to that ($md_w \leqslant 1/\sqrt{3}$) for the inclined straight-line segments.

Fig. 5.6(*a*) shows $\P_v$ in function of $md_w$ for the three analytical representations of gauze geometry. At low porosities (of the order of 0.5) there is some 10% discrepancy between the crossed-rods model and the sinusoidal. *Assuming the latter to be the more reliable, the figure translates to a 10% error in calculated regenerator dead space.* At more usual values of $\P_v$ – say, 0.7, there is still some 3–4% error. Such values are inconsistent with the level of accuracy currently sought in computer simulation of the Stirling gas circuit.

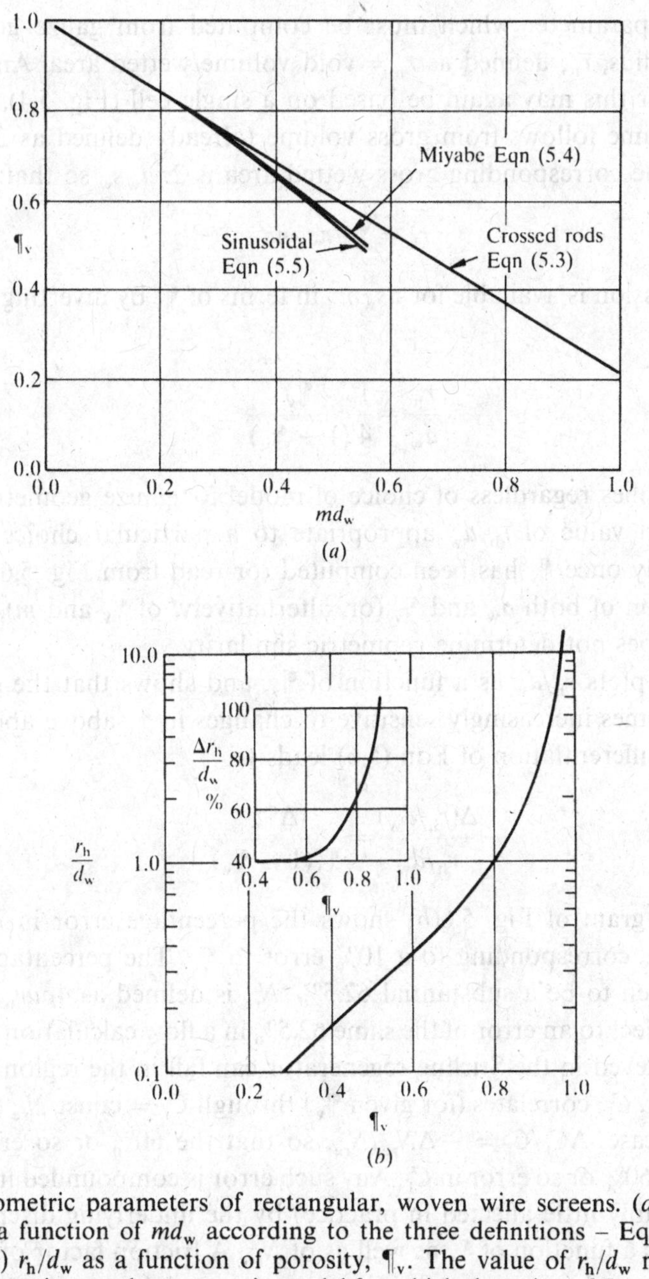


Fig. 5.6 Geometric parameters of rectangular, woven wire screens. (*a*) Volumetric porosity as a function of $md_w$ according to the three definitions – Eqns (5.3), (5.4) and (5.5). (*b*) $r_h/d_w$ as a function of porosity, $¶_v$. The value of $r_h/d_w$ read from the curve is appropriate to the geometric model for which the value of $¶_v$ was calculated. Inset: error in $r_h/d_w$ for 10% error in $¶_v$.

A further parameter which must be computed from gauze geometry is hydraulic radius, $r_h$, defined as $r_h$ = void volume/wetted area. An algebraic expression for this may again be based on a single cell (Fig. 5.4), for which the void volume follows from gross volume (already defined as $2d_w/m^2$) as $2\P_v d_w/m^2$. The corresponding gross wetted area is $2\pi d_w s_w$ so that:

$$r_h = \P_v/\pi s_w m^2$$

But an expression is available for $\pi s_w m^2$ in terms of $\P_v$ by inverting Eqn (5.2). Substituting:

$$\frac{r_h}{d_w} = \frac{1}{4}\frac{\P_v}{(1 - \P_v)} \tag{5.6}$$

Eqn (5.6) applies regardless of choice of model for gauze geometry, so that the numerical value of $r_h/d_w$ appropriate to a particular choice of model follows readily once $\P_v$ has been computed (or read from Fig. 5.6(*a*)). Since $r_h$ is a function of both $d_w$ and $\P_v$ (or, alternatively, of $\P_v$ and $m$), the value of $r_h$ alone does not determine geometric similarity.

Fig. 5.6(*b*) plots $r_h/d_w$ as a function of $\P_v$, and shows that the dependent variable becomes increasingly sensitive to changes in $\P_v$ above about 0.7. In fact, simple differentiation of Eqn (5.6) leads to

$$\frac{\Delta(r_h/d_w)}{r_h/d_w} = \frac{\Delta\P_v}{\P_v(1 - \P_v)} \tag{5.7}$$

The inset diagram of Fig. 5.6(*b*) shows the percentage error in $r_h/d_w$ as a function of $\P_v$ corresponding to a 10% error in $\P_v$. The percentage error at $\P_v = 0.8$ is seen to be a substantial 62.5%. $N_{re}$ is defined as $4\rho u r_h/\mu$, and so would be subject to an error of the same 62.5% in a flow calculation. Reynolds numbers achieved in the Stirling regenerator can fall in the region for which friction factor, $C_f$, correlates (for given $\P_v$) through $C_f$ = const/$N_{re}$ (see Chap. 7). For this case, $\Delta C_f/C_f = -\Delta N_{re}/N_{re}$, so that the 60% or so error in $N_{re}$ becomes a −60% or so error in $C_f$. Any such error is compounded in principle (though possibly little affected in practice) by the underlying discrepancy in $\P_v$, since $C_f$ is a function of $\P_v$ as well as of $N_{re}$. A friction factor of only 40% of the correct value goes some way towards explaining the discrepancy referred to in Section 5.1.

Any such error exists only to the extent of any mismatch between the hydraulic radius of the regenerator modelled and that used experimentally in deriving the working correlation $C_f = C_f(N_{re})$ (see Section 5.6). However, insofar as many of the data of Kays and London (for example) are for crossed rods (appropriately correlated using Eqn (5.3)), application of these correlations to woven screens incurs the risk of the problem in question. Other of the correlations by the same authors (e.g., their Fig. 7.9) are for mixed matrices, some woven, some of crossed rods. The curves cross in a fashion inconsistent with expectations of a properly constituted function of the form $C_f = C_f(\P_v, N_{re})$. Kays' and London's unique work has served its intended purpose in exemplary fashion – but that purpose was never the intimate analysis of the Stirling cycle machine: as a starting point the latter calls for clearly understood and consistent usage of such parameters as $\P_v$ and $r_h$.

In those regimes of flow involving the effects of compressibility an important parameter is the ratio (free-flow area/frontal area) – $\alpha_\P$. The crossed-rod picture allows two definitions:

$$\alpha_\P = (1 - md_w) \quad \text{or} \quad \alpha_\P = (1 - md_w)^2 \tag{5.8}$$

The second or the first alternative is used depending upon whether it is considered that flow responds simultaneously in the two directions orthogonal to the flow direction, or serially. Two square-woven gauzes, simplified as in Fig. 5.4(*b*), and having the same value of $\alpha_\P$ are geometrically similar.

For the sinusoidal case Pinker and Herbert offer the following expression, without derivation, for free-flow area ratio, $\alpha_\P$:

$$\alpha_\P = 1 - \frac{2}{\pi^2}\left\{(1 + 2N^2)\left[\frac{\pi}{2} - \cos^{-1}(N)\right] + 3N\sqrt{(1 - N^2)}\right\} \tag{5.9}$$

In Eqn (5.9) (only) $N = \pi md_w/2$.

It is clear that $\alpha_\P$ is a function only of $md_w$. Two rectangular, sinusoidal screens with the same $\alpha_\P$ are again geometrically similar.

Summarizing the geometric considerations raised to this point:

(*a*) Any of the $\P_v$ or $\alpha_\P$ determine similarity of a specific geometry between single gauzes, (namely, the geometry underlying the particular definition of $\P_v$ or $\alpha_\P$) but, strictly speaking, not between stacks of gauzes. This is because neither variable takes account of the relative orientation of successive gauzes in a stack.

(*b*) Heat transfer and flow friction correlations based on a given set of definitions of $\P_v$ and $\alpha_\P$ (and thus of $r_h$) are reliable for use in the context of the Stirling regenerator only if they are used in terms of $N_{re}$ etc. based on the original, respective definitions.

The foregoing treatment has given rise to a characteristic quantity, $\P_v$, by which geometric similarity is determined, and to a characteristic length, $r_h$, which is required in the formation of the defining dimensionless group or groups (e.g., $N_{re}$). It is worth noting that $r_h$ is not the sort of linear dimension one measures directly with a micrometer. This suggests that it may be possible to identify a length quantity to characterize porous materials of random geometry. The treatment of the next section follows Masha, Beavers and Sparrow[8] and Green and Duwez,[9] and is not restricted to any particular matrix structure.

## 5.3 Resistance coefficient of a porous medium

In general, for very low Reynolds number (creeping flow)

$$dp/dx = -\text{const}\ \mu U/\delta^2 \tag{5.10a}$$

In which $\delta$ is a length characterizing the pore openings. For flows where inertia effects dominate:

$$dp/dx = -\text{const}\ \rho U^2/\delta \tag{5.10b}$$

The transition between the two regimes is more gradual than in the case of pipe flow, so that at any one time the defining equations must contain terms proportional to both $U$ and $U^2$. Using the terminology of Green and Duwez the resulting (Forchheimer) law for pressure gradient in terms of combined viscous and localized acceleration effects becomes

$$-\frac{dp}{dx} = \frac{\mu U}{k_\alpha} + \frac{\rho U^2}{k_\beta} \tag{5.11}$$

$k_\alpha$ and $k_\beta$ have dimensions of [length]$^2$ and [length] respectively, and characterize the structure of the porous material. $k_\alpha$ [L$^2$] is equivalent to the Darcy permeability coefficient. $k_\beta$ [L] might be considered to be a measure of the tortuousness of the flow channels.

For a permeable material with pore space uniformly and randomly distributed, and with all porosity contributing to permeability (but see Green's and Duwez' comment on the matter) superficial velocity, $U$, may be related to pore velocity, $u$:

$$U = \P_v u \tag{5.12}$$

The conservation equations for steady, irreversible, *compressible* adiabatic flow through the porous solid then become:

mass continuity: $$d(\rho u) = 0 \tag{5.13}$$

momentum $$\rho u \frac{du}{dx} + \frac{dp}{dx} + \frac{\mu u}{k_\alpha} + \frac{\rho u^2}{k_\beta} = 0 \tag{5.14}$$

energy $$c_p\, dT + u\, du = 0 \tag{5.15}$$

Eqns (5.13)–(5.15) are the steady-flow equivalent of the general form of the conservation laws which constitute the statement of conditions in the flow passages of the regenerator of the working machine. For consistency the correlation of data obtained in steady-flow tests for use in Stirling cycle modelling must be based on these equations.

Using Eqn (5.12) the momentum equation may be written:

$$\gamma N_{ma}{}^2 \frac{du}{u} + \frac{dp}{p} + \gamma N_{ma}{}^2 \P_v{}^2 \left( \frac{1}{N_{re}{}^*} + 1 \right) \frac{dx}{k_\beta} = 0 \tag{5.16}$$

In Eqn (5.16) $N_{re}{}^*$ is a Reynolds number based on the permeability constant, $k$, *viz*:

$$N_{re}{}^* = \rho U k_\alpha / \mu k_\beta \tag{5.17}$$

$N_{ma}$ is Mach number, $u/a$, and $\gamma$ the specific heat ratio, so that Eqn (5.16) has the form:

$$\frac{dp}{p} = \frac{dp}{p}(\P_v, \gamma, N_{re}{}^*, N_{ma}) \text{ over given } dx \tag{5.16a}$$

Except where single gauzes are being tested, experimental correlations for porous materials tend to involve a specimen of finite flow length, $L$, $\gg dx$. Eqn (5.16) thus requires to be integrated over the test length, $L$. Omitting this process introduces the assumption that the pressure distribution is linear over the test length. It will be demonstrated that such an assumption is not always valid. The integration process parallels that for compressible flow with friction in a regular-shaped duct and results in[10]

$$\P_v{}^2 \left( \frac{1}{N_{re}{}^*} + 1 \right) \frac{x}{k_\beta} = \frac{1}{\gamma M_i{}^2} \left( 1 - \frac{M_i{}^2}{M^2} \right) + \frac{\gamma - 1}{2\gamma} \ln \frac{p^2}{p_i{}^2} \tag{5.18}$$

In Eqn (5.18) $x$ may be any distance from an upstream point, $i$, in the matrix where conditions are known to a point further downstream, subject to $x \leqslant L$. The equation permits a plot of $\Delta p/p$ vs $x$ (or, better, vs $x/L$). Masha *et al.* present such plots showing excellent agreement between experimentally determined and computed values over a range of $N_{re}{}^*$. It is significant that the curves become markedly non-linear at $N_{ma}$ less than 0.1.

Masha *et al.* remark that, if the assumption that flow is adiabatic is replaced with the assumption that it is isothermal, then Mach number, $M$, can be eliminated as an explicit parameter of Eqn (5.18). It is difficult to see how this can be, except for the case in which the heating effect of viscous dissipation exactly counterbalances the cooling effect of expansion. Otherwise the energy equation requires a term for heat input (or output) to account for the difference, and no such term carries over to Masha's 'isothermal' equation.

Eqn (5.18) serves to isolate the various components of static pressure drop observed in a steady-flow test. *It is important to be clear that usage of the Mach number term to this point reflects only part of the compressibility phenomenon.* It remains to decide how to deal with the rest.

## 5.4 Compressibility

Fig. 5.7 is taken from the study by Su[11] and shows pressure coefficient, $\Delta p/\frac{1}{2}\rho u^2$, as a function of Mach number for a single gauze. It is seen that the onset of choking ($\Delta p/\frac{1}{2}\rho u^2 \approx$ infinity) is gradual, becoming measurable in the region of $M = 0.05$. In an air-charged machine this corresponds to a $u$ of about 18 m/s at the cold end of the regenerator. At 3000 rpm with a crank radius of 25 mm peak piston speed is of the order of 7.5 m/s. With a ratio of piston area to regenerator area greater than unity, and free-flow/frontal area ratio proportional to $\P_v^2$, values of $M$ in excess of 0.05 are a possibility.

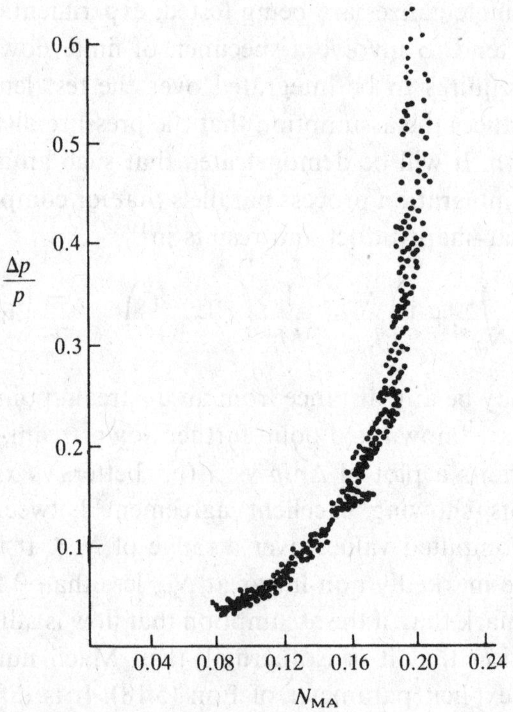


Fig. 5.7 Pressure loss coefficient as a function of approach Mach number for single square woven gauze of 0.714 porosity. (After Su.[11])

The compressibility effect here is that of the flow-area reduction of the single gauze acting as a convergent–divergent nozzle. It is difficult to imagine that the effect can be anything but a penalty on performance. The matter remains an open question pending further, comprehensive tests. In the meantime there is a choice of two ways in which to proceed:

(*a*) to acquire steady-flow correlations based on an equation such as Eqn (5.18), to embody in an appropriate simulation and to compare with experiment, or
(*b*) to monitor the regenerator computation for peak $N_{ma}$: if values arise in excess of 0.05, to redesign for operation at a lower maximum and thus to be sure of avoiding any penalty.

The former course calls for correlations for $C_f$ in terms of geometry, $N_{re}$ and $N_{ma}$. Extensive experimentation is implied, and it becomes important to take advantage of any available method of making the correlation more compact in presentation and in use. The concept of the 'true' Reynolds number[9] offers promise here.

## 5.5 The 'true' Reynolds number

If the definition of a Reynolds number is the ratio of the inertia forces in a flow system to the viscous forces, then for the flow system defined by Eqn (5.11) $N_{re}{}^*$ is given by Eqn (5.17). Friction factor, $C_f$, is defined as the ratio of dissipative forces to inertia forces, viz,

$$C_f{}^* = \frac{-k_\beta(\mathrm{d}p/\mathrm{d}x)}{\frac{1}{2}\rho U^2} \tag{5.19}$$

By definition:

$$C_f{}^* = 2/N_{re}{}^* \text{ in the Darcy regime}$$
$$= 2 \text{ in the inertial regime}$$

Over the entire range of flows, therefore (including compressible[8])

$$C_f{}^* = \frac{2}{N_{re}{}^*} + 2 \tag{5.20}$$

Fig. 5.8 is from the paper by Green and Duwez, and demonstrates the successful correlation of friction factor with Reynolds number for a range of porous metals having different geometries. This form of correlation calls for

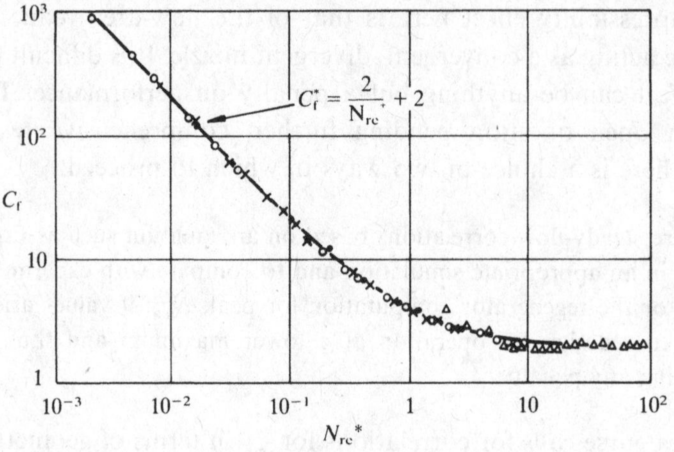


Fig. 5.8 Curve of 'true' friction factor, $C_f^*$, against 'true' Reynolds number, $N_{re}^*$, showing how the geometric parameter is absorbed. (From Green and Duwez.[9])

values of *two* characteristic length quantities, $k_\alpha$ and $k_\beta$, as opposed to the single geometric quantity ($\P_v$) used in the Kays and London type of presentation where there is, in general, a different curve of $C_f(\P_v)$ vs $N_{re}(r_h)$ for each $\P_v$. The coefficients in question are acquired by plotting Eqn (5.11) and determining the $k_\alpha$ and $k_\beta$ which provide a fit. The technique is standard in the field of flow through porous media,[12] and it is perhaps surprising that it has not so far received the attention of those working on Stirling machines. The fact that Miyabe *et al.*[6] ignore the implications must cast some doubt on the validity of their correlation based on the single length parameter $1/m - d_w$.

The principal advantage of a correlation which absorbs the geometric parameter is that it affords the possibility of presenting the effects of geometry, $N_{re}^*$ and $N_{ma}$ on a single chart. When the relevant data are eventually available, they may appear as suggested in Fig. 5.9, which plots $C_f^*$ against $N_{re}^*$ with $N_{ma}$ as parameter.

Pending acquisition of such data, the immediate option is to use correlations such as those of Kays and London, doing so with regard to the way in which the data were originally obtained, and noting that they are almost certainly invalidated by instances of local $u/a > 0.05$. The supposed problem is probably confined to designs based on air, $N_2$ or other gases of relatively high molecular mass, in which $a$ is relatively low. On the other hand, a successful air-charged machine offers certain potential advantages over the alternatives, and any set of general design rules for Stirling machines needs to be able to cope with air.

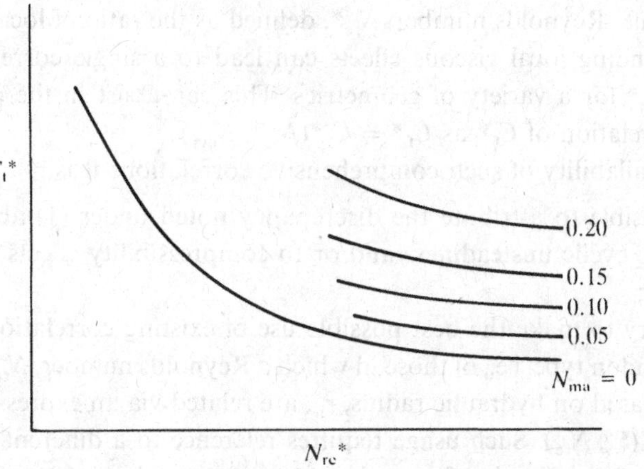


Fig. 5.9 Possible form of a correlation based on $C_f$* and $N_{re}$* with Mach number, $N_{ma}$ as parameter.

## 5.6 Summary

(1) Analysis and simulation based on steady-flow correlations between friction factor and Reynolds number generally lead to an underestimate of instantaneous heat exchanger and regenerator pressure gradient. The underestimate is popularly attributed to the cyclic unsteadiness of the gas processes within the Stirling machine.

(2) When friction factor and Reynolds number are based on a single length parameter (such as hydraulic radius, $r_h$) different geometries give rise to different curves of $C_f$ vs $N_{re}$.

(3) $¶_v$ and $r_h$ calculated on the assumption of crossed-rod geometry have numerical values different from those resulting from the sinusoidal assumption. The uncompensated discrepancy appears capable of causing significant error in flow loss computations.

(4) Steady-flow correlations of the type most frequently used in Stirling machine simulation *embody indeterminate values of Mach number*. Therefore they do not completely correlate the friction factor with all the flow variables. The literature on compressible flow (principally through single gauzes) is probably as extensive as that covering incompressible flow through gauze stacks, but has yet to be taken into account in Stirling cycle modelling.

(5) If steady-flow correlations are to be used in conjunction with the (unsteady flow) laws of mass, momentum and energy conservation, then those correlations require to be based on the steady-flow equivalent of the more general laws. When they are, it is found that Mach number is a parameter, and that the gradient of pressure through a porous medium of finite extent in the flow direction is increasingly non-linear with increasing Mach number – even at values as low as 0.1.

(6) Use of a 'true' Reynolds number, $N_{re}*$, defined as the ratio of local inertia effects to corresponding local viscous effects can lead to a single correlation between $C_f*$ and $N_{re}*$ for a variety of geometries. This can assist in the presentation of the full correlation of $C_f*$ as $C_f* = C_f*(N_{re}*, N_{ma})$.

(7) Pending availability of such comprehensive correlations it is

(*a*) not possible to attribute the discrepancy noted under (1) above authoritatively to cyclic unsteadiness and/or to compressibility effects and/or to heat transfer;

(*b*) necessary to make the best possible use of existing correlations of the Kays and London type, i.e., of those in which a Reynolds number, $N_{re}$, and a friction factor based on hydraulic radius, $r_h$, are related via an expression of the type $C_f = C_f(\P_v, N_{re})$. Such usage requires reference to a different curve for each different $\P_v$ *and each different definition of* $\P_v$, and some sort of independent check that, for any given $N_{re}$, corresponding $N_{ma} <$, say, 0.05.

# 6
# The equivalent Stirling cycle machine

## 6.1 Thermodynamically-similar volume variations

According to the discussion in Chap. 4 of the thermodynamic similarity conditions, specific cycle work is given by:

$$\zeta = \zeta\left( N_\tau, \kappa, \alpha, \mu_v, \begin{matrix} \mu_{dce} \\ \mu_{de} \\ \mu_{dr} \\ \mu_{dc} \\ \mu_{dcc} \end{matrix}, \begin{matrix} \alpha_{xe} \\ \alpha_{xr} \\ \alpha_{xc} \end{matrix}, \begin{matrix} \lambda_{he} \\ \lambda_{hr} \\ \lambda_{hc} \end{matrix}, \gamma, N_{MA}, N_{SG}, N_H, N_F \right) \tag{6.1}$$

Indicated thermal efficiency is a function of the same variables:

$$\eta_{th} = \eta_{th}\left( N_\tau, \kappa, \alpha, \mu_v, \begin{matrix} \mu_{dce} \\ \mu_{de} \\ \mu_{dr} \\ \mu_{dc} \\ \mu_{dcc} \end{matrix}, \begin{matrix} \alpha_{xe} \\ \alpha_{xr} \\ \alpha_{xc} \end{matrix}, \begin{matrix} \lambda_{he} \\ \lambda_{hr} \\ \lambda_{hc} \end{matrix}, \gamma, N_{MA}, N_{SG}, N_H, N_F \right) \tag{6.2}$$

It was an assumption underlying the theoretical development, and it is clear from Eqns (6.1) and (6.2) that, all other things being equal, fluid motion in the gas circuit is controlled by expansion and compression space volume variations in conjunction with dead volume distribution. If only two variable-volume spaces are involved, and if the variation is simple harmonic in both cases, then the volume variations are completely described when numerical values of *volume ratio*, $\kappa$, and of *phase angle*, $\alpha$, are given. (If the machine has a compression space comprising variable volumes separated by a heat exchanger, then an additional dimensionless parameter is required, but the principle is otherwise unaffected.)

Fig. 6.1 shows, in schematic form, the most common configurations involving the two variable volumes. In the opposed-piston layout (Fig. 6.1(*a*)), volume phase angle is the same as piston (or crank) phase angle, and volume

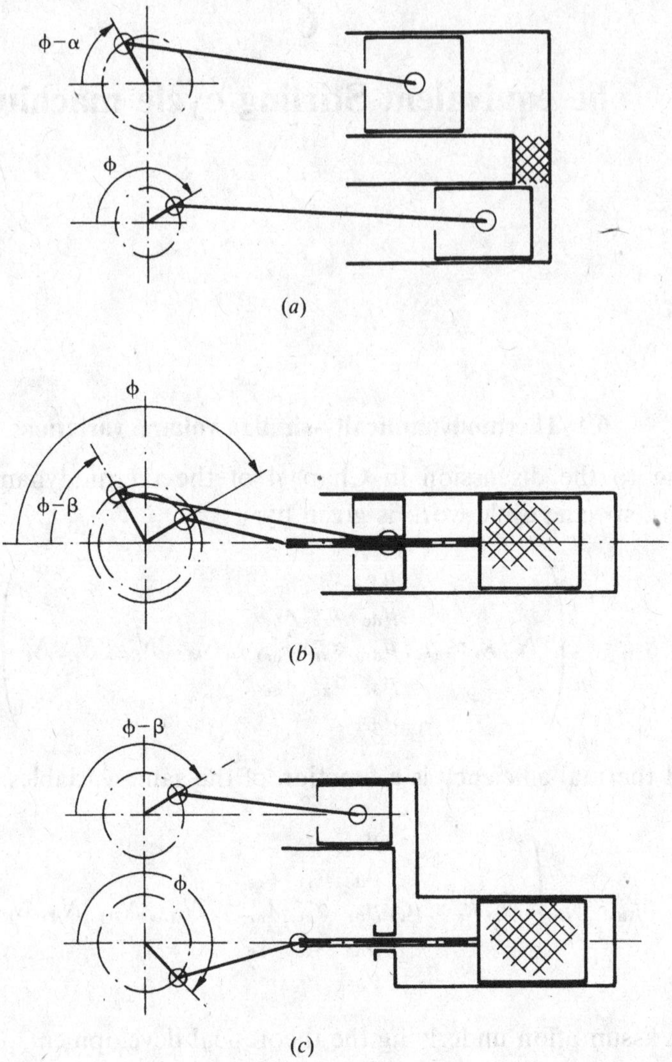


Fig. 6.1 The three, basic types of two-working-space Stirling machine: (*a*) Simple, parallel, opposed-piston ('alpha') type. (*b*) Uniform-bore coaxial ('beta') type. (*c*) Parallel, split compression space type. Note distinction between this *two working-space* machine and the *three-space* variant which is distinguished by having formal heat exchanger surfaces between the two regions of the compression space.

ratio, $\kappa$, is simply $V_C/V_E$. In the cases shown in Figs 6.1(*b*) and 6.1(*c*), volume phase angle, $\alpha$, is not generally equal to crank phase angle, $\beta$, and the ratio, $\lambda$, of volume swept by the piston to that swept by the displacer is generally different from volume ratio, $\kappa$. The matter of defining $\kappa$ and $\alpha$ for the study of a particular design thus has two aspects:

where the machine is not of opposed-piston type, equivalent $\kappa$ and $\alpha$ have to be defined corresponding to actual $\lambda$ and $\beta$;

independently of type of machine, the $\kappa$ and $\alpha$ (or $\lambda$ and $\beta$) are to be determined which are the approximate simple harmonic equivalent of the volume variations produced by the real crank mechanism.

## 6.2 Equivalent κ and α for sinusoidal volume variation

The treatment is that due to Finkelstein.[1] In Fig. 6.1(*b*) the variation in *total* volume is controlled by the compression-space piston alone. Denoting by $V_D$ the sum of all dead space, including that unswept in the variable-volume spaces at the dead-centre positions, enables total working fluid volume to be written down immediately:

$$V_e(\phi) + V_c(\phi) + V_D = \text{constant term} + \tfrac{1}{2}\lambda V_E[1 + \cos(\phi - \beta)] \quad (6.3)$$

Developing the corresponding expression for the opposed-piston machine of Fig. 6.1(*a*) will permit a comparison of terms from which the relationship between $\kappa$, $\alpha$, $\lambda$ and $\beta$ may be inferred. For the opposed piston case:

$$V_e(\phi) = \tfrac{1}{2}V_E[1 + \cos(\phi)] \quad (6.4)$$

$$V_c(\phi) = \tfrac{1}{2}V_C[1 + \cos(\phi - \alpha)] \quad (6.5)$$

The total working fluid circuit volume at any crank angle, $\phi$, then becomes:

$$\begin{aligned} V_e(\phi) + V_c(\phi) + V_D &= \tfrac{1}{2}V_E\{1 + \cos(\phi) + \kappa[1 + \cos(\phi - \alpha)] + 2\mu_D\} \\ &= \tfrac{1}{2}V_E[(1 + \kappa + 2\mu_D) + \cos(\phi) + \kappa\cos(\phi - \alpha)] \quad (6.6) \end{aligned}$$

In Eqn (6.6) $\mu_D = V_D/V_E$.

Expanding the term $\cos(\phi - \alpha)$, forming the sum of like trigonometric terms and then recombining into the form $\cos(A + B)$ gives

$$V_e(\phi) + V_c(\phi) + V_D = \tfrac{1}{2}V_E\left((1 + \kappa + 2\mu_D) + \sqrt{[1 + \kappa^2 + 2\kappa\cos(\alpha)]} \times \cos\left\{\phi - \arctan\left[\frac{\kappa\sin(\alpha)}{1 + \kappa\cos(\alpha)}\right]\right\}\right) \quad (6.7)$$

By comparison with Eqn (6.3)

$$\tan(\beta) = \frac{\kappa\sin(\alpha)}{1 + \kappa\cos(\alpha)} \quad (6.8)$$

$$\lambda = \sqrt{[1 + \kappa^2 + 2\kappa\cos(\alpha)]} \quad (6.9)$$

Eqns (6.8) and (6.9) may be inverted to permit expression of $\kappa$ and $\alpha$ in terms of $\lambda$ and $\beta$:

$$\tan(\alpha) = \frac{\lambda \sin(\beta)}{\lambda \cos(\beta) - 1} \tag{6.10}$$

$$\kappa = \surd[1 + \lambda^2 - 2\lambda \cos(\beta)] \tag{6.11}$$

By means of these last four equations, a displacer-type machine may be converted to the thermodynamically equivalent opposed-piston type. Conversely, after computation of the performance of an opposed piston machine, the values of $\lambda$ and $\beta$ may be determined for which a coaxial machine (Fig. 6.1(*b*) or (*c*)) will have an identical, computed ideal thermodynamic cycle.

In taking the inverse tangents called for by Eqns (6.8) and (6.10) it is necessary to choose the appropriate value from the multiple angles. Fig. 6.2 is a chart originally constructed by Finkelstein[1] which will give an approximate value of the angle required (and, indeed, of the volume ratio) against which may be checked the value obtained from tables or by use of a calculator.

It is not necessary to carry out algebraic conversion of values of dead space: these are merely the enclosed volumes not swept by either piston – regardless of configuration. Nevertheless, it is interesting to calculate explicitly the amount by which compression-end dead volume of the machine shown at Fig. 6.1(*c*) exceeds that of the other displacer-type machine (at Fig. 6.1(*b*)) by virtue of the fact that displacer and piston strokes have no overlap.

For this case

$$\begin{aligned} V_c(\phi) &= \tfrac{1}{2}\lambda V_E[1 + \cos(\phi - \alpha)] + \tfrac{1}{2}V_E[1 - \cos(\phi)] \\ &= \tfrac{1}{2}V_E\Bigg(1 + \lambda + \surd[1 + \lambda^2 - 2\lambda \cos(\beta)] \\ &\qquad \times \cos\left\{\phi - \arctan\left[\frac{\lambda \sin(\beta)}{\lambda \cos(\beta) - 1}\right]\right\}\Bigg) \end{aligned}$$

This last expression may be written

$$V_c = \tfrac{1}{2}V_E[1 + \lambda + \kappa \cos(\phi - \alpha)]$$

The minimum of $V_c$ represents an additional dead space amounting to $\frac{1}{2}V_E\{1 + \lambda - \surd[1 + \lambda^2 - 2\lambda \cos(\beta)]\}$. This is finite and positive for realistic values of $\lambda$ and $\beta$, so if two machines, one based on Fig. 6.1(*b*), the other on Fig. 6.1(*c*), have identical $\lambda$ and $\beta$, the latter will suffer the additional dead space which will reduce compression ratio and, all other things being equal, specific cycle work. However, the apparent advantage of the coaxial con-

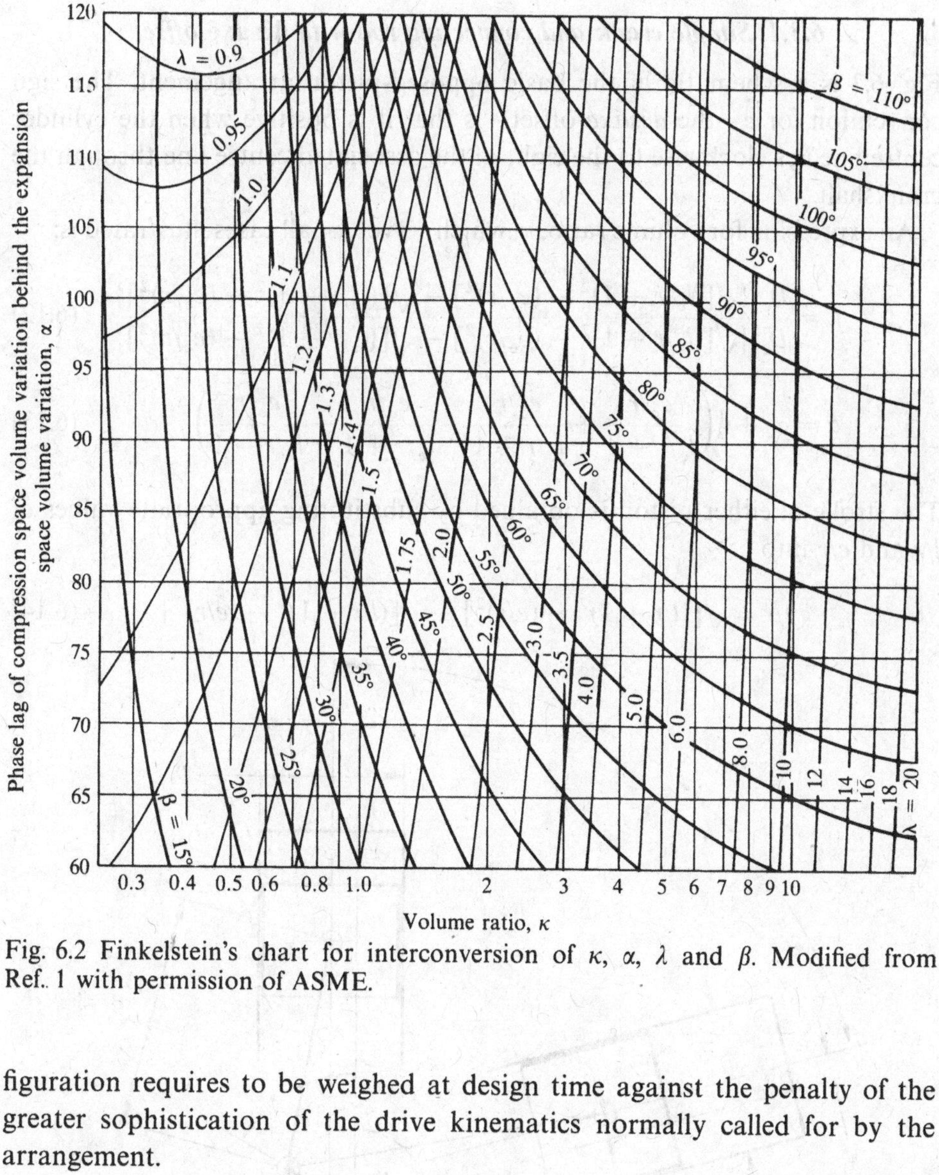


Fig. 6.2 Finkelstein's chart for interconversion of κ, α, λ and β. Modified from Ref. 1 with permission of ASME.

figuration requires to be weighed at design time against the penalty of the greater sophistication of the drive kinematics normally called for by the arrangement.

## 6.3 Values of phase angle and volume ratio for some common mechanisms

Appendix V deals in detail with the kinematics of a number of drive mechanisms, and includes kinematically-exact expressions for volume variations in each case. For present purposes, it is necessary to know only the respective κ and α, or, alternatively, to know the λ and β and to be able to convert.

### 6.3.1 *Simple crank and connecting rod with* désaxé *offset*

Fig. 6.3 is a schematic of the basic opposed-piston arrangement. The sign convention for $e$ – the *désaxé* offset – is that it is positive when the cylinder centre line lies clockwise to the right of the associated centre line through the crankshaft.

An expression for volume ratio, $\kappa$, which serves for all cases illustrated is:

$$\kappa = \frac{D_c{}^2\{\sqrt{[(l_c/r+1)^2-(e_c/r)^2]}-\sqrt{[(l_c/r-1)^2-(e_c/r)^2]}\}}{D_e{}^2\{\sqrt{[(l_e/r+1)^2-(e_e/r)^2]}-\sqrt{[(l_e/r-1)^2-(e_e/r)^2]}\}} \tag{6.12}$$

$$\alpha = \alpha_c + \tfrac{1}{2}\left(\frac{e_c/r}{l_c/r+1}+\frac{e_c/r}{l_c/r-1}-\frac{e_e/r}{l_e/r+1}-\frac{e_e/r}{l_e/r-1}\right) \tag{6.13}$$

The stroke of either piston is obtained by substituting appropriate values of $l/r$ and $e/r$ into

$$s/r = \sqrt{[(l/r+1)^2-(e/r)^2]}-\sqrt{[(l/r-1)^2-(e/r)^2]} \tag{6.14}$$

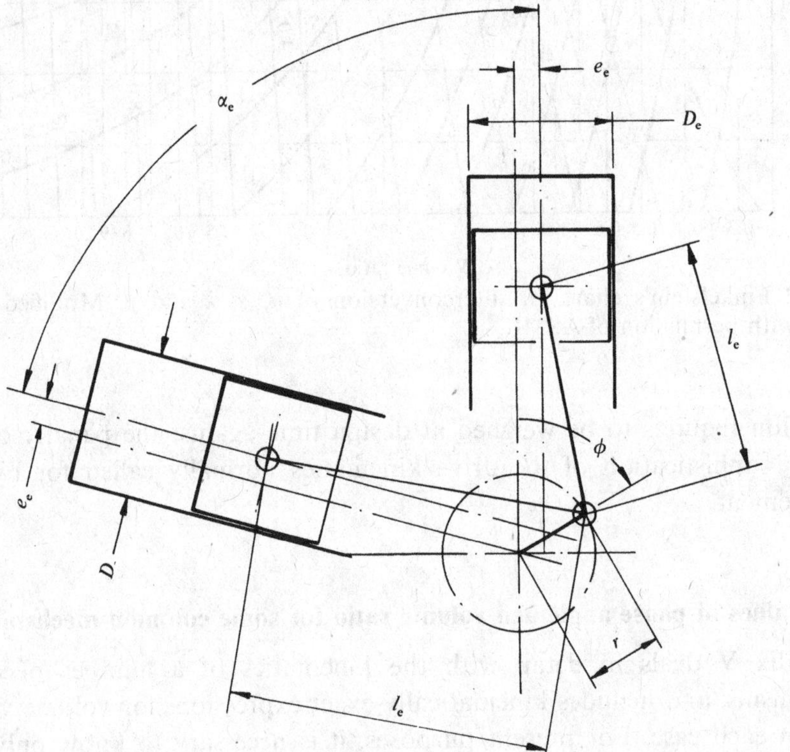


Fig. 6.3 Terminology for the practical, opposed-piston ('alpha') machine.

Net swept volume, required for computation of specific cycle work, is the simple sum of the swept volumes of individual cylinders, obtained in turn from Eqn (6.14) by a multiplication by the areas of respective bores.

In multi-cylinder embodiments an alternative opposed-piston arrangement is common: a variable-volume space *above* one piston communicates via heat exchangers and regenerator with the second variable-volume space *below* the next piston in the sequence. Appendix V explains how Eqns (6.12)–(6.14) are applied to this case.

### 6.3.2 The symmetrical rhombic drive

The rhombic drive offers the possibility of complete dynamic balance and of the elimination of side thrust on the reciprocating members. Achievement of these features calls for close (and therefore expensive) manufacturing tolerances, and results in a mechanism which is a bulky adjunct to the compact, coaxial gas circuit.

Fig. 6.4 is a schematic. Attention focusses on the *left-hand* crank, which rotates clockwise for the prime-mover mode, for which $e$ is positive as drawn, and which to this extent will be seen to correspond directly with Fig. 6.3 for the offset-cylinder opposed-piston case.

For uniform bore $D_e = D_c$, so for the symmetrical mechanism:

$$\text{piston swept volume ratio, } \lambda, = 1 \tag{6.15}$$

Dead centre positions differ by piston phase angle, $\beta$:

$$\beta = \arcsin\left(\frac{e/r}{l/r + 1}\right) + \arcsin\left(\frac{e/r}{l/r - 1}\right) \tag{6.16}$$

Equivalent $\kappa$ and $\alpha$ are obtained from Eqns (6.15) and (6.16) via the conversion expressions (Eqns (6.10) and (6.11)). Alternatively, it may be noted that for $D_e = D_c$, volume ratio, $\kappa$, is equal to the ratio of appropriate linear distances. Thus the ratio $V_C/V_E$ is equal to the ratio (maximum–minimum separation of the little-end eyes)/(stroke, $s$). Appendix V shows that there are two cases to be considered corresponding to $e \geqslant r$ and $e < r$, and that these give:

$$\kappa_{e \geqslant r} = \frac{\sqrt{[(l/r)^2 - (1 - e/r)^2]} - \sqrt{[(l/r)^2 - (1 + e/r)^2]}}{s/2r} \tag{6.17a}$$

$$\kappa_{e<r} = \frac{l/r - \sqrt{[(l/r)^2 - (1 + e/r)^2]}}{s/2r} \tag{6.17b}$$

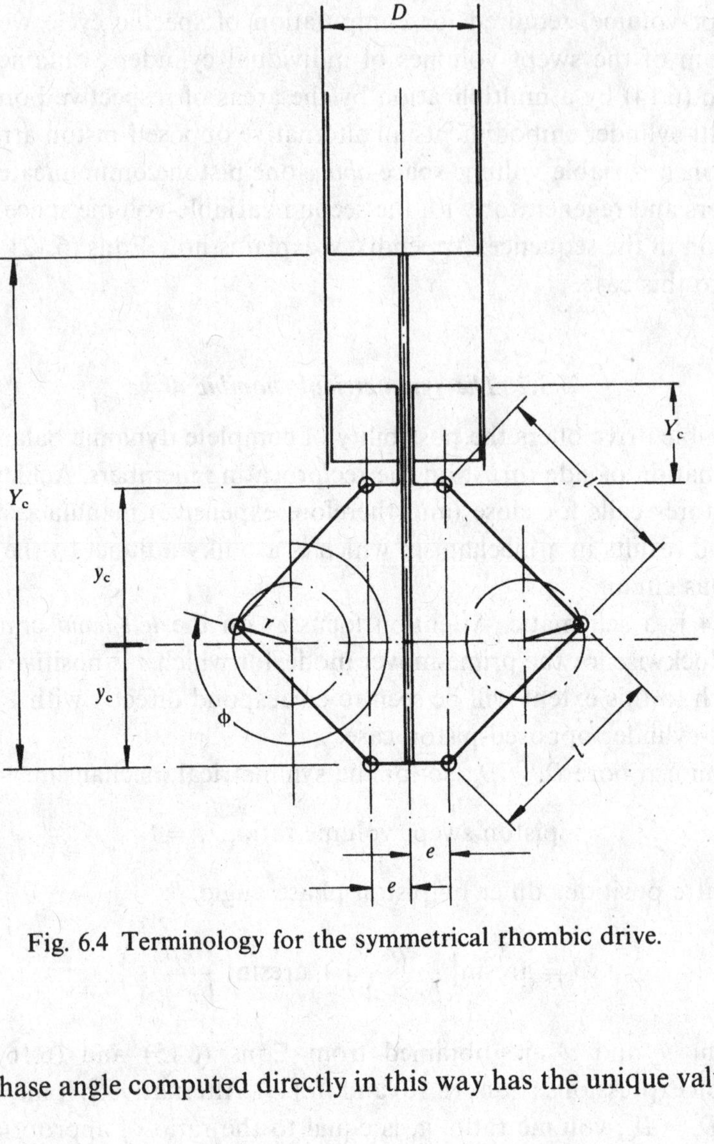


Fig. 6.4 Terminology for the symmetrical rhombic drive.

Volume phase angle computed directly in this way has the unique value given by:

$$\alpha = \pi/2 + \tfrac{1}{2}\left[\arcsin\left(\frac{e/r}{l/r+1}\right) + \arcsin\left(\frac{e/r}{l/r-1}\right)\right] \qquad (6.18)$$

Net swept volume, i.e., the volume swept out by piston and displacer with account of overlap, is:

$$\frac{V_{sw}}{\pi D^2 r/4} = 2\{\surd[(l/r)^2 - (e/r+1)^2] - \surd[(l/r-1)^2 - (e/r)^2]\} + Y_{c1} \qquad (6.19)$$

### 6.3.3 Ross drive mechanism

In the schematic of Fig. 6.5 it is assumed that the angularity of the rods connecting points $C$ and $E$ to the pistons is sufficiently small that the respective vertical components of piston motion may be assumed to be equal to those of $C$ and $E$. The angle $CFP$ is a right angle.

The kinematics are described by a large number of variables, and convenient expressions to enable computation of a $\lambda$ and an $\alpha$ are not readily written down. On the other hand, approximate $\kappa$ and $\alpha$ may be had direct by considering the simplified form of the mechanism discussed by Urieli and Berchowitz.[2] This is shown and discussed in Appendix V. The algebra leads to:

$$\kappa \approx \frac{D_c^{\,2} b_c}{D_e^{\,2} b_e} \tag{6.20}$$

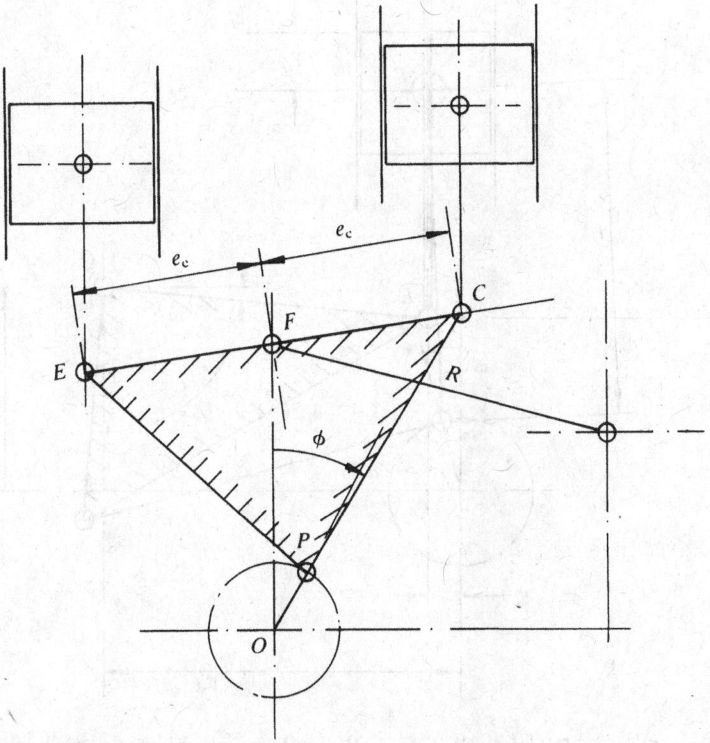


Fig. 6.5 Terminology for the symmetrical Ross drive mechanism.

A volume phase angle is estimated by noting the analogy with Fig. AV.2(*c*):

$$\alpha \approx \tfrac{1}{2}\left(\frac{e_c^*/r}{l_c^*/r+1}+\frac{e_c^*/r}{l_c^*/r-1}-\frac{e_e^*/r}{l_e^*/r+1}-\frac{e_e^*/r}{l_e^*/r-1}\right) \tag{6.21}$$

(NB: $e_e^*$ *is negative and* $\alpha_c = 0$).

### 6.3.4 Philips' bell-crank mechanism

Fig. 6.6 is a schematic of a general form of the mechanism. There is a greater number of variables even than in the case of the Ross drive. The prospective

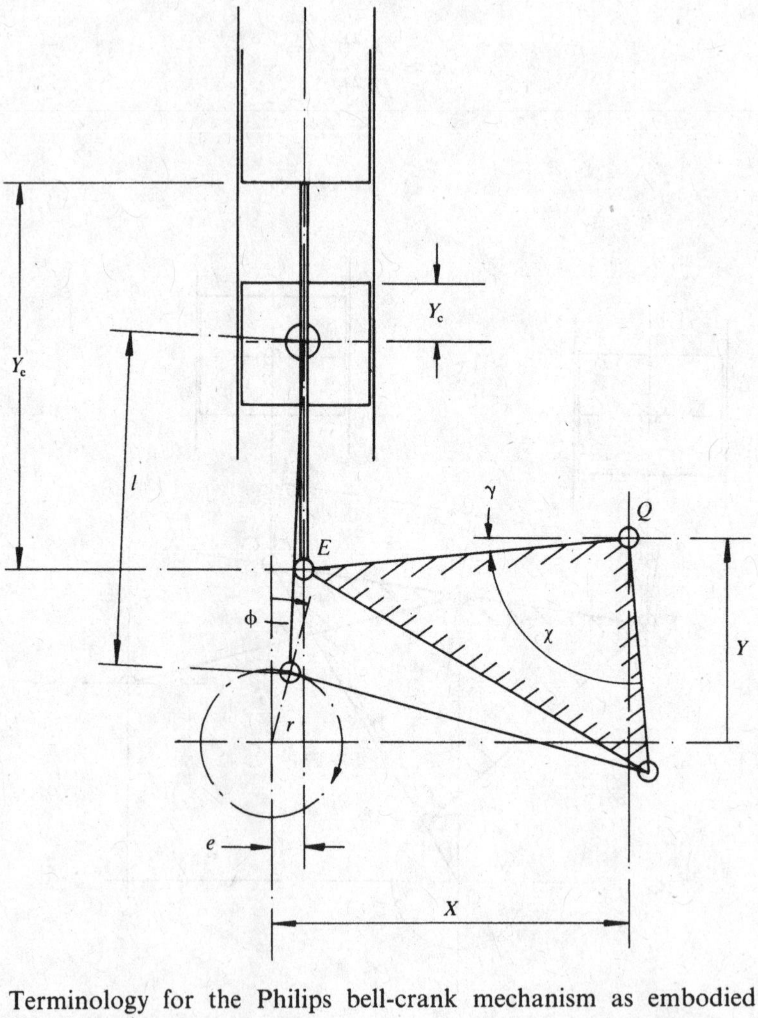


Fig. 6.6 Terminology for the Philips bell-crank mechanism as embodied in the MP1002CA air engine.

user of an existing mechanism may prefer to work by drawing the essential features to suitable scale. Crank positions corresponding to top and bottom dead centre positions of the displacer are easily constructed graphically. The compression-piston stroke is nearly enough $2r$. Rough values of $\lambda$ and $\beta$ follow, and may be converted to $\kappa$ and $\alpha$ in the now familiar way.

On the other hand, a user seeking the maximum thermodynamic benefit from the flexibility of the bell-crank arrangement will probably program for computer the expressions for volume variation in Appendix V. In this case, there will be a ready way of inspecting for the values of crank angle, $\phi$, corresponding to top and bottom dead-centre positions, *viz*, of $\phi_{y\min}$ and $\phi_{y\max}$. From such values may be calculated $\lambda$:

$$\lambda = \frac{\sqrt{[(l/r+1)^2 - (e/r)^2]} - \sqrt{[(l/r-1)^2 - (e/r)^2]}}{(QE/r)\{\sin[\gamma(\phi_{y\min})] - \sin[\gamma(\phi_{y\max})]\}} \tag{6.22}$$

Appendix V defines $\gamma$.

Piston phase angle, $\beta$, may be defined as the mean of the angular differences between top and bottom dead-centre positions:

$$\beta = \tfrac{1}{2}(\beta_{\text{tdc}} + \beta_{\text{bdc}}) \tag{6.23}$$

$$\beta_{\text{tdc}} = 2\pi + \arcsin\left(\frac{e_c/r}{l_c/r + 1}\right) - \phi_{y\max}$$

Based on conditions at bottom dead centre:

$$\beta_{\text{bdc}} = \phi_{y\min} - \arcsin\left(\frac{e_c/r}{l_c/r - 1}\right)$$

Values of $\kappa$ and $\alpha$ for use in the equivalent machine are obtained from $\lambda$ and $\beta$ by use of Eqns (6.10) and (6.11).

Net swept volume is the sum of individual volumes swept by piston and displacer minus any overlap. It may be calculated from the distance $Y_{sw}$ by substracting any stroke overlap from the sum of individual strokes:

$$Y_{sw}/r = (Y_e - Y_c)/r - (QE/r)\sin[\gamma(\phi_{y\max})] - \sqrt{[(l_c/r - 1)^2 - (e/r)^2]} \tag{6.24}$$

$(Y_e - Y_c)/r$ is found from the condition that the distance between lowermost displacer face and piston top face has a minimum value equal to $Y_{c1}$, and that this clearance is achieved when the two faces are at their closest:

$$(Y_e - Y_c)/r = Y_{c1}/r + \cos(\phi^*) + \sqrt{\{(l_c/r)^2 - [\sin(\phi^*) - e/r]^2\}} - Y/r - (QE/r)\sin[\gamma(\phi^*)] \tag{6.25}$$

$\phi^*$ is the value of $\phi$ for which piston face separation is a minimum. It cannot be found by visual examination of the equations for $y_e$ and $y_c$, and is best inspected for during a run of Eqns (AV.26) and (AV.3a) programmed for computer.

# 7

# A three-conservation law cycle model

## 7.1 Choice of analytical approach

The preceding chapters have established principles for the formulation of a cycle description. Illustrating the application of all of these principles will occupy the rest of this book. The present chapter will build up a description based on *thermodynamic similarity*. To keep the algebra manageable, involvement of the fourth conservation law (internal energy balance for the wall) is deferred. The functional form of the solution based on thermodynamic similarity (Chap. 4) is thereby reduced from that of Eqn (6.1) to:

$$\zeta = \zeta\left(\begin{matrix} & & & & \mu_{dce} & & & & & \\ & & & & \mu_{de} & \alpha_{xe} & \lambda_{he} & & & \\ N_\tau, & \kappa, & \alpha, & \mu_v, & \mu_{dr}, & \alpha_{xr}, & \lambda_{hr}, & \gamma, & N_{MA}, & N_{SG} \\ & & & & \mu_{dc} & \alpha_{xc} & \lambda_{hc} & & & \\ & & & & \mu_{dcc} & & & & & \end{matrix}\right) \tag{7.1}$$

with an analogous expression for indicated thermal efficiency.

At this stage there is a choice between, on the one hand, working directly in terms of the conservation laws in (non-linear) differential form and, on the other, introducing linearization or perturbation. The functional form of the solution is unaffected by the choice, and numerical results will be substantially the same for those operating conditions for which the linearization assumptions remain valid. The attraction of linearization is the opportunity it provides of isolating the effects of individual loss mechanisms. An arguable disadvantage is that it promises to distort performance in those regions of the operating envelope where an individual loss is high and where conditions therefore depart from the assumption of linearity.

The treatment which follows was first proposed by Organ[1] and uses the linearized approach. It works in terms not of absolute temperatures and pressures, but of local, instantaneous *departures* from ideal, reversible behaviour. Corresponding losses are calculated and integrated over a complete cycle. These are then subtracted from the ideal cycle work to yield net work.

The individual losses are evaluated in terms of the concepts of Availability Theory. Here, a lost opportunity for work production is expressed as $T_0\,dS$, where $T_0$ is the lowest temperature to which the device could realistically reject heat – in this case the compression exchanger value, $T_c$. The approach offers advantages amongst which are that:

(1) The expressions defining lost work under conditions of unsteady, compressible flow are simpler than those which define the flow itself. A way of looking at this is to note that unsteady effects are not inherently irreversible. Other loss mechanisms, such as conduction within the solid elements of the machine, also tend to have simple expression in terms of entropy creation rate.
(2) Individual contributions to reduced performance (i.e., individual sources of irreversibility) may be pinpointed. By contrast, in the conventional, non-linear formulation, a localized increase in, say, flow resistance normally means increased pumping loss. The latter may, however, be offset by enhanced heat transfer. The *net* effect usually remains unresolved, because the resulting change in overall machine performance is the result not only of the change in question, but also of secondary changes induced elsewhere.

The major simplifying assumptions are those discussed in Chapter 4 under thermodynamic similarity, but certain matters of detail bear confirmation.

## 7.2 Simplifying assumptions

It is assumed that:

(1) Flow is one-dimensional as suggested by Fig. 7.1. There is no diffusion (thermal or molecular) in the flow direction.
(2) The machine operates under conditions such that local, instantaneous values of lost work are small by comparison with corresponding useful energy fluxes. This is equivalent to the assumption that the machine is operating reasonably efficiently overall, and that there are no factors leading to disproportionate local irreversibilities.
(3) The local, instantaneous Stanton number, $N_{st}$, may be expressed as a function of local, instantaneous Reynolds number, $N_{re}$ and Prandtl number, $N_{pr}$, *viz*:

$$N_{st} = N_{st}(N_{re}, N_{pr})$$

Correspondingly,

$$C_f = C_f(N_{re})$$

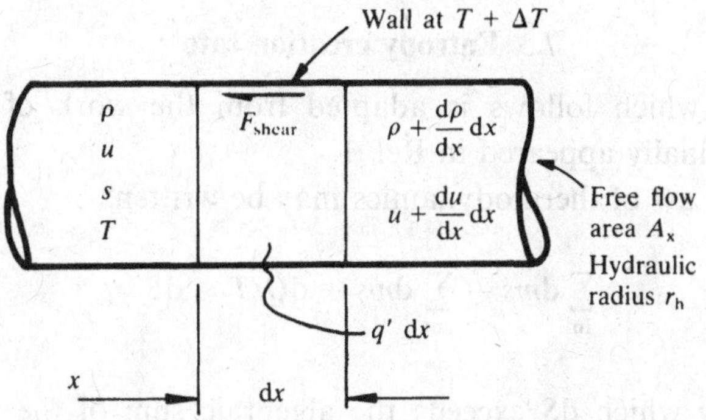


Fig. 7.1 Control volume for specifying rate of entropy creation in slab flow. (After Organ.[1]).

There is no restriction on the complexity of the functional form. When full dynamic similarity is being observed, the correlations must reflect the geometry of the heat transfer surface. To the extent that different correlations are required for the smooth passages of the heat exchangers vs the interrupted passage of the wire mesh regenerator, the requirement is true also of the formulation in terms of thermodynamic similarity. It will be argued later, however, that as long as attention is confined to regenerators made from stacks of wire screens and to exchanger passages of circular or rectangular cross-section, two pairs of correlations, one pair for the regenerator, one for the heat exchangers, will serve the purposes of all preliminary design work.

The goal of cycle analysis is to be able to identify changes in machine specification which lead to improved performance – i.e., to a reduction in the effects of irreversibilities. Each successive redesign thus improves the assumptions on which linearization is based. To this extent the assumptions are self-fulfilling. One is not really interested in a design in which losses are irremediably high, and so the linearization approach will always be a workable one.

The algebra which follows will reflect the additional assumptions of thermodynamic similarity. These do not represent a restriction on the Availability method, which may be applied equally where full dynamic similarity is allowed:

(4) Volume variations are simple harmonic for both expansion and compression spaces, with a constant phase angle between respective variations. Instantaneous volumes and volume change rates are therefore described completely when volume ratio, $\kappa$, and volume phase angle, $\alpha$ are defined.

## 7.3 Entropy creation rate

The treatment which follows is adapted from the work of Bejan.[2] The adaptation originally appeared in Ref. 1.

The Second Law of thermodynamics may be written:

$$\sum_{\text{in}} \mathrm{d}ms - \sum_{\text{out}} \mathrm{d}ms + \mathrm{d}Q/T \leqslant \mathrm{d}S$$

The amount by which $\mathrm{d}S$ exceeds the algebraic sum of the terms on the left-hand side measures the degree of irreversibility accompanying the transfer process. The excess of entropy over that involved in the reversible process may be called the *entropy generated*, $S'$. This is not a thermodynamic property. $S'$ may be expressed:

$$S' = \partial S/\partial t - Q'/T + \sum_{\text{out}} m's - \sum_{\text{in}} m's$$

In Fig. 7.1 the elemental length of duct, $\mathrm{d}x$, has cross-sectional area $A_x$ and hydraulic radius $r_h$. Fluid flows at the local, instantaneous rate $m'$ kg/s. There is heat transfer between fluid and unit length of wall at the instantaneous local rate $q_L'\,\mathrm{d}x$.

Bejan's method for defining the entropy generation rate within the control volume is adapted for present purposes in Appendix VII, and leads to:

$$S' = \left( q_L' \frac{\Delta T}{T^2} + \rho u \frac{F A_x}{T} \right) \mathrm{d}x \qquad (7.2)$$

The simplicity of this expression belies the fact that it derives from equations in differential form for energy and mass continuity for compressible flow with unsteady effects.

Now, $q_L'$ may be written $q_L' = hP\Delta T$, and $N_{st} = h/Gc_p$; $F$ has been defined under Notation, and so Eqn (7.2) becomes

$$S' = \left( \frac{q_L'^2 r_h}{T^2 m' c_p N_{st}} + \frac{m'^3 C_f}{2 A_x^2 T r_h \rho^2} \right) \mathrm{d}x$$

The term $u/|u|$ has been dropped, since $uu/|u|$ is always positive, consistent with irreversibility leading to positive $S'$. Recalling the definition of $N_{re}$ and eliminating $\rho$ via the ideal gas equation permits the previous expression to be rewritten:

$$S' = \left( \frac{q_L'^2 r_h}{T^2 m' c_p N_{st}(N_{re}, N_{pr})} + \frac{m'^3 C_f(N_{re}) R^2 T}{2p^2 A_x^2 r_h} \right) dx \tag{7.3}$$

$$= S_q' + S_f' \tag{7.4}$$

$S' = S'(x, t)$, and for evaluation it is necessary to have estimates of $q'(x, t)$ and $m'(x, t)$. By hypothesis $S'$ is small and so $q_L'$ and $m'$ are taken from an idealized model of the gas processes. The obvious choices are the 'isothermal' (or temperature-determined) and 'adiabatic' cycle models.

It appears to be taken for granted – possibly on account of a more plausible value for indicated thermal efficiency – that the adiabatic model is the more realistic. However, it will be seen from Appendix VIII that the assumption whereby the working fluid in the heat exchangers is at exchanger temperature at all times implies an instantaneous step change of temperature at both exchanger inlets over most of the inflow period. This is perhaps even harder to justify than the assumption of constant working space temperature on which the isothermal model is based.

The computer code which carried out the calculations shortly to be described has the facility of calling up either the isothermal or the adiabatic model to give the required, ideal values of $m'(x, t)$, from which corresponding $q_L'(x, t)$ follow. The computation of $m'(x, t)$ from the reference cycle analysis amounts to carrying out a cycle simulation which, in the case of the isothermal treatment, is represented by

$$\zeta = \zeta \begin{pmatrix} \mu_{dce} \\ \mu_{de} \\ N_\tau, \kappa, \alpha, \mu_v, \mu_{dr} \\ \mu_{dc} \\ \mu_{dcc} \end{pmatrix} \tag{7.1a}$$

In the adiabatic case, the formulation includes the variable $\gamma$:

$$\zeta = \zeta \begin{pmatrix} \mu_{dce} \\ \mu_{de} \\ N_\tau, \kappa, \alpha, \mu_v, \mu_{dr}, \gamma \\ \mu_{dc} \\ \mu_{dcc} \end{pmatrix} \tag{7.1b}$$

The computational work involved in the adiabatic model exceeds that of the isothermal counterpart by some orders. Initial optimization work involves mechanized cycling of the simulation through many permutations of the dimensionless parameters. On the other hand, according to Eqn. (7.1b),

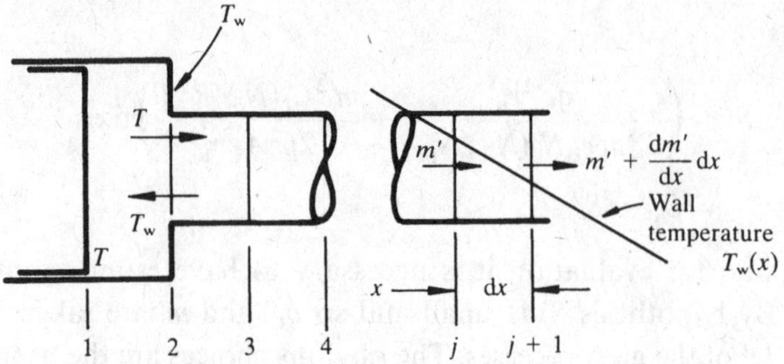


Fig. 7.2 Control volume as one of a series forming a Stirling cycle machine. (After Organ.[1])

normalized mass flow rates, $\sigma'$, pressures, $\psi$, and heat transfers are functions of crank angle, $\phi$, and of specific heat ratio, $\gamma$, alone when a machine of fixed geometry operates at given temperature ratio, $N_\tau$. This means that, for a given machine at given $N_\tau$ there are only two solutions to the adiabatic analysis, one for monatomic, the other for diatomic gases. For the temperature-determined analysis, there is evidently one solution only, applicable to all gases. Available computing power may thus dictate the choice of reference model – adiabatic or isothermal – depending upon whether the simulation is being used for analysis or for optimization.

Both the temperature-determined and adiabatic analyses assume that fluid temperature in the dead spaces is equal to that of the immediately adjacent wall. Fig. 7.2 shows schematically the unsteady flow of a compressible fluid in which the temperature gradient in the flow direction is the same as that in the wall, namely, $\mathrm{d}T_w/\mathrm{d}x$. For this case, Appendix IX shows that instantaneous mass flow rate, $m'$, at location $j + 1$ is given by

$$m_{j+1}' = -m_e' - \frac{1}{R}\sum_{2}^{n}\frac{1}{T_{wj+1}} A_j \,\mathrm{d}x \frac{\mathrm{d}p}{\mathrm{d}t} \qquad \text{for } n = j + 1 \qquad (7.5)$$

The negative sign before $m_e'$ reflects the fact that the ideal reference cycle analysis assumes flows *into* the variable-volume spaces to be positive.

The same appendix derives the expression for ideal heat transfer rate per unit length of duct:

$$q_L' = -A_x \frac{\mathrm{d}p}{\mathrm{d}t} + R\frac{\gamma}{\gamma - 1}\left[m'\frac{\mathrm{d}T_w}{\mathrm{d}x} + U(-\dot{m}_{ws})m_{ws}'\frac{T - T_w}{\mathrm{d}x}\right] \qquad (7.6)$$

In Eqn (7.6) $m'$ is given by Eqn (7.5) and the subscript ws signifies 'working space'. The appendix explains why the term $U(-m_{ws})m_{ws}'$ is evaluated only when the flow passage element under consideration immediately adjoins a variable-volume space.

Normalizing the expressions for $m'$ and $q_L'$ and letting the prime, ', refer to crank angle, $\phi$:

$$\sigma_{\phi j+1}' = -\sigma_{\phi e}' - \sum_{2}^{j+1} \frac{1}{\tau_{w_{j+\frac{1}{2}}}} \mathrm{d}\mu_j \frac{\mathrm{d}\psi}{\mathrm{d}\phi} \tag{7.7}$$

$\phi = \omega t$ and $\mathrm{d}\mu_j = \mathrm{d}V_{dj}/V_E$, where $\mathrm{d}V_{dj}$ is the dead volume element between points $j-1$ and $j$.

Substituting Eqn (7.7) into Eqn (7.6) and normalizing the remaining terms:

$$\frac{q_\phi}{MRT_{ref}} = -\mathrm{d}\mu_j \frac{\mathrm{d}\psi}{\mathrm{d}\phi} + \frac{\gamma}{\gamma - 1} \times \text{Eqn (7.7)} \times \text{Eqn. (7.9)} \tag{7.8}$$

$$\text{Eqn (7.9)} = \mathrm{d}\tau - U(-\sigma_{ws})(\tau - \tau_w) \tag{7.9}$$

$q_\phi'$ is the heat transferred within control volume $\mathrm{d}\mu_j$ per unit increment in crank angle, $\phi$. For a machine of given geometry there are only two values for normalized heat transferred in crankshaft interval $\mathrm{d}\phi$ according to Eqn (7.8) – one for monatomic, the other for diatomic gases.

It is now necessary to substitute Eqn (7.8) for heat transfer and Eqn (7.7) for mass flow rate into Eqn (7.3) for entropy generation rate. To do this, normalized expressions are required for $N_{pr}$ and $N_{re}$.

### 7.3.1 Prandtl number, $N_{pr}$ and Reynolds number $N_{re}$

The rôle of the Prandtl number has already been discussed in Chapter 4, where it was dismissed as a constant and dropped from the list of parameters of the solution. This remains the recommended approach, but the opportunity arises here of showing that the parameter may be retained if desired, with negligible penalty in terms of increased computation. It is assumed that

$$N_{pr} = \mu c_p/k = \mu_{ref} c_{p\,ref}/k_{ref} = N_{PR} \tag{7.10}$$

By definition,

$$N_{re} = 4r_h|m'|/\mu A_x$$

(The expression in the source[1] incorrectly omits the area term, but takes it into account in the subsequent development.)

The wide range of temperature involved means that corresponding variations in $\mu$ must be accounted for. This may be accomplished in terms of the

Sutherland temperature, $T_{su}$, and a reference value of dynamic viscosity coefficient, $\mu_{ref}$:

$$\frac{\mu}{\mu_{ref}} = \left(\frac{1 + T_{su}/T_{ref}}{T/T_{ref} + T_{su}/T_{ref}}\right) N_\tau^{3/2} = f(\tau, \tau_{su}, N_\tau) \tag{7.11}$$

With this definition

$$N_{re} = 8\pi N_{RE} \frac{\lambda_h |\sigma'|}{\alpha_x \times \text{Eqn (7.11)}} \tag{7.12}$$

It will be recalled that $\lambda_h = r_h/L_{ref}$, where, to achieve thermodynamic similarity, the choice of $L_{ref}$ is $R_H$ (Chap. 4), defined in turn as $V_{ref}/A_{ref} = MRT_{ref}/p_{ref}A_{wetted}$. Eqn (7.12) expresses local, instantaneous $N_{re}$ as the product of two types of quantity – those which vary with location and crank angle and those which are fixed.

### 7.3.2 Evaluation of entropy generation, $S_q'$, due to heat transfer

Fig. 7.2 labelled the boundaries, $j$, of subdivisions of the gas circuit. It is now convenient to make use of the same subscript to refer to the control volumes formed between successive subdivisions. Thus, control volume $j$ is that which lies between subdivisions $j - 1$ and $j$. In the analysis which follows, a quantity, $dS_j$, of entropy generated during time or crankshaft interval $dt$ or $d\phi$, means the entropy generated in control volume $j$ due to transfer of heat element $dq_j$ (taken as the mean of values calculated (Eqn (7.8)) at planes $j - 1$ and $j$) or to friction (based on the mean of flows (Eqn (7.7)) again evaluated at planes $j - 1$ and $j$).

Normalizing Eqn (7.3) by $MRT_{ref}$ and changing the independent variable from $t$ to $\phi$ ($=\omega t = 2\pi N_s t$) permits the left-hand term to be written:

$$\frac{T_0\, dS_{q_j}}{MRT_{ref}} = \frac{(\gamma - 1)}{\gamma} \frac{q_\phi'^2 \lambda_{h_j} \alpha_{x_j}}{\tau_j^2 \sigma_\phi' N_{st}\, d\mu_j} d\phi \tag{7.13}$$

Eqn (7.13) has made use of the fact that normalized volume, $d\mu_j$, = normalized cross-sectional area, $\alpha_{x_j}$, multiplied by the appropriate elemental length of channel, $d\lambda_j$. An area$_j$ is defined at a discrete point, $j$, a volume$_j$ or a length$_j$ between points $j - 1$ and $j$. The usage is self-explanatory, but some care is necessary in programming to avoid meaningless combinations of the subscripted quantities.

To proceed it is necessary to have $N_{st}$ as a function of $N_{re}$ and $N_{pr}$. In Fig. 7.3($a$) have been superimposed Kays' and London's correlations[3] for smooth

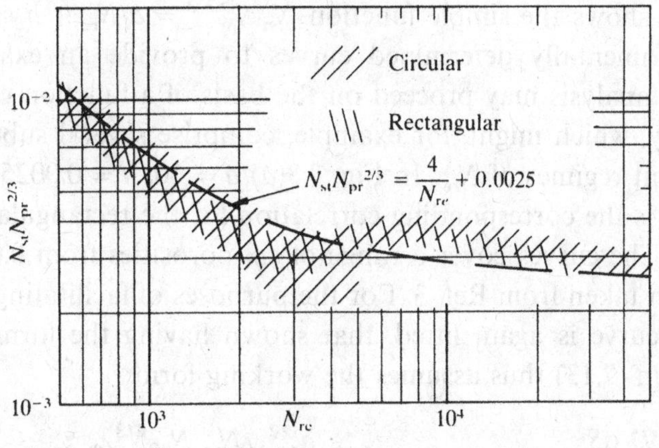


(*a*)

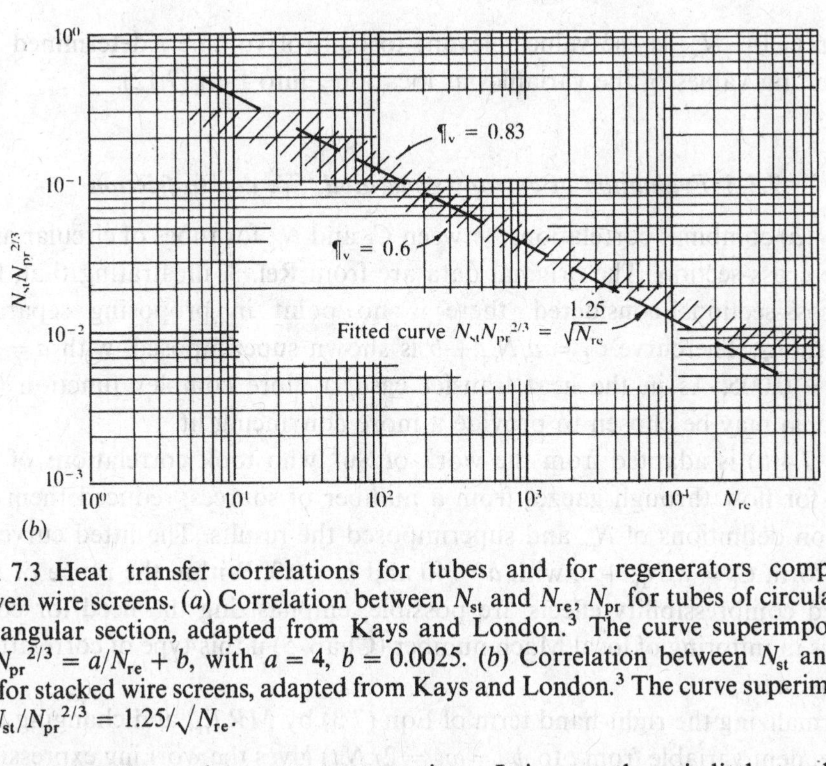


(*b*)

Fig. 7.3 Heat transfer correlations for tubes and for regenerators comprising woven wire screens. (*a*) Correlation between $N_{st}$ and $N_{re}$, $N_{pr}$ for tubes of circular and rectangular section, adapted from Kays and London.[3] The curve superimposed is $N_{st}N_{pr}^{2/3} = a/N_{re} + b$, with $a = 4$, $b = 0.0025$. (*b*) Correlation between $N_{st}$ and $N_{re}$, $N_{pr}$ for stacked wire screens, adapted from Kays and London.[3] The curve superimposed is $N_{st}/N_{pr}^{2/3} = 1.25/\sqrt{N_{re}}$.

ducts of circular and square cross-sections. It is seen there is little to choose between them, and $r_h$ as the defining linear dimension thus permits a unique correlation between $N_{st}$ and $N_{re}$. Reference to the source[3] will show minimal difference between the correlations of Fig. 7.3(*a*) and that for slots of rectangular cross-section having aspect ratio of 3. By the time the aspect ratio reaches 8 there is some divergence, but this is well within the uncertainty associated with the numerical values of other variables.

The figure shows the simple function $N_{st}N_{pr}^{2/3} = a/N_{re} + b$ superimposed on the experimentally determined curves to provide an example of the method. The analysis may proceed on the basis of a function of any degree of complexity, which might, for example, comprise several subfunctions for fitting different regimes of $N_{re}$. In Fig. 7.3(*a*) $a = 40$, $b = 0.0025$.

Fig. 7.3(*b*) is the corresponding correlation for the rectangular wire gauze regenerator. The curves cover a range of six porosities from 0.602 to 0.832, and are again taken from Ref. 3. For the purposes of facilitating illustration, a simplified curve is again fitted, that shown having the form $N_{st}N_{pr}^{2/3} = 1.25/\sqrt{N_{re}}$ Eqn (7.13) thus assumes the working form:

$$\frac{T_0\, dS_{q_j}}{MRT_{ref}} = d\zeta_{s_q} = (\gamma - 1)/\gamma\, \frac{q_\phi'^2 \sqrt{N_{re_j}} N_{PR}^{2/3} \lambda_{h_j} \alpha_{x_j}}{1.25\tau_j^2 |\sigma_\phi'|\, d\mu_j}\, d\phi \qquad (7.14)$$

In Eqn (7.14) $N_{re_j}$ is the value relevant to control volume $j$ determined by substituting values of the variables at location $j$ into Eqn (7.12).

### 7.3.3 Evaluation of entropy generated, $S_f'$, due to friction

Fig. 7.4(*a*) combines correlations between $C_f$ and $N_{re}$ for tubes of circular and square cross-section. The original data are from Ref. 3, illustrating that, for the cross-sections considered, there is no point in proposing separate correlations. The curve $C_f = a/N_{re} + b$ is shown superimposed, with $a = 16$ and $b = 0.005$. As in the heat transfer case, a more complex function (or functions) may be chosen to provide a more convincing fit.

Fig. 7.4(*b*) is adapted from the work of Su[4] who took correlations of $C_f$ vs $N_{re}$ for flow through gauzes from a number of sources, reduced them to common definitions of $N_{re}$ and superimposed the results. The fitted curve is of the form $C_f = a/N_{re} + b$, with $a = 40$ and $b = 0.3$. Within the range of $N_{re}$ covered compressibility effects are possible, emphasizing the need for continuous monitoring of local Mach number (Chap. 5) if this type of correlation is to be used.

Normalizing the right-hand term of Eqn (7.3) by $MRT_{ref}$ and changing the independent variable from $t$ to $\phi$ ($= \omega t = 2\pi N_s t$) gives the working expression for normalized entropy generated in control volume $d\mu_j$ over crank angle increment $d\phi$:

$$\frac{T_0\, dS_{f_j}}{MRT_{ref}} = d\zeta_{sf} = \frac{2\pi^2 N_{MA}^2 \tau_j |\sigma_{\phi j}'^3| (a/N_{re_j} + b)}{\lambda_{h_j} \alpha_{x_j}^3 \psi^2}\, d\mu_j\, d\phi \qquad (7.15)$$

in which $N_{re_j}$ is again the value relevant to control volume $j$ determined by

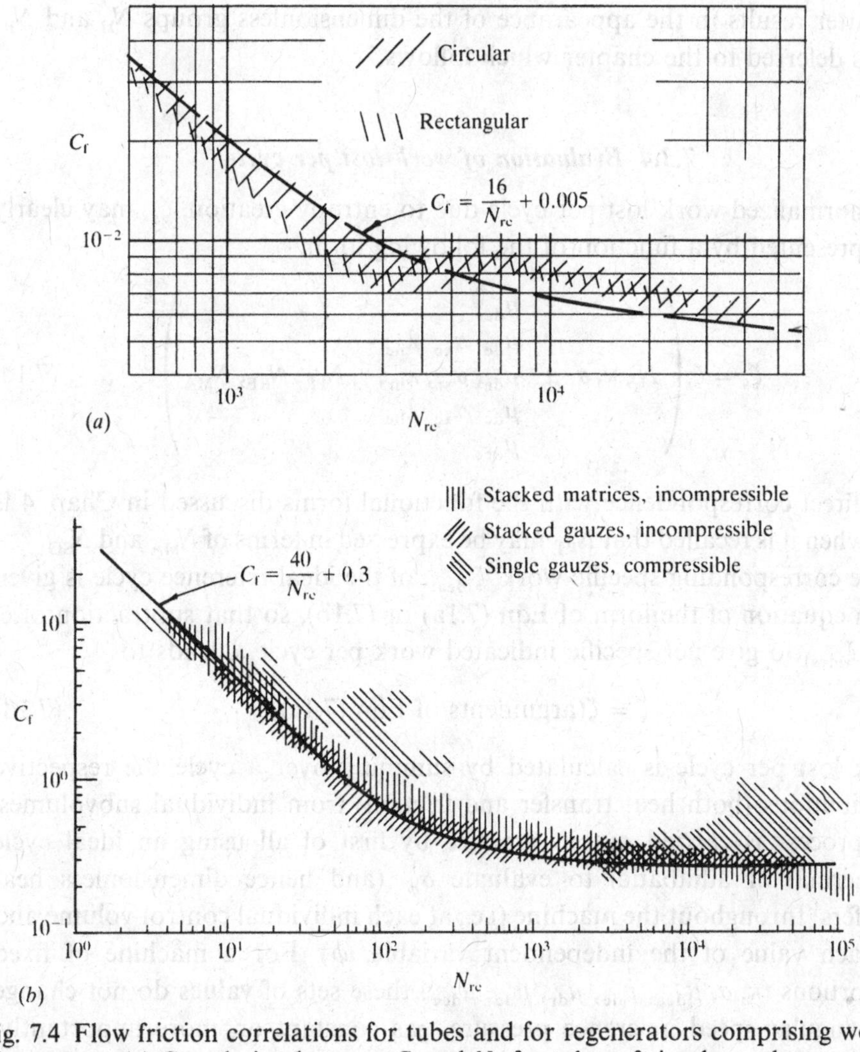


Fig. 7.4 Flow friction correlations for tubes and for regenerators comprising woven wire screens. (*a*) Correlation between $C_f$ and $N_{re}$ for tubes of circular and rectangular section, adapted from Kays and London.[3] The curve superimposed is $C_f = a/N_{re} + b$, with $a = 16$, $b = 0.005$. (*b*) Superposition of correlations between $C_f$ and $N_{re}$ for stacked wire screens *on the assumption of incompressible flow*, after Su.[4] (Adapted from Kays and London.[3]) The curve superimposed is $C_f = a/N_{re} + b$, with $a = 40$, $b = 0.3$.

substituting values of the variables at location $j$ into Eqn (7.12). The equation is readily separated into terms which, for a machine of fixed dimensions ($\kappa$, $\alpha$, $\alpha_x$, $\mu_d$, $\lambda_h$) do not vary for gases having the same $\gamma$.

Analogous expressions may be derived to define entropy generated at the irreversibilities arising as a result of sudden expansion or contraction or of heat flow due to a temperature gradient in the solid walls. Consideration of the latter results in the appearance of the dimensionless groups $N_H$ and $N_F$, and is deferred to the chapter which follows.

### 7.3.4 Evaluation of work lost per cycle

Net, normalized work lost per cycle due to entropy creation, $\zeta_s$, may clearly be represented by a function of the following form:

$$\zeta_s = \zeta_s\left( N_\tau, \kappa, \alpha, \mu_v, \begin{matrix} \mu_{dce} \\ \mu_{de} \\ \mu_{dr}, \\ \mu_{dc} \\ \mu_{dcc} \end{matrix} \begin{matrix} \alpha_{xe} \\ \alpha_{xr}, \\ \alpha_{xc} \end{matrix} \begin{matrix} \lambda_{he} \\ \lambda_{hr}, \\ \lambda_{hc} \end{matrix} \gamma, N_{PR}, N_{RE}, N_{MA} \right) \tag{7.1c}$$

The direct correspondence with the functional forms discussed in Chap. 4 is seen when it is recalled that $N_{RE}$ may be expressed in terms of $N_{MA}$ and $N_{SG}$.

The corresponding specific work, $\zeta_{ideal}$, of the ideal reference cycle is given by an equation of the form of Eqn (7.1a) or (7.1b), so that subtraction of $\zeta_s$ from $\zeta_{ideal}$ to give net specific indicated work per cycle, $\zeta$ leads to

$$\zeta = \zeta(\text{arguments of Eqn (7.1c)}) \tag{7.1d}$$

Work lost per cycle is calculated by summing over a cycle the respective contributions (both heat transfer and friction) from individual subvolumes. The process is carried out in practice by first of all using an ideal cycle (isothermal or adiabatic) to evaluate $\sigma_\phi'$ (and hence dimensionless heat transfers) throughout the machine (i.e., at each individual control volume and for each value of the independent variable, $\phi$). For a machine of fixed proportions ($\kappa$, $\alpha$, $\mu_{dce}$, $\mu_{de}$, $\mu_{dr}$, $\mu_{dc}$, $\mu_{dcc}$) these sets of values do not change with angular speed, reference pressure, gas constant or, more importantly, with the dimensionless groups $N_{PR}$, $N_{RE}$ or $N_{MA}$. They may therefore be stored and picked up by a second computer program for systematic multiplication by the variable terms (those involving $N_{RE}$ and/or $N_{MA}$). If a machine of fixed $\kappa$, $\alpha$, $\mu_{dce}$, $\mu_{de}$, $\mu_{dr}$, $\mu_{dc}$, $\mu_{dcc}$ is being investigated only the second program requires to be systematically rerun.

Fig. 7.5 is a schematic of a machine in which the expansion exchanger is divided into $n_e$ subvolumes, the regenerator into $n_r$, and the compression exchanger into $n_c$. The total number of subvolumes is thus $n_e + n_r + n_c$, and there are $1 + n_e + n_r + n_c$ node points at which flow rates $\sigma_\phi'$ are evaluated. Numerical values of the various $n$ are best selected by trial and error: suitable values are those above which net lost available work is independent of the particular $n$ in each case.

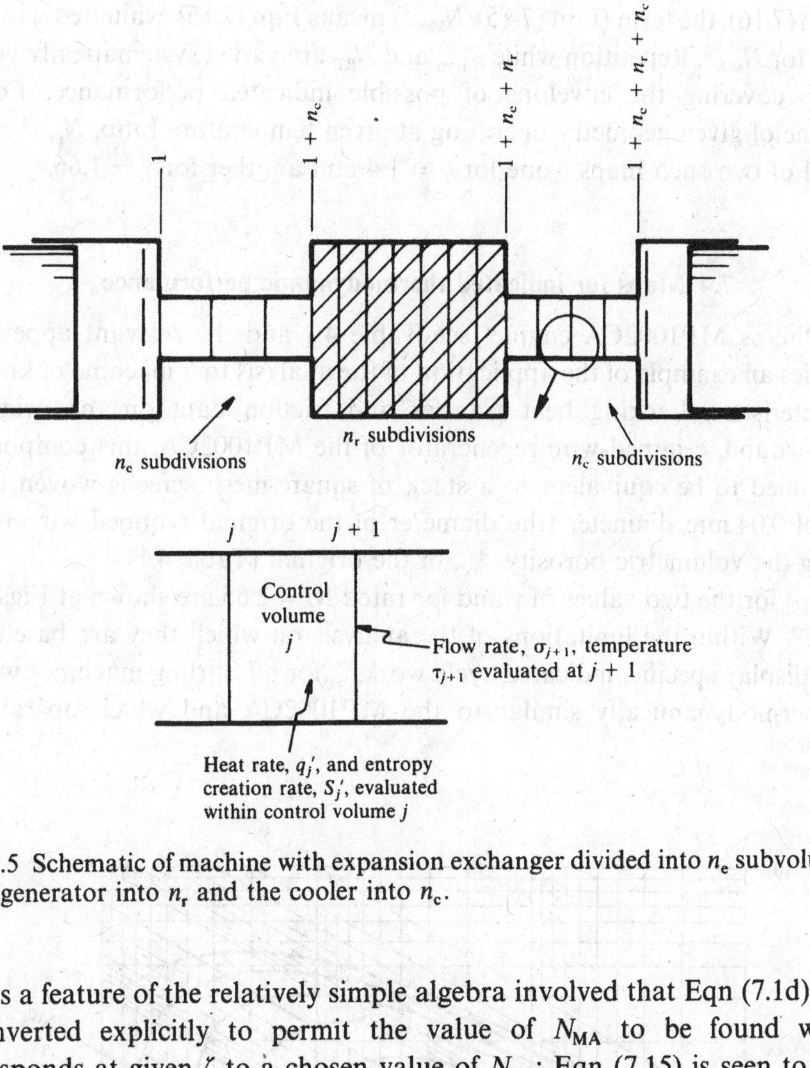


Fig. 7.5 Schematic of machine with expansion exchanger divided into $n_e$ subvolumes, the regenerator into $n_r$ and the cooler into $n_c$.

It is a feature of the relatively simple algebra involved that Eqn (7.1d) may be inverted explicitly to permit the value of $N_{MA}$ to be found which corresponds at given $\zeta$ to a chosen value of $N_{RE}$: Eqn (7.15) is seen to be a linear function of the 'constant' term $N_{MA}^2$. (From the definition of $N_{MA}$ this is the same as saying that, for a machine of given $R$, $R_H$ and $T_{ref}$, loss due to flow friction is proportional to $N^2$.) $\zeta_{sf}$ for the cycle may thus be found by first summing the variable terms of Eqn (7.15) and multiplying the result by $N_{MA}^2$. A range of nominal values of net, indicated specific cycle work, $\zeta_{nom}$, is selected, e.g., 0.1, 0.2, 0.3, 0.4 . . . and corresponding $\zeta_{sq}$ and (Eqn (7.15))/$N_{MA}^2$ calculated. The value of $N_{MA}$, say, $N_{MA_{0.1}}$ which results in a particular value – say, 0.1 – of $\zeta_{nom}$ is then found by the elementary process:

$$N_{MA_{0.1}} = \sqrt{\left(\frac{0.1 - \zeta_{sq}}{\text{Eqn (7.15)}/N_{MA}^2}\right)} \qquad (7.16)$$

In Eqn (7.16), the term (Eqn (7.15)/$N_{MA}^2$) means Eqn (7.15) evaluated without the factor $N_{MA}^2$. Repetition while $\zeta_{nom}$ and $N_{RE}$ are varied systematically yields a map covering the envelope of possible indicated performance. For a machine of given geometry operating at given temperature ratio, $N_\tau$, there is a total of two such maps – one for $\gamma = 1.4$ and another for $\gamma = 1.66$.

## 7.4 Maps for indicated thermodynamic performance

The Philips MP1002CA engine (see Table 4.1 and the relevant appendix) provides an example of the application of the analysis to a machine of known characteristics. Lacking heat transfer and friction data on the original, spiral-wound, crimped-wire regenerator of the MP1002CA, this component is assumed to be equivalent to a stack of square-mesh screens woven from wire of 0.04 mm diameter (the diameter of the original crimped wires) and having the volumetric porosity, $¶_v$, of the original (Table 4.1).

Maps for the two values of $\gamma$ and for rated $N_\tau = 2.65$ are shown at Figs. 7.6 and 7.7. Within the limitations of the analysis on which they are based the maps display specific, indicated cycle work, $\zeta$, for *all* Stirling machines which are thermodynamically similar to the MP1002CA and which operate at

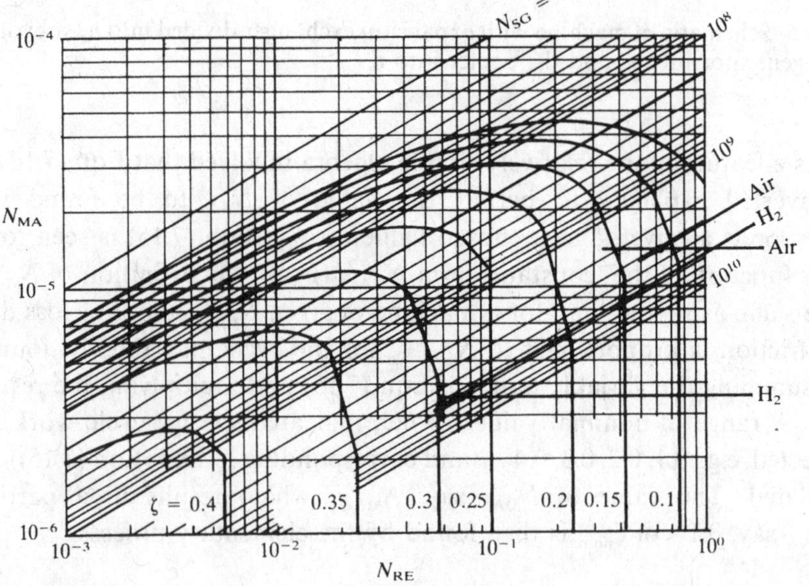


Fig. 7.6 Performance map of Philips MP1002CA engine at rated $N_\tau$ (Table 4.1) and for diatomic working fluid ($\gamma = 1.4$) showing operating points for air and $H_2$.

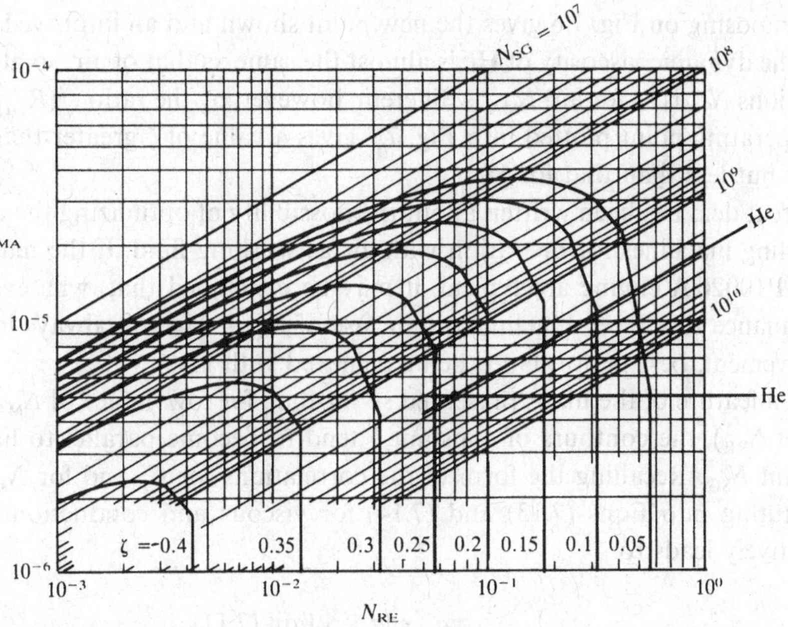


Fig. 7.7 Performance map of Philips MP1002CA engine at rated $N_\tau$ (Table 4.1) and for monatomic working fluid ($\gamma = 1.66$) showing operating points for air and He.

$N_\tau = 2.65$. The ordinates and abscissae are respectively $N_{MA}$ and $N_{RE}$. These two parameters are related to $N_{SG}$ via the relationship $N_{RE} = N_{MA}{}^2 N_{SG}$, permitting lines of constant $N_{SG}$ to be superimposed on the map as an alternative input scale. (Note that the definitions of $N_{RE}$ and $N_{MA}$ differ by constant factors from those reported in the review chapter (Chap. 2, Figs. 2.16, 2.17) and that the specification of the MP1002CA machine has been refined since the data for those earlier diagrams were computed.)

Entered on Fig. 7.6 are lines corresponding to rated $N_{SG}$ and $N_{MA}$ for the MP1002CA (Table 4.1) for the normal working fluid, i.e., for air. A value of $\zeta$ of approximately 0.12 is read at the intersection. This is for a $\zeta$ defined as $W_{cyc}/MRT_{ref} = W_{cyc}/p_{ref}V_{ref} = \mu_v W_{cyc}/p_{ref}V_E$. With $\mu_v = 0.514$ (Table 4.1) $W_{cyc}/p_{ref}V_E = 0.335$. The MP1002CA drives its own compressor, has a complex linkage, and might be expected to have relatively low mechanical efficiency – say 0.7. This would give $\zeta$ based on shaft output equal to 0.235 – substantially in excess of the Beale number value of 0.15. The excess is consistent with the fact that a number of losses have still to be taken into account. Multiplying up by $p_{ref}V_E$ gives $W_{cyc} = 30$ J, corresponding to some 745 W indicated.

$H_2$ has the same value of $\gamma$ as does air, but the different $\mu$ and $R$ lead to different $N_{MA}$ and $N_{SG}$ at rated $p_{ref}$ and $N_s$. Calculating the values and superimposing on Fig. 7.6 gives the new point shown and an improved value of $\zeta$. The dynamic viscosity of He is almost the same as that of air, so at rated conditions $N_{SG}$ is as for air. $N_{MA}$ is different, however, by the ratio $\sqrt{(R_{air}/R_{He})}$. The operating point plotted into Fig. 7.7 gives a value of $\zeta$ greater than that for air but less than that for $H_2$.

A great deal has been written about the possibility of optimizing the design of Stirling machines for operation with air as working fluid. If the maps for the MP1002CA engine are typical, it may be concluded that, whatever the performance of a given machine on air, use of $H_2$ or He will always lead to improvement, best performance being achieved with $H_2$.

It is a feature of the maps presented so far that, for low values of $N_{MA}$ (and thus of $N_{RE}$), the contours of constant $\zeta$ tend to become parallel to lines of constant $N_{SG}$. Recalling the form of the correlations for $C_f$ and for $N_{st}$, and substituting into Eqns (7.13) and (7.14) for viscous and conduction losses respectively leads to:

$$d\zeta_{s_q} + d\zeta_{s_f} = 0 + \frac{\pi a}{4N_{SG}} \frac{\tau_j |\sigma_{\phi j}'^3| \text{ Eqn (7.11)}}{\lambda_{h_j}{}^2 \alpha_{x_j}{}^2 \psi^2} d\mu_j \, d\phi \qquad (7.17)$$

In Eqn (7.11) $a$ is the constant from the $C_f$–$N_{re}$ correlation. The equation confirms that, at these low Reynolds numbers, losses are due entirely to friction, that they are inversely proportional to $N_{SG}$, as the maps confirm, and that the performance of a machine of fixed thermodynamic geometry at given $N_\tau$ is measured in terms of $N_{SG}$ alone.

Tentatively supposing that a map for a machine of different thermodynamic geometry would have roughly the same form as that of Figs. 7.6 and 7.7 it is seen that there is a lower limit of $N_{SG}$ below which $\zeta = 0$, and no work is produced for *any* possible value of $N_{RE}$. For the MP1002CA machine this appears to be $10^7$. Corresponding to realistic $N_{RE}$, a lowest limit on $N_{SG}$ might be judged to be $10^8$. Low $N_{SG}$ arises at low pressure and/or high rpm. Low pressures, at 1 atm ($=10^5$ Pa), are encountered in 'hot-air' engines – a degenerate form of Stirling engine frequently having only a vestigial regenerator. Assuming $\mu = 2 \times 10^{-5}$ and inverting the definition of $N_{SG}$ for $p_{ref} = 10^5$ gives a limiting $N_s$ of 25, or 1500 rpm. Although extrapolation of the MP1002CA map to such cases is precarious, it is well known that the net output of the typical, unpressurized, small hot-air engines is achieved at or below 1000 rpm. Certain designs achieve higher rpm under no-load conditions, suggesting finite indicated work above 1000 rpm, but the 3000 or 4000 rpm of pressurized machines is unknown.

Conversely, to produce useful output, a machine must operate with $N_{SG}$

suitably in excess of the critical minimum (but note from the maps that, above a certain point an increase in $N_{SG}$ at constant $N_{MA}$ gives reduced $\zeta$). An interesting instance is afforded by the Malone engine[5,6] which used water as the working fluid at high pressure and the relatively low rpm of 25–250. Taking $\mu$ for water as $10^{-3}$ and using for consistency the lower of the pressures quoted (7 MPa) as the reference value gives a range of $N_{SG}$ of $1.7 \times 10^9 \leqslant N_{SG} \leqslant 1.7 \times 10^{10}$ – i.e., suitably above the nominal minimum. The acoustic speed for water is some 4.5 times that for air, so that for an identical value of the product $N_s R_H$, the operating point in Fig. 7.6 would be at the intersection of $N_{MA} = 3 \times 10^{-6}$ and the chosen value of $N_{SG}$. This will be seen to be a region of the map where calculated losses are low. The reduced $N_s$ requires $R_H$ to be increased in proportion to maintain constant $N_s R_H$ – an increase consistent with the superior heat transfer of the liquid working fluid.

Mechanically the MP1002CA is quite unsuited for operation at elevated pressures with water as the working fluid. Moreover, its physical layout differs substantially from what is known of the Malone engine. (Insufficient data are available to compute values of the $\mu_d$, $\lambda_h$ etc. which would indicate how similar the machines were thermodynamically.) Nevertheless, the performance maps suggest that a Stirling cycle machine capable of the relevant wall stresses and bearing loads, and possessing the thermodynamic geometry of the MP1002CA engine, would produce power at the pressures and angular speeds claimed for the Malone engine, and do so with a favourable level of internal losses. The conclusion underlines the importance of the parameters $N_{SG}$ and $N_{MA}$ in the thermodynamic design of these machines.

Fig. 7.8 is a map for the GPU3 engine specified in terms of the parameters listed in Table 4.1. The operating point gives $\zeta \approx 0.08$ based on working fluid circuit volume. Taking into account the ratio $\mu_v = 0.384$ gives $\zeta$ (based on $V_E$) = 0.21 indicated. Corresponding indicated power is 4.4 kW, which exceeds the experimentally determined indicated value of 3.96 kW[7] but by less of a margin that one might expect, bearing in mind that other internal losses have still to be taken into account. With an assumed 75% mechanical efficiency, Beale number, $N_B = 0.15$. At the low $p_{ref}$ for which these values are calculated, the machine is operating away from the original design point: a higher computed $N_B$ would be expected at rated rpm and charge pressure.

Fig. 7.9 is the corresponding design chart for $\gamma = 1.4$, allowing performance on the diatomic gases air and $H_2$ to be estimated. $\zeta$ for air lies close to zero, suggesting that at 42 bar charge pressure the engine would not run at 2500 rpm – even unloaded. However, at high values of internal loss the linearity assumption on which the computations are founded is suspect, and

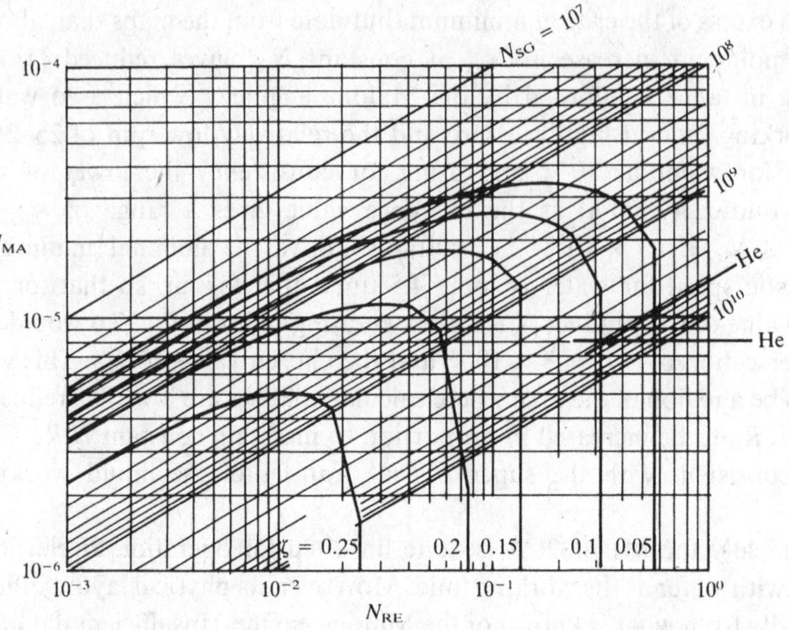


Fig. 7.8 Performance map of General Motors GPU3 engine at conditions given in Table 4.1 and for monatomic working fluid ($\gamma = 1.66$) showing the operating point for He.

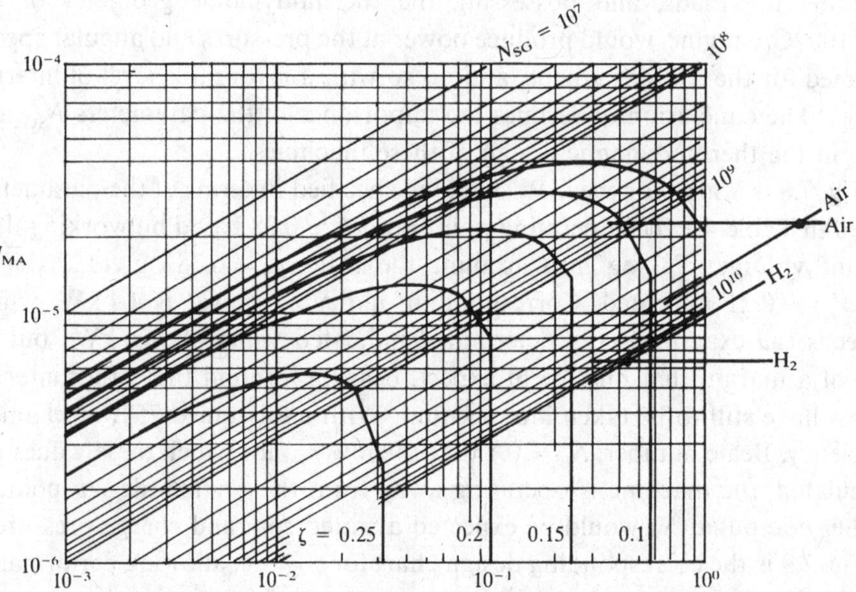


Fig. 7.9 Performance map of General Motors GPU3 engine at conditions given in Table 4.1 and for diatomic working fluid ($\gamma = 1.4$) showing the operating points for air and $H_2$.

the result is therefore inconclusive. The point for $H_2$ suggests some 60% greater cycle work than for helium at the same rpm and $p_{ref}$.

The maps afford a means of looking into a matter which arises frequently in discussion of the regenerator and of the qualities required of that component: the regenerators of many successful Stirling machines have been made from the finest wires commercially available (i.e., from mesh woven from wire of the order of 0.025 mm diameter). On this basis it might be thought that regenerator performance improves continuously with decreasing wire size, and that commercial availability is thus a limitation on performance.

Hydraulic radius is independent of porosity, and the functional form of the thermodynamic similarity approach thus permits $\lambda_h$ for the regenerator to be varied whilst keeping regenerator dead space and all other geometric and operational parameters constant. This is tantamount to varying wire diameter. Fig. 7.10 shows the effect of reducing $\lambda_h$ for the MP1002CA regenerator by a factor of 10. Output with air as the working fluid and at rated conditions has fallen below what would be the contour of $\zeta = 0$ as increased losses due to flow resistance outweigh gains due to enhanced heat transfer area. It is tempting to undertake an exercise in selecting optimum

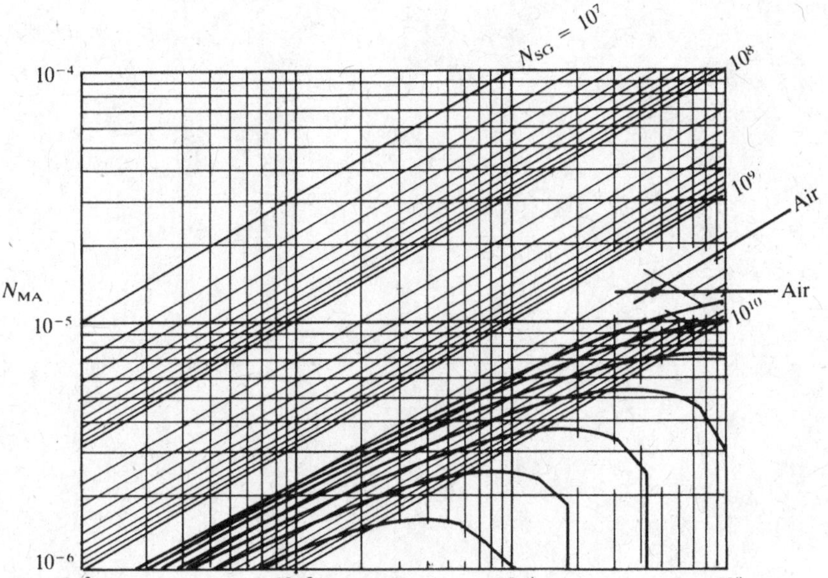


Fig. 7.10 Performance map for Philips MP1002CA engine as for Fig. 7.6, but with the dimensionless hydraulic radius of regenerator, $\lambda_{h_r}$, reduced by a factor of 10. Computed losses exceed indicated cycle work despite the apparently favourable increase in heat transfer area.

porosity and hydraulic radius. However, the matter of thermal conduction loss internal to the wires has yet to be looked into, and will almost certainly be a function of regenerator geometry. This is the task of the chapter which follows.

Finally, in connection with use of a map as a design aid, there is little point in achieving a high $\zeta$ if the penalty is a value of $N_s$ such that net *power* falls below requirements. For a given working fluid the power of a machine of fixed geometry is proportional to the product of $\zeta$ with $N_{MA}$. The map presentation permits rapid homing in on the combination of $\zeta$ and $N_s$ giving the required power output. There is no risk of inadvertently choosing an inefficient design or operating point, since low efficiencies are inevitably tied to high losses (low values of net $\zeta$), and may thus be spotted immediately.

# 8

# Extension to four conservation laws

## 8.1 Thermal conduction

The ideal cycle envisages a continuous, invariant distribution of gas temperature between $T_e$ and $T_c$, corresponding exactly to the invariant temperature profile of the adjacent solid wall. (Regenerator temperature distribution need not have the usual linear form.) It also assumes that heat flow is confined to a direction perpendicular to working fluid motion. Any departure in practice from these ideal conditions is accompanied by flow of heat down a gradient in temperature. This represents an irreversibility, and a corresponding contribution to reduced thermal efficiency.

Losses due to heat flow within the gas have already been discussed and dealt with in an approximate, one-dimensional treatment. In addition there is thermal conduction within the solid components, and the effect might be expected to be most marked in the regenerator where heat flux rates are highest. This chapter will focus on conduction within the regenerator, but the treatment may be considered a model for dealing with the heat exchangers.

For the practical cycle to approach the ideal, the regenerator should:

have high thermal capacity to minimize temperature swing;

have high thermal conductivity to minimize temperature gradients;

offer large surface area to minimize temperature differences between solid and working fluid;

have small dead volume to maximize pressure swing;

offer minimum resistance to flow.

It is clear that all of these requirements cannot be satisfied simultaneously, and that regenerator design must be a compromise. For a rational choice to be possible it is evidently necessary to be able to predict in some detail the transient thermal response of the matrix. This means looking at the cyclic variation of temperature within wire elements of the order of 0.025 mm

(0.001 in) in diameter. Recalling that Jakob[1] has classed the thermal analysis of the regenerator as among the most difficult analytical problems in engineering, and bearing in mind that he was referring to a model of the regenerator simplified to the flow of a gas past a plane, solid wall, the magnitude of the design problem becomes apparent.

Authors who have recognized the problem include Kim and Qvale[2] and Miyabe, Takahashi and Hamaguchi.[3] The former authors model an individual wire as a pin fin and make use of the standard analytical expressions for spanwise temperature distribution and for efficiency of such fins. The span direction for the pin fin lies in the plane of an individual gauze disc, and the idea of a temperature distribution in this coordinate direction conflicts with the notion of one-dimensional flow.

Miyabe *et al.* assume that conditions in a specimen wire from a given disc represent those in all wires of the same disc, and look qualitatively at the effect on cyclic variation of the radial temperature distribution of extreme instances of thermal capacity and conductivity. Their discussion is made with reference to a figure reproduced here as Fig. 8.1, and bears repetition:

> If the wire diameter, $d_w$, is too large (*c*), heat does not penetrate to the centre of the wire within the blow time. Thus, some domain of the sectional area does not contribute to the heat storage. If $d_w$ is too small (*a*), heat penetrates to the centre of the wire before the blow time expires. In this case, effective heat storage volume is insufficient, although the total volume of wire is used effectively. Therefore, to augment the heat storage capacity, (the) number of screens must be increased. This leads to the increase of flow losses.

The authors proceed to look at the wire as a thin column, and present the relationship between the temperature on the axis of the column and that of the surrounding fluid in terms of the Biot modulus, $B_i$, $(=(hd/k)_{\text{matrix}})$ and a Fourier number, $F_o$ $(=\alpha t_b/d^2)$ based on blow time, $t_b$ with parameter $T_r$ $(=T_{\text{wire}}/T_{\text{fluid, mean}})_{r=0}$. Fig. 8.2 is the plot of the relationship presented by

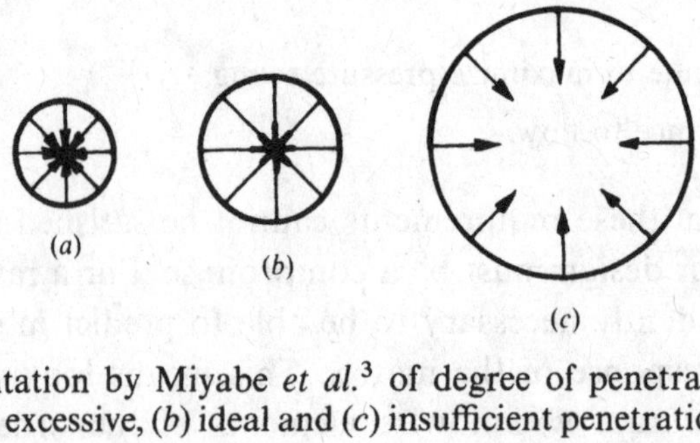


Fig. 8.1 Representation by Miyabe *et al.*[3] of degree of penetration of temperature wave showing, (*a*) excessive, (*b*) ideal and (*c*) insufficient penetration. (© 1982 IEEE.)

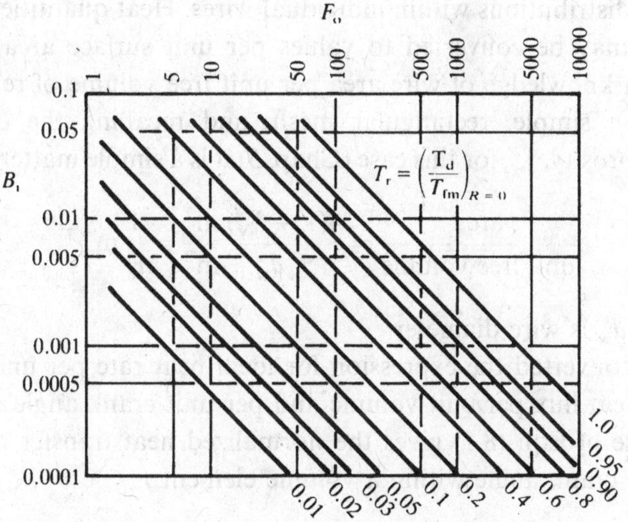


Fig. 8.2 Relationship plotted by Miyabe *et al.*[3] between Biot number $hd_w/2k_{wire}$ and Fourier number ($4\alpha t_{blow}/d_w^2$ in their notation). They recommend trial-and-error selection of $d_w$ to give $T_r$ of between 0.95 and 1.0. (© 1982 IEEE.)

Miyabe *et al.* They recommend selection of a value of wire diameter by trial and error such that $0.95 \leqslant T_r \leqslant 1.0$.

The indicated power of an engine tested experimentally with a variety of regenerator packings was highest for the packing for which $T_r$ was closest to 1.0.[3] On this basis the design criterion of Miyabe *et al.* appears plausible and useful. It would nevertheless be reassuring to have a means of carrying out explicit computation of the losses per cycle in terms of temperature transients in order to see how they relate.

The treatment which follows is intended for incorporation into the perturbation/availability approach of the preceding chapter. It is emphasized, however, that the method proposed for determining the temperature history within the wires is equally appropriate to a (non-linear) treatment in terms of absolute variables.

## 8.2 Heat transfer rates per wire

Appendix IX developed an expression for the rate of heat exchange of the ideal cycle per unit of free volume of the exchanger or regenerator. For the case of a machine operating efficiently (the condition being designed for) ideal, local instantaneous heat fluxes between wall and gas will approximate actual values, and may be used as a basis for estimating heat flow, and, from this, temperature distributions within individual wires. Heat quantities per unit of free volume may be converted to values per unit surface area of the wire element via a knowledge of wire area per unit free volume of regenerator.

Assuming a simple, rectangular mesh, and recalling the definition of volumetric porosity, $\P_v$, for this case (Chap. 5) it is a simple matter to arrive at:

$$\frac{\text{area}}{\text{unit free volume}} = \frac{4(1 - \P_v)}{\P_v d_w} \frac{m^2}{m^3} = \frac{1}{r_h} m^{-1} \tag{8.1}$$

In Eqn (8.1) $d_w$ is wire diameter.

Eqn (7.8) converted the expression for ideal heat rate per unit volume to normalized heat flux per unit volume and per unit crank angle. Multiplying by the inverse of Eqn (8.1) gives the normalized heat transfer rate per unit area per unit crank angle within a volume element $j$:

$$\frac{q_\phi'}{MRT_{ref}} = \frac{r_h}{V_{ref}} \left( -\frac{d\psi}{d\phi} + \frac{\gamma}{\gamma - 1} \sigma_j' \frac{d\tau_w}{d\mu_j} \right) \tag{8.2}$$

The conditional term has been omitted from Eqn (8.2), but is required for the end control volumes (only) if account is being taken of any discontinuity between the temperature of the incoming gas and regenerator wall temperature at that instant. As before, $d\tau_w$ is the change in dimensionless temperature between the end faces of the control volume.

It would be possible in principle to choose the number of regenerator subdivisions sufficiently large to give one control volume per gauze, so that Eqn (8.2), with a slightly different set of numerical values in each case, would be evaluated for each gauze. In practice it is expedient to have a smaller number of subdivisions, in which case all gauzes in a given control volume – and therefore all individual wires – are assumed to be subject to a common instantaneous heat flux per unit area and per unit crank angle.

The surface of the individual wire of a woven gauze is not uniformly subject to the same flow conditions. Actual heat flux per unit area is therefore not uniform, and there are corresponding axial and circumferential temperature gradients and components of heat flow. Since the degree of non-uniformity is not known at this stage, it cannot be taken into account quantitatively. In the circumstances one might argue that, because the surface heat flux rate is non-uniform, it must be greater locally than the value calculated on the assumption of uniformity. This suggests multiplication of losses eventually to be calculated by a factor (>1) as and when means of assessing the degree of non-uniformity come to light. For the moment the treatment proceeds on the basis of uniform, radial heat flow.

## 8.3 The diffusion equation

### 8.3.1 Internal points

The thermal diffusion equation in polar coordinates is[4]

$$\frac{\partial^2 T}{\partial r^2} + \frac{1}{r}\frac{\partial T}{\partial r} = \frac{1}{\alpha}\frac{\partial T}{\partial t} \tag{8.3}$$

In anticipation of numerical solution, and of the desirability of being able to specify convenient radial subdivisions for the wire, normalization is provisionally carried out with diameter, $d_w$, as reference length. With $\lambda$ for dimensionless radial coordinate, $r/d_w$, Eqn (8.3) converts to:

$$\frac{\partial^2 \tau}{\partial \lambda^2} + \frac{1}{\lambda}\frac{\partial \tau}{\partial \lambda} = \frac{\omega d_w{}^2}{\alpha}\frac{\partial \tau}{\partial \phi} \tag{8.4}$$

Recalling the definitions of the dimensionless groups $N_H$ and $N_F$, and that the reference length adopted for thermodynamic similarity is $R_H$ it is seen that:

$$1/N_F = \omega R_H{}^2/2\pi\alpha$$

Eqn (8.4) becomes:

$$\frac{\partial^2 \tau}{\partial \lambda^2} + \frac{1}{\lambda}\frac{\partial \tau}{\partial \lambda} = \frac{2\pi}{N_F}\frac{d_w{}^2}{R_H{}^2}\frac{\partial \tau}{\partial \phi} \tag{8.4a}$$

At the centre line $\lambda = 0$, but

$$\frac{1}{\lambda}\frac{\partial \tau}{\partial \lambda} \to \frac{\partial^2 \tau}{\partial \lambda^2}$$

so that, at the centre line

$$\frac{\partial^2 \tau}{\partial \lambda^2} = \frac{\pi}{N_F}\frac{d_w{}^2}{R_H{}^2}\frac{\partial \tau}{\partial \phi} \tag{8.4b}$$

### 8.3.2 At the boundary

At the wire surface, heat per unit time by conduction per unit area is $-k_r\,\partial T/\partial r$, so that the corresponding rate per radian is $-(k_r/\omega)\,\partial T/\partial r$. This must equal the heat rate per radian per unit area given by Eqn (8.2):

$$\frac{k_r}{\omega}\frac{\partial T}{\partial r} = MRT_{ref}\ \text{Eqn (8.2)}$$

$$\frac{\partial \tau}{\partial \lambda} = 2\pi N_H \frac{d_w}{R_H}\frac{r_h}{R_H}\left(-\frac{d\psi}{d\phi} + \frac{\gamma}{\gamma - 1}\sigma_{\phi j}{}'\frac{d\tau_w}{d\mu_j}\right) \tag{8.5}$$

## 8.4 Discretization

Numerical solution requires the differential coefficients to be put into the form of difference relationships with finite values for the increments, $\Delta\lambda$, $\Delta\phi$ in the independent variables $\lambda$ and $\phi$. In general, the increments may not be chosen arbitrarily because of the effect of the ratio $\Delta\lambda/\Delta\phi$ on numerical stability.

Using the subscripts $\phi$ and $\lambda$ to denote the angular and radial coordinate directions respectively, and $\Delta$ to denote a change in a particular coordinate direction, Eqn (8.4b), for example, becomes:

$$\Delta_\lambda{}^2\tau = \tfrac{1}{2}\left\{\frac{2\pi}{N_F}\frac{d_w{}^2}{R_H{}^2}\frac{\Delta\lambda^2}{\Delta\phi}\right\}\Delta_\phi\tau$$

$$= \tfrac{1}{2}\Lambda\Delta_\phi\tau \tag{8.4c}$$

$\Lambda$ represents the term between the curly braces, and $1/\Lambda$ is equivalent to the Fourier number for the interval which, in terms of absolute variables, is $\alpha\Delta t/\Delta x^2$.[4] The condition for numerical stability for the explicit formulation (difference form of the term $\Delta_\lambda{}^2\tau$ based on values at current time rather than on the mean during the integration step) is $2/\Lambda \leqslant 0.5$. The magnitude of $\Delta\lambda$ is $0.5/(n-1)$, where $n$ is the number of radial subdivisions. $\Delta\phi$ must be chosen to satisfy the stability criterion. It is unlikely that the $\Delta\phi$ used in calculating the discrete volumes, pressures and local, instantaneous mass and heat flow rates from the ideal reference cycle would be suitable, since considerations of stability did not apply to that stage of the computation, and a relatively coarse step size was appropriate. It might accordingly be necessary to provide for a large number of cycles of the heat diffusion solution corresponding to each increment in crankshaft angle.

With $\alpha \approx 6 \times 10^{-6}$ for stainless steel, a wire of 0.025 mm diameter subdivided into 10 to give $\Delta r = 0.0025 \times 10^{-3}$ m gives maximum permissible $\Delta t$ from the stability criterion ($\Delta t \leqslant 0.5(\Delta r)^2/\alpha$) of $0.52 \times 10^{-6}$. At 3000 rpm one revolution takes 0.02 s, and one degree about $55 \times 10^{-6}$ s. Use of a time step corresponding to one degree of crankshaft rotation would thus exceed by a factor of 100 the criterion for satisfactory solution of the radial temperature profile in the wire by an explicit scheme.

By contrast, implicit numerical schemes (spatial differences formulated as the averages of differences at the beginning and end of the current time step) are unconditionally stable. This advantage is offset to some extent by the fact that the method generates equations simultaneous in the unknowns (the temperatures after a time step). If the number of subdivisions of the conduction path is such that simple algebraic means for inversion cannot

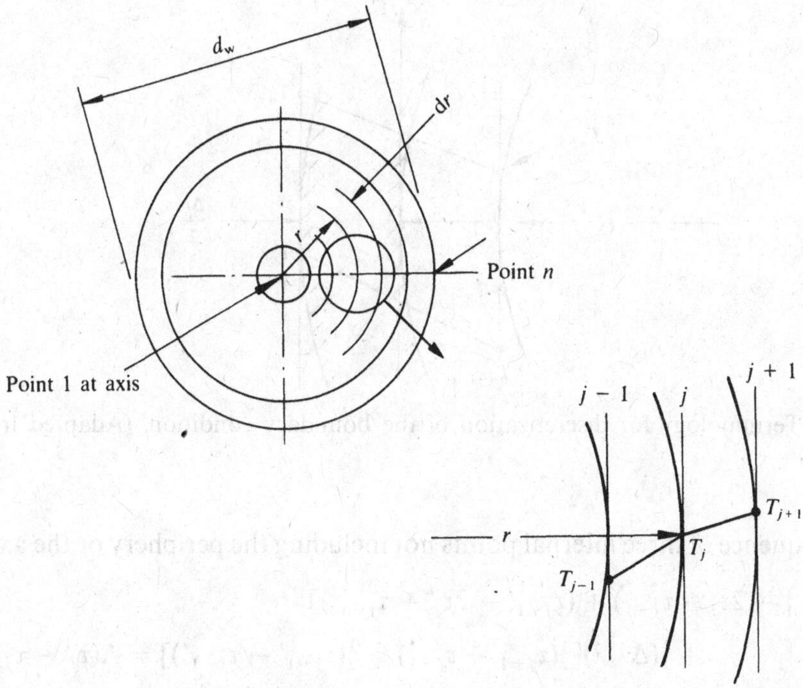


Fig. 8.3 Internal radial subdivision of a cylindrical wire element as the basis for a numerical discretization scheme.

cope, then numerical schemes for the inversion of matrices have to be invoked. The stability of the implicit formulation permits the increment in $\Delta\phi$ for the internal temperature solution to be set equal to the crankshaft increment of the main simulation. The natural choice for the discretization scheme is thus the implicit formulation.

Fig. 8.3 represents the cross-section through a single wire, showing $n$ annular subdivisions. There is a case[5] for non-linear variation of the radial increments, corresponding to the non-linear influence of the $1/r$ term of the diffusion equation. In practice the convenience of equal subdivisions is the deciding factor. With the prime (′) to denote (unknown) temperatures after the increment $\Delta\phi$ in $\phi$, the discrete form of Eqn (8.4c) for the axial location is:

$$\tfrac{1}{2}[(\tau_2 - \tau_1) - (\tau_1 - \tau_2) + (\tau_2' - \tau_1') - (\tau_1' - \tau_2')] = \Lambda(\tau_1' - \tau_1)/2$$

Rearranging:

$$\tau_1'(-1 - \Lambda/2) + \tau_2' = \tau_1(1 - \Lambda/2) - \tau_2 \qquad (8.6)$$

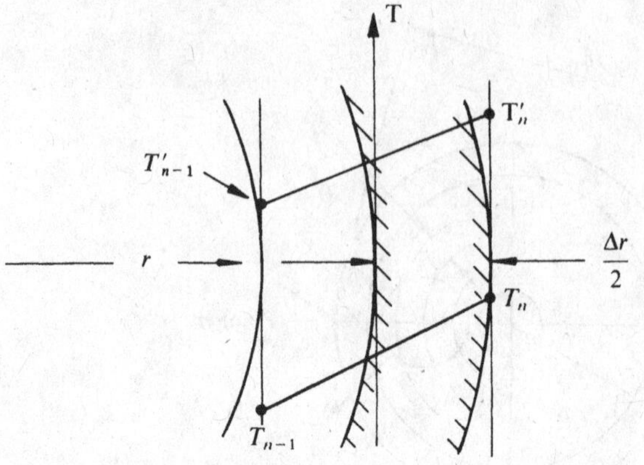


Fig. 8.4 Terminology for discretization of the boundary condition. (Adapted from Arpaci.[5])

For a sequence of three internal points not including the periphery or the axis:

$$\tfrac{1}{2}[(\tau_{j+1} - 2\tau_j + \tau_{j-1}) + (\tau_{j+1}' - 2\tau_j' + \tau_{j-1}')] + \tfrac{1}{2}(\Delta\lambda/\lambda)[\tfrac{1}{2}(\tau_{j+1} - \tau_{j-1}) + \tfrac{1}{2}(\tau_{j+1}' - \tau_{j-1}')] = \Lambda(\tau_j' - \tau_j)$$

Collecting like terms and rearranging:

$$\tau_{j-1}'(1 - \Delta\lambda/2\lambda) + \tau_j'(-2 - 2\Lambda) + \tau_{j+1}'(1 + \Delta\lambda/2\lambda) = \tau_{j-1}(-1 + \Delta\lambda/2\lambda) + \tau_j(2 - 2\Lambda) + \tau_{j+1}(-1 - \Delta\lambda/2\lambda) \quad (8.7)$$

Discretization of the boundary condition (Eqn (8.4)) does not exactly follow the analytical form. The scheme advocated by Arpaci[5] is applied to points $n-1$ and $n$. Rate of heat flow per unit area by conduction into the left-hand face of the shaded control volume of Fig. 8.4 minus rate of heat flow by convection to the gas through the right-hand face must equal the rate of energy storage. The finite temperature gradient appropriate to the conduction term is that at $\Delta r/2$, i.e., $(T_n - T_{n-1})/\Delta r$. With $q'$ for surface heat flux rate per unit area:

$$-k\frac{\Delta T}{\Delta r} - q' = \frac{\Delta r}{2}\rho c\frac{\Delta T_{3\Delta r/4}}{\Delta t}$$

$q'$ is the right-hand side of Eqn (8.2) multiplied by $MRT_{\text{ref}}$. Substituting Eqn (8.2), working with the means of the temperature gradients between points $n-1$ and $n$ (on the boundary):

$$\tfrac{1}{2}[(\tau_n - \tau_{n-1}) + (\tau_n' - \tau_{n-1}')] - 2\Delta\lambda[\text{RHS of Eqn (8.5)}] = (\Lambda/2)[\tfrac{1}{4}(\tau_{n-1}' - \tau_{n-1}) + \tfrac{3}{4}(\tau_n' - \tau_n)]$$

Collecting like terms and rearranging:

$$\tau_{n-1}'(1 - \Lambda/4) + \tau_n'(-1 - 3\Lambda/4) = 2\Delta\lambda[\text{RHS of Eqn (8.5)}] + \tau_{n-1}(-1 - \Lambda/4) + \tau_n(1 - 3\Lambda/4) \quad (8.8)$$

The coefficients of Eqns (8.6), (8.7) and (8.8) form a tridiagonal matrix which may be inverted by an appropriate computer software routine to yield $n$ values of $\tau_j'$ – i.e., the dimensionless temperatures after the crank angle increment. These become the 'known' temperatures which are substituted into the right-hand sides of Eqns (8.6), (8.7) and (8.8) for the next integration step.

## 8.5 Implementation

The foregoing relationships are coded in FORTRAN as an integral part of a suite of simulation programs developed by the author and based on the thermodynamic similarity principle. Matrix inversion is handled by the relatively unsophisticated but well-known and well-tried routine SIMQX. At the time of writing the programs are run on Hewlett Packard 9000 series 300/800 computers, graphics calls using the GRASP library and being handled by the X system.

The regenerator calculation starts with wire temperatures graded linearly between expansion and compression end values, the radial distribution in individual wires being uniform. After two complete cycles of the wire temperature computation conditions are essentially steady state.

At rated operating conditions (Table 4.1) there is a cyclic swing of +/− some 10 K at the outer surface of wires near the expansion end. At rated wire diameter (see the specification in Appendix AVI) radial temperature gradients are negligible at all locations and at all points around the cycle. In fact, to obtain a pictorial impression of the radial temperature wave it is necessary to assume a decrease in characteristic Fourier number, $N_F$ of some three orders of magnitude and an increase of the same order in regenerator thermal load (i.e., in $N_H$). The result is shown in Fig. 8.5 – again for a wire at the expansion end. The vertical plane represents a slice through the wire centre line, so that one symmetrical half of the diametral distribution is illustrated. The figure emphasizes a phenomenon not widely publicised, whereby the maximum and minimum of the cyclic temperature swing occur during the cold blow, and are separated by much less than the intuitively expected 180° of crankshaft rotation, the actual value being some 30°.

It might be asked why such extreme modifications to $N_H$ and $N_C$ do not

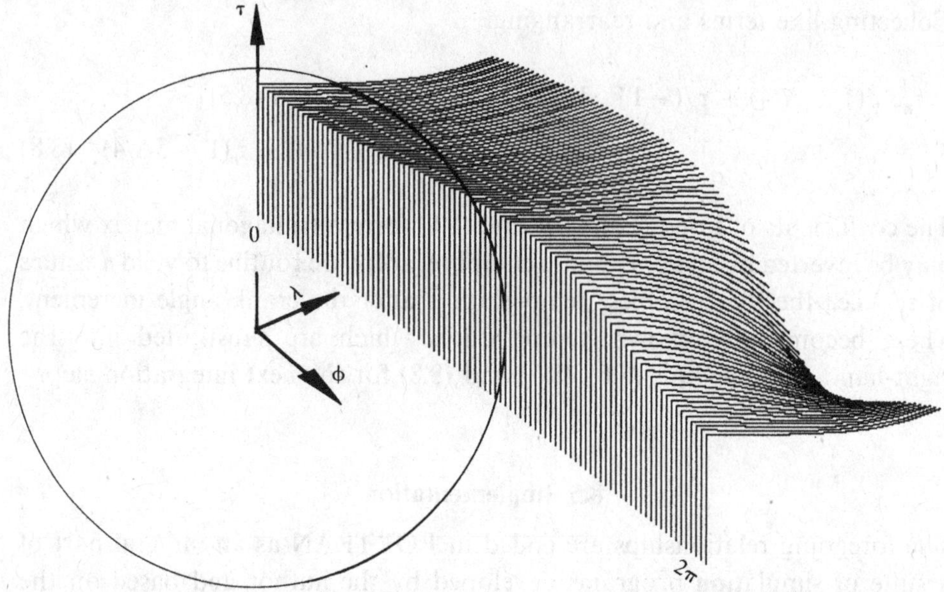


Fig. 8.5 Radial distribution of dimensionless temperature, $T/T_{ref}$ against crank angle for a typical wire at the expansion end. Regenerator thermal capacity ratio is as for Philips MP1002CA at rated conditions (Table 4.1), but thermal diffusivity is set to an artificially low value (by reducing $N_F$) to exaggerate radial temperature profile.

lead to a completely unrepresentative result. By way of explanation, note that

$$N_H N_C = \frac{p_{ref} N_s L_{ref} R_H}{k_r T_{ref}} \frac{\alpha_r}{N_s {R_H}^2}$$

From the definition of $\alpha_r$ as $k_r/\rho_r c_r$.

$$N_H N_C = \frac{p_{ref}}{T_{ref} \rho_r c_r} = \frac{MR}{V_{ref} \rho_r c_r} = \frac{\rho_{ref} c_p}{\rho_r c_r} \frac{(\gamma - 1)}{\gamma}$$

The product of the two dimensionless groups is thus equivalent to the ratio of the total, nominal heat capacity of the gas to that of the matrix. Decreasing one of the groups by a certain factor whilst increasing the other by the same factor thus preserves the original ratio, so in Fig. 8.5 the bulk periodic temperature swing remains characteristic of that of the regenerator wire at nominal values of $N_H$ and $N_C$, while only the *gradients* (the $\partial T/\partial r$) in spanwise temperature distribution are exaggerated.

A rough calculation of $\rho_{ref} c_p/\rho_r c_r$ for the MP1002CA leads to the value 1/250. If one quarter of the total mass of gas passes through the regenerator changing its temperature between 60 and 600 °C one might guess at a temperature swing at of some 4(600 – 60)/250, or about 10 °C – a value not far from that determined by the simulation.

The fact that a 1000-fold decrease in thermal conductivity is required before radial temperature gradients become significant suggests that, for machines thermodynamically similar to the MP1002CA at least, the search for alternative regenerator materials may concentrate on specific thermal capacity ($\rho_r c_r$), cost, high temperature performance and manufacturing considerations to the virtual exclusion of thermal conductivity. There seems, in other words, little chance of insufficient temperature wave penetration (right-hand illustration of Fig. 8.1) with any metallic wire of the diameter in question.

A look at Eqn (8.4c) reveals that the temperature calculation is influenced by $(d_w/R_H)^2$, and that it is thus more sensitive to a given percentage increase in wire diameter than to the corresponding percentage decrease in $N_F$. Accordingly, increasing $d_w/R_H$ by a factor of $\sqrt[3]{1000}$ produces the exaggerated temperature profiles of Fig. 8.5.

The use of a specific regenerator geometry (the screen woven from cylindrical wires) has to some extent gone against the spirit of thermodynamic similarity, having called for geometric description in terms of two dimensionless variables $¶_v$ and $d_w/R_H$ over and above those anticipated in Chaps. 4 and 7. It will be shown shortly that a choice of regenerator hydraulic radius, $\lambda_{hr}$, together with a value of *either* $¶_v$ *or* $d_w/R_H$ fixes the geometry of the wire screen, so that only one additional dimensionless variable has actually resulted from the choice of a specific form of regenerator. The temperature profile of Fig. 8.5 thus applies to all Stirling machines having the same values of the parameters used to process the ideal reference cycle ($\alpha$, $\kappa$, $N_\tau$, the various $\mu$), and of $\lambda_{hr}$, $N_H$, $N_F$, and $¶_v$ (or $d_w/R_H$). It is unaffected by changes in $N_{MA}$ and $N_{SG}$.

## 8.6 Available work lost due to conduction within the regenerator

The expression in cylindrical polar coordinates for rate of entropy creation per unit volume, $S_v'$ when heat flow is confined to the radial direction is:

$$S_v' = \frac{k}{T^2}\left(\frac{\partial T}{\partial r}\right)^2$$

Normalizing by reference to wire diameter as previously:

$$S_v' = \frac{k}{\tau^2 {d_w}^2}\left(\frac{\partial \tau}{\partial \lambda}\right)^2 \tag{8.9}$$

The volume of an annulus of unit axial length is $2\pi r\,\mathrm{d}r$. Entropy generation rate per unit length of wire, $S_l'$, thus becomes

$$S_l' = \frac{2\pi k}{d_w{}^2} \sum \frac{\lambda}{\tau^2}\left(\frac{\partial \tau}{\partial \lambda}\right)^2 \mathrm{d}\lambda \qquad (8.10)$$

Using the reasoning which led to Eqn (8.1), the length of wire per unit volume of regenerator housing is:

$$4(1 - ¶_v)/\pi d_w{}^2$$

Correspondingly, the length $l_v$ per unit of free regenerator volume (the gas-side formulation is entirely in terms of free volumes) is:

$$l_v = 4(1 - ¶_v)/¶_v \pi d_w{}^2 \qquad (8.11)$$

Substituted into Eqn (8.10), Eqn (8.11) gives rise to an expression for entropy generation rate per unit volume of the control volume used to evaluate the heat transfers which were the basis for determining the temperature gradients in the wires. Forming finite differences and normalizing by the now familiar $MRT_0$ gives specific lost work per unit increment in crank angle, $\mathrm{d}\phi$:

$$\mathrm{d}\zeta_{sk} = \frac{4}{\pi\,\mathrm{d}\lambda^2} \frac{1 - ¶_v}{¶_v} \frac{1}{(d_w/R_H)^2 N_H} \sum_1^{n-1} \lambda_j \frac{(\tau_{j+1} - \tau_j)^2}{(\tau_{j+1} + \tau_j)^2} \mathrm{d}\phi\, \mathrm{d}\mu_j \qquad (8.12)$$

It is easy to show (see Chap. 5) that for the square mesh gauze $d_w = 4r_h(1 - ¶_v)/¶_v$. With this substitution, and recalling that $r_h/R_H$ for the regenerator is denoted $\lambda_h$, the final expression for available work lost during crank increment $\mathrm{d}\phi$ in dimensionless volume element $\mathrm{d}\mu_j$ is:

$$\mathrm{d}\zeta_{sk} = \frac{1}{\pi\,\mathrm{d}\lambda^2} \frac{¶_v}{1 - ¶_v} \frac{1}{4\lambda_{hr}{}^2 N_H} \sum_1^{n-1} \lambda_j \frac{(\tau_{j+1} - \tau_j)^2}{(\tau_{j+1} + \tau_j)^2} \mathrm{d}\phi\, \mathrm{d}\mu_j \qquad (8.12a)$$

The squaring of the temperature difference terms is consistent with the fact that entropy is generated, and available work lost, for both possible directions of heat flow. Eqns (8.12) and (8.12a) confirm that work lost in this way is, like the temperature profiles themselves, a function of a dimensionless parameter extra to those identified in Chap. 4, namely, $¶_v$ (or $d_w/R_H$). With regenerator conduction loss taken into account, the functional expression for specific cycle work (Eqn (4.9)):

$$\zeta = \zeta\left(\begin{array}{llll|l} & \mu_{dce} & & & \\ & \mu_{de} & & \alpha_{xe} & \lambda_{he} \\ N_\tau, \kappa, \alpha, \mu_v, & \mu_{dr}, & \gamma, & \alpha_{xr}, & \lambda_{hr}, \P_v, N_{MA}, N_{SG}, N_H, N_F \\ & \mu_{dc} & & \alpha_{xc} & \lambda_{hc} \\ & \mu_{dcc} & & & \end{array}\right) \quad (8.13)$$

parameters of ideal reference cycle | parameters of loss computation

The parameters of Eqn (8.13) have been grouped in a form which emphasizes those required for computation of the ideal reference cycle, and those required for the computation of losses. The parameters required for loss computation (only) may be varied at will without need for recomputation of the reference cycle. This means that indicated performance over all possible rpm, charge pressures, working fluids, exchanger hydraulic radii, regenerator materials and mesh numbers may be investigated by varying the eleven parameters of the loss calculation. In practice, the task reduces to varying the values of nine such parameters, and producing the corresponding map with the remaining two ($N_{MA}$, $N_{RE}$ – or $N_{SG}$) as independent variables.

## 8.7 Conclusions

At the outset an intention was expressed of looking for support for the design guidelines proposed by Miyabe *et al.*[3] The latter are formulated in terms of a Biot and Fourier numbers – respectively $hd_w/2k$ and $4\alpha t_{blow}/d_w{}^2$ as defined by Miyabe.

There is no unique value of $h$ for the regenerator, but a characteristic value such as a mean or a maximum may be calculated. The local, instantaneous Stanton number, $N_{st} = h/Gc_p$, so that $hd_w/2k = Gc_p N_{st} d_w/2k$. The $N_{st}$ are calculated at all locations within the regenerator during the course of the lost work computation, so it is a simple matter to store the maximum value of $abs(Gc_p N_{st})$ for eventual multiplication by the constant terms to give corresponding maximum Biot number, $N_{Bi}$. In fact, computation is carried out in dimensionless variables, in terms of which

$$N_{Bi} = 2\pi N_{st}(\text{location}, \phi) N_H \sigma_\phi{}' \frac{1}{\alpha_x} \frac{d_w}{R_H} \frac{\gamma}{\gamma - 1} \quad (8.14)$$

The maximum calculated in this way for the MP1002CA engine under rated conditions is 0.008 – a value close to the 0.00877 of Miyabe's 'optimum' regenerator. The corresponding value of Fourier number (Miyabe's definition) is 800 – some $2\frac{1}{2}$ times that for Miyabe's optimum case. Entering these values in the design chart (Fig. 8.2) leads to a $T_r$ (Miyabe's definition) with a value somewhat greater than the recommended 0.95–1.0. Either a reduction in $N_F$ or an increase in $d_w/R_H$ would be required to bring the design into line with the Miyabe criterion.

The computer simulation based on the analysis of this and the previous chapter offers an opportunity to isolate the effects of such design changes. In Fig. 8.6 $d_w/R_H$ is varied over a factor of about 10 above and below the

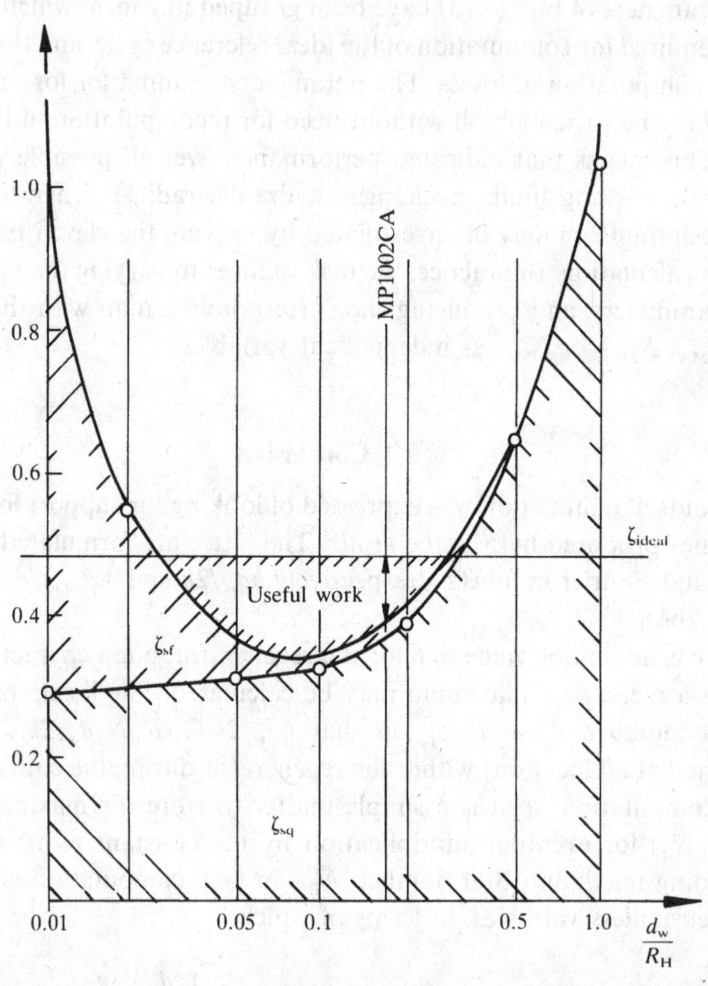


Fig. 8.6 The effect on total, lost, available specific work of variation in dimensionless wire diameter, $d_w/R_H$. The small area between the upper curve and the horizontal line at $\zeta_{ideal}$ represents the range of $d_w/R_H$ for which the cycle produces useful indicated work. The rated value for the MP1002CA machine is somewhat in excess of the apparent optimum.

nominal, and total lost specific work plotted. Over the range in question available work lost due to thermal conduction within the wires is negligible. The two principal losses are thus those due to convective heat transfer and flow friction. It is seen that decreasing $d_w/R_H$ brings about a reduction in losses due to convective heat transfer (the $\zeta_{sq}$) but a rapid increase in flow losses (the $\zeta_{sf}$). At high $d_w/R_H$ the contribution from friction tends to the very low value of the slotted heat exchangers alone. At low $d_w/R_H$ the convective heat transfer losses are asymptotic to the value calculated for the slotted heat exchangers. At these extremes, the assumption of linearity on which the computations are based cannot be justified.

The horizontal line at the value 0.48 is the specific work of the ideal reference cycle under rated conditions. For values of total lost work below 0.48 there is useful cycle work. The rated value of $d_w/R_H$ for the MP1002CA machine determines the operating point, which is seen to lie above the point at which useful cycle work would be a maximum. On this criterion, $d_w/R_H$ would ideally be some 65% of that of the wire actually installed.

It has been shown that the diameter of the regenerator wire may be increased and/or the conductivity of the wire material reduced substantially without penalty in terms of excessive internal temperature gradients and corresponding losses. Within wide limits, therefore the choice of regenerator wire diameter has little to do with ensuring adequate temperature wave penetration, since this is virtually assured, but everything to do with attaining a heat transfer area which is large enough to keep down convective heat transfer losses without creating excessive flow losses.

The finding indicates that a vast range of materials is suitable for use as wires for a Stirling regenerator, given suitable mechanical properties and a product $\rho_r c_r$ comparable with that of stainless steel.

The fact that internal conduction losses are found to be negligibly small over all reasonable choices of regenerator wire does not mean that the analysis formulated to deal with the case is redundant. It will be required again, virtually unchanged, to deal with the unsteady heat flows in the tubular heat exchangers, where conduction losses are substantial.

# 9

# Full dynamic similarity

## 9.1 Implications of full dynamic similarity – scaling

Chap. 4 applied the principles of dynamic similarity to the formulation of the one-dimensional equations governing fluid behaviour in the gas circuit, and isolated the resulting defining dimensionless groups. Chap. 8 argued that, with the condition of geometric similarity relaxed, an approximate similarity of gas circuit working fluid processes may be specified in terms of dimensionless groups which determine thermodynamic similarity.

The full dynamic similarity treatment relies on detailed specification of drive mechanism kinematics and on a comprehensive geometric description of all heat exchanger passages. Heat transfer and flow friction correlations are for the specified heat exchanger geometry. Scaling from a design of one physical size to another is by changing the value of reference length, and applies only between machines whose gas circuit geometry and drive kinematics are identical. The conditions for full dynamic similarity are thus more restrictive than those for thermodynamic similarity. It is tempting to think that they are also more rigorous than the alternative, until it is appreciated *inter alia* that computations based on both presuppose that all phenomena are one-dimensional, whereas in reality they are three-dimensional.

Because heat exchanger geometry is specified, it becomes possible in selected cases to propose a treatment of the losses associated with transient thermal conduction within the exchanger walls. The obvious instance is that of the exchanger comprising cylindrical tubes: for this case the equations of one-dimensional transient thermal conduction are of the same form as those for the regenerator wire (dealt with in Chap. 8).

The 'equivalent' Stirling machine has now evolved to its final state: Fig. 9.1 shows the isothermal source and sink communicating with the working fluid through the walls of exchangers, the radial distribution of temperature which is allowed to vary over a cycle corresponding to local, instantaneous heat flux rate. It is possible to provide for similar variation of wall temperature

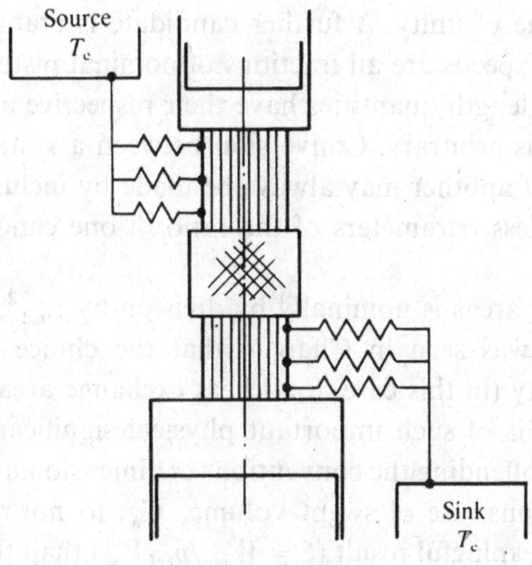


Fig. 9.1 Equivalent, one-dimensional Stirling cycle machine, showing indirect contact with source and sink via the thermal resistance (capacitance not indicated) of heat exchanger walls.

in the variable-volume spaces, but it can be inferred from the findings of Section 9.3 that the additional computational work is likely to be out of proportion to the benefits. The transient temperature within the solid material of the regenerator matrix is accounted for, and piston motions (and thus volume variations) determined by drive mechanism kinematics. If it is required to take into account conduction in the direction of the machine axis, or of displacer 'shuttle' heat transfer, these are calculated separately, and added to or subtracted once per cycle from the quantities computed by integration around the cycle. Losses due to friction in seals and at the bearings of the drive mechanism may be estimated in terms of local, instantaneous loads (pressure plus inertia) and corresponding friction coefficients. This computation may be carried out independently of that for the thermodynamic cycle, in the manner explained in Ref 1, and is not treated here.

## 9.2 Choice of reference length

In order to normalize the numerous length quantities which describe the gas circuit and drive mechanism it is necessary to select a reference length, $L_{ref}$ – preferably a quantity which leads to dimensionless groups having physical meaning. Thus, one possibility (in view of the gas dynamics treatment of Chapter 11 and Appendix XI) is the nominal gas path length from piston face to piston face. The notional length of a single wave passage thus assumes the convenient value of unity. A further candidate is crank radius, $r$. With this choice, particle speeds are all fractions of nominal piston face speed, $\omega r$. Other fundamental length quantities have their respective attractions, but in the end the choice is arbitrary. Conversion between a system involving one reference length and another may always be made by inclusion in the list of defining dimensionless parameters of the ratio of one candidate $L_{ref}$ to the other.

Normalization of areas is nominally by division by $L_{ref}^2$, and volumes by $L_{ref}^3$. However, it was seen in Chap. 4 that the choice of an area as a normalizing quantity (in this case, total heat exchange area) can lead to the quantity $R_H$ which is of such important physical significance that it seems worthwhile risking offending the conventions of dimensional analysis to retain it. For similar reasons, use of swept volume, $V_{sw}$, to normalize cycle work produces a more meaningful result ($\zeta = W_{cyc}/p_{ref} V_{sw}$) than the more pedantic alternatives (say, $W_{cyc}/p_{ref} r^3$). Provided there are sufficient length quantities, $\lambda_j = l_j/L_{ref}$, to specify machine geometry completely, geometric (and thus dynamic) similarity is preserved despite these liberties.

In a cycle simulation formulated so as to exploit dynamic similarity principles, one would expect to keep separate the two arrays of dimensionless lengths which define crank kinematics on the one hand and the gas circuit on the other. The separate grouping can be reflected in the functional expression for computed, indicated cycle by using subscript k for the lengths of kinematic members, and gc for length quantities which describe the gas circuit:

$$\zeta = \zeta\left( N_\tau,\ V_E/r^3,\ \begin{matrix} \lambda_{k_1} & \lambda_{gc_1} \\ \lambda_{k_2} & \lambda_{gc_2} \\ \lambda_{k_3}, & \lambda_{gc_3}, \\ \vdots & \vdots \\ \lambda_{k_n} & \lambda_{gc_n} \end{matrix}\ \P_v,\ \gamma,\ N_{MA},\ N_{SG},\ \begin{matrix} N_{H_e} & N_{F_e} \\ N_{H_r}, & N_{F_r} \\ N_{H_c} & N_{F_c} \end{matrix} \right) \tag{9.1}$$

Regenerator porosity, $\P_v$, in Eqn (9.1) is a further example of a variable which has not been normalized directly by (reference length)$^3$, the indirect normalization by an intermediate volume – in this case that of the regenerator capsule – resulting in a more familiar and more usable form. Normalized wire diameter, $d_w/r$ is included among the various $\lambda_{gc_n}$. Separate Fourier and surface conductance numbers, $N_H$ and $N_F$ have been specified for the expansion end exchanger, regenerator and cooler in anticipation of calculation of transient conduction effects in each. An alternative presentation of the various $N_F$ would be possible by specifying one $N_F$ only (for regenerator, say)

Table 9.1. *Variables defining the kinematics of the drive mechanism of the Philips MP1002CA air engine*

| Variable | Definition | Value |
|---|---|---|
| $r$ | crank-pin offset | 13.5 mm |
| $e/r$ | (*désaxé* offset)/$r$ | 0.0 |
| $h/r$ | (displacer con-rod centres)/$r$ | 4.16 |
| $QF/r$ | (bell-crank arm length)/$r$ | 5.08 |
| $QE/r$ | (bell-crank arm length)/$r$ | 4.82 |
| $X/r$ | (bell-crank fulcrum $x$ offset)/$r$ | 4.82 |
| $Y/r$ | (bell-crank fulcrum $y$ offset)/$r$ | 3.10 |
| $l/r$ | (piston con-rod centres)/$r$ | 6.28 |
| $X_{VE}$ | $V_E/r^3$ | 25.10 |

and two thermal diffusivity ratios – heater to regenerator and cooler to regenerator. By the same token, alternative presentations are possible for the various $N_H$.

Specification of heat exchanger geometry is straightforward in respect of the regenerator having a geometrically regular packing, and of the parallel flow passages themselves, but manifolds and connecting ducts generally pose a problem. This may be dealt with by adjusting the diameter of the duct to that of the adjacent working space or exchanger tube, and then pro-rating the length to keep the enclosed volume at the original value. This approach has been taken in Tables 9.1 and 9.2 which list the dimensionless parameters defining drive mechanisms and gas circuit of the Philips MP1002CA engine. Reference length, $L_{ref}$ is crank-pin offset, $r$. For definitions of kinematic quantities, reference may be made to Section AV.4.

*Any* analysis of the Stirling cycle machine may be cast in the functional form of Eqn (9.1), whether subject to transformation to Characteristics form, whether perturbed before solution, whether in cartesian coordinates or Lagrange. The sole reservation is that, for those solutions based on fixed grid schemes, the eventual numerical answer is additionally a function of one or more Courant numbers, $N_C$.

Individual expressions for normalized losses due to heat transfer and friction are essentially those already developed for the thermodynamic similarity case, but there are differences of detail due to the different base for normalization of the input data: in the definition of $N_{re}$ in terms of $N_{RE}$ (Eqn (4.4)) the ratio $R_H/L_{ref}$ is unity for the thermodynamic similarity case, but a fractional constant when the length term in $N_{RE}$ is other than $R_H$ (as is generally the case when dynamic similarity is being observed).

Table 9.2. *Variables defining the gas circuit of the Philips MP1002CA air engine*

| Variable | Definition | Value |
|---|---|---|
| Hydraulic radii | | |
| $\lambda_{he}$ | $r_{he}/r$ | 0.0127 |
| $\lambda_{hr}$ | $r_{hr}/r$ | 0.003 |
| $\lambda_{hc}$ | $r_{hc}/r$ | 0.0127 |
| $r/R_H$ (for conversion between $_{ts}$ and $_{ds}$) | | 13.5/0.2166 |
| Flow passage lengths | | |
| $\lambda_{ce}$ | (equiv. length exp. sp. dead vol.)/$r$ | 0.24 |
| $\lambda_e$ | (length exp. exchanger)/$r$ | 2.76 |
| $\lambda_r$ | (length regen. housing)/$r$ | 2.15 |
| $\lambda_c$ | (length comp. exchanger)/$r$ | 2.76 |
| $\lambda_{cc}$ | (equiv. length comp. sp. dead vol.)/$r$ | 0.28 |
| Free-flow (cross-sectional) areas | | |
| $\alpha_{xce}$ | (exp. end area)/$r^2$ | 13.0 |
| $\alpha_{xe}$ | (exp. exchr. free-flow area)/$r^2$ | 1.12 |
| $\alpha_{xr}$ | (regen housing x-sect. area)/$r^2$ | 3.93 |
| $\alpha_{xc}$ | (comp. exchr. free-flow area)/$r^2$ | 1.12 |
| $\alpha_{xcc}$ | (comp. end area)/$r^2$ | 13.0 |
| Regenerator | | |
| ¶$_v$ | volumetric porosity | 0.8 |
| $\lambda_w$ | $d_w/r$ | 0.0030 |
| Operating parameters at rated conditions | | |
| $N_{MA}$ | $N_s r/\sqrt{(RT_{ref})}$ | 0.0012 |
| $N_{SG}$ | $p_{ref}/N_s\mu$ | 3.11E + 09 |
| $N_{RE}$ | $N_{SG}N_{MA}{}^2$ | 4.5E + 03 |
| $N_H$ | $p_{ref}N_s r^2/kT_{ref}$ (regenerator) | 1.06 |
| $N_F$ | $\alpha_r/N_s r^2$ (regenerator) | 0.0013 |

Eqn (7.13) for elemental lost work due to surface film conductance loss was:

$$\frac{T_0\,dS_{q_j}}{MRT_{ref}} = \frac{(\gamma-1)}{\gamma}\,\frac{q_\phi{}'^2\lambda_{h_j}\alpha_{x_j}}{\tau_j{}^2\sigma_\phi{}'N_{st}\,d\mu_j}\,d\phi \tag{7.13}$$

The variables whose numerical values depend on a geometric normalizing quantity are $\lambda_{h_j}$, $\alpha_{x_j}$ and $d\mu_j$, viz, $r_{h_j}/R_H$, $A_{x_j}/A_w$, $dV_j/V_{ref}$. From the definition of $R_H$ it is seen that the normalizing quantities cancel between the three terms, so that the value of the expression is unaffected when the variables are normalized by $r$, $r^2$ and $r^3$ respectively. Provided $N_{st}$ is deduced from an appropriately evaluated local, instantaneous $N_{re}$ Eqn (7.13) may be used for the dynamic similarity case.

Slight modification is called for to the expression for incremental work lost to flow friction. Eqn (7.15) may be written:

$$\frac{T_0\,\mathrm{d}S_{\mathrm{f}_j}}{MRT_{\mathrm{ref}}} = \mathrm{d}\zeta_{\mathrm{sf}} = \frac{2\pi^2 N_{\mathrm{MA}}{}^2 \tau_j |\sigma_{\phi j}{}'^3| (a/N_{\mathrm{re}_j} + b)}{\lambda_{\mathrm{h}_j} \alpha_{\mathrm{x}_j}{}^2 \psi^2}\,\mathrm{d}\lambda_j\,\mathrm{d}\phi \qquad (7.15)$$

The numerical difference between $N_{\mathrm{MA}}$ for the thermodynamic similarity (ts) and dynamic similarly (ds) cases may be expressed:

$$N_{\mathrm{MA}_{\mathrm{ds}}} = N_{\mathrm{MA}_{\mathrm{ts}}} \frac{L_{\mathrm{ref}}}{R_{\mathrm{H}}}$$

Consideration of the ratio $\mathrm{d}\mu_j/(\lambda_{\mathrm{h}_j}, \alpha_{\mathrm{x}_j}{}^2)$ reveals that the geometric normalizing quantities do not cancel, and that there is an unbalanced $(\text{area})^2$ term implying a numerical factor $r^4/A_{\mathrm{w}}{}^2$ between Eqn (7.15) evaluated for the thermodynamic and dynamic similarity cases. The term $V_{\mathrm{E}}/r^3$ $(=X_{\mathrm{VT}} = V_{\mathrm{E}}/L_{\mathrm{ref}}{}^3)$ is chosen (Eqn (9.1)) to relate volumes to lengths on the basis that one does not wish to be tied to the notional volume $L_{\mathrm{ref}}{}^3$. With this definition, Eqn (7.15) for the thermodynamic similarity case becomes, for dynamic similarity:

$$\frac{T_0\,\mathrm{d}S_{\mathrm{f}_j}}{MRT_{\mathrm{ref}}} = \mathrm{d}\zeta_{\mathrm{sf}} = \frac{2\pi^2 (N_{\mathrm{MA}})_{\mathrm{ds}}{}^2 \tau_j |\sigma_{\phi j}{}'^3| (a/N_{\mathrm{re}_j} + b)}{(\lambda_{\mathrm{h}_j} \alpha_{\mathrm{x}_j}{}^2)_{\mathrm{ds}} \psi^2 (\mu_{\mathrm{v}}/X_{\mathrm{VE}})^2} (\mathrm{d}\lambda_j)_{\mathrm{ds}}\,\mathrm{d}\phi \qquad (9.2)$$

In Eqn (9.2) $\mu_{\mathrm{v}} = V_{\mathrm{E}}/V_{\mathrm{wfc}}$ as before, and the subscript ds signifies that the variables in question have been normalized by reference to the length quantity $r$. The equation now returns the same numerical value of lost work increment as did Eqn (7.15) when both are applied to the same control volume of the same machine.

## 9.3 Performance maps and scaling

For comparison with the thermodynamic similarity treatment Fig. 9.2(*a*) is a performance map for the Philips MP1002CA air engine, showing the dependence of normalized, computed indicated cycle work, $\zeta$, on $(N_{\mathrm{MA}})_{\mathrm{ds}}$ and $(N_{\mathrm{RE}})_{\mathrm{ds}}$ $(=N_{\mathrm{SG}}(N_{\mathrm{MA}})_{\mathrm{ds}}{}^2)$. Transient conduction losses have been set at zero. To obtain $\zeta$ at rated conditions, the chart is entered with the $N_{\mathrm{MA}}$ and $N_{\mathrm{RE}}$ from Table 9.2, giving $\zeta = 0.275$. (A difference is to be expected from the thermodynamic similarity case due to the use of heat transfer and friction correlations specific to the slotted heat exchangers, and to the revised volume variations.)

Before looking into the principal rôle of the dynamic similarity approach

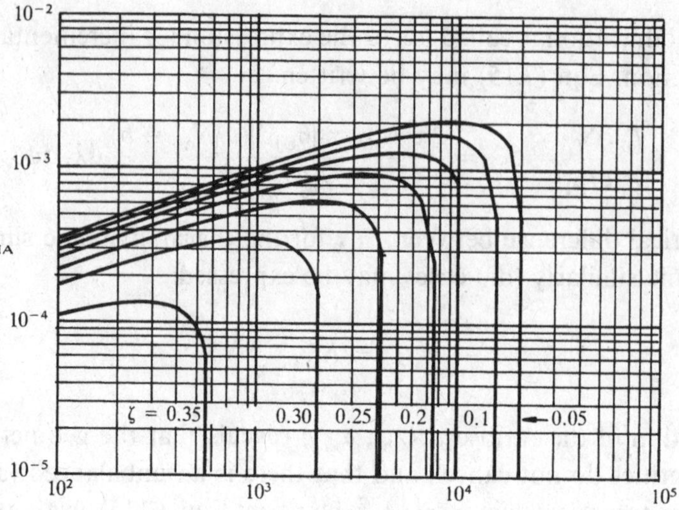


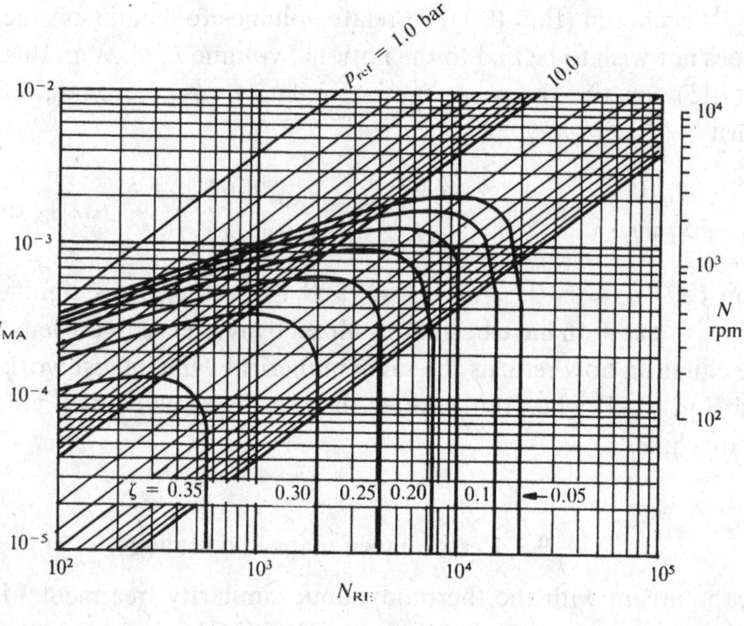


Fig. 9.2 (*a*) and (*b*) Performance maps for *all* Stirling engines geometrically similar to the Philips MP1002CA machine and operating on diatomic working fluid (e.g., air) at rated $N_\tau$. Both maps apply over the entire operating envelope (i.e., are valid for all $N_{MA}$, $N_{RE}$), but map (*b*) has superimposed speed and pressure scales valid for the original MP1002CA operating on air at rated $N_\tau$.

(that of scaling) it is worth noting that *for a given machine with a given working fluid* ($\gamma$, $R$, $\mu$) a simple modification to the map enables performance to be read off directly in terms of reference pressure, $p_{ref}$ and rpm, $N$. From the definition of $N_{MA}$, $N_s = N_{MA}\sqrt{(RT_{ref})}/L_{ref}$, so that, with $r$ for $L_{ref}$, rpm $= N = 60N_{MA}\sqrt{(RT_{ref})}/r$. With $r$ known (and fixed for the given machine) the rpm scale lies parallel to the $N_{MA}$ scale (Fig. 9.2(*b*)).

Substituting the expression for $N$ into that for $N_{RE}$ gives:

$$N_{MA} = \frac{\mu_{ref}\sqrt{(RT_{ref})}}{L_{ref}\,p_{ref}} N_{RE} = (\text{constant for specific machine})\frac{N_{RE}}{p_{ref}}. \quad (9.3)$$

Lines of constant $p_{ref}$ may thus be drawn on the map as shown in Fig. 9.2(*b*). The intersection of $p_{ref} = 15$ atm (14 gauge nominal + atmospheric) with $N = 1500$ rpm will be found to coincide with the operating point previously determined in terms of $N_{RE}$ and $N_{MA}$.

The earlier map of Fig. 9.2(*a*) may be used to infer performance, at original $N$ and $p_{ref}$, of a machine scaled linearly from the MP1002CA. Suppose, for example, that it is proposed to increase linear dimensions by a factor of 2, giving a machine of $2^3 = 8$ times the original swept volume ($= 8 \times 60 = 480$ cm$^3$). $N_{RE}$ increases by the factor $2^2$, $N_{MA}$ by 2. Entering the map at the new values predicts a reduction in $\zeta$ by a factor of about 4. (It is noteworthy that, in the operating range under consideration, the increase in $N_{MA}$ has little influence, the performance reduction being due almost entirely to the increased $N_{RE}$.) Within the range of size variation investigated by the author it has proved possible to restore the original $\zeta$ in a linearly-scaled machine by modifying heat exchanger and regenerator geometry. The process calls for a rerun of the thermodynamic design package and results in a new map, since the result of the changes is a new machine (i.e., geometric and thus dynamic similarity have been lost). However, it would be unwise to carry out the extrapolation too far on the basis of $N_{RE}$ and $N_{MA}$ only, since the process ignores the effect of linear scale on regenerator performance and on heat exchanger wall thickness and corresponding conduction losses.

## 9.4 Available work lost in heat exchanger transient conduction

The analysis presented relates specifically to cylindrical tubular heat exchangers with flow in the direction of the tube axis. In the usual case of multiple tubes in parallel it is assumed that flow in one tube is representative of that in all. All tubes have the same, uniform wall thickness, and the external surface of all tubes is at constant, uniform temperature – $T_e$ at the expansion end, $T_c$

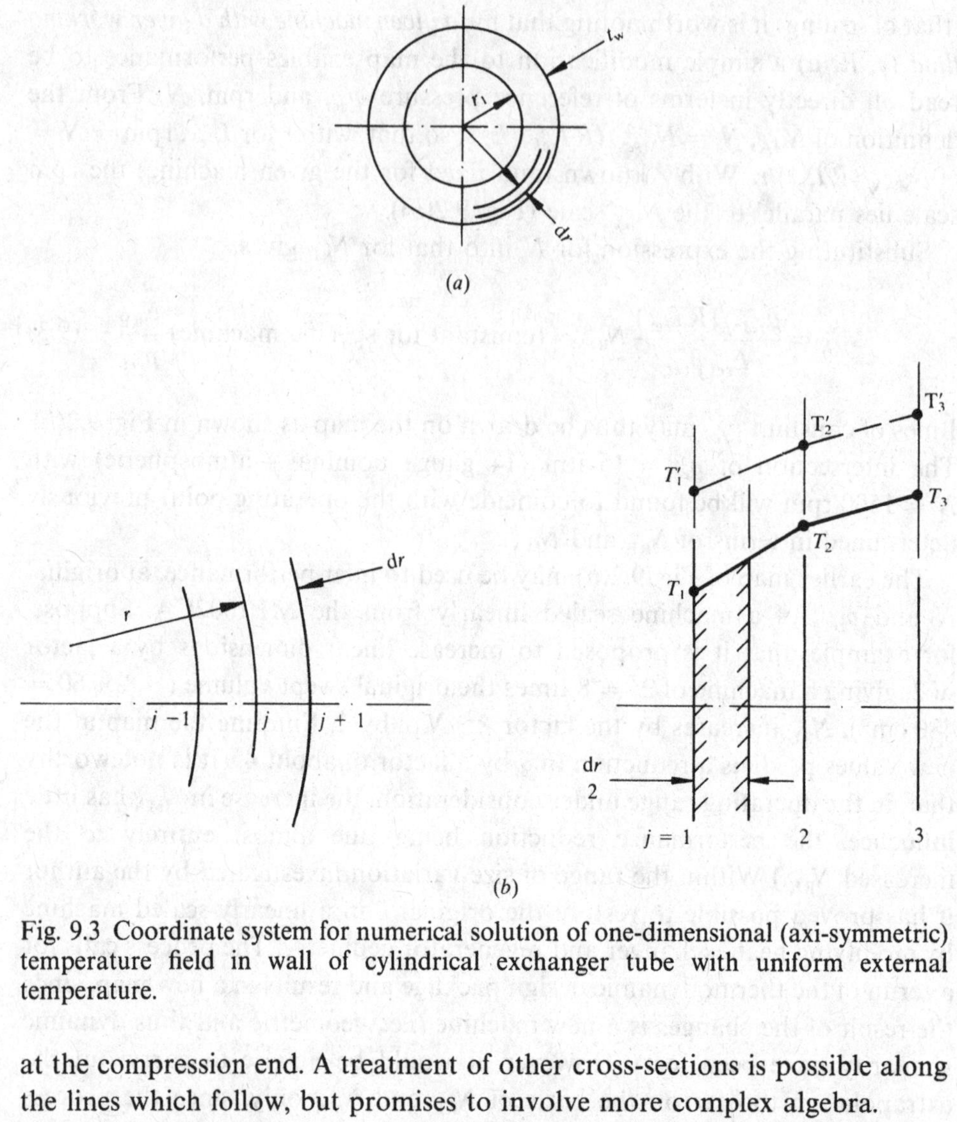


Fig. 9.3 Coordinate system for numerical solution of one-dimensional (axi-symmetric) temperature field in wall of cylindrical exchanger tube with uniform external temperature.

at the compression end. A treatment of other cross-sections is possible along the lines which follow, but promises to involve more complex algebra.

Fig. 9.3(*a*) represents a cross-section through a typical tube of wall thickness $t_w$. Internal to the tube wall heat flow is governed by the diffusion equation in polar coordinate form. This is precisely the equation (Eqn (8.4)) already used to describe heat flow in the regenerator wire. With radial dimensions internal to the wall normalized with respect to wall thickness, $t_w$, Eqn (8.4) becomes:

$$\frac{\partial^2 \tau}{\partial \lambda^2} + \frac{1}{\lambda}\frac{\partial \tau}{\partial \lambda} = \frac{2\pi}{N_F}\frac{t_w{}^2}{r^2}\frac{\partial \tau}{\partial \phi} \qquad (9.4)$$

$$N_F = \frac{\alpha}{N_s r^2} \text{ (with } \alpha_e \text{ to form } N_{Fe},\ \alpha_c \text{ to form } N_{Fc}\text{)}$$

At the outer boundary, dimensionless temperature has the constant value $N_\tau$ at the expansion end and unity at the compression end. At the innermost radius, the boundary condition parallels that for the outer surface of the regenerator wire, although the expression for the dimensionless heat transfer rate at the surface becomes so simplified with the loss of the temperature gradient term that it is now convenient to write the condition directly as:

$$\frac{\partial\tau}{\partial\lambda} = -2\pi N_{\mathrm{H}} \frac{r_{\mathrm{h}}}{r} \frac{t_{\mathrm{w}}}{r} \frac{\mathrm{d}\psi}{\mathrm{d}\phi} \tag{9.5}$$

Discretization proceeds as for the case of the regenerator wire (Fig. 9.3(*b*)). For internal points:

$$\begin{aligned}\tau_{j-1}'(1 - \mathrm{d}\lambda/2\lambda) + \tau_j'(-2 - 2\Lambda) + \tau_{j+1}'(1 + \mathrm{d}\lambda/2\lambda)\\ = \tau_{j-1}(-1 + \mathrm{d}\lambda/2\lambda) + \tau_j(2 - 2\Lambda) + \tau_{j+1}(-1 - \mathrm{d}\lambda/2\lambda)\end{aligned} \tag{9.6}$$

$$\Lambda = \frac{2\pi}{N_{\mathrm{F}}}\left(\frac{t_{\mathrm{w}}}{r}\right)^2 \frac{\mathrm{d}\lambda^2}{\mathrm{d}\phi}$$

Heat flow *to* the gas is now radially *inwards*, leading to an arrangement of subscripts which differs in sense from those of the regenerator case:

$$\tau_1'(1 + \tfrac{3}{4}\Lambda) + \tau_2'(-1 + \tfrac{1}{4}\Lambda) = 2\mathrm{N}\,\mathrm{d}\lambda\,\mathrm{d}\psi/\mathrm{d}\phi + \tau_1(-1 + \tfrac{3}{4}\Lambda) + \tau_2(1 + \tfrac{1}{4}\Lambda) \tag{9.7}$$

$$\mathrm{N} = 2\pi N_{\mathrm{H}} \frac{r_{\mathrm{h}}}{r} \frac{t_{\mathrm{w}}}{r}$$

At the outer circumference:

$$\tau_n = N_\tau \text{ (expansion end)} \tag{9.8a}$$

$$\tau_n = \text{unity (compression end)} \tag{9.8b}$$

When written out for a succession of $n$ stations, $j$, at radial locations $\mathrm{d}\lambda$ $(= d_r/t_{\mathrm{w}})$ apart, the coefficients of Eqns (9.6), (9.7) and (9.8) form a tridiagonal matrix. As in the case of the regenerator solution, this may be inverted numerically to yield $n$ values of $\tau_j'$ – i.e., the dimensionless temperatures after the crank angle increment. These become the 'known' temperatures which are substituted into the right-hand sides of Eqns (9.6), (9.7) and (9.8) to provide the starting point for the next integration step.

The integration process is initiated at $\phi = 0$ with uniform values of $\tau_j$ equal at the expansion end to $N_\tau$ and at the compression end to unity. Cyclic steady state is reached in a single revolution, and computation of transient conduction losses is withheld to a second, complete revolution.

The analysis has been applied to a machine identical with the Philips MP1002CA engine with the exception of the expansion and compression end exchangers. These are assumed modified to an arrangement in which the original 160 slots are replaced by $n_t$ cylindrical tubes. $n_t$ is chosen so as to maintain the original free-flow area and hydraulic radius. With $d_i$ for wall internal diameter $\frac{1}{4}\pi d_i^2 n_t = A_{x_e}$ for equivalence of free-flow area. Hydraulic radius of a cylindrical tube is $d_i/4$, so that for compatibility $d_i/4 = r_{h_e}$. Combining these two conditions:

$$n_t = A_{x_e}/4\pi r_{h_e}^2$$

Now, $A_{x_e} = \alpha_{x_e} r^2$, $r_{h_e} = \lambda_{h_e} r$ for the dynamic similarity case, so that

$$n_t = \alpha_{x_e}/4\pi\lambda_{h_e}^2$$

Using the values of Table 9.2 it is seen that 410 cylindrical tubes are required for equivalence with the 160 original rectangular slots.

Each tube has the same internal wetted length as the original slots. Conductivity and diffusivity are taken to be as for the stainless steel of the regenerator. Tube wall thickness is 0.5 of internal diameter. (In designing a comparable engine from scratch, one would almost certainly arrive at different proportions from these, and both material and proportions would be different between expansion and compression exchangers.)

Solution of the cyclic distribution of wall temperature along the length of expansion and compression space exchangers permits calculation of total entropy creation per cycle and of corresponding lost available work. The analysis parallels that for the case of the regenerator. Expressing differential quantities as finite differences:

$$d\zeta_{sk} = n_t \frac{d\lambda}{(V_{ref}/r)^3 N_H} \sum_{1}^{n-1} \lambda_j \frac{(\tau_{j+1} - \tau_j)^2}{(\tau_{j+1} + \tau_j)^2} d\phi \qquad (9.9)$$

$N_{H_e}$ or $N_{H_c}$ substitutes $N_H$ depending on which of the two exchangers, expansion or compression, is being analysed. Some care is necessary in the interpretation of Eqn (9.9), in which $d\lambda$ in the numerator is the length of a control volume measured in the axial direction of the exchanger tube bundle, while $\lambda_j$ is the (dimensionless) radial coordinate of temperature $\tau_j$ in the wall of an individual tube. The summation of Eqn (9.9) is carried out over a crankshaft revolution and over the lengths of expansion and compression exchangers. The latter process is facilitated by noting that it is necessary only to compute for a typical station at the expansion end and to multiply by the number of stations. The same is true of the compression end (but not of the regenerator!).

Fig. 9.4(*a*) shows the computed variation of the distribution of dimensionless temperature as a function of radius and crank angle. For improved visual effect the value of $N_H$ has been arbitrarily increased above the value of Table 9.2 by a factor of 10. The location is the control volume immediately adjacent to the expansion end, but since for the constant-temperature exchanger the ideal cycle surface heat flux rate is independent of location (Eqn (7.8)) the profile is typical of all stations between cylinder outlet and regenerator inlet. The plot confirms the isothermal condition at the outer radius (heat source).

The ideal reference cycle behind the simulation which produced Fig. 9.4(*a*) is the isothermal (rather than either of the alternative candidates – adiabatic or Finkelstein's 'generalized thermodynamic analysis'). The reason for the choice is to make a point: net heat transferred in the constant volume exchangers according to the isothermal analysis is *zero*. The horizontal straight line superimposed over the variation at the inner wall highlights the variation of surface temperature above and below the datum (nominal) value at that point, and draws attention to the following features:

(*a*) There is about as much area above the curve as below. (Heat transfer rate is not proportional to temperature, but the picture is qualitatively consistent with zero net heat transfer.)

(*b*) The surface temperature profile is asymmetrical. With qualification as in (*a*) above, this is in sympathy with the asymmetry of heat flux over a cycle (Ref. 2, Fig. 3).

(*c*) The instants (values of crank angle, $\phi$) at which the wall temperature gradient immediately below the surface is zero might be expected to correspond to the point at which surface temperature equals nominal 'isothermal' gas temperature. It is clear from the plots that they do not, indicating that computed heat transfers differ somewhat from those which would result from calculation in terms of heat transfer coefficients and temperature differences.

(*d*) Most significantly of all, *even though net heat transferred over the cycle is zero* there is, from Eqn. (9.9), a cyclic loss of available work. The calculated loss may be expected to increase when the effects are included of heat conduction rate computed as the mean for the cycle.

It is seen from Eqn (9.9) that losses due to transient conduction do not compound the steady component linearly. It is nevertheless interesting to isolate the magnitude of exchanger conduction loss for the hypothetical steady heat flow case. Considering the expansion end exchanger, the mean heat transfer rate at $N_s$ cycles/s is $Q_E N_s$. The heat transfer rate per unit area is:

$$Q_E N_s / n_t \pi d l = \partial T_w / \partial r$$

Substituting into the expression for entropy creation rate per unit volume

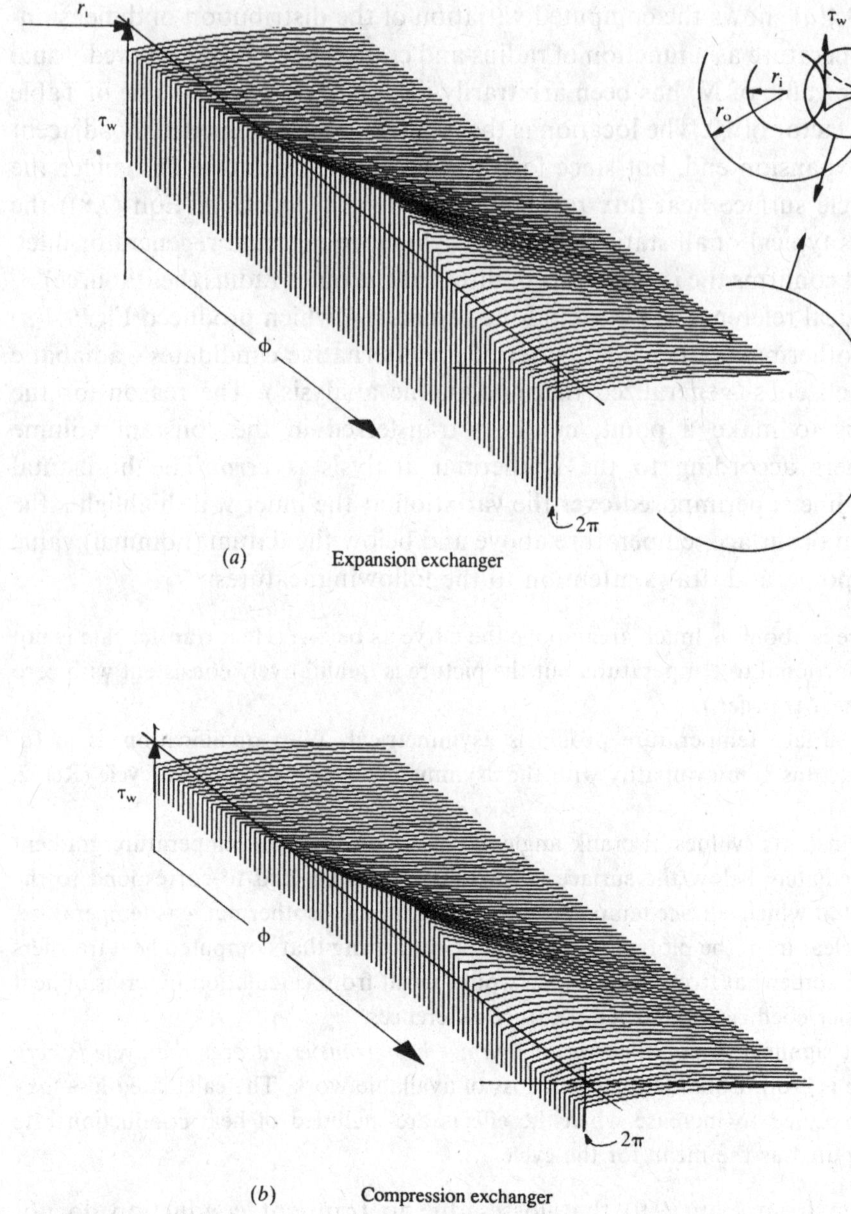


Fig. 9.4 Computed variation of dimensionless wall temperature as a function of radius and crank angle for a Philips MP1002CA engine modified to tubular heat exchangers as explained in the text. (*a*) Expansion exchanger. (*b*) Compression exchanger.

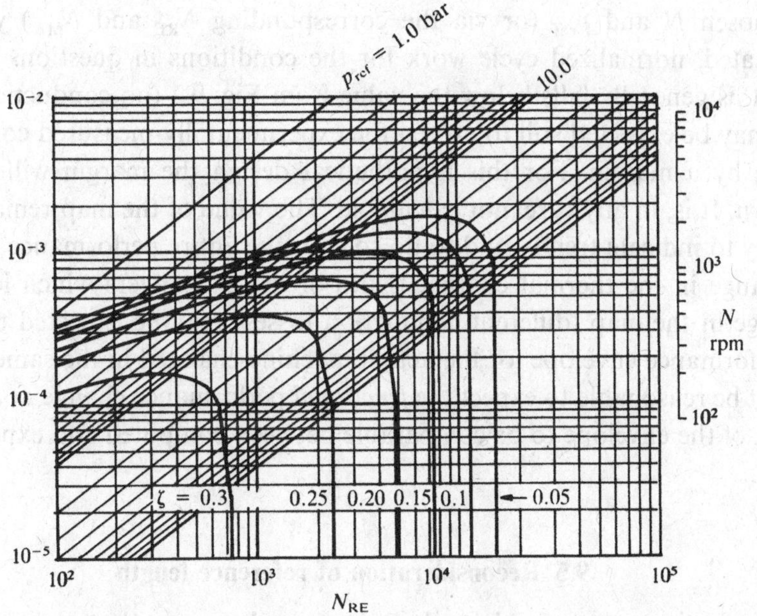


Fig. 9.5 Performance map for a Philips MP1002CA machine modified as described in text. Note the difference in performance between the modified machine (in which account is taken of transient conduction losses) and that predicted by the maps of Fig. 9.2, where conduction losses were set to zero.

(Section 8.6), while assuming for algebraic convenience that wall surface area is characterized by a single diameter, $d$, (rather than by both inner, $d_i$ and outer, $d_o$) gives

$$S' = \frac{k}{T^2 r}(Q_E N_s)^2 \frac{t_w}{\pi d l k_w n_t}$$

Subjecting the foregoing equation to the now familiar process of normalization:

$$\zeta_{Skcycle} = \frac{Q_E{}^2}{(MRT_o)^2} \frac{1}{N_t{}^2} \frac{N_H}{dlk_w n_t} \frac{t_w/r}{V_{de}/V_{wfc}} \frac{r_{he}}{R_H} \tag{9.10}$$

The value of $Q_E{}^2/(MRT_0)^2$ corresponding to conditions which produced the temperature profiles was 0.65, giving a value of $\zeta_{Skcycle}$ from Eqn (9.10) of 0.0374. The corresponding cyclic loss due to the transient effect was 0.52E-03.

The conduction loss calculation may be integrated into the process of performance map generation. The map of Fig. 9.5 is computed for the conditions applying to that of Fig. 9.2, but taking account of work lost in transient conduction in regenerator and heat exchangers. Entering the map at chosen $N$ and $p_{ref}$ (or via the corresponding $N_{RE}$ and $N_{MA}$) yields the estimated, normalized cycle work for the conditions in question. It is seen that this generally falls below the value from Fig. 9.2 (no conduction losses) but may be expected still to exceed the experimentally-measured counterpart value by a margin. For this hypothetical design the margin will never be known. It is, in any case, not calculable. The value of the map remains in its ability to indicate trends, and to do so over the entire performance envelope: a change in the thermal characteristics of the exchangers which leads to a change in the map (different conduction losses) may be expected to lead to a performance envelope for the actual machine changed in the same sense. It might be reasonable to expect the *fractional* performance change at any given point of the envelope to be comparable between computed and experimental cases.

## 9.5 Reconsideration of reference length

It is sometimes the case that what appears obvious and straightforward in concept is less so in practice. Thus, while the algebra for the thermodynamic similarity case is readily coped with in parallel with the dynamic similarity counterpart, the requirement for different sets of numerical data for the two cases has been found to be confusing in use and conducive to error. A further matter which has been found problematical is the requirement for two, similar, but not identical, suites of computer code, some elements of which are interchangeable, others of which are not.

The difficulty may be avoided by noting that the conditions for dynamic similarity may be achieved through the use of *any* consistent set of normalizing quantities – including that used in developing the thermodynamic similarity, viz, $R_H$ for length, $A_{wetted}$ for areas and $V_{wfc}$ for volumes.

Conversion to any other system of reference variables, $L_{ref}$, $A_{ref}$, $V_{ref}$ may be applied retrospectively to computed results rather than to defining equations and to input data: the process is simply that of multiplying the length result of interest by $R_H/L_{ref}$, the area result by $A_{wetted}/A_{ref}$ etc.

Geometric similarity is still a precondition of dynamic similarity. Moreover, it is necessary to use heat transfer and friction factor correlations specific to the exchanger geometry under consideration, otherwise the scaling process developed in Chap. 7 is not valid. The fact that the reference length, $R_H$, is not a quantity determined by measurement with a micrometer is no obstacle: one computes in terms of dimensionless lengths, etc., and on settling on the final thermodynamic configuration, calculates from $\zeta$ (net dimensionless cycle work) the $M$ (or $V_{ref}$) to give required (absolute) performance. A numerical value of $R_H$ follows via $N_{MA}$ when $N$ is specified, and $A_{wetted} = V_{wfc}/R_H$. Absolute values of the rest of the gas circuit data and operating conditions follow.

The variable $r$ (crank-pin eccentricity) is not redundant: it is the logical reference length for normalizing the algebra of the crank kinematics. Except for cycle analysis in terms of the Method of Characteristics (Chap. 11 and Appendix XI) the rôle of the kinematic calculation is to generate volume variations. For greatest flexibility in program design, volume variations are best computed separately from the main process of thermodynamic simulation, and in terms of lengths normalized by $r$. The resulting values of volume variation are readily interfaced with the gas circuit code via a single conversion factor $r^3/V_{wfc}$.

# 10

# Lumped-parameter analysis

## 10.1 The lumped-parameter interpretation

There are no formal definitions distinguishing between nodal, finite-cell and lumped-parameter formulations. However, the latter term is appropriate to those treatments in which substantial sections of the gas circuit are dealt with (*a*) as if fluid and/or wall properties were uniform, and/or (*b*) by spatial integration where integration would ideally be with respect to both time and distance, and/or (*c*) the inverse of (*b*). In Chap. 13 a method will be described whereby different numerical formulations may be compared with reasonable objectivity. In the meantime, there is no quantitative evidence that a suitable lumped-parameter analysis gives a less accurate prediction of overall performance than do finite-cell schemes with their substantially greater demands on computing time. The *genre* is thus worth considering in detail.

The pioneer was Finkelstein, and his first lumped-parameter[1] model provides a suitable introductory example.

## 10.2 Finkelstein's generalized thermodynamic analysis

The treatment is essentially the adiabatic analysis (Appendix VIII) modified to take account of finite heat transfer in the variable-volume spaces (Fig. 10.1). In the regenerator and heat exchangers, the working fluid temperature is that of the immediately adjacent solid wall. Momentum effects are not considered, so at any instant, pressure throughout the gas circuit is uniform. The working fluid behaves as an ideal gas. In the original formulation volume variations were simple harmonic with constant phase angle, but the algebra is readily modified to cope with the volume variations produced by realistic crank mechanisms.

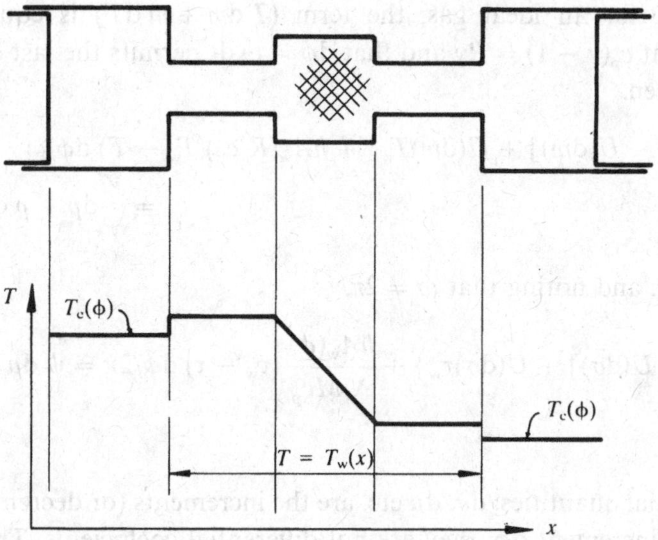


Fig. 10.1 Schematic indicating the processes considered during analysis in terms of lumped parameters.

### 10.2.1 The energy equation for the variable-volume spaces

The energy equation in 'engineering thermodynamics' form (Chap. 2) for a variable-volume space is:

$$\mathrm{d}Q + \mathrm{d}H - p\,\mathrm{d}V = \mathrm{d}U \tag{10.1}$$

With heat $\mathrm{d}Q$ exchanged by convection during time interval $\mathrm{d}t$ given by $\mathrm{d}Q = hA_\mathrm{w}(T_\mathrm{w} - T)\,\mathrm{d}t$, Eqn (10.1) becomes

$$hA_\mathrm{w}(T_\mathrm{w} - T)\,\mathrm{d}t + \mathrm{d}H - p\,\mathrm{d}V = c_\mathrm{v}(T\,\mathrm{d}m + m\,\mathrm{d}T)$$

$A_\mathrm{w}$ is the exposed or wetted area, and is thus a function of piston position, viz, $A_\mathrm{w}(\phi)$. During inflow, $\mathrm{d}H$ is equal to $\mathrm{d}mc_pT_\mathrm{w}$, while during outflow it is $\mathrm{d}mc_pT$. Otherwise, the above form of the energy equation applies to both cases. Finkelstein combines inflow and outflow cases by defining a function of $\mathrm{d}m$ – a step function, $U(\mathrm{d}m)$, such that

$$U(\mathrm{d}m) = \begin{cases} 1 & \text{for } \mathrm{d}m \geqslant 0 \quad \text{(inflow)} \\ 0 & \text{otherwise} \quad \text{(outflow)} \end{cases}$$

Inflow and outflow cases may thus be combined, giving:

$$\mathrm{d}mc_p\{T[1 - U(\mathrm{d}m)] + U(\mathrm{d}m)T_\mathrm{w}\} + hA_\mathrm{w}(T_\mathrm{w} - T)\,\mathrm{d}t - p\,\mathrm{d}V = c_\mathrm{v}(T\,\mathrm{d}m + m\,\mathrm{d}T)$$

Noting that, for an ideal gas, the term $(T\,\mathrm{d}m + m\,\mathrm{d}T)$ is equivalent to $\mathrm{d}(pV)/R$, that $c_p(\gamma - 1) = R\gamma$ and that $\mathrm{d}\phi = \omega\,\mathrm{d}t$ permits the last expression to be rewritten:

$$R\,\mathrm{d}m\{T[1 - U(\mathrm{d}m)] + U(\mathrm{d}m)T_w\} + hA_w(R/c_p)(T_w - T)\,\mathrm{d}\phi/\omega = \frac{V}{\gamma}\,\mathrm{d}p + p\,\mathrm{d}V \quad (10.2)$$

Normalizing, and noting that $\omega = 2\pi N_s$:

$$\mathrm{d}\sigma\{\tau[1 - U(\mathrm{d}\sigma)] + U(\mathrm{d}\sigma)\tau_w\} + \frac{hA_w(\phi)}{N_s M c_p}(\tau_w - \tau)\,\mathrm{d}\phi/2\pi = \psi\,\mathrm{d}\mu + \frac{\mu}{\gamma}\,\mathrm{d}\phi \quad (10.3)$$

The differential quantities, $\mathrm{d}\sigma$, $\mathrm{d}\mu$ etc. are the increments (or decrements) over crank angle increment $\mathrm{d}\phi$; they are not differential coefficients. There is one equation of the form of Eqn (10.3) for each of the two working spaces. As usual, $\psi = pV_{ref}/MRT_{ref}$, $\mu = v/V_{ref}$.

Before proceeding it is worth examining the dimensionless heat transfer term, $hA_w(\phi)/N_s M c_p$. This may be written $(hA_{wo}/N_s M c_p)a_w(\phi)$, in which $a_w(\phi)$ is the dimensionless wetted area, and attention is focussed on the term in parentheses. This has the form of a Stanton number, since $N_{st} = h/c_p G$, and might be written $N_{ST} a_w(\phi)$. Finkelstein treated the term as a parameter to which he gave a selection of constant values during solution – 0 (corresponding to the adiabatic case) *via* 0.1, 0.5, 1.0 to infinity (for the 'isothermal' case). The use of arbitrary constants gives the impression that there is no means of evaluating the term. On the contrary, much is known about heat transfer in the cylinders of reciprocating machines, the authoritative account being that of Annand.[2]

### 10.2.2 Evaluation of heat transfer coefficient

Annand identifies two modes of heat transfer – convection and radiation. It is probable that the convective mode dominates in the Stirling cycle machine, in which case the correlation proposed by Annand is

$$N_{nu} = hD/k = aN_{re}b \quad (10.4)$$

Fig. 10.2 is taken from Annand's paper and indicates the large number of reciprocating engines and individual experimental results justifying the correlation. The slope of the curves gives $b = 0.7$, while the value of $a$ is found to vary between $0.35 \leqslant a \leqslant 0.8$ depending on the intensity of gas motion.

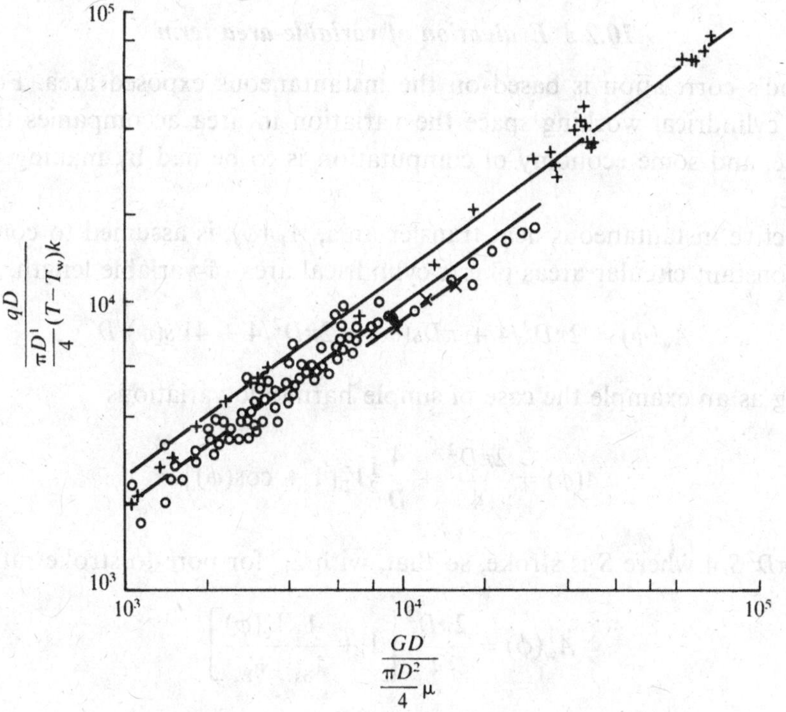


Fig. 10.2 Data collected and reinterpreted by Annand[2] illustrating the correlation of Eqn (10.4). The label on the horizontal scale is $N_{re}$; $G$ = charge mass flow rate = $\pi D^2 g/4$. (Courtesy Institution of Mechanical Engineers.).

In Eqn (10.4) $D$ is cylinder diameter. Annand's definition of $N_{re}$ is based on $D$ and instantaneous piston speed, $u_p$, $\approx 2\pi N_s r \sin(\phi)$, where $r$ is the crank-pin offset. The correlation is embodied in the present treatment by noting first of all that the Reynolds number of the correlation is $N_{re} = \rho u_p D/\mu$. With $p/RT$ for $\rho$ and with $u_p$ written in full, $N_{re}$ converts to:

$$N_{re} = 2\pi \frac{\psi}{\tau} N_{RE} \left[ \frac{r}{L_{ref}} \frac{D}{L_{ref}} |\sin(\phi)| \right] \tag{10.5}$$

In Eqn (10.5) $N_{RE}$ ($= N_{MA}{}^2 N_{SG}$) is the now familiar characteristic Reynolds number. In those formulations using crank radius, $r$, as the reference length the term in square brackets reduces to $[2\lambda_{bs}|\sin(\phi)|]$ where $\lambda_{bs}$ is the bore-to-stroke ratio of the working space under consideration. $N_{re}$ may now be evaluated in terms of the (dimensionless) variables of the solution as computation proceeds. Corresponding $N_{nu}$ and hence $h$ follow via Annand's correlation.

### 10.2.3 Evaluation of variable-area term

Annand's correlation is based on the instantaneous exposed area. For the usual, cylindrical working space the variation in area accompanies that of volume, and some economy of computation is to be had by making use of the fact.

Effective instantaneous heat transfer area, $A_W(\phi)$, is assumed to comprise two, constant circular areas plus a cylindrical area of variable length, $s(\phi)$:

$$A_w(\phi) = 2\pi D^2/4 + \pi D s(\phi) = 2\pi D^2/4 + 4Vs(\phi)/D$$

Taking as an example the case of simple harmonic variations

$$A(\phi) = \frac{2\pi D^2}{4} + \frac{4}{D}\tfrac{1}{2}V_E[1 + \cos(\phi)]$$

$V_E = \pi D^2 S/4$ where $S$ is stroke, so that, with $\lambda_{bs}$ for bore-to-stroke ratio

$$A_w(\phi) = \frac{2\pi D^2}{4}\left[1 + \frac{1}{\lambda_{bs}}\frac{V_e(\phi)}{V_E}\right]$$

$$= A_{wo}a(\phi) \tag{10.6}$$

$$A_{wo} = 2\pi D^2/4$$

$$a(\phi) = 1 + \frac{1}{\lambda_{bs}}\frac{V_e(\phi)}{V_E}$$

An efficient simulation program will precompute an array of values of dimensionless volumes from the trigonometric function $f_{ve}(\phi)$ rather than evaluate at each call. These values may be used directly in the variable part of the area term. Such economies become essential in the context of optimization, which requires the thermodynamic cycle to be recomputed some hundreds or thousands of times (see Chap. 14).

### 10.2.4 Discretization

As for the adiabatic analysis (Appendix VIII) total gas circuit mass content, $M$, is the sum of $m_e$ and $m_c$ with the mass instantaneously occupying the dead space. Normalizing

$$\psi = \frac{1 - \sigma_e - \sigma_c}{\nu} \tag{10.7}$$

Differentiating

$$d\psi = \frac{-d\sigma_e - d\sigma_c}{\nu} \tag{10.8}$$

Dimensionless temperatures, $\tau$, may be expressed in terms of pressure, $\psi$, and volumes, $\mu$:

$$\tau = \psi\mu/\sigma \tag{10.9}$$

Eqn (10.3) may thus be reduced to a difference expression in terms of the variable $\sigma$ only. Two such difference expressions are formed, one each for expansion and compression spaces. Simultaneous solution leads to values of $\sigma_e(\phi + d\phi)$ and $\sigma_c(\phi + d\phi)$ for all $\phi$. Back substitution then gives $\psi(\phi + d\phi)$, $\tau_e(\phi + d\phi)$ and $\tau_c(\phi + d\phi)$.

The discretization scheme is as for the adiabatic analysis, being an expansion about current (known) values of the dependent variable, $\phi$, to corresponding mid-interval ($\frac{1}{2}\,d\phi$) values, i.e., $\psi_\phi + \frac{1}{2}\,d\psi$, $\sigma_{e_\phi} + \frac{1}{2}\,d\sigma_e$ and $\sigma_{c_\phi} + \frac{1}{2}\,d\sigma_c$. The treatment differs slightly from that of Appendix VIII in that $\psi_\phi$ is retained in the differential relationships (i.e., $\sigma_{e_\phi}$ and $\sigma_{c_\phi}$ are not substituted). With **H** as an abbreviation (it is *not* a true parameter) for $N_{ST}\,d\phi/2\pi$ the difference relationship for either space becomes:

$$\psi_\phi\left(\frac{\mu}{\sigma_\phi + \frac{1}{2}\,d\sigma}\{d\sigma[1 - U(d\sigma)] - \mathbf{H}a(\phi)\} - d\mu\right)$$
$$+ d\psi\left(\frac{\frac{1}{2}\mu}{\sigma_\phi + \frac{1}{2}\,d\sigma}\{d\sigma[1 - U(d\sigma)] - \mathbf{H}a(\phi)\} - \tfrac{1}{2}\,d\mu - \mu/\gamma\right)$$
$$+ d\sigma U(d\sigma)\tau_w + \mathbf{H}a(\phi)\tau_w = 0 \tag{10.10}$$

Substituting for $d\psi$ and noting that $1/(\sigma_\phi + \frac{1}{2}\,d\sigma) \approx (1 - \frac{1}{2}\,d\sigma/\sigma_\phi)/\sigma_\phi$ gives

$$\psi_\phi\left(\frac{\mu}{\sigma_\phi}\{d\sigma[1 - U(d\sigma)] - \mathbf{H}a(\phi)\}(1 - \tfrac{1}{2}\,d\sigma/\sigma_\phi) - d\mu\right)$$
$$- \frac{(d\sigma_e + d\sigma_c)}{\nu}\left(\frac{\frac{1}{2}\mu}{\sigma_\phi}\{d\sigma[1 - U(d\sigma)] - \mathbf{H}a(\phi)\}(1 - \tfrac{1}{2}\,d\sigma/\sigma_\phi) - \tfrac{1}{2}\,d\mu - \mu/\gamma\right)$$
$$+ d\sigma U(d\sigma)\tau_w + \mathbf{H}a(\phi)\tau_w = 0 \tag{10.11}$$

In Eqns (10.10) and (10.11) unsubscripted dependent variables may take the subscript e, in which case the expression describes the expansion space, or c to relate to the compression space. Dealing with the expansion space by applying subscripts as appropriate, expanding and grouping separately the multipliers of $\sigma_e$ and $\sigma_c$:

$$\mathrm{d}\sigma_e\left[\frac{\psi_\phi \nu \mu_e}{\sigma_{e\phi}}\{[1 - U(\mathrm{d}\sigma_e)] + \mathbf{H}_e a_e(\phi)/2\sigma_{e\phi}\} + \frac{\mu_e}{2\sigma_{e\phi}}\mathbf{H}_e a_e(\phi) + \frac{\mathrm{d}\mu_e}{2} + \frac{\mu_e}{\gamma} + U(\mathrm{d}\sigma_e)\nu\tau_{w_e}\right] + \mathrm{d}\sigma_c\left[\frac{\mathrm{d}\mu_e}{2} + \frac{\mu_e}{\gamma} + \frac{\mu_e}{2\sigma_{e\phi}}\mathbf{H}_e a_e(\phi)\right]$$

$$= -\mathbf{H}_e a_e(\phi)\nu\tau_{w_e} + \psi_\phi \nu[\mathrm{d}\mu_e + \mu_e \mathbf{H}_e a_e(\phi)/\sigma_{e\phi}]$$

$$+ \text{terms of higher order in the differences} \quad (10.12)$$

For the compression space:

$$+ \mathrm{d}\sigma_e\left[\frac{\mathrm{d}\mu_c}{2} + \frac{\mu_c}{\gamma} + \frac{\mu_c}{2\sigma_{c\phi}}\mathbf{H}_c a_c(\phi)\right]$$

$$+ \mathrm{d}\sigma_c\left[\frac{\psi_\phi \nu \mu_c}{\sigma_{c\phi}}\{[1 - U(\mathrm{d}\sigma_c)] + \mathbf{H}_c a_c(\phi)/2\sigma_{c\phi}\} + \frac{\mu_c}{2\sigma_{c\phi}}\mathbf{H}_c a_c(\phi) + \frac{\mathrm{d}\mu_c}{2} + \frac{\mu_c}{\gamma} + U(\mathrm{d}\sigma_c)\nu\tau_{w_c}\right]$$

$$= -\mathbf{H}_c a_c(\phi)\nu\tau_{w_c} + \psi_\phi \nu[\mathrm{d}\mu_c + \mu \mathbf{H}_c a_c(\phi)/\sigma_{c\phi}]$$

$$+ \text{terms of higher order in the differences} \quad (10.13)$$

Eqns (10.12) and (10.13) are of the form:

$$a_{11}\,\mathrm{d}\sigma_e + a_{12}\,\mathrm{d}\sigma_c = b_1 \quad (10.14a)$$

$$a_{21}\,\mathrm{d}\sigma_e + a_{22}\,\mathrm{d}\sigma_c = b_2 \quad (10.14b)$$

The solutions are clearly:

$$\mathrm{d}\sigma_e = (b_1 a_{22} - b_2 a_{12})/\Delta \quad (10.15a)$$

$$\mathrm{d}\sigma_c = (b_2 a_{11} - b_1 a_{21})/\Delta \quad (10.15b)$$

$$\Delta = a_{11}a_{22} - a_{21}a_{12}$$

Finally, the new values of $\sigma_e$ and $\sigma_c$ after the integration step are:

$$\sigma_{e\phi+\mathrm{d}\phi} = \sigma_{e\phi} + \mathrm{d}\sigma_e \quad (10.16a)$$

$$\sigma_{c\phi+\mathrm{d}\phi} = \sigma_{c\phi} + \mathrm{d}\sigma_c \quad (10.16b)$$

The boundary conditions that $\sigma_{e2(n+1)\pi} = \sigma_{e(2n\pi)}$, $\sigma_{c2(n+1)\pi} = \sigma_{c(2n\pi)}$ are achieved by cycling the computation through a sufficient number of revolutions to give convergence.

It is desirable to have as many means as possible of checking for correct functioning of the computer code which implements the solution. One such check is to set high values of $N_{ST_e}$ and $N_{ST_c}$ (e.g., via arbitrarily large $\mathbf{H}_e$ and $\mathbf{H}_c$) and compare computed $\sigma$, $\tau$ and $\psi$ with values predicted by the 'isothermal' analysis fed with the same data ($N_\tau$, $\kappa$, $\alpha$, $\nu$). For the check to be convincing it has been found necessary to include in the right-hand sides of Eqns (10.12) and (10.13) the higher order terms identified earlier. These are (to be included in $b_1$)

$$\frac{\psi_\phi \nu \mu_e (d\sigma_e)^2}{2\sigma_{e_\phi}{}^2}[1 - U(d\sigma_e)]$$

$$+ (d\sigma_e + d\sigma_c)\left[\frac{\frac{1}{2}\mu_e}{\sigma_{e_\phi}}\{d\sigma_e[1 - U(d\sigma_e)](1 - \tfrac{1}{2}\, d\sigma_e/\sigma_{e_\phi}) + \tfrac{1}{2}\mathbf{H}_e a_e(\phi)\, d\sigma_e/\sigma_{e_\phi}\}\right]$$

(to be included in $b_2$)

$$\frac{\psi_\phi \nu \mu_c (d\sigma_c)^2}{2\sigma_{c_\phi}{}^2}[1 - U(d\sigma_c)]$$

$$+ (d\sigma_e + d\sigma_c)\left[\frac{\frac{1}{2}\mu_c}{\sigma_{c_\phi}}\{d\sigma_c[1 - U(d\sigma_c)](1 - \tfrac{1}{2}\, d\sigma_c/\sigma_{c_\phi}) + \tfrac{1}{2}\mathbf{H}_c a_c(\phi)\, d\sigma_c/\sigma_{c_\phi}\}\right]$$

At the first pass through an integration step the $d\sigma_e$, $d\sigma_c$ used to compute the higher-order terms are those from the previous step. By carrying out each integration step at least twice, current values of $d\sigma_e$ and $d\sigma_c$ may be incorporated. In this way it has been found possible to achieve agreement with the results of the 'isothermal' analysis to within some 0.01%.

### *10.2.5 Some specimen results*

Fig. 10.3 shows computed variation with crank angle of dimensionless pressure, $\psi$, for a heat pump described by $N_\tau = T_e/T_c = 0.5$, $\kappa = 1.0$, $\alpha = 90°$, $\nu = 0.5$ and $\gamma = 1.4$ for comparison with the results of Finkelstein's[1] original account. (Finkelstein used $\nu = 1.0$; his definitions of $\nu$ and $\psi$ are respectively twice and half of those used here). Working space volume variations are simple harmonic (as per Finkelstein). The parameter is $N_{ST} = N_{STe} = N_{STc}$, and is constant over a cycle. Fig. 10.4 shows variations in dimensionless working space temperatures, $\tau_e$ and $\tau_c$.

While the description of the working space processes is sophisticated, even by comparison with that of the most recent one-dimensional treatments, the neglect of flow losses and of limitations on heat transfer in the regenerator

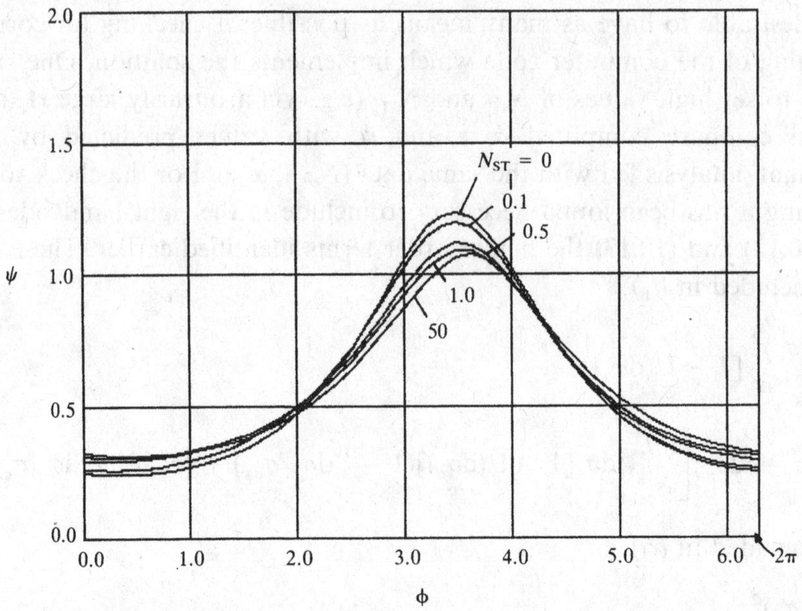


Fig. 10.3 Variation of dimensionless pressure, $\psi$, with crank angle, $\phi$. The parameter is $N_{ST} = N_{STe} = N_{STc}$. The model is essentially Finkelstein's generalized thermodynamic analysis.[1] (NB: With the definition of dimensionless pressure adopted in this text, $\psi$ has twice the numerical value of Finkelstein's $\psi$.).

and exchangers means that this simple lumped-parameter model returns a grossly overoptimistic estimate of cycle work over a wide range of the operating envelope. Finkelstein's second cycle description is inherently capable of a more accurate representation of the gas processes.

## 10.3 Finkelstein's improved cycle description

Fig. 10.5(*a*) shows seven points in the gas circuit at which properties are evaluated. The treatment[3] provides for the walls of the cylinder and the adjacent heat exchanger to be at different, constant temperatures. However, the principles involved may be illustrated without compromise on the assumption that the expansion cylinder and exchanger are at the common temperature, $T_{we}$, the compression cylinder and associated exchanger both being at $T_{wc}$.

### *10.3.1 Defining equations*

The working space description is identical to that for the earlier analysis, except that the temperature of the gas during inflow is no longer $T_w$. Instead,

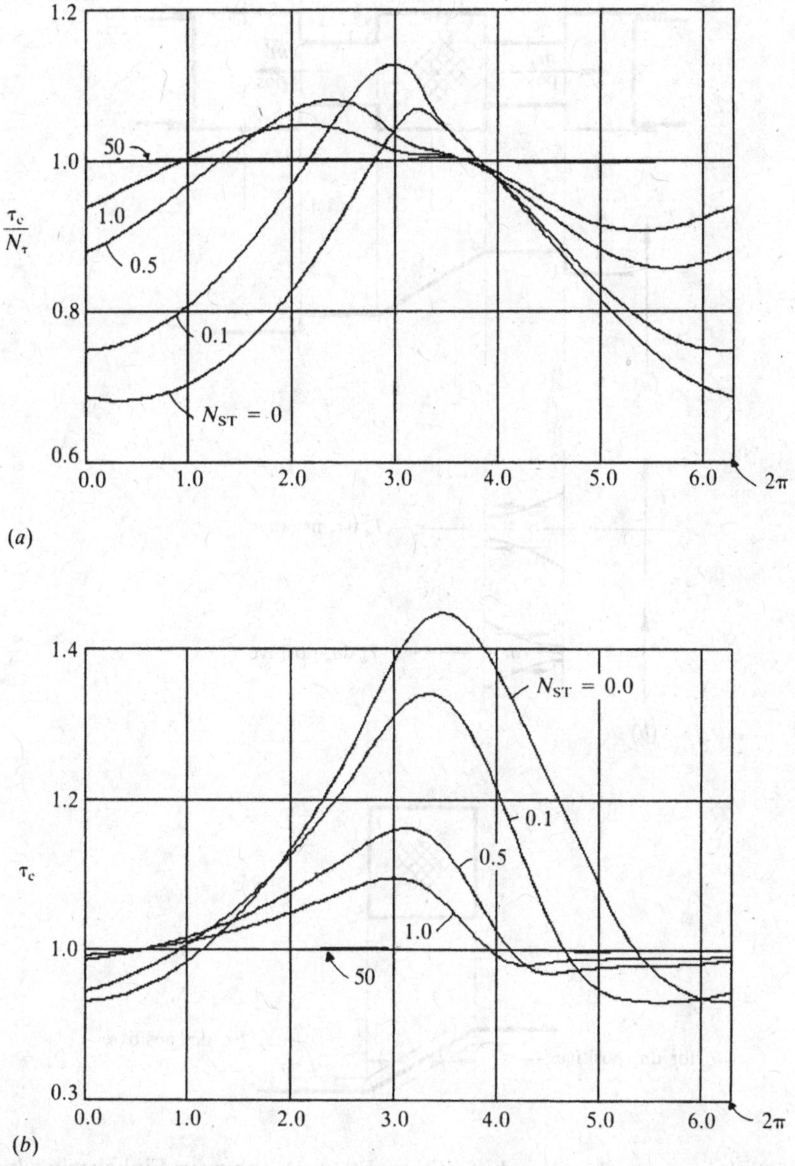


Fig. 10.4 Variation of dimensionless working space temperature according to Finkelstein's generalized thermodynamic analysis.[1] The parameter is $N_{ST} = N_{STe} = N_{STc}$. (*a*) Expansion space (machine acting as heat pump). $\tau_e$ divided by $N_\tau$ so that base-line value becomes unity. (*b*) Compression space (machine acting as heat pump).

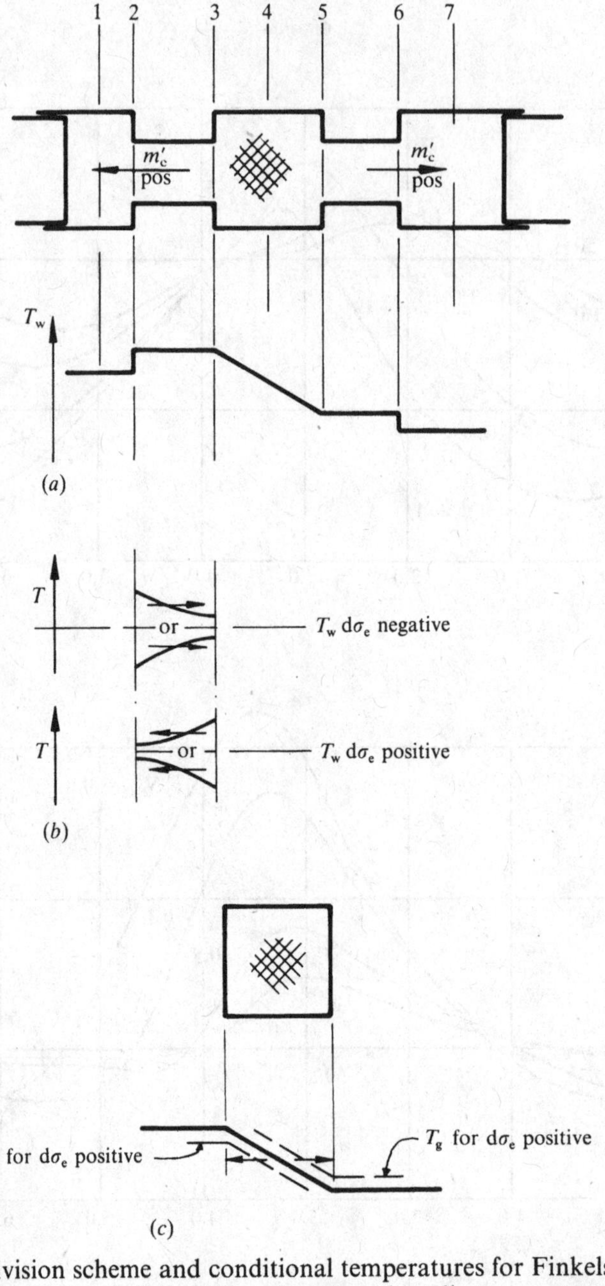


Fig. 10.5 Subdivision scheme and conditional temperatures for Finkelstein's lumped parameter analysis. (*a*) Subdivision scheme and wall temperature distribution. (*b*) Gas temperature distribution in exchangers for the two possible flow directions. (*c*) Regenerator gas outlet temperature for the two possible flow directions.

a temperature $T_2$ is used at the expansion end ($T_6$ at the compression end) where $T_2$ is the variable temperature achieved at point 2 by gas flowing from regenerator outlet towards the working space.

Eqn (10.3) is rewritten for the expansion space with this modification

$$\mathrm{d}\sigma_e\{\tau_1[1 - U(\mathrm{d}\sigma_e)] + U(\mathrm{d}\sigma_e)\tau_2\} + \mathbf{H}_e a_e(\phi)(\tau_{we} - \tau_1) = \psi_1\,\mathrm{d}\mu_e + \mu_e\,\mathrm{d}\psi_1/\gamma \tag{10.17}$$

Correspondingly, for the compression space

$$\mathrm{d}\sigma_c\{\tau_7[1 - U(\mathrm{d}\sigma_c)] + U(\mathrm{d}\sigma_c)\tau_6\} + \mathbf{H}_c a_c(\phi)(\tau_{wc} - \tau_7) = \psi_7\,\mathrm{d}\mu_c + \mu_c\,\mathrm{d}\psi_7/\gamma \tag{10.18}$$

The process of algebraic reduction employed in the generalized thermodynamic analysis is not possible here, so solution will be in terms of 11 variables – $\psi_1$, $\psi_4$, $\psi_7$, $\sigma_e$, $\sigma_c$, $\tau_1$, $\tau_2$, $\tau_3$, $\tau_5$, $\tau_6$ and $\tau_7$. This calls for 11 equations, and a further two of these are provided by a statement of mass in the three principal spaces – two of variable volume and one combined dead space:

$$\sigma_e = \psi_1\mu_e/\tau_e \tag{10.19}$$

$$\sigma_c = \psi_7\mu_c/\tau_c \tag{10.20}$$

The expression for dead-space mass content is an approximation based on mid-plane pressure, $\psi_4$ – i.e., without account of temperature variation. It is thus identical to the comparable term used in the isothermal and adiabatic analyses:

$$\psi_4 = \frac{1 - \sigma_e - \sigma_c}{\nu}$$

In the analysis the expression takes the differential form:

$$\mathrm{d}\psi_4 = \frac{-\mathrm{d}\sigma_e - \mathrm{d}\sigma_c}{\nu} \tag{10.21}$$

Flow losses are dealt with in terms of two pressure differences, one between $p_1$ and $p_4$, the other between $p_4$ and $p_7$. This calls for the rather uncomfortable lumping together of the flow characteristics of a complete heat exchanger and half of a regenerator. Had the formulation been in terms of four pressures, one for each cylinder and one each at the interfaces of the expansion exchanger and regenerator, and of the compression exchanger and regenerator, the flow loss model would have been much more plausible. Provisionally following Finkelstein's original approach, the assumption is made that equivalent values of friction factor, $C_f$, exchanger length, $L$, wetted perimeter, $P$ and cross-sectional area, $A_x$, may be specified, thus permitting, for the expansion end

$$\Delta p = p_4 - p_1 = -\tfrac{1}{2}\rho u^2 C_f LP(u/|u|)/A_x$$

The term $u/|u|$ ensures that the sign of $\Delta p$ always opposes that of $u$. Now, $\rho u^2 = (\rho u)^2/\rho = m'^2/\rho A_x = m'^2 RT/pA_x \approx m'^2 RT/\tfrac{1}{2}[(p_1 + p_4)A_x]$. Substituting, noting that $A_x/P = r_h$ and using for $m'$ the value at expansion space inlet/outlet (point 1) leads to:

$$\tfrac{1}{2}(p_4 - p_1)(p_4 + p_1) = -\tfrac{1}{2}\frac{m_e'^2 RTC_f L}{A_x^2 r_h}\frac{-m_e'}{|-m_e'|}$$

The negative sign with $m_e'$ takes account of the fact that $m_e'$ is *positive* for flow *into* the cylinder (i.e., for the 'wrong' direction – right to left in Fig. 10.5). Normalizing, and taking into account that $m'$ to this point has been mass rate with *time* (whereas the independent variable of the normalized equations is crank angle, $\phi$) gives:

$$(\psi_4 - \psi_1)(\psi_4 + \psi_1) = -\mathrm{d}\sigma_e^2 \tau_{2\text{-}4}\mathbf{F}_e \frac{-\mathrm{d}\sigma_e}{|-\mathrm{d}\sigma_e|}$$

$\tau_{2\text{-}4}$ is some suitable mean temperature over the relevant length of flow passage. The matter of the sign of the right-hand side may be simplified by replacing $\mathrm{d}\sigma_e^2$ by $-\mathrm{d}\sigma_e|\mathrm{d}\sigma_e|$, giving, finally:

$$(\psi_4 - \psi_1)(\psi_4 + \psi_1) = +\mathrm{d}\sigma_e|\mathrm{d}\sigma_e|\tau_{2\text{-}4}\mathbf{F}_e \qquad (10.22)$$

$$\mathbf{F}_e = \frac{4\pi^2 C_f(N_{re})L/L_{ref}V_{ref}^2}{\mathrm{d}\phi^2(r_h/L_{ref})A_x^2 L_{ref}^2}N_{MA}^2; \left(N_{MA} = \frac{N_s L_{ref}}{\sqrt{(RT_{ref})}}\right)$$

(because of the $\mathrm{d}\phi$ component $\mathbf{F}_e$ is not a true parameter – just a convenient abbreviation (cf. $\mathbf{H}_e$)). The corresponding expression for the compression end is

$$(\psi_7 - \psi_4)(\psi_7 + \psi_4) = -\mathrm{d}\sigma_c|\mathrm{d}\sigma_c|\tau_{4\text{-}6}\mathbf{F}_c \qquad (10.23)$$

$$\mathbf{F}_c = \frac{4\pi^2 C_f(N_{re})L/L_{ref}V_{ref}^2}{\mathrm{d}\phi^2(r_h/L_{ref})A_x^2 L_{ref}^2}N_{MA}^2$$

In the definitions of $\mathbf{F}$, $r_h$, $A_x$ etc. take appropriate local values. In view of other approximations, suitable values for $\tau_{2\text{-}4}$ and $\tau_{4\text{-}6}$ are respectively $N_\tau$ $(T_e/T_{ref} = T_e/T_c)$ and 1.0 $(T_c/T_c)$.

Heat transfer in the exchangers (Fig. 10.5(*b*)) is dealt with by assuming that instantaneous temperature distribution is that which would apply under steady-flow conditions at the current, instantaneous value of flow rate into or out of the adjacent variable-volume space. With subscript i for inlet (whichever of the two ends that may be) and o for outlet:

$$T_w - T_o = (T_w - T_i)\exp\left\{-\frac{hA_{wh}}{m'c_p}\right\}$$

The form of the equation presupposes that flow is positive in the direction inlet-to-outlet. Since $m_e'$, for example, is computed as negative when it corresponds to outflow (left–right in Fig. 10.5) the signs used in the equation must reflect this. Considering the expansion end and normalizing:

$$\tau_{we} - \tau_3 = (\tau_{we} - \tau_2)\exp(+\mathbf{H}_{he}/d\sigma_e) \qquad (10.24)$$

$$\mathbf{H}_{he} = N_{ST_{he}}\, d\phi/2\pi$$

$$N_{ST_{he}} = h_{he}A_{whe}/N_s c_p M$$

At compression end

$$\tau_{wc} - \tau_6 = (\tau_{wc} - \tau_5)\exp(-\mathbf{H}_{hc}/d\sigma_c) \qquad (10.25)$$

$$\mathbf{H}_{hc} = N_{ST_{hc}}\, d\phi/2\pi$$

$$N_{ST_{hc}} = \mathrm{h}_{hc}A_{whc}/N_s c_p M$$

These steady-flow relationships clearly represent a drastic rationalization of reality: on the other hand, one may see clearly what assumption has been made, and the nature of the solution to which it leads. This much may not be claimed of the more 'rigorous' fixed-grid schemes – however many subdivisions they employ.

Nine relationships have been so far stated. The remaining two equations relate temperatures either side of the regenerator on the assumption of a constant temperature recovery factor, $\eta_r$. The functioning of the regenerator with short blow times, varying inlet conditions and varying pressure is so complex as to deny confidence in *any* attempt to define empirical constants of operation. Nevertheless, the assumption here leads for the expansion end (Fig. 10.5(*c*)) to

$$m_e' > 0: \quad T_3 = T_5 + \eta_r(T_{we} - T_{wc})$$

Normalizing

$$d\sigma_e > 0: \quad \tau_3 = \tau_5 + \eta_r(\tau_{we} - \tau_{wc})$$

For the compression end:

$$d\sigma_c > 0: \quad \tau_5 = \tau_3 + \eta_r(\tau_{wc} - \tau_{we})$$

These two equations do not apply when $d\sigma_e$ or $d\sigma_c$ (or both) $< 0$. To cover these cases:

$$d\sigma_e < 0: \quad \tau_2 = \tau_1$$

$$d\sigma_c < 0: \quad \tau_6 = \tau_7$$

Each pair of cases may be combined using the step function, $U$:

$$(\tau_2 - \tau_1) - U(d\sigma_e)[\tau_3 - \tau_5 - \eta_r(\tau_{we} - \tau_{wc}) - (\tau_2 - \tau_1)] = 0 \quad (10.26)$$

$$(\tau_6 - \tau_7) - U(d\sigma_c)[\tau_5 - \tau_3 - \eta_r(\tau_{wc} - \tau_{we}) - (\tau_6 - \tau_7)] = 0 \quad (10.27)$$

### 10.3.2 Discretization and solution scheme

As in the case of the previous solution, the defining equations (11 in number this time) are expanded about their values at the current value, $\phi$, of the independent variable, to the new value $\phi + d\phi$. Dependent variables which are not differentials thus assume mid-interval forms exemplified by $\psi_1 + \frac{1}{2}d\psi_1$, $\sigma_e + \frac{1}{2}d\sigma_e$, $\tau_1 + \frac{1}{2}d\tau_1$, etc. The coefficients of the 11 resulting unknowns, $d\psi_1$, $d\psi_4$, $d\psi_7$, $d\sigma_e$, $d\sigma_c$, $d\tau_1$, $d\tau_2$, $d\tau_3$, $d\tau_5$, $d\tau_6$ and $d\tau_7$ form an $11 \times 11$ matrix which can be solved by standard numerical means. To provide an example, Eqn (10.17) is expanded. For conciseness, the $\psi$, $\sigma$ and $\tau$ taking values which apply at the start of the integration interval will not be subscripted as such. Volumes, differentials of volumes and areas are evaluated at mid-interval. For example, $a_e$ signifies $a_e(\phi + \frac{1}{2}d\phi)$, $\mu_e$ signifies $\mu_e(\phi + \frac{1}{2}d\phi)$, and that part of the computer program which generates the array of values of volumes, areas and their differences must calculate accordingly.

$$d\sigma_e\{(\tau_1 + \tfrac{1}{2}d\tau_1)[1 - U(d\sigma_e)] + U(d\sigma_e)(\tau_2 + \tfrac{1}{2}d\tau_2)\} + \mathbf{H}_e a_e(\tau_{we} - \tau_1 - \tfrac{1}{2}d\tau_1)$$
$$= \mu_e\, d\psi_1/\gamma + (\psi_1 + \tfrac{1}{2}d\psi_1)\, d\mu_e$$

Grouping terms gives

$$a_{1,1}\,d\psi_1 + a_{1,2}\,d\psi_4 + a_{1,3}\,d\psi_7 + a_{1,4}\,d\sigma_e + a_{1,5}\,d\sigma_c + a_{1,6}\,d\tau_1 + a_{1,7}\,d\tau_2$$
$$+ a_{1,8}\,d\tau_3 + a_{1,9}\,d\tau_5 + a_{1,10}\,d\tau_6 + a_{1,11}\,d\tau_7 = b_1 \quad (10.28)$$

In Eqn (10.28)

$$a_{1,1} = -\mu_e/\gamma - \tfrac{1}{2}\,d\mu_e \qquad a_{1,2} = 0$$

$$a_{1,3} = 0 \qquad a_{1,4} = \tau_1[1 - U(d\sigma_e)] + U(d\sigma_e)\tau_2$$

$$a_{1,5} = 0 \qquad a_{1,6} = -\mathbf{H}_e a_e$$

$$a_{1,7} = 0 \qquad a_{1,8} = 0$$

$$a_{1,9} = 0 \qquad a_{1,10} = 0$$

$$a_{1,11} = 0$$

$$b_1 = -\mathbf{H}_e a_e(\tau_{we} - \tau_1) - \tfrac{1}{2}\,d\sigma_e\{d\tau_1[1 - U(d\sigma_e)] + U(d\sigma_e)\,d\tau_2\} + \psi_1\,d\mu_e$$

Discretization of the remaining ten equations produces ten further sets of coefficients. For brevity, only the non-zero terms will be cited. Discretizing the energy balance for the compression space (Eqn (10.18)) gives the following non-zero coefficients for row 2 of the matrix:

$$a_{2,3} = -\mu_c/\gamma - \tfrac{1}{2}\,d\mu_c \qquad a_{2,5} = \tau_7[1 - U(d\sigma_c)] + U(d\sigma_c)\tau_6$$

$$a_{2,7} = -\mathbf{H}_c a_c$$

$$b_2 = -\mathbf{H}_c a_c(\tau_{wc} - \tau_7) - \tfrac{1}{2}\,d\sigma_c\{d\tau_7[1 - U(d\sigma_c)] + U(d\sigma_c)\,d\tau_6\} + \psi_7\,d\mu_c$$

The following non-zero coefficients result from discretizing the expression for total mass content (Eqn (10.21)) and belong in row 3:

$$a_{3,2} = \nu \qquad a_{3,4} = 1$$

$$a_{3,5} = 1 \qquad (b_3 = 0)$$

From Eqn (10.19) for expansion space mass content:

$$a_{4,1} = -\mu_e \qquad a_{4,4} = \tau_1$$

$$a_{4,6} = \sigma_e \qquad b_4 = \psi_1\,d\mu_e$$

From Eqn (10.20) for compression space mass content:

$$a_{5,3} = -\mu_c \qquad a_{5,5} = \tau_7$$

$$a_{5,7} = \sigma_c \qquad b_5 = \psi_7\,d\mu_c$$

The expressions for pressure drop (Eqns (10.22), (10.23)) are non-linear in the flow rate ($d\sigma$) terms. This is dealt with by lumping the $|d\sigma|$ component with the coefficient and relying on the second pass of the integration interval to insert it as a 'known' quantity. From Eqn (10.22):

$$a_{6,1} = -\psi_1 \qquad a_{6,2} = \psi_4$$

$$a_{6,4} = -N_\tau|\mathrm{d}\sigma_e|\mathbf{F}_e \qquad b_6 = -\psi_4{}^2 + \psi_1{}^2 + \tfrac{1}{4}(\mathrm{d}\psi_1{}^2 - \mathrm{d}\psi_4{}^2)$$

Non-zero coefficients from the corresponding equation applied to the compression end are

$$a_{7,2} = -\psi_4 \qquad a_{7,3} = \psi_7$$

$$a_{7,5} = +1.0 \times |\mathrm{d}\sigma_c|\mathbf{F}_e \qquad b_7 = \psi_4{}^2 - \psi_7{}^2 + \tfrac{1}{4}(\mathrm{d}\psi_4{}^2 - \mathrm{d}\psi_7{}^2)$$

The equations for exchanger heat flow (Eqns (10.24) and (10.25)) remain implicit in the d$\sigma$ terms after expansion. The former, for the expansion end, gives

$$a_{8,7} = \tfrac{1}{2}\exp(\mathbf{H}_{he}/\mathrm{d}\sigma_e) \qquad a_{8,8} = -\tfrac{1}{2}$$

$$b_8 = (\tau_{we} - \tau_2)\exp(\mathbf{H}_{he}/\mathrm{d}\sigma_e) - (\tau_{we} - \tau_3)$$

For the compression end:

$$a_{9,9} = \tfrac{1}{2}\exp(-\mathbf{H}_{hc}/\mathrm{d}\sigma_c) \qquad a_{9,10} = -\tfrac{1}{2}$$

$$b_9 = (\tau_{wc} - \tau_5)\exp(-\mathbf{H}_{hc}/\mathrm{d}\sigma_c) - (\tau_{wc} - \tau_6)$$

There is little practical difference between including the two foregoing sets of terms in the matrix, and omitting them (leaving a 9 × 9 matrix) in favour of evaluating the source equations (Eqns (10.24), (10.25)) in steady-state form after each integration step. If this course is taken, the $\tau_2$ term of $a_{1,4}$ is replaced by $N_\tau$, the $\tau_6$ term of $a_{2,5}$ by 1.0, and the d$\tau_2$ of $b_1$ and the d$\tau_6$ of $b_2$ disappear. Similar considerations apply to the set of conditional equations for the temperatures across the regenerator (Eqns (10.26), (10.27)), offering the possibility of reducing the matrix to 7 × 7. From Eqn (10.26):

$$a_{10,6} = +\tfrac{1}{2}[1 + U(\mathrm{d}\sigma_e)] \qquad a_{10,7} = +\tfrac{1}{2}[1 + U(\mathrm{d}\sigma_e)]$$

$$a_{10,8} = -\tfrac{1}{2}U(\mathrm{d}\sigma_e) \qquad a_{10,9} = +\tfrac{1}{2}U(\mathrm{d}\sigma_e)$$

$$b_{10} = (\tau_2 + \tau_1)[1 + U(\mathrm{d}\sigma_e)] - U(\mathrm{d}\sigma_e)[\tau_3 - \tau_5 - \eta_r(\tau_{we} - \tau_{wc})]$$

From Eqn (10.27):

$$a_{11,8} = +\tfrac{1}{2}U(\mathrm{d}\sigma_c) \qquad a_{11,9} = -\tfrac{1}{2}U(\mathrm{d}\sigma_c)$$

$$a_{11,10} = +\tfrac{1}{2}[1 + U(\mathrm{d}\sigma_c)] \qquad a_{11,11} = -\tfrac{1}{2}[1 + U(\mathrm{d}\sigma_c)]$$

$$b_{11} = (\tau_6 + \tau_7)[1 + U(\mathrm{d}\sigma_c)] - U(\mathrm{d}\sigma_c)[\tau_5 - \tau_3 - \eta_r(\tau_{wc} - \tau_{we})]$$

At each integration step the solution program first of all sets the entire array to zero. It then evaluates each non-zero array element and feeds the value into the appropriate array location for numerical reduction by a suitable library subroutine. A simple Gauss or Gauss-Seidel method such as that embodied in SIMQ or SIMQX is suitable for obtaining specimen results. If the program (or a derivative of it) is to be used to produce performance maps (see below) or in optimization studies then it is worth carrying out the extra coding required to take advantage of the greater efficiency of, say, the Nag™ library programs F01BRF (de-composition) and F04AXF.

### 10.3.3 Specimen output from the simulation

It was shown under Section 10.2 that $N_{ST_e}$, $N_{ST_c}$ are functions only of $N_{RE}$ (in reality also of $N_{pr}$ and thus of $N_{PR}$), certain (normalized) physical dimensions of the machine, and of the dependent variables, the $\psi$ and the $\tau$. It can be shown that the same is true of the $N_{ST_e}$, $N_{ST_c}$ and of the $N_{ST_{he}}$, $N_{ST_{hc}}$ of the more comprehensive analysis of Section 10.3. The friction parameters embodied in the abbreviations $\mathbf{F}_e$ and $\mathbf{F}_c$ are clearly functions only of $N_{RE}$, of $N_{MA}$, of (normalized) gas circuit dimensions (hydraulic radius, cross-sectional area and length (or volume)) and of dimensionless mass rates. Assuming simple harmonic volume variations and constant phase angle the functional form of the simulation model as an expression of dimensionless cycle work, $\zeta_{MRT}$, may therefore be written

$$\zeta_{MRT} = \zeta_{MRT}\left[ N_\tau, \kappa, \alpha, \nu(N_\tau), \begin{pmatrix} \text{various dimen-} \\ \text{sionless lengths} \\ \text{and areas} \end{pmatrix}, \gamma, \eta_r, N_{PR}, N_{RE}, N_{MA} \right] \quad (10.29)$$

Overlooking the incorporation of the empirical $\eta_r$, the functional form may be said to be identical in essence to that of the three-conservation law (CL-3) perturbation analysis of Chap. 7. The practical difference may be expected to be in the respective sensitivities to variations in the parameters – particularly in $N_{RE}$, $N_{MA}$. In principle, the divergence between the two models could be illustrated by producing performance maps from Eqn (10.19) for the same machines as the maps of Chap. 7 and over the same range of $N_{RE}$ and $N_{MA}$. With the lumped-parameter simulation in its present form, however, (flow losses computed over a heat exchanger plus one half of a regenerator) this would be both inconvenient and unfair – inconvenient because of the difficulty in specifying appropriate $C_f$–$N_{re}$ correlations for the composite flow channels, and unfair because the three-pressure-point arrangement does not permit the best possible modelling of the gas circuit. With two pressure points at the

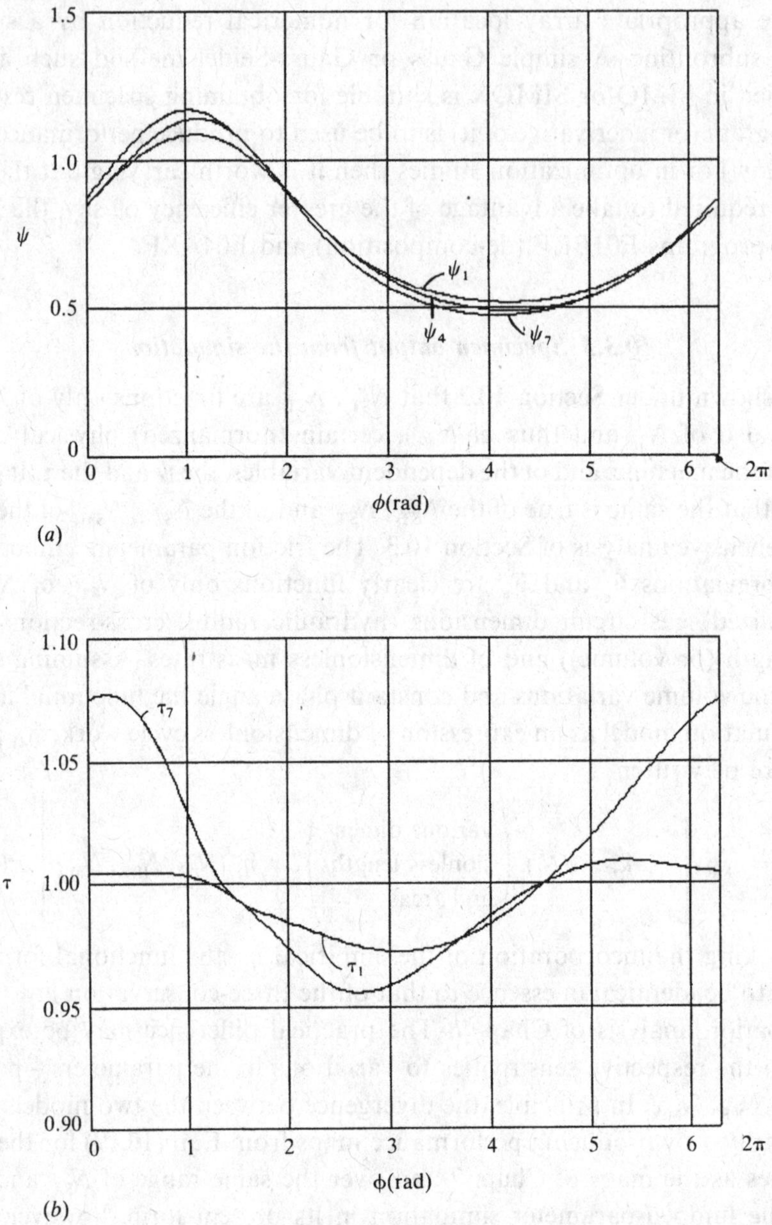


Fig. 10.6 Sample output from lumped-parameter model based on Finkelstein's analysis.[3] (*a*) Variation of pressure with crank angle at locations 1, 4 and 7. (*b*) Variation with crank angle of working space temperatures.

regenerator end faces replacing the single mid-plane point standard $C_f$–$N_{re}$ correlations could be used at a cost of a marginal increase in array size and consequent computation. The presentation of maps is accordingly deferred pending modification to four pressure points.

In the interim, Fig 10.6 shows a selection of curves produced to compare results of the present formulation with originals by Finkelstein.[3] Finkelstein used $N_\tau = 2.0$ (prime-mover mode), $\gamma = 1.4$, $\kappa = 1.0$, $\alpha = \pi/2$, $N_{STe}/2\pi = N_{STc}/2\pi = N_{ST_{he}}/2\pi = N_{ST_{hc}}/2\pi = 0.5$, $a_e(\phi) = a_c(\phi) = 1.0$. His individual values of dead space ratio convert to $v = 0.68$. (In this later analysis the definitions of both $v$ and $\psi$ are precisely those preferred here.) The values of $\mathbf{F}_e\,d\phi^2$ and $\mathbf{F}_c\,d\phi^2$ used to achieve the results shown in Fig. 10.6(*a*) were unity as opposed to the 0.05 used for Finkelstein's somewhat different flow/pressure drop relationship.

To achieve temperature swings comparable to those obtained by Finkelstein, the various $N_{ST}/2\pi$ had to be increased from 0.5 to 0.8. Otherwise, and noting the slightly different pressure swing, the curves of Figs 10.6 correspond completely to the originals obtained with the use of an early digital computer running a program simulating a differential analyser.[3]

# 11

# Implementation of the Method of Characteristics

## 11.1 State of the art

The reader who has not used the Method of Characteristics may wish to study the background treatment of Appendix XI before embarking on the current chapter.

At the time of writing, work to implement a comprehensive computer simulation of the Stirling cycle based on the Method of Characteristics is on a far smaller scale than modelling by finite-cell and nodal techniques, and attracts the attention of a minority of analysts. Other than that the Characteristics method gives an *appearance* of greater difficulty, there is no good reason why a method which offers authentic solution of the defining equations should continue to take second place to *ad hoc* numerical schemes which, outside the field of Stirling machines, would not be contemplated for solution of even the simplest of problems of gas dynamics.

It is true that a construction which is straightforward when carried out graphically is not always readily implemented on the computer, as those will know who have written – or even used – computer-based packages for carrying out mechanical draughting. This may be the reason for a great deal of effort (outside the field of Stirling machines) to develop a fixed-grid scheme equivalent to the Method of Characteristics. The name usually associated with such schemes is Hartree,[1] the essence of whose treatment is shown in Fig. 11.1: the $t$–$x$ plane is subdivided into a rectangular grid. Assuming that fluid properties (expressed in terms of $a$ and $u$) are known at time $t$, those at $t + \Delta t$ are estimated by projecting elements of **I**, **II** and path characteristics backwards from the fixed $x$ at $+\Delta t$. Two arguable shortcomings of the scheme are that (*a*) in general, each step forward in time calls for interpolation between (known) property values at the points fixed along the $x$ direction, and (*b*) the resulting solution – numerical or graphical – loses the pictorial quality (see Fig. 2.12) of the counterpart integrated along continuous Mach lines. A development of the Hartree method – the so-called lambda scheme

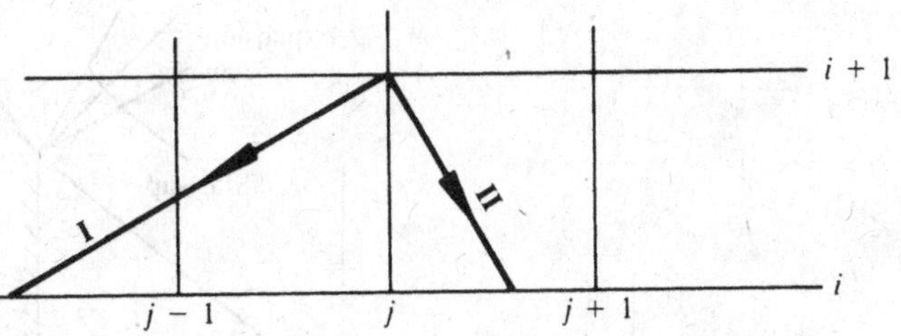


Fig. 11.1 Schematic of Hartree method.[1]

as espoused by Rispoli[2] for Stirling modelling work – checks whether a Mach line projected back from a point at $x, t + \Delta t$ could actually have reached this latter point on its forward travel, given the intervening $a$ and $u$ and the possibility of local shocks or other discontinuities. An enlightened implementation of the Hartree method would contain such provision anyway, so the methods are essentially equivalent.

Fig. 11.2 represents a simplified Stirling machine at the start of a Characteristics solution by the traditional (graphical) process. For the moment, discontinuities in cross-sectional area and in fluid properties (the cause of reflected waves) are not considered. To keep computing time within reasonable bounds it is necessary (as with the case of finite-cell schemes) to contrive instantaneous start-up from rest. From the material of Appendix XI it will have been deduced that the impulsive *compressive* start-up of a piston generates a shock. The Method of Characteristics can cope with shocks, but it would be a pity to have to do so, given that shocks do not occur in the real machine following conventional start-up.

A piston driven by a crank starting instantaneously to angular speed $\omega$ from a dead-centre position initially has zero velocity, and so does not generate a shock. An instantaneous *expansive* start-up generates an expansion fan – but an expansion poses no problem. Thus it is possible to specify a satisfactory start-up procedure: the left-hand (expansion-space) piston is set at the outer dead-centre position. Instantaneous linear velocity corresponding to start-up to full angular speed is zero. There can thus be no possibility of a shock forming at the expansion end. The time of traverse of a wave is now estimated corresponding to still gas conditions. Speed locally is $1/a$, since $u$ is initially zero. An approximate value will suffice which, in terms of normalized coordinates ($\theta = t/(L/a_0)$) is unity – a clear demonstration of the merits of working in terms of normalized variables. The compression space crank is now 'wound forward' to an angle taking into account both the fixed phase angle, $\alpha$, *and* the angle to be turned through at rated crank speed in the time for a wave pass. For reasonable values of the parameters of operation,

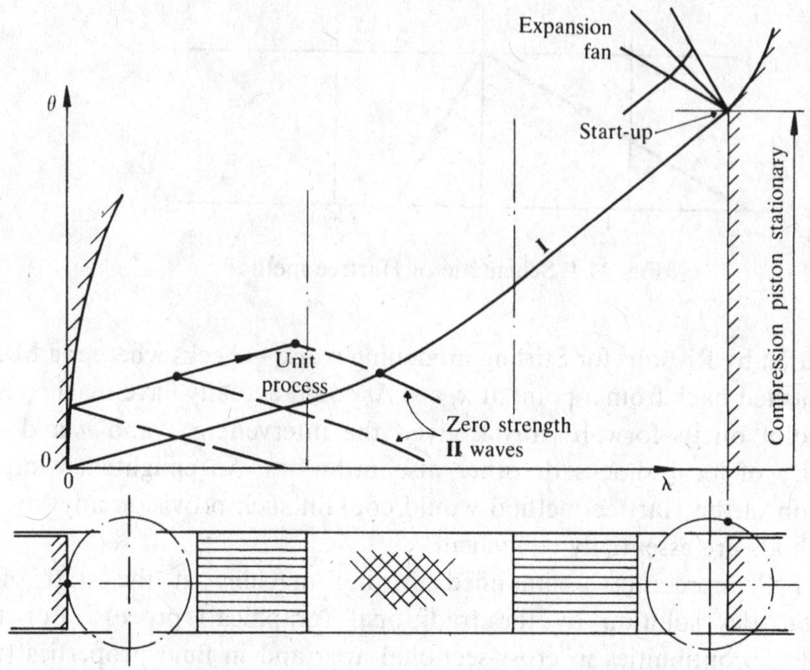


Fig. 11.2 First sweep and unit process in the Method of Characteristics applied to the equivalent Stirling cycle machine.

this angle is such that compression piston start-up causes an *expansion*, which is free of the problem of shocks. The expansion piston is now put into motion. The phase angle between the two cranks is correct at the instant of arrival at the compression piston face of the first (compressive) **I** Mach wave. The compression piston is started at this point. The same effect can be achieved by reversing the rôles of the pistons.

The first Mach wave is sketched in Fig. 11.2. The next wave is initiated at the left piston face at a suitable time after the first, and with the aid of an imaginary (zero strength) **II** wave from the still gas region. A mesh of labelled points is built up along the second **I** wave on the basis of regularly-spaced **II** waves. The figure shows the development of a typical unit process. There is no difference between a **II** Mach line which originates from the right-hand piston face and a **I** wave which is reflected at the same instant. The total number of waves being handled at any one time may therefore be kept constant by allowing as **II** waves only the reflection of **I** waves. If there are $n$ unit processes on the first left-to-right scan of the **I** wave, then there will be $n$ unit processes to be dealt with on subsequent scans. If the Mach net is to be plotted, it is necessary to store $t$ and $x$ (or $\theta$ and $\lambda$) coordinates of the $n$ intersections on two successive **I** Mach lines.

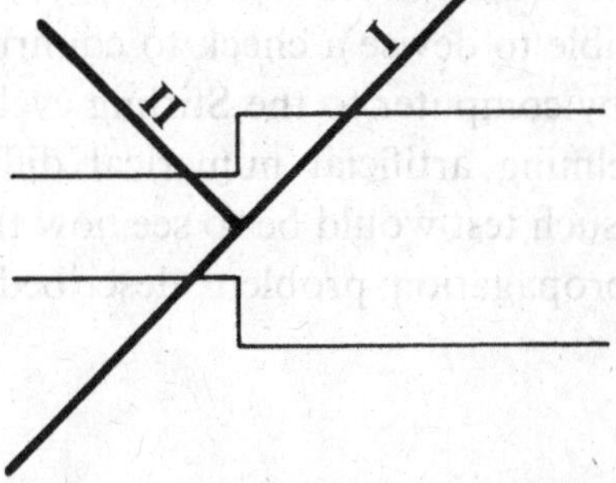


Fig. 11.3 **II** Mach wave reflected from area discontinuity on passage of **I** wave.

The individual unit process is standard for all integration steps, and the storage of a predetermined number of *x* and *t* coordinates for each set of *a* and *u* no more complex than the storage of the same number of *t* and *x* for the fixed grid. Where, then, is the factor which makes the notion of the Method of Characteristics in computer implementation so daunting?

It is this: the discussion preceding skated round the matters of discontinuity in cross-sectional area and fluid properties. Both give rise to a transmitted and a reflected wave component (Fig. 11.3) so that an incident **I** wave generates a new **II** wave, and an incident **II** wave generates a new **I** wave. Each newly-arising wave generates further new waves on encountering a discontinuity. Whatever the computational storage capacity available, the prospect of an indefinitely increasing number of mesh points is an unattractive one.

The problem of *fluid property* discontinuities is easily dealt with by agreeing that these, if they exist, will be small enough not to cause reflections. (The *possibility* of the existence of such discontinuities has been argued by Organ,[3] but rather as a vehicle for the comparison of different methods of numerical solution than as a reflection site for the finite pressure wave.) Area changes, however, cannot be argued away. There are at least two possibilities:

The generalized treatment of unsteady, one-dimensional, compressible flow in a duct (Ref. 4 and Appendix XI) permits gradual change of cross-sectional area. One might explore to what extent the idea of a 'gradual' change can be stretched to cover the rates of area change encountered in the Stirling cycle machine.

The unit process might be looked at to see whether the effects of the new (reflected) wave generated at the discontinuity can be adequately taken into account by merging it with an adjacent wave (or waves) of the original complement. Given that the original choice of *n* waves was arbitrary, and that, within limits, the flow field is equally well described by more or fewer waves, this approach must offer possibilities.

Before investing the substantial effort threatened by the proposed investigation, it would seem sensible to devise a check to confirm that the Method of Characteristics applied by computer to the Stirling cycle machine is likely to be free of the overwhelming artificial numerical diffusion which plagues finite-cell schemes. One such test would be to see how the method copes with the temperature-wave propagation problem described at the beginning of Appendix XI.

## 11.2 A test for the Method of Characteristics

The artificial numerical dispersion of temperatures occurs independently of that of pressures, and so may be examined in a frictionless flow-field at constant pressure. The equations for the characteristic *directions* (Eqns (AXI.8a)–(AXI.8c) are unaffected by this simplification, but those for the *changes along* the characteristic directions simplify. The path line equation (Appendix XI, Section AXI.13) is

$$\mathrm{d}p_{\text{path}} = \frac{p}{a}\left[\frac{2\gamma}{\gamma - 1}\,\mathrm{d}a_{\text{path}} - \frac{\gamma}{a}(uF + q^*)\,\mathrm{d}t_{\text{path}}\right] \tag{11.1}$$

$F$ has been assumed zero, and $q^*$ is heat transfer per unit mass and per unit time, which, for the tube with wall at temperature $T_w$, may be written:

$$q^* = \frac{N_{\text{st}}}{r_h} u c_p (T_w - T) \tag{11.2}$$

With $a$ for $\sqrt{(\gamma R T)}$ and $\mathrm{d}\theta$ for $\mathrm{d}t/(L/a_0)$ the path line equation may be written

$$\frac{\mathrm{d}a}{a} = \frac{\gamma - 1}{2\gamma}\frac{\mathrm{d}p}{p} + \frac{N_{\text{st}}}{2r_h/L}\frac{u}{a_0}\left[\frac{1}{(a/a_0)^2} - 1\right]\mathrm{d}\theta_{\text{path}}$$

or, since there is no pressure change:

$$\frac{\mathrm{d}a}{a} = \frac{N_{\text{st}}}{2r_h/L}\frac{u}{a_0}\left[\frac{1}{(a/a_0)^2} - 1\right]\mathrm{d}\theta_{\text{path}} \tag{11.3}$$

For an element between points 1 and 2 of a path line labelled $k$:

$$\ln(a_k)_2 - \ln(a_k)_1 = \frac{N_{\text{st}}}{2r_h/L}\frac{\underline{u}_{1-2}}{a_0}\left\{\frac{1}{(\underline{a}_{1-2}/a_0)^2} - 1\right\}\mathrm{d}\theta_{\text{path}} \tag{11.3a}$$

In Eqn (11.3a) the underlined symbols take their respective mean values between points 1 and 2.

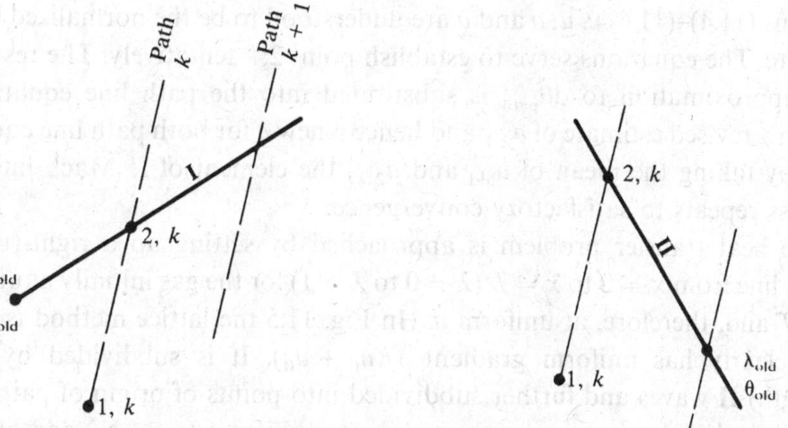


Fig. 11.4 Typical unit process based on intersection of a Mach line with a path line.

Inspection of the equations for change of state along the Mach lines (Eqns (AXI.9a) and (AXI.9b)) reveals that, for the contrived case of uniform pressure and velocity, no change in state is predicted – the waves just travel at local gradients corresponding to the $a$ and $u$ deduced from change along the path lines. This focusses attention on the fact that the *unit integration process* (in the case of variable pressure as well as in the present, special instance) generally involves not, as might be inferred from standard treatments, a pair of Mach lines, but one Mach line and one path line. Fig. 11.4 is a schematic of typical unit processes involving the $k$th path line and a **I**- or **II**-running Mach line. The **II**-running case may be used to indicated the numerical algorithm used to proceed from the pair of known points to the new point.

Dimensionless time and location are respectively $\theta_{\text{old}}$ and $\lambda_{\text{old}}$ at the right-most known point, and this allows the point to be a Mach line intersection or the intersection with an earlier path line. The coordinates, $\theta_{2,k}$ and $\lambda_{2,k}$ are sought of the new point, 2, $k$, on the basis that the path line gradient between 1, $k$ and 2, $k$ is the inverse, $1/\underline{u}$, of local mean particle speed, and that Mach line gradient is correspondingly $1/(\underline{u} - \underline{a}_{\text{II}})$. Solving the two, simultaneous, linear equations for the straight line elements, noting from the starting assumptions that $\underline{u}$ must be equal to the constant $u_0$ gives:

$$\theta_{2,k} = \frac{-\lambda_{1,k} + \lambda_{\text{old}} + u_0\theta_{1,k} - (u_0 - \underline{a}_{\text{II}})\theta_{\text{old}}}{\underline{a}_{\text{II}}} \tag{11.4}$$

$$\lambda_{2,k} = \lambda_{1,k} + u_0\, \mathrm{d}\theta_{\text{path}} \tag{11.5}$$

$$\mathrm{d}\theta_{\text{path}} = \theta_{2,k} - \theta_{1,k} \tag{11.6}$$

In Eqns (11.4)–(11.6) $u$, $\underline{u}$, $a$ and $\underline{a}$ are understood to be the normalized forms, $u/a_0$ etc. The equations serve to establish point $2, k$ tentatively. The resulting, first approximation to $d\theta_{path}$ is substituted into the path line equation to obtain a revised estimate of $a_{2,k}$ and hence a new $\underline{a}$ for both path line equation and, by taking the mean of $a_{old}$ and $a_{2,k}$, the element of **II** Mach line. The process repeats to satisfactory convergence.

The heat transfer problem is approached by setting up a right-running Mach line from $x = 0$ to $x = L$ ($\lambda = 0$ to $\lambda = 1$) for the gas initially at uniform $u$, $p$, $T$ and, therefore, at uniform $a$. (In Fig. 11.5 the lattice method is used.) This clearly has uniform gradient $1/(u_0 + a_0)$. It is subdivided by (zero strength) **II** waves and further subdivided into points of origin of path lines. Initially, all **II** waves travel right-to-left at gradient $1/(u_0 - a_0)$ and all path lines left-to-right at gradient $1/u_0$. At the origins of all intersections between the original **I** wave with the **II** waves and the path lines $a = a_0$ and $u = u_0$ (normalized form again). The step change in temperature (or, if required, a gradual change) is initiated in the form of a new value of $a$ for all path lines originating from the left at $\theta > 0$. The unit processes are applied from left-to-right, building on the original **I** wave, and for as many sweeps as are required for the temperature wave to pass through the exchanger.

The Mach plane does not display temperature (or acoustic speed) directly, although an impression may be had by observing the increased gradient of the Mach lines to the left of the particle path carrying the initial discontinuity, the increase confirming reduced values of $|u_0 + a|$ and $|u_0 - a|$, and hence of $a$ (or $T$). However, plots of $T/T_0$ vs $\lambda$ may be produced from numerical values computed intermediate to the construction of the Mach net. Fig. 11.6 contains examples, confirming that the only difference in the profile from one instant of time to the next is the extent to which the front has propagated. The plots (and the Mach plane) confirm the total absence of numerical diffusion about the temperature front.

The analytical solution (Eqn (AXI.2)) does not superimpose exactly on the curve obtained numerically, but precise correspondence is hardly to be expected, given the coarseness of the integration mesh. To check that numerical and analytical results must, in the limit, coincide, the path line equation (Eqn (11.3a)) may be converted back to a form involving $T$ rather than $a$. With $da/a = \frac{1}{2} dT/T$ (from differentiation of $a = \sqrt{(\gamma RT)}$) the equation becomes

$$\frac{dT}{T} = \frac{N_{st}}{r_h/L} u(T_w - T)\, dt_{path} \qquad (11.3b)$$

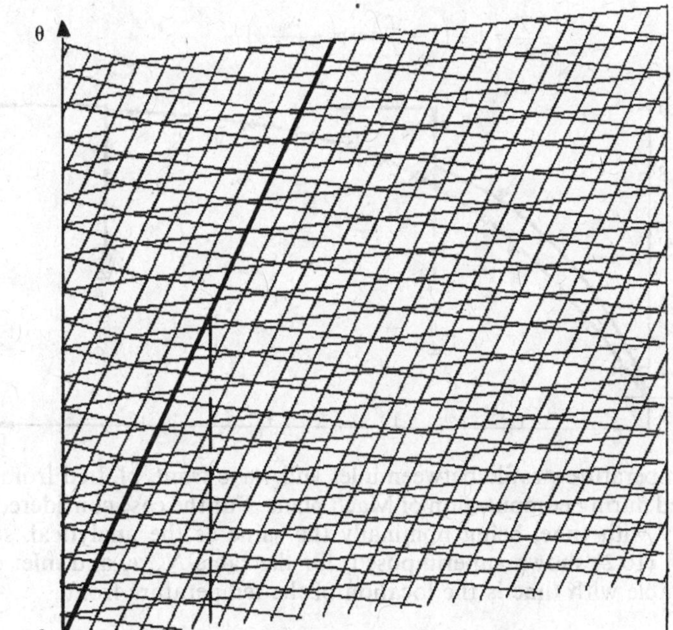

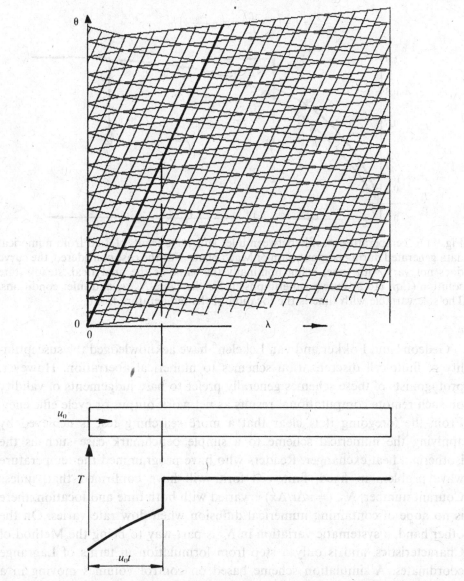

Fig. 11.5 Solution in physical (Mach) plane of early stages of the constant pressure, constant velocity heat transfer problem with step change in temperature at inlet. The heavy **I** Mach line denotes the passage of the temperature front. Increasing gradient of Mach lines behind the front corresponds to a reduction in acoustic speed at the lower temperatures.

Noting that $u\,dt$ is the $dx$ travelled by a specific particle, integration of the path line equation in this form leads to the analytical solution given as Eqn (10.2). To this extent, the Characteristics method with sufficiently fine integration mesh is bound to yield the appropriate numerical answer.

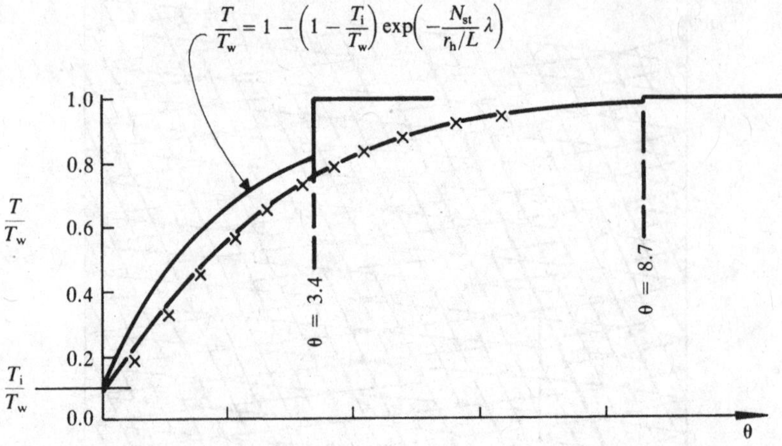


Fig. 11.6 Temperature profile between inlet and wave front, plotted from numerical data generated during computation of Mach plane. For the case considered, the curve does not vary with time, being nominally the same as the analytical, steady-state solution (Eqn (10.2) shown superimposed) for the same $N_{st}$, $r_h$ and inlet conditions. The sole variable with time is the location of the temperature front.

Gedeon[5] and Fokker and van Eckelen[6] have acknowledged the susceptibility of finite-cell discretization schemes to numerical aberration. However, protagonists of these schemes generally prefer to base judgements of validity on such remote computational results as net work output or cycle efficiency. From the foregoing it is clear that a more searching test is achieved by applying the numerical scheme to a simple benchmark case such as the isothermal heat exchanger. Readers who have programmed the temperature wave problem in fixed, finite-cell form will have confirmed that, unless Courant number, $N_C$, $(= u\Delta t/\Delta x)$ is varied with both time and location, there is no hope of containing numerical diffusion when flow rate varies. On the other hand, a systematic variation in $N_C$ is part-way to being the Method of Characteristics, and is only a step from formulation in terms of Lagrange coordinates. A simulation scheme based on control volumes moving in a Lagrange reference frame has been formulated by Organ (Chap. 13) and implemented by Rix[7] for application to those regimes of Stirling machine operation where pressure waves are considered unimportant, and application of the Method of Characteristics not justified.

That the Method of Characteristics copes appropriately with a temperature discontinuity suggests that there will be no problem with the continuous temperature distributions assumed in normal operation of the gas circuit. There remains the matter of discontinuities in cross-sectional area and of the associated reflections. It is Mach waves, and not particle paths, which have reflections, and so it would be possible – not to mention convenient – to have available for experiment a form of the Method of Characteristics in which integration could proceed in unit processes involving Mach lines alone rather than the intricate combination of Mach lines and particle paths.

The homentropic Characteristic[4, 8] relationships would, in principle serve. However, a disadvantage is that the case excludes heat transfer and flow friction. Cycle work would be zero, and there would thus be no indication that the essential nature of the Stirling gas processes was being appropriately modelled. Use will be made instead of a Characteristics formulation in which fluid temperature is determined at all times by the temperature of the immediately adjacent solid wall. The treatment, fully described in Ref. 9, might be loosely described as a dynamic equivalent to the 'isothermal' analysis. Friction and area change are modelled, but fixing the temperature means that solution can proceed without consideration of particle paths.

## 11.3 The temperature-determined compressible flow-field

The control volume of Fig. 11.7 is essentially that for the generalized case (Fig. AXI.2) except that heat transfer is dealt with implicitly by the condition of zero temperature difference between working fluid and immediately adjacent wall.

The equation of conservation of mass is as for the general case:

$$\frac{\partial \rho}{\partial t} + \rho \frac{\partial u}{\partial x} + u \frac{\partial \rho}{\partial x} + \frac{\rho u}{A_x} \frac{\mathrm{d}A_x}{\mathrm{d}x} = 0 \qquad (11.7)$$

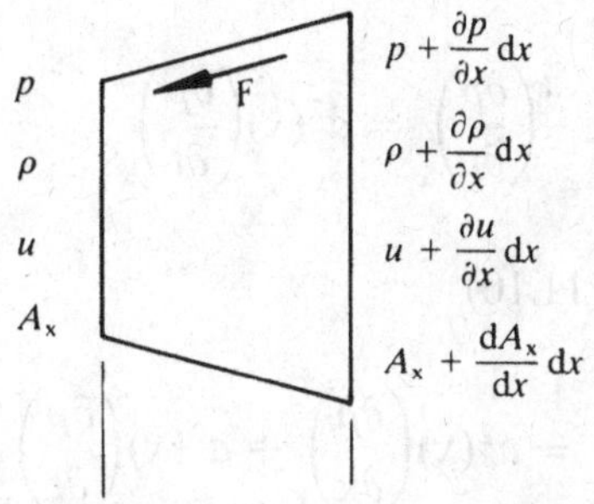


Fig. 11.7 Control volume for flow in which the temperature of the fluid is always equal to that of the adjacent wall.

as is the equation for conservation of momentum after subtraction of Eqn (11.7)

$$\frac{\partial u}{\partial t} + u\frac{\partial u}{\partial x} + \frac{1}{\rho}\frac{\partial p}{\partial x} + F = 0 \tag{11.8}$$

$F$ is the wall friction term already defined as

$$F = \frac{C_f}{r_h}\frac{u^2}{2}\frac{u}{|u|}$$

If a third equation can be found corresponding to the energy equation of the general treatment, then it should be possible, in principle, to deal with the set of three equations using Shapiro's determinant form and thus to establish characteristic directions and equations for change of state for the temperature-determined case.

As might be suspected from experience of the 'static' temperature-determined (or 'isothermal') model, the outstanding equation linking $p$, $\rho$ and $T$ is the ideal gas equation, $p = \rho RT$. Differentiating with respect to $x$ at constant $t$:

$$\left(\frac{\partial p}{\partial x}\right)_t = RT\left(\frac{\partial \rho}{\partial x}\right)_t + \rho R\frac{\mathrm{d}T}{\mathrm{d}x}$$

$$= a^2(x)\left(\frac{\partial \rho}{\partial x}\right)_t + \rho R\frac{\mathrm{d}T}{\mathrm{d}x} \tag{11.9}$$

In Eqn (11.9) $a$ is now the *isothermal* sound speed, $\sqrt{(RT)}$.

Differentiating now with respect to $t$ at constant $x$:

$$\left(\frac{\partial p}{\partial t}\right)_x = a^2(x)\left(\frac{\partial \rho}{\partial t}\right)_x \tag{11.10}$$

Adding Eqns (11.9) and (11.10)

$$\left(\frac{\partial p}{\partial x}\right)_t + \left(\frac{\partial p}{\partial t}\right)_x = a^2(x)\left(\frac{\partial \rho}{\partial x}\right)_t + a^2(x)\left(\frac{\partial \rho}{\partial t}\right)_x + \rho\frac{\mathrm{d}a(x)^2}{\mathrm{d}x} \tag{11.11}$$

Eqns (11.7), (11.8) and (11.11) may now be laid out as were the corresponding equations in Appendix XI:

$$
\begin{array}{llllllll}
\mathrm{d}x\,\dfrac{\partial u}{\partial x} & +\,\mathrm{d}t\,\dfrac{\partial u}{\partial t} & & & & & = & \mathrm{d}u \\[2ex]
 & & \mathrm{d}x\,\dfrac{\partial p}{\partial x} & +\,\mathrm{d}t\,\dfrac{\partial p}{\partial t} & & & = & \mathrm{d}p \\[2ex]
 & & & & \mathrm{d}x\,\dfrac{\partial \rho}{\partial x} & +\,\mathrm{d}t\,\dfrac{\partial \rho}{\partial t} & = & \mathrm{d}\rho \\[2ex]
\rho\,\dfrac{\partial u}{\partial x} & & & & u\,\dfrac{\partial \rho}{\partial x} & +\,\dfrac{\partial \rho}{\partial t} & = & \dfrac{\rho u\,\mathrm{d}A_x}{A_x\,\mathrm{d}x} \\[2ex]
u\,\dfrac{\partial u}{\partial x} & +\,\dfrac{\partial u}{\partial t} & +\,\dfrac{1}{\rho}\dfrac{\partial p}{\partial x} & & & & = & -F \\[2ex]
 & & u\,\dfrac{\partial p}{\partial x} & +\,\dfrac{\partial p}{\partial t} & -\,a^2\dfrac{\partial \rho}{\partial x} & -\,a^2\dfrac{\partial \rho}{\partial t} & = & \rho\,\dfrac{\mathrm{d}a(x)^2}{\mathrm{d}x}
\end{array}
$$

As before, the solution for any chosen partial differential coefficient may be expressed in determinant form. The following example represents the solution for $\partial u/\partial x$:

$$
\frac{\begin{vmatrix}
\mathrm{d}u & \mathrm{d}t & 0 & 0 & 0 & 0 \\
\mathrm{d}p & 0 & \mathrm{d}x & \mathrm{d}t & 0 & 0 \\
\mathrm{d}\rho & 0 & 0 & 0 & \mathrm{d}x & \mathrm{d}t \\
\dfrac{\rho u\,\mathrm{d}A_x}{A_x\,\mathrm{d}x} & 0 & 0 & 0 & u & 1 \\
-F & 1 & 1/\rho & 0 & 0 & 0 \\
\rho\,\mathrm{d}a(x)^2/\mathrm{d}x\!: & 0 & u & 1 & -a^2 & -a^2
\end{vmatrix}}{\begin{vmatrix}
\mathrm{d}x & \mathrm{d}t & 0 & 0 & 0 & 0 \\
0 & 0 & \mathrm{d}x & \mathrm{d}t & 0 & 0 \\
0 & 0 & 0 & 0 & \mathrm{d}x & \mathrm{d}t \\
\rho & 0 & 0 & 0 & u & 1 \\
u & 1 & 1/\rho & 0 & 0 & 0 \\
0 & 0 & u & 1 & -a^2 & -a^2
\end{vmatrix}} = \frac{\partial u}{\partial x} \qquad (11.12)
$$

The condition that the determinant of the coefficients be zero leads to the following expression for the local gradients of the Mach lines:

$$\frac{dt}{dx_{I}} = \frac{1}{u + \sqrt{[RT(x)]}} \tag{11.13a}$$

$$\frac{dt}{dx_{II}} = \frac{1}{u - \sqrt{[RT(x)]}} \tag{11.13b}$$

In Eqns (11.13) $a$ has been replaced by $\sqrt{[RT(x)]}$ since it is in terms of values of this quantity that solution will proceed. Setting the numerator of Eqn (11.12) equal to zero gives the equations for change in fluid properties along the respective characteristic directions:

Along the **I** characteristic direction:

$$du_{I} = -\sqrt{(RT)}\frac{dp}{p} + \left[-F + u\frac{\sqrt{(RT)}}{T}\frac{dT}{dx} + u\sqrt{(RT)}\frac{1}{A_x}\frac{dA_x}{dx}\right]dt_{I} \tag{11.14a}$$

Along the **II** direction:

$$du_{II} = \sqrt{(RT)}\frac{dp}{p} + \left[-F - u\frac{\sqrt{(RT)}}{T}\frac{dT}{dx} + u\sqrt{(RT)}\frac{1}{A_x}\frac{dA_x}{dx}\right]dt_{II} \tag{11.14b}$$

It is seen that Eqns (11.14) are equivalent to Eqns (AXI.9) with $\gamma$ replaced by unity, and with the term for heat exchange by temperature difference replaced by a term corresponding to Eqn (7.6) for heat exchanged as the fluid moves up or down the imposed temperature gradient. A significant difference from the earlier treatment is that particle trajectories, $dt/dx = 1/u$ are not characteristic directions, and it is not necessary to construct them in order to be able to proceed with a solution.

The state change equations contain terms accounting for the rate of area change, $dA_x/dx$, and for suitable values solution proceeds precisely as already explained for the homentropic case (Appendix XI). Fig. 2.12 was obtained from a computer program embodying Eqns (11.13) and (11.14) and using the lattice variant of the integration sequence. There are no moving temperature discontinuities in this case, but the solution displays an alternative physical phenomenon which eludes finite-cell schemes, but which the Method of Characteristics treats faithfully. This is the expansion wave resulting from impulsive start of the right-hand piston: the components of the fan diverge while the expansion wave dissipates (as must be the case with a real impulsive expansion), but then settle down to reflect local cyclic expansion and compression as do the other Mach lines. The non-standard spacing of the lines from the original fan is preserved indefinitely, indicating that the

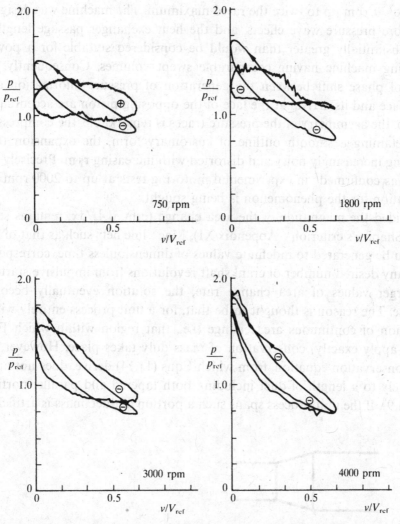

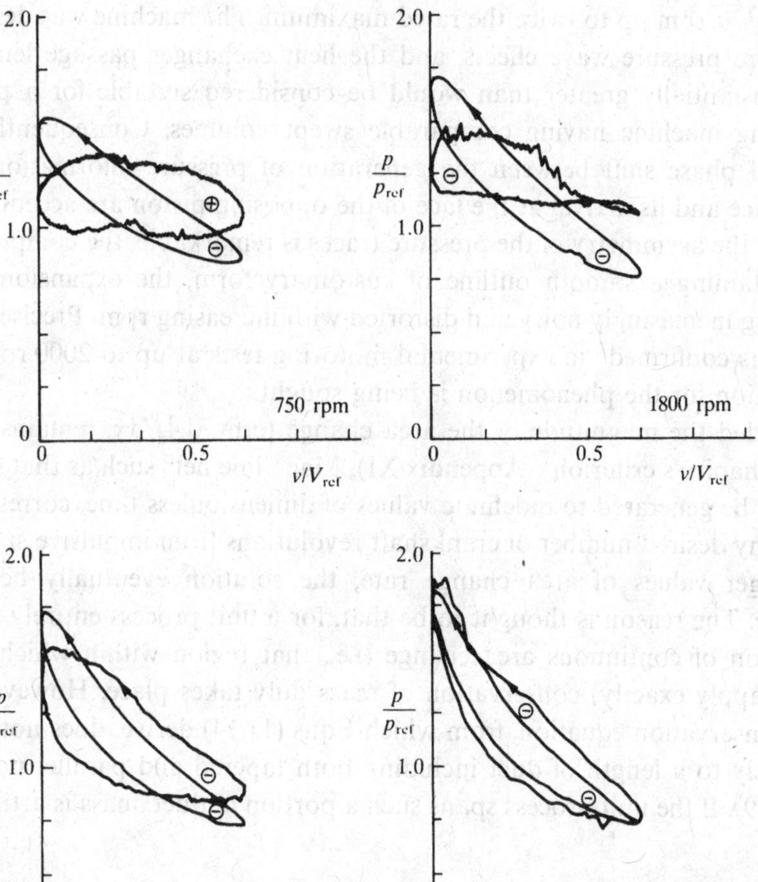


Fig. 11.8 Indicator diagrams computed by the Method of Characteristics for the SM1[7] laboratory engine for various crankshaft speeds (after Organ[9]). The results provide theoretical confirmation of a phenomenon noted in experiment, whereby the expansion traces become increasingly 'noisy' with increasing rpm, while the traces for the compression space remain relatively unaffected, except for an increase in amplitude.

thermodynamic information being carried by the wave components is being properly transported and is not subject to averaging or dissipative effects at successive steps in the integration process.

Indicator diagrams corresponding to Fig. 2.12 have been obtained for a range of crankshaft speeds,[9] and reveal a phenomenon which has not, to the knowledge of the author, arisen in other forms of Stirling simulation: Fig. 11.8 shows indicator diagrams corresponding to motoring of the SM1 machine[7] at rpm up to twice the rated maximum. The machine was designed to explore pressure wave effects, and the heat exchanger passage length is thus substantially greater than would be considered suitable for a power-producing machine having comparable swept volumes. Consequently, the effects of phase shift between the generation of pressure information at a piston face and its arrival at the face of the opposing piston are accentuated. Even so, the asymmetry of the pressure traces is remarkable, the compression trace retaining a smooth outline of customary form, the expansion trace becoming increasingly noisy and distorted with increasing rpm. Precisely this trend was confirmed[7] in experimental motoring tests at up to 2000 rpm. An explanation for the phenomenon is being sought.

Provided the magnitude of the area change term, $dA_x/dx$, remains small (recall Shapiro's criterion – Appendix XI), Mach line nets such as that of Fig. 2.12 can be generated to indefinite values of dimensionless time, corresponding to any desired number of crankshaft revolutions from impulsive start-up. For larger values of area change rate, the solution eventually becomes unstable. The reason is thought to be that, for a unit process entirely within the region of continuous area change (i.e., that region within which Eqns (11.14) apply exactly) conservation of mass duly takes place. However, the mass conservation equation, from which Eqns (11.14) derive, does not apply accurately to a length of duct including both tapered and parallel portions (Fig. 11.9). If the unit process spans such a portion of duct, mass is artificially

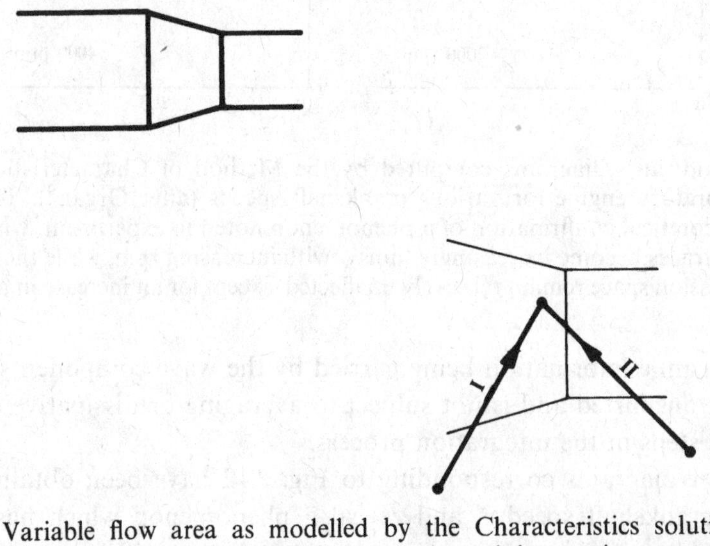


Fig. 11.9 Variable flow area as modelled by the Characteristics solution which produced Fig. 2.12. At least some unit processes in each integration sweep span part or all of the area transition.

lost or gained, and the cumulative effect after a number of integration steps upsets the integration sequence.

Experimentation with integration step size (initial Mach line spacing) suggests that increasing the number of Mach lines does not cure the problem; after all, in any one scan of the exchanger system shown at the base of Fig. 2.12 there will, in the general case, be eight unit processes each spanning a discontinuity in area gradient. It appears necessary to develop a unit process capable of coping with an abrupt area discontinuity.

## 11.4 A unit process for area discontinuity

Fig. 11.10 indicates the four cases of a single Mach line, **I** or **II**, incident upon an area discontinuity, positive or negative. In each case, an incident **I** wave gives rise to a new **II** wave, and *vice versa*. All cases are amenable, in both lattice and field interpretation, to treatment by standard means[4] provided the generation of the extra wave can be tolerated; several cycles of the Stirling

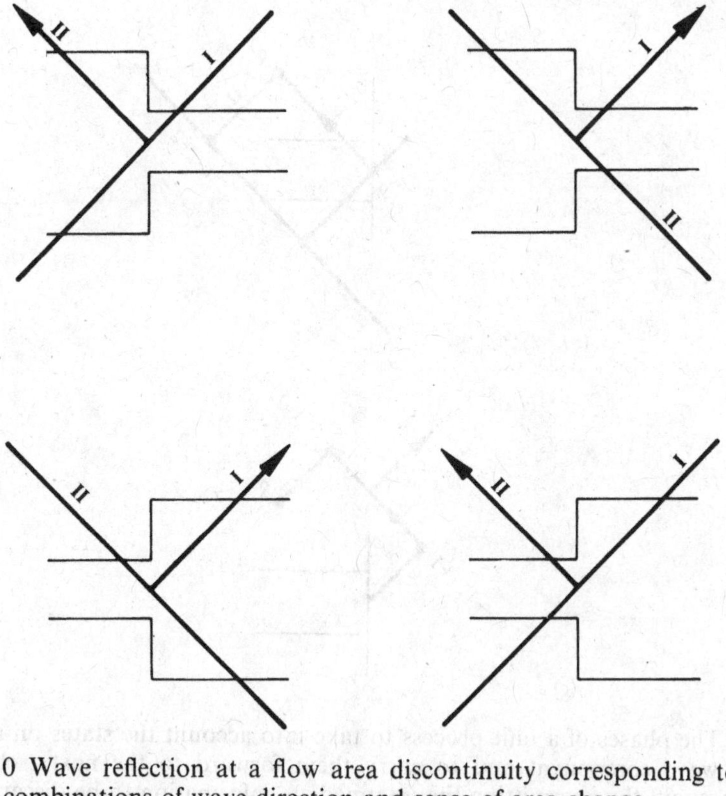


Fig. 11.10 Wave reflection at a flow area discontinuity corresponding to the four possible combinations of wave direction and sense of area change.

machine are necessary for attainment of steady state, and it is preferred not to have to handle the geometric increase in number of Mach lines over such an interval.

Fig. 11.11($a$) shows a unit process, between two points of known states

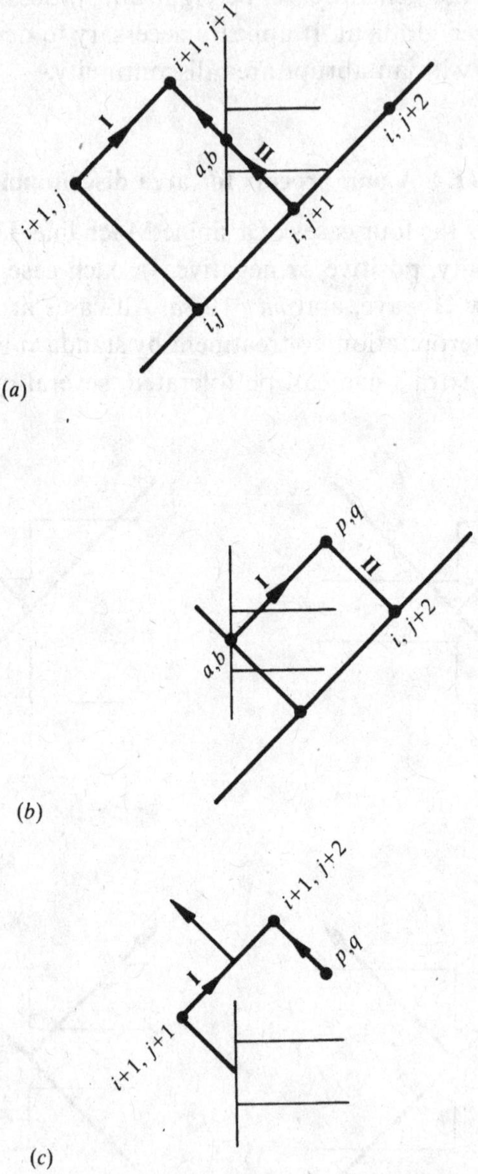


Fig. 11.11 The phases of a unit process to take into account the states on an extra (reflected) wave component and to carry them forward to the next (scheduled) integration point, thus permitting the intermediate information to be discarded.

and coordinates, $i + 1, j$ and $i, j + 1$ and one new point, $i + 1, j + 1$. The process is superimposed over an area discontinuity, and gives rise to a new **I** Mach line, originating from Mach plane coordinates here labelled $a, b$. The extra wave generated eventually forms part of a separate unit process which will also span the area discontinuity. If, as has so far been the norm, the integration scan is from left to right, the next unit process has to cope with the extra Mach line. This may be achieved by applying the new unit process twice (Fig. 11.11(*b*)), first from known points $a$, $b$ and $i, j + 2$ to a new, intermediate point at $p, q$, and a subsequent unit process (Fig. 11.11(*c*)) to $i + 1, j + 2$. This last step will generate an extra Mach line, **I** or **II** depending upon which of the pair of Mach lines reaching point $i + 1, j + 2$ strikes the area discontinuity first.

Nothing has been compromised in calculating the states at $i + 1, j + 2$ and the location of this point, which is thus determined with the same precision as that of all other points to the current stage. If the **I** Mach line is now forgotten, the only information lost in proceeding with the rightward scan on the basis of the new point (plus the next 'previous' point, $i, j + 3$) is the extra resolution which would have been gained by the reduction in integration step size. The unit process must set a flag to alert the unit process based on $i + 1, j + 1$, at the next sweep there is now an extra wave to be taken account of, namely that (Fig. 11.11(*c*)) generated in the last stages of the current unit process.

A computer simulation implementing the foregoing steps is under development and test, but is not yet ready for use. Thus for the time being the only functioning simulation based comprehensively on the Method of Characteristics, with account of friction and heat transfer, remains that of Sirett.[10] His presentation of the Mach plane benefited from a feature of the particular pen plotter used, whereby the starting point of a line segment had slightly increased intensity over the rest. This feature leads to the effect seen in Fig. 11.12, which plots the paths of particles between Mach lines, but not the Mach lines themselves. The latter nevertheless show up by virtue of the feature of the plotter described.

The work advanced as far as the achievement of indicator diagrams and reliefs of temperature and pressure in function of location and time (Figs 11.13 and 11.14), but was regrettably not pursued.

While it has here been argued, and the work of Sirett has demonstrated, that the Method of Characteristics may be adapted to accommodate wave reflection sites, what has been achieved, in a sense, is merely to unearth a further latent area of difficulty: the wire mesh regenerator is a sequence of hundreds – or in some instances – thousands of wave reflection sites. If

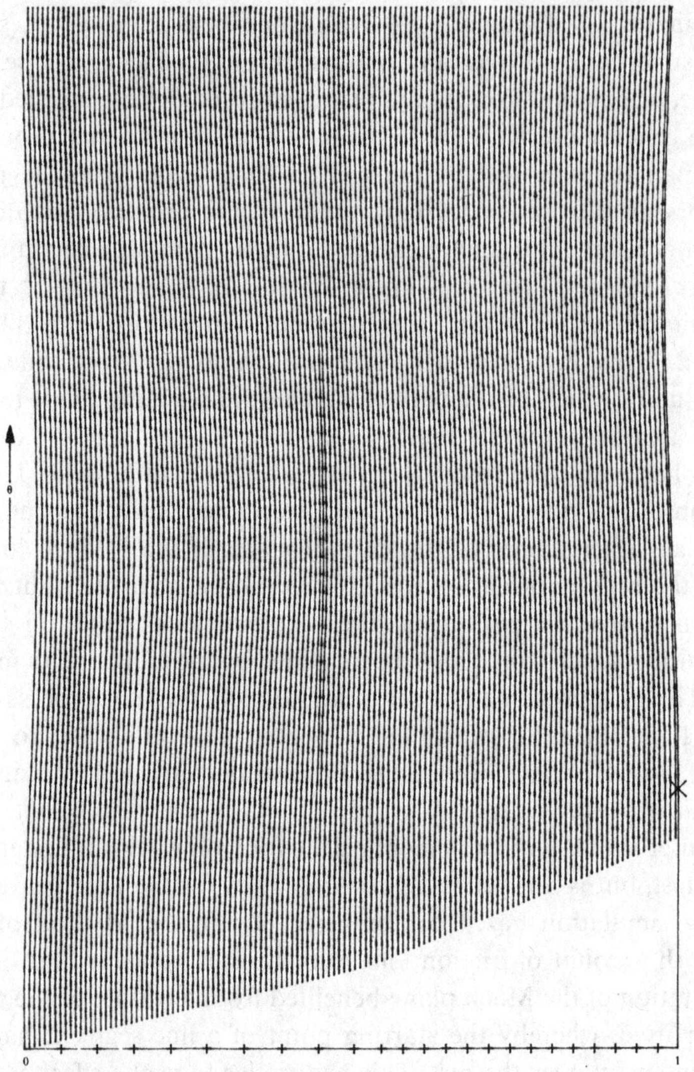


Fig. 11.12 Mach plane plot produced by Sirett's implementation of the Method of Characteristics. (After Sirett.[10]).

computation of cycle states must take into account waves reflected from the basic area discontinuities between working spaces and heat exchangers, it follows that it must account also for the infinitely larger number of reflections per pass which arise in the regenerator.

A thousand or so individual area discontinuities separated by some 0.025 mm is not at first sight an attractive proposition for any process of numerical integration. The chapter which follows, however, shows that, with certain simplifications, the problem becomes straightforward.

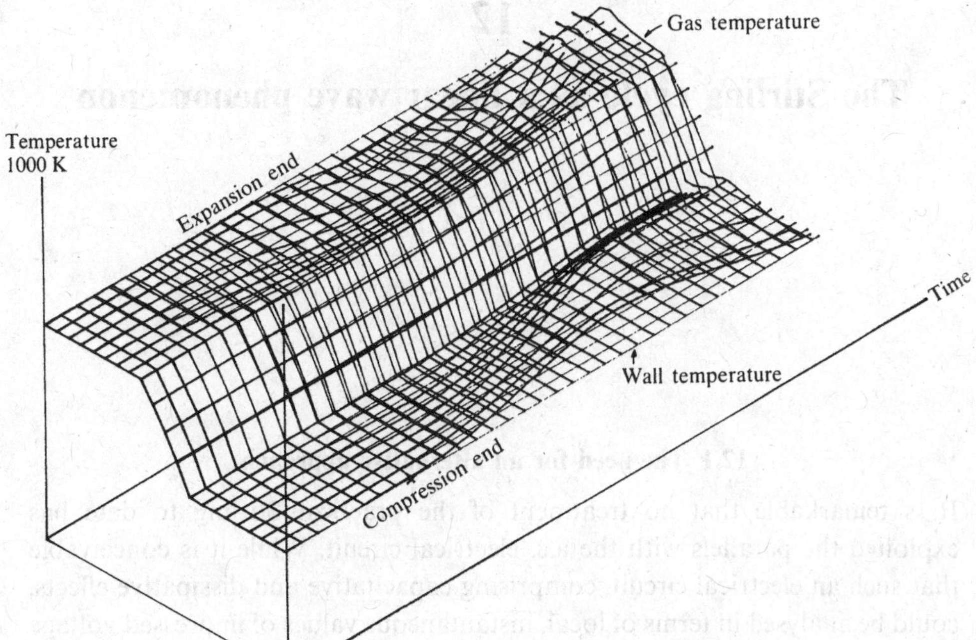


Fig. 11.13 Relief of temperature as a function of location and crank angle produced by Sirett's Characteristics simulation.

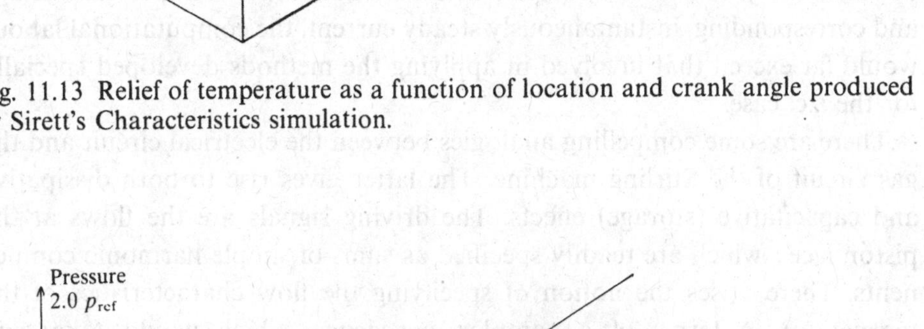

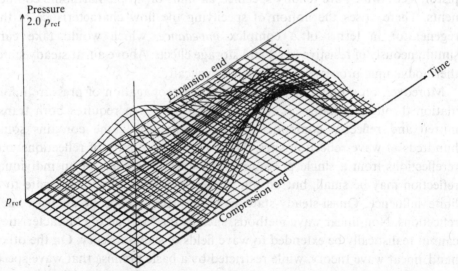

Fig. 11.14 Relief of pressure as a function of location and crank angle produced by Sirett's Characteristics simulation.

# 12

# The Stirling cycle as a linear wave phenomenon

## 12.1 The need for an alternative approach

It is remarkable that no treatment of the practical Stirling to date has exploited the parallels with the a.c. electrical circuit. While it is conceivable that such an electrical circuit, comprising capacitative and dissipative effects, could be analysed in terms of local, instantaneous values of impressed voltage and corresponding, instantaneously steady current, the computational labour would far exceed that involved in applying the methods developed specially for the a.c. case.

There are some compelling analogies between the electrical circuit and the gas circuit of the Stirling machine. The latter gives rise to both dissipative and capacitative (storage) effects. The driving signals are the flows at the piston faces, which are readily specified as sums of simple harmonic components. There arises the notion of specifying the flow characteristics of the regenerator in terms of a complex *impedance*, which would take care simultaneously of resistive (d.c.) and storage effects. Above all, at steady state, thermodynamic processes are repeated cyclically.

Moreover, on a one-dimensional view, the propagation of pressure information through a change in cross-sectional flow area requires both transmitted and reflected components. The regenerator alone contains some hundreds of wave reflection sites. These imply an infinity of reflections and rereflections from a single pressure disturbance. The effect of an individual reflection may be small, but an infinity of small effects may aggregate to a finite influence. Quasi-steady-state methods cannot cope at all with such reflections. Non-linear wave methods, such as the Method of Characteristics, cannot realistically be extended to wave fields of such intricacy. On the other hand, linear wave theory, while restricted by a basic premise that wave speed is equal to local acoustic speed, can readily handle an unlimited number of wave reflection sites.

This chapter applies the methods of linear wave analysis to computation of the instantaneous pressure distribution in one-dimensional, compressible flow through a regenerator modelled as a series (some hundreds) of flow area discontinuities. The approach predicts the experimentally observed asymmetry between compression and expansion space traces. As far as it goes the method is completely free of the need to carry out numerical integration. It thus avoids problems of integration mesh size and of associated convergence and stability. It thus represents an alternative starting point for cycle analysis which, if heat transfer effects can eventually be taken into account, may lead to more tractable formulations than the quasi-steady-flow treatments which have become the norm.

## 12.2 Simplifying assumptions

(1) The working fluid is an ideal gas.

(2) It adopts the temperature of the immediately adjacent solid wall.

(3) The $j$th gauze in a stack of $n$ regenerator gauzes (Fig. 12.1($a$)) is uniformly at temperature $T_j$:

$$T_j = T_e + \frac{T_c - T_e}{n} j$$

(4) Fluid particle velocity is always negligible by comparison with local, instantaneous acoustic speed, $a$. Pressure information therefore propagates at $+a$ and $-a$.

(5) A given harmonic of piston motion may be represented by the real (or imaginary) part of the simple harmonic velocity $\omega r\, e^{i\omega t}$ at a point. The approximation is a reasonable one where $r$ is substantially less than total heat exchanger passage length, but is not as good when the condition is not met.

(6) A single regenerator gauze may be modelled as a flow passage with an area discontinuity (Fig. 12.1($b$)). The length of the individual parallel portions is equal to wire diameter, $d_w$, in each case. The ratio of the smaller to the larger cross-sectional area is equal to free flow area ratio, $\alpha_{\P}$, for the gauze (and, equivalently, for the matrix as a whole).

(7) Where necessary in order to expedite analysis, $\rho = p_{ref}/RT$, where $T$ is local, absolute temperature and $p_{ref}$ is constant.

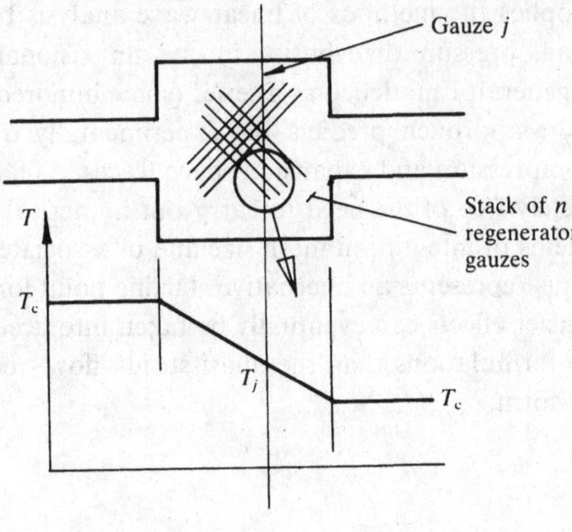


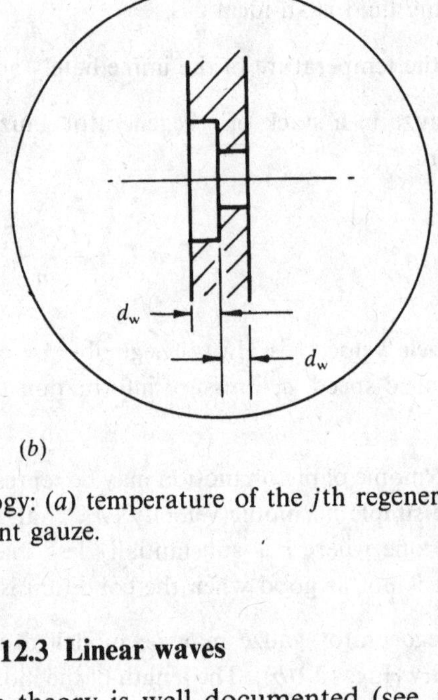


Fig. 12.1 Regenerator terminology: (*a*) temperature of the $j$th regenerator gauze in a stack of $n$ gauzes; (*b*) equivalent gauze.

## 12.3 Linear waves

One-dimensional linear wave theory is well documented (see, for example, Ref. 1) and leads to

$$\frac{\partial^2 p}{\partial t^2} = a^2 \frac{\partial^2 p}{\partial x^2} \tag{12.1}$$

In Eqn (12.1) $p$ is the pressure excess above the mean (or hydrostatic) value and $a^2$ is defined as $\partial p/\partial \rho$ which, under assumption (2) above, is equivalent to $RT$.

The following satisfy Eqn (12.1):

$$p = f\left(t - \frac{x}{a}\right), \qquad u = \frac{p}{\rho a} = \frac{f(t - x/a)}{\rho a}$$

(wave in direction of $x$ increasing)

$$p = g\left(t + \frac{x}{a}\right), \qquad u = \frac{p}{\rho a} = \frac{g(t + x/a)}{\rho a}$$

(wave in direction of $x$ decreasing)

Anticipating a geometric model of the Stirling machine in the form of a series of ducts of different cross-sectional areas, the local instantaneous volume flow rate, $J$, is defined as

$$J = A_x u$$

At any instant, $t$, and location, $x$, a parallel element of duct will contain two wave components, one running from left to right, the other from right to left:

$$p(x, t) = f\left(t - \frac{x}{a}\right) + g\left(t + \frac{x}{a}\right)$$

$$J(x, t) = \frac{A_x[f(t - x/a) + g(t + x/a)]}{\rho a}$$

The left-running wave may embody components that are reflections from a right-running wave which passed earlier, and *vice versa*.

The task is to specify $p(x, t)$ and $J(x, t)$ at all $t$ and all $x$ throughout a crankshaft revolution. The method adopted will be to suppose the possibility of Fourier analysis of the waves into simple harmonic components. The linearity property then permits the components to be summed.

## 12.4 Description of the flow-field in terms of linear waves

Fig. 12.2($a$) is a schematic of the expansion space with its contents at absolute temperature $T_e$. The piston face motion is modelled as a simple harmonic velocity at $x = 0$. A superposition of several such harmonics will permit the fundamental and higher harmonics of any piston motion to be specified.

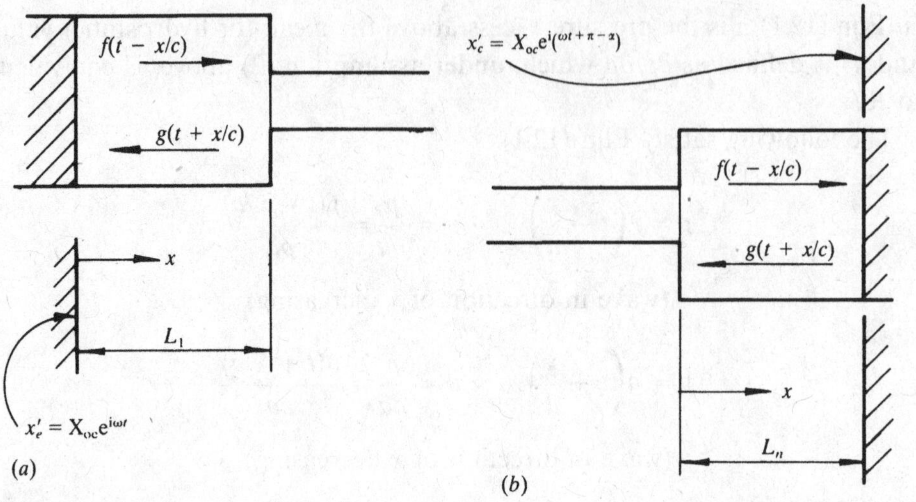


Fig. 12.2 Terminology for expansion and compression spaces: (*a*) expansion end; (*b*) compression end.

The right-running wave, $f$, is the resultant of the wave generated at $x = 0$ by the piston harmonic under consideration and any component of an earlier, left-running wave reflected from the piston 'face'. Wave $g$, currently left, is the aggregate of that part of the left-running wave that emerged at $x = L_1$.

Now consider a specific harmonic component of the general wave. At any point in the expansion space ($0 \leqslant x \leqslant L_1$):

$$p(x, t) = F_1\, e^{i\omega(t - x/a)} + G_1\, e^{i\omega(t + x/a)}$$

Subscript 1 has been chosen to indicate that the expansion space is the first in a sequence of $n$ interconnected spaces, each of uniform cross-sectional area, which receive the same treatment:

$$J(x, t) = A_{x1} \frac{F_1\, e^{i\omega(t - x/a)} - G_1\, e^{i\omega(t + x/a)}}{\rho a}$$

At $x = 0$, $J$ is specified in terms of piston face velocity:

$$J(0, t) = A_{x1} \frac{F_1\, e^{i\omega t} - G_1\, e^{i\omega t}}{\rho a} = A_{x1}\omega r_e\, e^{i\omega t}$$

whence

$$\omega r_e \rho a = F_1 - G_1 \tag{12.2}$$

Pressure at the piston face is:

$$p(0, t) = F_1\, e^{i\omega t} + G_1\, e^{i\omega t} \tag{12.3}$$

The comparable expressions for the compression piston face (Fig. 12.2(*b*)) are derived from:

$$J(L_n, t) = A_{xn}\omega r_c \, e^{i(\omega t + \pi - \alpha)}$$

$$\omega r_c \rho a \, e^{i(\pi - \alpha)} = F_n \, e^{-i\omega L_n/a} - G_n \, e^{i\omega L_n/a} \qquad (12.4)$$

$$p(L_n, t) = F_n \, e^{i\omega(t - L_n/a)} + G_n \, e^{i\omega(t + L_n/a)} \qquad (12.5)$$

In the foregoing expressions $\rho$ and $a$ assume values appropriate to $T$ and thus to location.

The four equations, (12.2)–(12.5) contain between them the four unknown (complex) quantities $F_1$, $G_1$, $F_n$, and $G_n$. If there were no discontinuities in cross-sectional area or variation in fluid properties between the faces of expansion and compression pistons, then $F_1$ would be the same function as $F_n$, $G_1$ the same as $G_n$, and Eqns (12.2) and (12.4) would be sufficient to determine them. Then $p(x, t)$ could be expressed for all $x$ and $t$. The solution of this special case (isothermal, constant area tube) as $\omega$ tends to zero will be checked against the predictions of the 'isothermal' analysis in a later section.

Changes in cross-sectional area do exist, however, and a one-dimensional model of compressible flow requires wave reflections at each to conserve mass and momentum. The changes of flow area at the variable-volume spaces and heat exchangers are obvious instances. In addition, each regenerator gauze causes a contraction and reexpansion of the flow. In the Stirling machine to be modelled there are 330 gauzes. This number of reflecting sites gives rise to an infinity of reflections and rereflections of a single incident wave. On experience with non-linear approaches such as the Method of Characteristics it might be thought that there would be no hope of coping. Formulated in terms of linear waves, the problem is, on the contrary, straightforward.

Fig. 12.3 indicates the $j$th area transition in a sequence of any number of such transitions. The flow sections on either side are assumed to be of constant cross-sectional area. At any instant there are forward ($F_j$, $F_{j+1}$) waves on both sides of the area discontinuity, and respective reverse-running waves, $G_j$, $G_{j+1}$. Equating pressures at the area discontinuity ($u$ is assumed sufficiently small that the steady-flow momentum change can be ignored):

$$A_j \, e^{-i\omega L_j/a} + B_j \, e^{i\omega L_j/a} = A_{j+1} + B_{j+1} \qquad (12.6)$$

Equating flows:

$$A_{x_j}(F_j \, e^{i\omega(t - L_j/a_j)} - G_j \, e^{i\omega(t + L_j/a_j)})/\rho_j a_j = A_{x_{j+1}}(F_{j+1} \, e^{i\omega t} - G_{j+1} \, e^{i\omega t})/\rho_{j+1} a_{j+1}$$

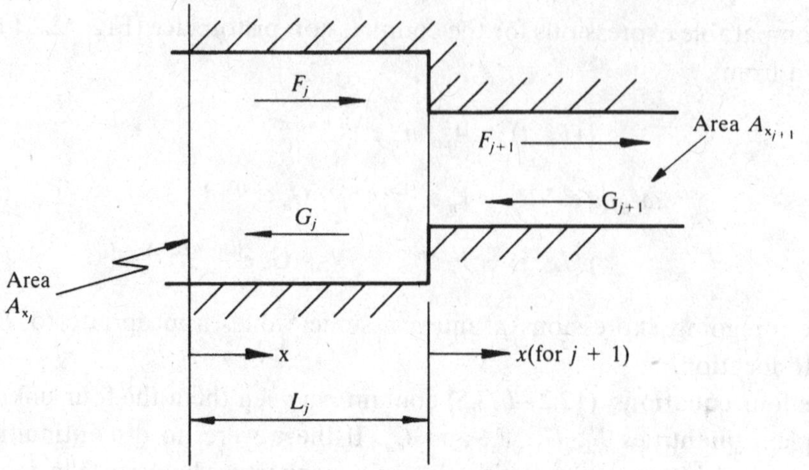


Fig. 12.3 Conditions at typical area discontinuity.

For computational purposes $\rho_j$ and $a_j$, $\rho_{j+1}$ and $a_{j+1}$ take their mean respective values over the lengths $L_j$ and $L_{j+1}$. At the interface the values are common, so that

$$\left(\frac{F_j\,\mathrm{e}^{-\mathrm{i}\omega L_j/a_j} - G_j\,\mathrm{e}^{\mathrm{i}\omega L_j/a_j}}{A_{\mathrm{x}j+1}}\right)A_{\mathrm{x}_j} = F_{j+1} - G_{j+1} \tag{12.7}$$

For a system of $n - 2$ area transitions there are $n - 2$ equations such as Eqns (12.6) and (12.7), plus two flow equations at the boundary. A solution for all the $n$ complex $F$ and $G$ is thus possible. The intuitive expectation is for manipulation of a substantial matrix. In fact, it is possible to proceed with a sequence of almost trivial analytical steps:

Eqns (12.6) and (12.7) may be rearranged to express the relationship between $F_{j+1}$ and $F_j$, $G_{j+1}$ and $G_j$ in the form of a simple $2 \times 2$ matrix:

$$\begin{matrix} F_{j+1} \\ G_{j+1} \end{matrix} = \begin{vmatrix} a_{11} & a_{12} \\ a_{21} & a_{22} \end{vmatrix} \begin{matrix} F_j \\ G_j \end{matrix} \tag{12.8}$$

where

$$a_{11} = 0.5(1 + \alpha_{\mathrm{x}})\,\mathrm{e}^{-\mathrm{i}\omega L_j/a_j}$$
$$a_{12} = 0.5(1 - \alpha_{\mathrm{x}})\,\mathrm{e}^{\mathrm{i}\omega L_j/a_j}$$
$$a_{21} = 0.5(1 - \alpha_{\mathrm{x}})\,\mathrm{e}^{-\mathrm{i}\omega L_j/a_j}$$
$$a_{22} = 0.5(1 + \alpha_{\mathrm{x}})\,\mathrm{e}^{\mathrm{i}\omega L_j/a_j}$$
$$\alpha_{\mathrm{x}} = A_{xj}/A_{xj+1}$$

(The $\alpha$ with subscript x is an area ratio, not to be confused with the unsubscripted symbol for volume phase angle.)

Eqn (12.8) is of the form

$$\begin{matrix} F_{j+1} \\ G_{j+1} \end{matrix} = \mathbf{Z}_1 \begin{matrix} F_j \\ G_j \end{matrix} \tag{12.8a}$$

$$Z_1 = \begin{vmatrix} a_{11} & a_{12} \\ a_{21} & a_{22} \end{vmatrix}$$

If the interface between the parallel lengths of tube $j+1$ and $j+2$ is now dealt with the following equations are obtained

$$\begin{matrix} F_{j+2} \\ G_{j+2} \end{matrix} = \mathbf{Z}_2 \begin{matrix} F_{j+1} \\ G_{j+1} \end{matrix}$$

$$\mathbf{Z}_2 = \begin{vmatrix} b_{11} & b_{12} \\ b_{21} & b_{22} \end{vmatrix}$$

$F_{j+2}$ and $G_{j+2}$ may now be related to $F_j$ and $G_j$ via

$$\begin{matrix} F_{j+2} \\ G_{j+2} \end{matrix} = \mathbf{Z} \begin{matrix} F_j \\ G_j \end{matrix} \tag{12.9}$$

$$\mathbf{Z} = \begin{vmatrix} a_{11}b_{11} + a_{21}b_{12} & a_{12}b_{11} + a_{22}b_{12} \\ a_{11}b_{21} + a_{21}b_{22} & a_{12}b_{21} + a_{22}b_{22} \end{vmatrix} \tag{12.9a}$$

Thus, an equation of the form of Eqn (12.9) relates expansion space conditions directly to those in the compression space. **Z** is always a simple, 2 × 2 matrix obtained by successive multiplication in the manner of expression (12.9). Eqns (12.2) and (12.9) together determine $F_1$, $G_1$, $F_n$, $G_n$. From values of **Z** stored intermediate to the final value, any $F_j$, $G_j$ may be determined.

Knowing machine geometry, boundary conditions (piston amplitudes, phase angle, $\alpha$) and temperature distribution (to give the $a_j$) the distributions of pressure and flow are determined as a function of time and location throughout a cycle at angular frequency $\omega$. A specific numerical solution takes the form of the amplitude and phase of pressure and flow for given $\omega$. If piston motion derives from a mechanism with many harmonics, the linearity of the problem allows the solutions for the different frequency components to be added, giving net pressure and flow as a function of time and location.

## 12.5 Normalization

Under the assumption $a = \sqrt{(RT)}$ the term $\omega L/a$ may be written:

$$\frac{\omega L}{a} = 2\pi \frac{L}{L_{\text{ref}}} \frac{N_{\text{s}} L_{\text{ref}}}{\sqrt{(RT)}}$$
$$= 2\pi\lambda_j N_{\text{MA}}$$

$N_{\text{MA}}$ is the now familiar characteristic Mach number. Dividing Eqns (12.2), (12.4) and (12.9) by $p_{\text{ref}}$ and noting that $\rho = p/a^2$ amounts to expressing the solution as

$$\frac{p}{p_{\text{ref}}} = \frac{p}{p_{\text{ref}}}(\lambda_1, \lambda_2, \lambda_3, \ldots, \lambda_n, N_\tau, N_{\text{MA}}, \phi) \tag{12.10}$$

In Eqn (12.10), the $\lambda_1, \ldots, \lambda_n$ (which include the constant phase angle, $\alpha$) define the geometry of the machine. Temperature ratio, $N_\tau$, and characteristic Mach number, $N_{\text{MA}}$, complete the description of $p/p_{\text{ref}}$ as a function of crank angle, $\phi$. If there is more than one input harmonic (from either piston or both) then $p/p_{\text{ref}}$ has the value given by summing for each $N_{\text{MA}}$ corresponding to each harmonic.

Harmonic piston face pressure components may be integrated analytically with respect to simple harmonic displacement components to give net, specific cycle work, $\zeta$:

$$\zeta = \zeta(\lambda_1, \lambda_2, \lambda_3, \ldots, \lambda_n, N_\tau, N_{\text{MA}}) \tag{12.11}$$

## 12.6 Experimental results and comparison with computation

The Stirling machine described by Rix[2] has been modelled using the linear wave approach. An outline specification of the machine is included in Appendix VI.

The results to be referred to are those obtained by Rix and previously reproduced as Fig. 2.3. These showed pressures at the piston faces as functions of crank angle. The parameters were $N_{\text{MA}}$ and a second group, $N_{\text{SG}}$, which takes care of dynamic similarity for the case of a real fluid having a finite coefficient of dynamic viscosity. In the figure, expansion and compression traces are separated by an arbitrary vertical distance, and no significance attaches to the net vertical offset of one trace relative to the other. In a given column of traces, $N_{\text{MA}}$ and $N_{\text{SG}}$ are common regardless of working fluid.

The phenomenon of interest is the fact that, with $N_{\text{MA}}$ increasing, the expansion trace becomes increasingly noisy, increasingly distorted and reduced in amplitude by comparison with the compression trace. In order to achieve anything comparable by conventional computation Rix found it necessary to augment the steady-flow friction correlations for the wire screen by up to an order of magnitude.

Fig. 12.4(*a*) shows traces computed from the present analysis over the same values of $N_{MA}$ but with $N_{SG}$ at a high value to give the effect of vanishing viscosity. First and second harmonics of the Ross crank drive linkage are accounted for. As for the experimental traces, vertical separation is by an arbitrary amount.

In the experimental counterpart a small temperature difference (of the order of 10 °C) had been measured between the fluids external to expansion and compression exchangers. With heat being lifted from the expansion end and delivered to the compression end a temperature difference somewhat greater than this is implied in the mean temperatures of working fluid in the expansion and compression spaces. In recognition of this, a value of $N_\tau$ of 0.833 is assumed.

For $N_\tau = 1$, compression and expansion space pressure swing amplitudes are identical, as the theoretical development insists. For the small temperature difference supposed here, pressure excursions at the expansion end suffer a small reduction in amplitude by comparison with those at the compression end, but resulting asymmetry is slight. Phase shift between the two traces increases noticeably with increasing $N_{MA}$ but is again insufficient to account for the experimentally observed distortions. The inviscid fluid clearly does not behave like the real counterpart.

Viscous dissipation requires to be taken into account. The treatment (that described by Lighthill[3]) is analytically simple and is applied to each gauze individually. The effect of viscosity is to give rise to a velocity distribution across the flow which is a boundary-layer effect at high Reynolds number and a Poiseuille distribution at low values. The entire range may be taken into account by defining a complex acoustic speed which is a function of local Womersley parameter.[3] For a comprehensive treatment the reader is referred to the source, from which it will be seen in particular that use of empirical correlations such as that between $N_{re}$ and the friction factor, $C_f$, is completely avoided. Instead, the effects of viscosity are characterized by the dimensionless group $N_{SG}$.

Fig. 12.4(*b*) shows traces with the effects of viscosity taken into account. Individual traces are for the same combinations of $N_{MA}$ and $N_{SG}$ as the experimental counterparts. Trends in trace geometry are comparable, but amplitudes are not quite right. When the experimental traces were obtained, gain was adjusted so that the traces more or less filled the screen. Amplitude generally increased with increasing $N_{MA}$, calling for reduced gain. The computed traces are also scaled to fill the frame, but amplitudes when viscous effects are incorporated are somewhat less than experimental values, and increase with $N_{MA}$ at a lower rate. The equations modified for complex sound speed do not exactly conserve mass.

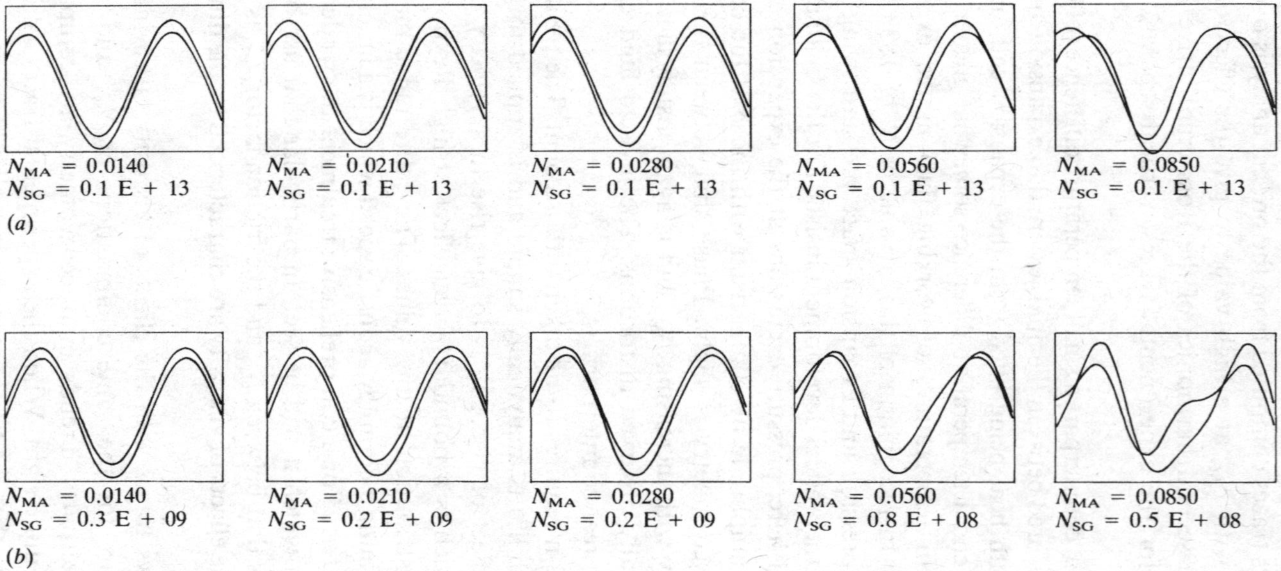


Fig. 12.4 Computed traces corresponding to experimental conditions Fig. 2.3. (*a*) No viscosity ($N_{SG} \rightarrow \infty$), $N_{MA}$ increasing. (*b*) Viscous effects included.

## 12.7 Comparison with the static 'isothermal' case

Fig. 12.5 shows a tube of length $L$. At the left (expansion) and right (compression) extremities the velocities, $u$, and corresponding volume flow

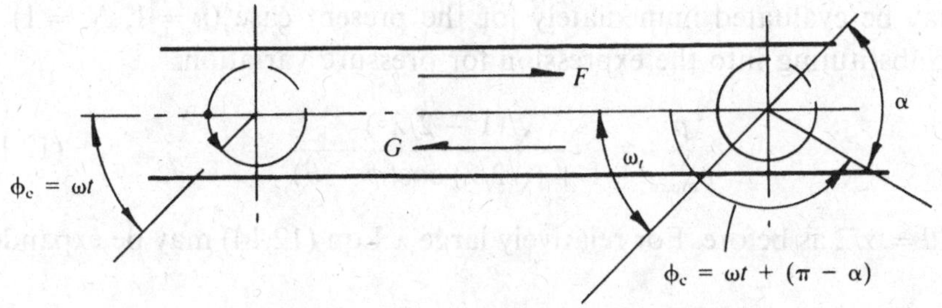


Fig. 12.5 Isothermal tube of uniform cross-sectional area.

rates, $J$, are as for the case already dealt with (Figs 12.2 and 12.3). With $F_1 = F_n$ and $G_1 = G_n$, $F$ and $G$ follow from Eqns (12.3) and (12.4) by simple elimination:

$$\frac{F}{p_{\text{ref}}} = \frac{2\pi N_{\text{MA}}(\kappa\, \mathrm{e}^{\mathrm{i}(\alpha - 2\pi\lambda N_{\text{MA}})} - 1)}{\mathrm{e}^{-4\mathrm{i}\pi\lambda N_{\text{MA}}}} \tag{12.12a}$$

$$\frac{G}{p_{\text{ref}}} = \frac{F}{p_{\text{ref}}} - 2\pi N_{\text{MA}} \tag{12.12b}$$

$$\lambda = L/r_{\text{e}}$$

The solution for $p(x, t)$ is formed by substituting Eqns (12.12a) and (12.12b) into Eqn (12.5). For low revolutions per minute $N_{\text{MA}}$ tends to zero and the solution equation assumes a factor $2\pi N_{\text{MA}}/\sin(\lambda 2\pi N_{\text{MA}})$, which tends to $1/\lambda$. The solution becomes:

$$\frac{p(t)}{p_{\text{ref}}} = \frac{2}{\lambda} \sin\left(\frac{\alpha}{2}\right) \sin\left(\phi + \frac{\alpha}{2}\right) \tag{12.13}$$

where $\phi = \omega t$.

This is the 'static' solution, in which the time for pressure to propagate between piston faces is so short relative to the time for a cycle that pressure at any instant may be considered uniform. In Eqn (12.13) the term $(2/\lambda)\sin(\alpha/2)$ has the form of an amplitude. Inspection of Fig. 12.5 confirms that amplitude is greatest for $\alpha = \pi$ $(\sin(\pi/2) = 1)$. It is also easy to confirm that maximum and minimum pressures occur at $\omega t = +\alpha/2$ and $-\alpha/2$ as Fig. 12.5 would predict.

The Schmidt analysis provides a check on the predictions of Eqn (12.13); Fig. 12.5 may be viewed as a cylinder closed by pistons whose volume variations are out of phase by $\alpha$ rad. Using Finkelstein's formulation[4] there is defined a dead space parameter, being the ratio of volume unswept (corrected for temperature) to volume swept by the expansion-space piston. This may be evaluated immediately for the present case ($\kappa = 1$, $N_\tau = 1$) as $\lambda - 2$. Substituting into the expression for pressure variation:

$$\frac{p}{p_{\text{mean}}} = \frac{\sqrt{(1 - 2/\lambda^2)}}{1 + (\sqrt{2}/\lambda)\cos(\phi - \theta)} \tag{12.14}$$

Where $\theta = \alpha/2$ as before. For relatively large $\lambda$ Eqn (12.14) may be expanded to give

$$\frac{p}{p_{\text{mean}}} = 1 - \frac{\sqrt{2}}{\lambda}\cos\left(\phi - \frac{\alpha}{2}\right) \tag{12.15}$$

The unity term on the right-hand side corresponds merely to the pressure offset, $p_{\text{mean}}$, from zero. There is thus a sinusoidal ($\alpha/2 = \pi/4$ and $\sin(\phi - \pi/4) = -\cos(\phi - \pi/4)$) variation of pressure with amplitude $\sqrt{2}/\lambda$. For the supposed value of $\alpha$ ($\pi/2$) the amplitude term of Eqn (12.13) is also $\sqrt{2}/\lambda$, and Eqn (12.15) becomes identical with Eqn (12.13).

The degenerate form of the wave formulation for the special case of $N_{MA} = 0$ thus gives the same results as the Schmidt equation provided $\lambda \gg 1$.

## 12.8 Acknowledgement

The treatment of the Stirling cycle machine in terms of linear waves was made possible by virtue of considerable help from Dr Ann Dowling of Cambridge University Engineering Department.

# 13

# Finite-cell solution schemes

## 13.1 Expectation *vs* reality

Most finite-cell integration schemes proposed to date have been of one-dimensional Eulerian type: the working fluid circuit is subdivided into fixed, contiguous control volumes. Corresponding to the 'slab flow' simplification, individual control volumes are rectangular. The laws of conservation of mass, momentum and energy are applied to a typical control volume. To this extent, the process of setting up the numerical model corresponds exactly to that of developing the conservation laws in analytical, differential form. Possibly for this reason it is often supposed that, by subdividing into sufficiently small control volumes and time steps, '. . . the solution becomes independent of the numerical parameters',[1] i.e., that it becomes equivalent to the analytical counterpart.

In fact, this is not so; a fixed-grid finite-cell solution scheme may, as demonstrated by Berchowitz, Urieli and Rallis,[2] show convergence towards some specific numerical result as interval sizes are progressively reduced. However, there is nothing to say that the numerical result in question is that which would be returned by analytical solution of the same problem. The discrepancy is due to a subprocess inherent in the numerical sequence, whereby fluid properties are linearized across each control volume at each step in the integration sequence. The arbitrarily imposed linearity is the feature which permits the fluid properties to be defined throughout the flow-field. On the other hand, flow-field properties may not be linear between fixed grid points and be capable at the same time of transporting information at particle speed (heat transfer and friction information) or at combined particle and acoustic speed (pressure information).

The issue has already been discussed at some length in the context of the Method of Characteristics (Chapter 11 and Appendix XI), where it was recommended that the reader carry out an exercise in numerical analysis to convince himself of the nature of the problem. If the exercise was tackled it

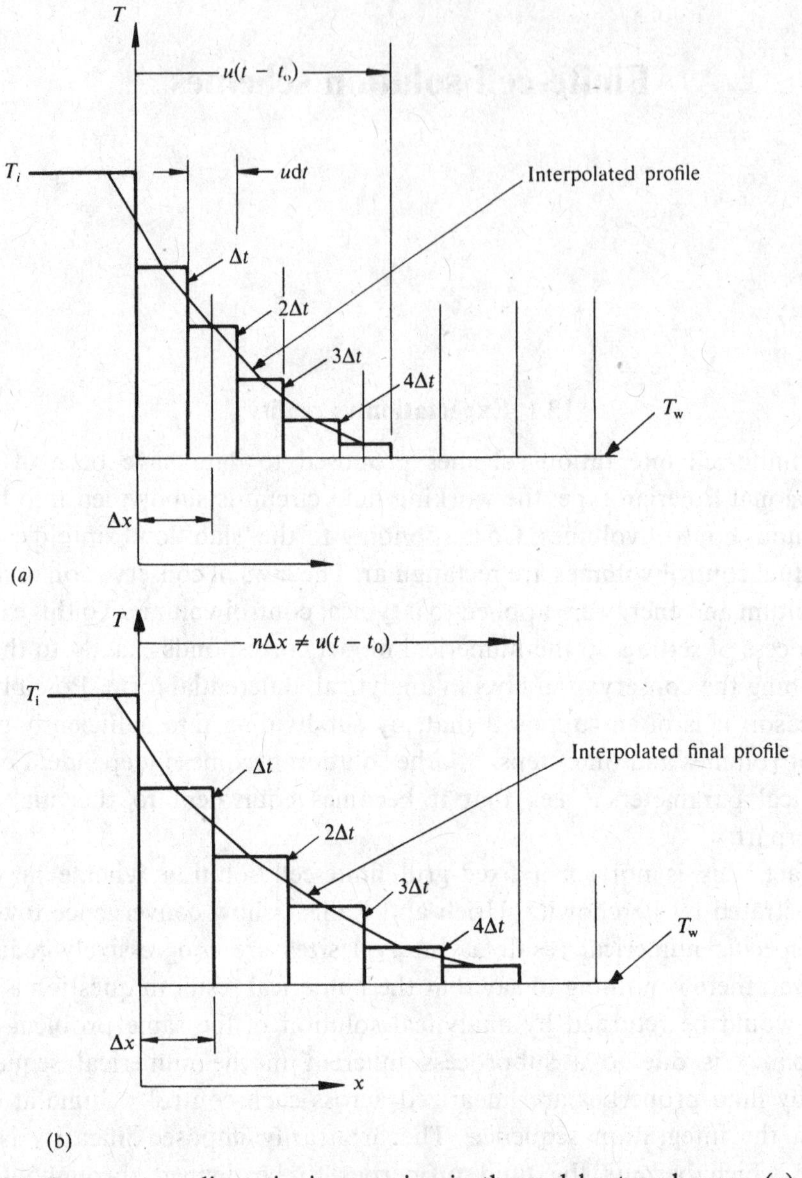


Fig. 13.1 Temperature discontinuity entering isothermal heat exchanger. (*a*) Temperature profile development consistent with the simplifications of slab flow. (*b*) The artificial diffusion introduced by the fixed-grid integration scheme.

will have revealed what is now shown schematically in Fig. 13.1: a discontinuity in temperature enters a heat exchanger at time $t = 0$. The true progress of the one-dimensional wave (no diffusion in the flow direction) is at particle velocity, $u$. A discontinuity persists at the head of the wave. At any given $t$

the incomplete temperature profile superimposes over the steady-state solution. The Eulerian numerical scheme, by contrast, causes the front to propagate at the speed of the integration process. Not only is this speed generally false, but the temperature history at any given point varies (not shown) as integration proceeds, in conflict with the prediction of the analytical solution.

As far as the author is aware, it has not been the custom to subject numerical simulations of the Stirling cycle to a test such as the imposition of temperature discontinuities, despite the fact that to do so is elementary. The reason may lie in the argument that such severe discontinuities do not arise in practice, and that the response of conventional simulations to such temperature profiles as *do* arise is plausible.

On the other hand, recent attempts at isolating and characterizing individual thermodynamic losses[3] focus on a phenomenon whereby a heat exchanger passage contains two or more 'slugs' of gas, each having its discrete level of turbulence depending on whether it resides permanently in the exchanger, or whether it enters once per cycle from an adjacent working space. The front which separates such 'slugs' travels essentially at local particle speed as does the temperature front (and, incidentally, coincides identically with the fronts investigated by Organ[4]). If the analyst is not persuaded of the desirability of having the numerical solution preserve temperature front information, perhaps the argument will be more convincing now that it is seen to apply equally to the notion of fronts between different levels of turbulence.

If finite-cell numerical schemes based in the Eulerian reference frame cannot cope, then it is necessary to look for an alternative which can. The Method of Characteristics affords *all* of the desired facilities without exception. On the other hand, it calls for a great deal of homework in advance of implementation, and its substantial demands on processing time are greatest at conditions of operation for which pressure wave propagation effects are least significant. What is required, therefore, is a scheme which fills the operational gap between 'static' conditions (e.g., in the limit, zero angular speed) and the point where wave propagation effects influence the outcome.

For the present, one-dimensional flow case a suitable scheme may be constructed in the Lagrange reference frame. In Lagrange coordinates, the computational frame moves at local particle speed. The transport phenomena of concern are thus accounted for correctly. (It is worth noting that Lagrange methods fail, or become excessively complicated, in two dimensions where there is substantial shear deformation of the flow net.[5])

## 13.2 The Lagrange reference frame

The energy equation in standard (Eulerian) form is[6]

$$q = (\partial/\partial t)[\rho A_x(c_v T + \tfrac{1}{2}u^2)] + (\partial/\partial x)[\rho u A_x(c_v T + p/\rho + \tfrac{1}{2}u^2)] \quad (13.1)$$
$$q = h(T_w - T)/\rho r_h$$

In Eqn (13.1) the appearance of $c_v T$ corresponds to the assumption of ideal gas behaviour. Following Shapiro's treatment, Eqn (13.1) is expanded, the equation of continuity (Eqn (AXI.3)) subtracted, and an outstanding term in $\partial p/\partial x$ eliminated by means of the momentum equation (Eqn (AXI.4)) to give

$$q + uF = \frac{\mathrm{D}(c_v T)}{\mathrm{d}t} + \frac{p}{\rho A_x}\frac{\partial}{\partial x}(A_x u) \quad (13.2)$$

In Eqn (13.2) D/d$t$ is the 'substantial' or 'material' derivative, and applies while following a particle at velocity $u$. It is thus part-way to Lagrange form, and in its present format is already a more compact expression of energy conservation than was the Eulerian counterpart: it states that the combined effect of heat exchange (which may have positive or negative value) and friction heating (always positive from the definition of $F$) is to change the terms on the right-hand side. The latter comprise internal energy, and a quantity based on $p\,\partial u/\partial x$. Examination of this last term shows it to be the rate of local reversible work done when the system boundaries dilate or contract against pressure $p$. Having the equation in this form facilitates comparison with other usages such as the more heuristic expression of Rix,[7] who discounts the effects of friction (see Section 13.4 for the phenomena missed when friction is omitted), and that of Crowley[8] who retains the friction term and extends the description to include the effects of mass transfer.

At this point there is a departure from Shapiro's treatment: the equation of mass conservation may be expanded:

$$\rho\frac{\partial(A_x u)}{\partial x} + A_x\left\{\frac{\partial \rho}{\partial t} + u\frac{\partial \rho}{\partial x}\right\} = 0$$

The term in the curly braces will be recognized as the substantial derivative, so that

$$\frac{1}{\rho A_x}\frac{\partial(A_x u)}{\partial x} + \frac{1}{\rho^2}\frac{\mathrm{D}\rho}{\partial t} = 0$$

Provisionally assuming ideal gas behaviour (the restriction will be reexamined later), the term $\rho^2$ may be replaced by $(p/RT)^2$ and the result substituted into Eqn (13.2) to give:

$$q + uF = \frac{\mathrm{D}(c_{\mathrm{v}}T)}{\mathrm{d}t} - \frac{RT^2}{p}\frac{\mathrm{D}}{\mathrm{d}t}\left(\frac{p}{T}\right) \tag{13.3}$$

Eqn (13.3):

has been formulated without neglecting *any single one* of the terms of the original partial differential equations of conservation of mass, momentum and energy. It is therefore consistent with all the physical and geometric phenomena embodied in those differential equations.

is in Lagrange form. It therefore applies to a particle, and, within limits, to a control volume moving with velocity $u$. The total differential form means that the integration process is uncluttered by consideration of artificial numerical convection etc.

is algebraically simpler than the corresponding expression in the Eulerian system.

places no restriction on the correlations used for $F$ and $q$. Therefore with realistic assumptions about the respective variations of $p$ and $u$ with crank angle, $\omega t$, and the variation with $x$ of $T_{\mathrm{w}}(x)$ (see definition of $q$) it *amounts to a one-line statement of the thermodynamic processes affecting a working fluid particle as it cycles within the heat exchanger system.*

Eventually, Eqn (13.3) will be embodied, together with the corresponding form of the mass and momentum conservation laws, in a numerical solution which will solve for $u$ and $p$, thus permitting modelling of a particular gas circuit under specified conditions of operation. In the meantime it may be solved in isolation to show its potential for illustrating representative aspects of gas circuit phenomena.

## 13.3 A one line description of gas circuit processes

The system of normalization used in this text will by now be familiar, except that little attention has been paid to the variables $u$ (particle velocity) and $\rho$. In the context of the Method of Characteristics, where high values of $u$ are encountered, the logical normalizing quantity for $u$ is reference acoustic speed, $a_0$, so that $\mathbf{u} = u/a_0$. In the present case, $u$ is assumed $\ll a_0$, so that a more realistic normalizing base is $N_{\mathrm{s}}L_{\mathrm{ref}}$. This leads to numerical values of $\mathbf{u} = u/N_{\mathrm{s}}L_{\mathrm{ref}}$ which tend to be in the hundreds when $L_{\mathrm{ref}}$ is $R_{\mathrm{H}}$, and which are of the order of unity when $L_{\mathrm{ref}}$ is crank-pin offset, $r$; $\rho$ is normalized by $M/V_{\mathrm{ref}}$, so that $\wp = \rho V_{\mathrm{ref}}/M$. For the ideal gas case, and from the definitions of $\psi$ and $\tau$, $\wp = \psi/\tau$.

In the Lagrange formulation $u = \mathrm{d}x/\mathrm{d}t$ (Fig. 13.2($a$)), and the independent variable is $\phi = \omega t$. Normalized distance, $x/L_{\mathrm{ref}}$ is denoted $\lambda$, so that the

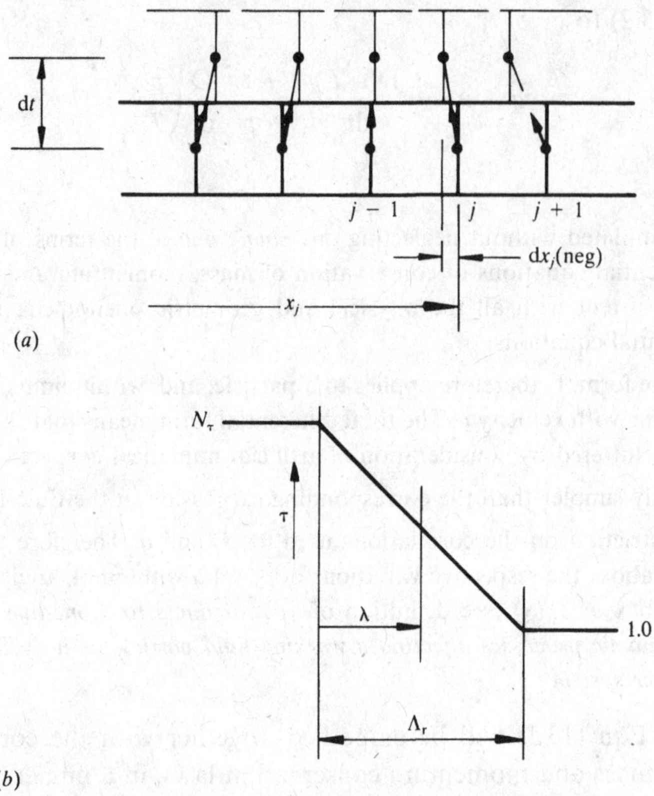


Fig. 13.2 Coordinate system for Lagrangian solution. (*a*) The Lagrangian reference frame, indicating the time-dependent motion of system boundaries. (*b*) Distribution of wall temperature to which a moving fluid particle is exposed.

relationship between $\boldsymbol{u}$ and $\lambda$ is:

$$\boldsymbol{u} = 2\pi(\mathrm{d}\lambda/\mathrm{d}\phi) \tag{13.4}$$

With the foregoing definitions, the normalized form of Eqn (13.3) with terms substituted for $q$ and $F$ is

$$-(\gamma - 1)\frac{\tau^2}{\psi}\frac{\mathrm{D}}{\mathrm{d}\phi}\frac{\psi}{\tau} + \frac{\mathrm{D}\tau}{\mathrm{d}\phi} = \frac{\gamma}{2\pi\lambda_\mathrm{h}} N_{\mathrm{ST}}\frac{\tau}{\psi}[\tau_\mathrm{w}(\lambda) - \tau] + \frac{(\gamma - 1)}{2\pi} N_{\mathrm{MA}}{}^2 \frac{\boldsymbol{u}^3 C_\mathrm{f}}{2\lambda_\mathrm{h}}\frac{\boldsymbol{u}}{|\boldsymbol{u}|} \tag{13.5}$$

As before, $N_{\mathrm{ST}} = hA_{\mathrm{ref}}/N_\mathrm{s}Mc_\mathrm{p}$.

The differential coefficients in Eqn (13.5) are equivalent to total coefficients when following a particle at velocity $\boldsymbol{u}$, so a simple numerical integration in uniform steps of $\mathrm{d}\phi$ yields particle location and corresponding variation in dimensionless temperature over a cycle $\phi = 0$ to $2\pi$. The procedure will be demonstrated for a specific case. It is assumed that, in a particular Stirling machine, a particle of working fluid traverses the regenerator during a cycle, just reaching the ends at the instants of flow reversal. Particle velocity and pressure amplitude are simple harmonic functions of crank angle with an arbitrary phase angle between the two. As the solution proceeds it will become apparent that it is a simple matter to extend the treatment to other amplitudes (e.g., particle entering an isothermal exchanger tube) and to velocity and pressure functions of any desired form.

Regenerator temperature distribution is assumed linear as indicated in Fig. 13.2(*b*), so that, in terms of the dimensionless distance between expansion end and instantaneous particle location, $\lambda$, the temperature of the wall element with which the particle is instantaneously exchanging heat is:

$$\tau_w(\lambda) = N_\tau + \frac{1.0 - N_\tau}{\Lambda_r}\lambda \tag{13.6}$$

For a particle undergoing simple harmonic motion, velocity, $u(\omega t) = U_0 \sin(\omega t)$. Applying the condition that the particle travels the full length of the regenerator in half a cycle leads to $u = \pi N_s L_r \sin(\omega t)$, where $L_r$ is regenerator flow passage length. From the definition of normalized velocity:

$$\boldsymbol{u}(\phi) = \pi\Lambda_r \sin(\phi), \qquad \Lambda_r = L_{reg}/L_{ref} \tag{13.7a}$$

By similar reasoning, particle location as a function of $\phi$ becomes:

$$\lambda(\phi) = -\tfrac{1}{2}\Lambda_r(\cos\phi - 1) \tag{13.7b}$$

The values of all variables in Eqn (13.3) are now specified except those of $\tau$; $N_{ST}$ and $C_f$ may be any desired functions of flow passage geometry and local, instantaneous Reynolds number. To complete the example, the 'thermodynamic similarity' correlations for the wire gauze (Figs 7.3(*b*), 7.4(*b*)) are supposed. With this arrangement, the dimensionless hydraulic radius, $\lambda_h$ is a separate parameter of the problem which may take arbitrary (but realistic) values.

Accounting for the fact that, for normal $N_{pr}$, $h$ (in $N_{ST}$) is $h(N_{re}) = h(N_{RE}, \wp, \boldsymbol{u})$, and that $C_f = C_f(N_{RE}, \wp, \boldsymbol{u})$ it comes as no surprise that the expression, and thus the solution, for $\tau$, compares with that for the cycle as a whole, namely:

$$\tau = \tau(N_\tau, \gamma, \lambda_h, \Lambda_r, N_{MA}, N_{SG}, (N_{PR}), \phi)$$

Additional parameters not mentioned explicitly are pressure swing amplitude and the phase between pressure and velocity. In a real machine these are functions of the basic parameters listed.

Eqn (13.3) must be discretized to make $D\tau$ explicit: it has been found possible to proceed by merely expanding the coefficient $D(\psi/\tau)\,d\phi$ and leaving the terms of the right-hand side untouched. However, if extreme values of $N_{ST}$ are to be investigated (as, e.g., to simulate 'isothermal' operation) then it is necessary to expand the two terms in $\tau$ on the right-hand side to the mid-interval point, viz, to $\tau + \frac{1}{2}\,d\tau$, and to take the $\frac{1}{2}\,d\tau$ terms into consideration. The range of parameter values over which the solution will function is further enhanced by retaining the higher order terms, the $(d\tau)^2$, which result from the expansion, which thus has the form

$$a\,d\tau^2 + b\,d\tau + c = 0 \tag{13.8}$$

The standard solution of a quadratic equation gives

$$d\tau = \frac{-b \pm \sqrt{[b^2 - 4ac]}}{2a}$$

and inspection of a range of numerical results establishes that the positive sign applies. Carrying out the expansion leads to the following values for the coefficients of Eqn (13.8):

$$a = \frac{(\gamma - 1)}{(\psi_\phi + \frac{1}{2}\,d\psi)\tau_0\,d\phi}(\psi_\phi + \tfrac{3}{4}\,d\psi) + \tfrac{1}{4}\frac{\gamma}{2\pi}\frac{N_{ST}}{\lambda_h} \tag{13.9a}$$

$$b = \frac{(\gamma - 1)}{(\psi_\phi + \frac{1}{2}\,d\psi)\,d\phi}\psi_0 + \frac{\gamma}{2\pi}\frac{N_{ST}\frac{1}{2}[\tau_w(\lambda) - 2\tau_0]}{\lambda_h(\psi_\phi + \frac{1}{2}\,d\psi)} + \frac{1}{d\phi} \tag{13.9b}$$

$$c = \frac{-(\gamma - 1)\tau_0\,d\psi}{(\psi_\phi + \frac{1}{2}\,d\psi)\,d\phi} - \frac{\gamma N_{ST}\tau[\tau_w(\lambda) - \tau_0]}{2\pi\lambda_h(\psi_\phi + \frac{1}{2}\,d\psi)} + \text{RHS} \tag{13.9c}$$

In Eqn (13.9c) for $c$, RHS is an abbreviation for the complete flow friction term at the right-hand side of Eqn (13.5), unmodified except for being evaluated at mid-integration interval, $\phi + \frac{1}{2}\,d\phi$. Terms with subscript 0 take the numerical value corresponding to the start of the integration step.

### 13.4 A numerical example

To provide a specific example, the parameters of Philips' MP1002CA engine are used (Table 4.1). Thus, the various numerical values will reflect a reference area of 0.55 m$^2$, a reference volume of 0.12E − 03 and, correspondingly, a reference length, $L_{ref} = R_H$ of 0.216E − 03. We fix $N_\tau$ at 2.65, while parameters

$\lambda_h$, $N_{SG}$ and $N_{MA}$ (and hence $N_{RE}$) are varied either side of their nominal values to investigate the effect of such variation.

A discontinuity is specified in terms of the properties of adjacent particles. Eqn (13.5) defines what happens to a particular particle. At this stage, therefore, we are looking not at discontinuities, but at the temperature history of a particular particle. It will have to be kept in mind that, at certain extreme values of the parameters the regenerator would have insufficient thermal capacity for the linear gradient to be maintained, and that at others, unrealistically high pressure gradients would arise.

For Fig. 13.3(*a*) all parameters have the standard values for the MP1002CA machine at rated conditions. The temperature history of the particle settles to cyclic equilibrium after a small number of reversals. As anticipated, temperature swing either side of nominal is slight, but it is noteworthy that such swing as occurs is not symmetrical.

In Fig. 13.3(*b*), $N_{MA}$ and $\lambda_h$ retain rated values, but $N_{SG}$ is increased by more than an order of magnitude. This effectively increases the thermal capacity of the working fluid particle and thus the temperature difference either side of nominal. Some two complete cycles are necessary for cyclic equilibrium to be attained, and thereafter the temperature excursions occur more or less symmetrically either side of the linear distribution for the matrix.

The Lagrange formulation promises a direct insight into the significance or otherwise of the phénomenon of friction reheating: in Eqn (13.5) the final term on the right-hand side is always positive, while the primary direction of heat flow clearly changes sign more or less in sympathy with flow direction. On this basis it might be anticipated that gas-to-matrix temperature difference would be higher on the hot blow ($\boldsymbol{u}$ positive) than on the cold blow: the reheating effect would add to the heat load to be dumped on the hot blow, while lightening the load on the cold blow. To see the effect it is necessary to set a value of $N_{MA}$ considerably outside the range achieved in the MP1002CA machine. Fig. 13.3(*c*) illustrates the effect of increasing $N_{MA}$ by an order of magnitude. On the hot-to-cold pass the reheating effect causes an increased temperature differential. On the reverse pass, it contributes to the process of heating by convection, thus bringing the temperature of the particle closer to that of the matrix. The asymmetry in Fig. 13.3(*c*) is probably only part of the phenomenon previously illustrated in Fig 2.3.

A comprehensive cycle description based on the Lagrange coordinate framework will shortly be discussed. In the meantime it will be worthwhile to define the qualities sought in a simulation which might be used for the design of a prototype machine costing perhaps £$\frac{1}{4}$ million to construct.

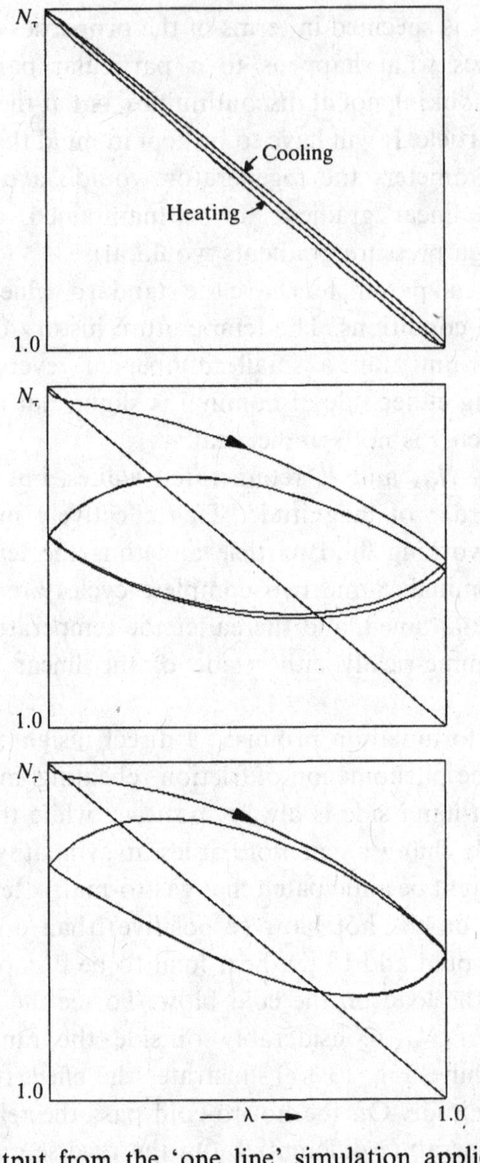


Fig. 13.3 Specimen output from the 'one line' simulation applied to the Philips MP1002CA engine, showing fluid particle history from start-up for (*a*) rated conditions, (*b*) $N_{SG}$ ($= p_{ref}/N_s\mu$) increased, thereby increasing working fluid thermal capacity and (*c*) rated $N_{SG}$ with $N_{MA}$ enhanced to exaggerate the effects of friction reheating.

## 13.5 Verifiability

A problem which afflicts *any* numerical formulation is that of verifying that the computer code correctly implements the intended numerical algorithms. Those who have followed the debate about flight safety in connection with the fly-by-wire A320 Airbus will be aware of a view, widespread among numerical analysts, that there is no such thing as guaranteed freedom from errors in computer code of any size. Implementation of the finite-cell scheme affords the chance of looking into the 'checkability' of simulation code. In this connection a priority will be to exploit every opportunity for modular program architecture (a very different approach from that of the authors of, e.g., Ref. 9, as reference to their program listing will show).

There is greatest potential for verification if the formulation permits interchange of blocs of code within a common main program so as to represent a range of models of the same gas circuit: for example, an 'isothermal' (CL-1) model such as the Schmidt is free of propagating discontinuities, so that a generalized finite-cell formulation with only the CL-1 (mass conservation) module active ought to return values identical to those of the closed-form solution. ('Generalized', because for this special case (only), an Eulerian formulation should also give values equal to those from the closed-form solution.) With the CL-3 (energy conservation) module active, results identical to those of the isothermal analysis should again result (from a Lagrange or other appropriate formulation) when $N_{ST}$ is set equal to infinity.

The required modular form has been implemented in the 'building-bloc' approach outlined in Chap. 2. A key feature is the *implicit* integration scheme. This gives rise to a matrix of the coefficients of the unknowns (the increments in the dependent variables) at each step in the solution. Predetermined blocs of the matrix are reserved for particular conservation laws (or for corresponding 'dummy' laws – see below). Within such individual blocs, subblocs are reserved for particular dependent variables – pressure, temperature, etc. Thus, simulations may be built from a selection of statements of conservation of mass and/or momentum and/or energy, but in each combination, the energy conservation law (for example) will be plugged into the same bloc of matrix. The scheme is not restricted to any particular number of gas circuit subdivisions or time (or crank angle) steps. A separate bloc is reserved for the gas law, so a ready comparison is available between the assumptions of ideal and, say, van der Waals' gas behaviour.

## 13.6 The 'building-bloc' approach

### *13.6.1 Subdivision of the gas circuit*

For convenience of presentation and discussion of the algebra, a form of equivalent machine will be modelled in which free-flow area is uniform between the piston faces. Extension of the treatment so that it deals with flow area discontinuities is straightforward, but is best handled separately at a later stage. On the other hand, any desired variation of hydraulic radii and of wall temperature distribution may be specified at this juncture.

Fig. 13.4(*a*) shows the gas circuit, with total gas mass, $M$, divided into an

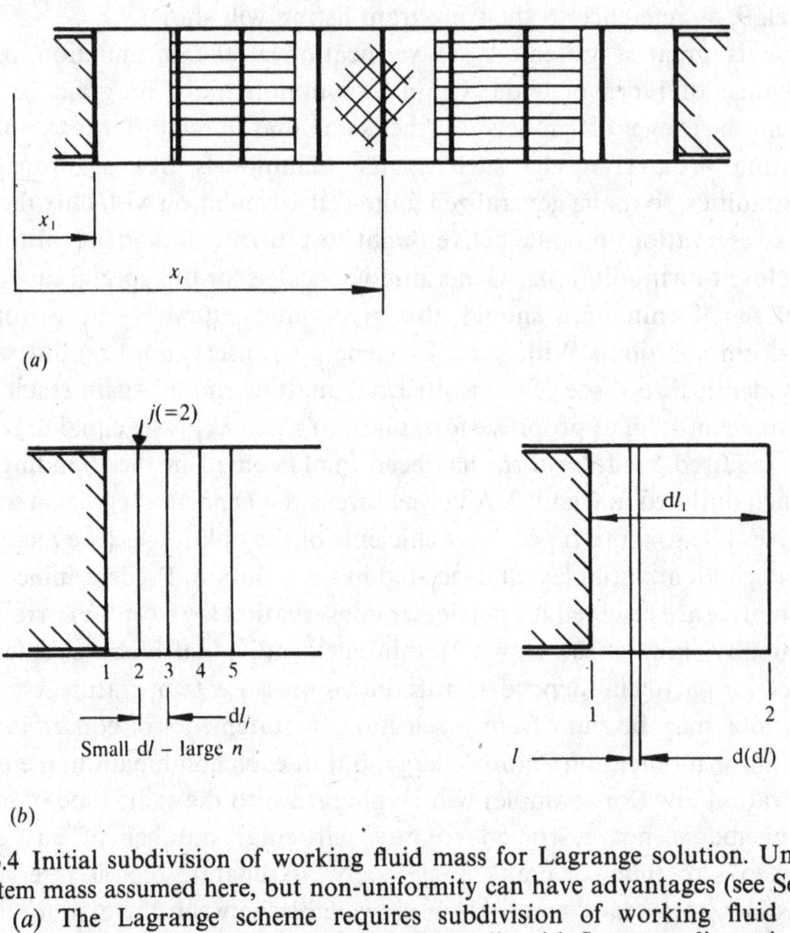


Fig. 13.4 Initial subdivision of working fluid mass for Lagrange solution. Uniform subsystem mass assumed here, but non-uniformity can have advantages (see Section 13.10). (*a*) The Lagrange scheme requires subdivision of working fluid mass. Subsystem boundaries correspond only coincidentally with flow area discontinuities. (*b*) Initial setting of subsystem boundaries requires stepping through from left to right using guessed values of mean subsystem density (i.e., of temperature). The process iterates to convergence (last calculated subsystem boundary coincides with known location of right-hand piston face).

arbitrary number, $n$, of masses, d$m$. The most convenient arrangement is for all the d$m$ to be equal, but the restriction is not essential. (Later it will be shown that there are occasions when non-uniform subdivision is almost essential.) Fluid properties are defined at the mid-points of the subsystems as indicated. $n$ may be increased essentially without limit, bringing the numerical model increasingly close to the spirit of the defining equations. The gas circuit shown represents the opposed-piston machine, but the method extends without complication to machines with compression space divided by a heat exchanger.

By contrast with the Eulerian formulation, the location, $\lambda_j$, of the interface between adjacent subsystems is a variable to be solved for as numerical solution proceeds. Starting values require to be set for all variables, including the $n - 1$ values of $\lambda_j$. This is most easily accomplished with the aid of the 'isothermal' analysis (Section AII.4), in which the dimensionless total mass is unity, so that individual (dimensionless) submasses are of magnitude $1/n$. Wall temperature distribution provides the required initial working fluid temperature distribution once the mid-points of the subsystems are fixed. This is achieved in one of two ways, depending essentially upon the magnitude of $n$, and is described (Fig. 13.4($b$)) in terms of absolute variables.

For $n$ sufficiently large, $\mathrm{d}m \approx \rho_j \, \mathrm{d}V_j \approx \rho_j A_x \, \mathrm{d}l_j$. Starting from the left-hand piston face, successive $\mathrm{d}l_j$ are calculated by inverting this latter equality, the required value of $\rho_j$ being set in terms of initial $p$ and the $T_w$ at $l_j$, where $T_w$ is expressed as per Eqn (13.6). A bit of iteration can find the $T_w$ at $l_j + \frac{1}{2}(l_{j+1} - l_j)$. For smaller values of $n$ and correspondingly larger lengths of individual subsystems, d$m$ is divided into, say, 20 subsubunits and the foregoing process applied 20 times, starting again at the left-hand piston face, and accumulating the elemental lengths occupied by the subsubvolumes. After this number of applications of the algorithm, the process has arrived at the interface between the first two subsystems – $l_2$. The process repeats as far as $l_{n+1}$ (compression piston face) and the result is checked against the (known) value of $l_{n+1}$.

In practice the process of initial set-up is carried out in terms of the dimensionless variables of the problem, $\psi$, $\tau$, $\wp$, $\lambda$. It is worth noting that the 'fixed' values of $\lambda$ which define exchanger boundaries may be calculated in terms of the various exchanger (dimensionless) cross-sectional areas, $\alpha_x$, and dead volumes, $\mu_d$. Moreover, instantaneous piston positions may be expressed in terms of the respective (dimensionless) volume variations. There is accordingly no need to set up a special data file for the program, which can make use of the input data ($\alpha_x$, $\mu_d$ etc.) for the simulations described in earlier chapters.

There are $n$ unknown values of each of $\psi$, $\tau$, $\wp$, and $n-1$ of $\lambda$. The size of the array of unknowns is $4n-1$. Since there is no essential connection between the number and siting of the nodes for the gas and those for the solid walls, discussion will proceed for the moment on the assumption that three conservation laws are to be taken into account, i.e., the matter of wall temperature variation is deferred.

### 13.6.2 A single conservation law model

A single conservation law model comprises (*a*) a geometric description of the gas circuit, (*b*) a statement of mass conservation and (*c*) an equation of state linking $p$, $\rho$, and $T$. Since CL-2, CL-3 and CL-4 models also contain these same three elements, formulation of the CL-1 model is an essential step in setting up any of the more comprehensive variants.

The law of mass conservation in the Lagrange system of Fig. 13.4 is

$$\rho_j A_x(x_{j+1} - x_j) = M/n$$

The equivalent normalized form is

$$\wp_j(\lambda_{j+1} - \lambda_j) = 1/n\alpha_x \tag{13.10}$$

Comparison with the equivalent expression in Eulerian coordinates will underline the elegant simplicity of the Lagrange form.

Eqn (13.10) and the other conservation laws have to be discretized for numerical solution. The equation is first of all expanded about $\phi = 0$ to $\phi = \phi + \mathrm{d}\phi$, giving:

$$(\wp_j + \mathrm{d}\wp_j)([\lambda_{j+1} - \lambda_j] + [\mathrm{d}\lambda_{j+1} - \mathrm{d}\lambda_j]) = 1/n\alpha_x$$

Multiplying out, subtracting Eqn (13.10) and retaining terms of higher order gives:

$$\mathrm{d}\wp_j[\lambda_{j+1} - \lambda_j] + \mathrm{d}\lambda_j[-\wp_j] + \mathrm{d}\lambda_{j+1}[+\wp_j] = -\mathrm{d}\wp_j(\mathrm{d}\lambda_{j+1} - \mathrm{d}\lambda_j) \tag{13.11}$$

The terms in square brackets are the coefficients of the unknowns which are to be fed into the array for numerical solution. The coefficients are entered according to the column reserved for the associated variable. The allocation of columns used here (which is arbitrary) is, first, $n$ columns for $\mathrm{d}\psi$, followed by $n$ each for $\mathrm{d}\tau$ and $\mathrm{d}\wp$ and $n-1$ for $\mathrm{d}\lambda$ in that order. A CL-4 model would follow with $n_w$ spaces for the coefficients of dimensionless wall temperature increment, $\mathrm{d}\tau_w$. The first of the $n$ equations, which follow the pattern of Eqn

(13.11), is that for the extreme left-hand subsystem, and it occupies the first row of the matrix. This and the $n$th such equation are special cases, however, since in the former case $d\lambda_j$ is known (being $d\lambda_1 = d\lambda_e$ of the expansion piston), while in the latter case it is $d\lambda_{j+1}$ which is known, being $d\lambda_{n+1} = d\lambda_c$ of the compression piston. These two special cases of Eqn (13.11) thus lead to coefficients given by:

$$d\wp_1[\lambda_2 - \lambda_1] + d\lambda_2[\wp_1] = d\lambda_1\wp_1 - d\wp_1(d\lambda_2 - d\lambda_1) \tag{13.12a}$$

$$d\wp_n[\lambda_{n+1} - \lambda_n] + d\lambda_n[-\wp_n] = -d\lambda_{n+1}\wp_j - d\wp_n(d\lambda_{n+1} - d\lambda_n) \tag{13.12b}$$

If it were envisaged that ideal gases only were to be modelled, Eqn (13.10) could have been formulated in terms of $\psi$ and $\tau$, the array of unknowns would have been smaller by a factor of some $0.75 \times 0.75 \approx 0.5$ and solution correspondingly speedier. The form of Eqns (13.12), however, affords the valuable option of being able to specify, e.g., a van der Waals' gas.

To relate the coefficients just defined to their respective locations in the matrix, the first row is considered. Coefficients of individual density changes occupy columns $2n+1$ to $3n$, those of length increments columns $3n+1$ to $4n-1$. Row 1 therefore contains all zeros but for the following coefficients:

$$a_{1,2n+1} = \lambda_2 - \lambda_1$$

$$a_{1,3n+1} = \wp_1$$

$$b_1 = d\lambda_1\wp_1 - d\wp_1(d\lambda_2 - d\lambda_e)$$

A typical row $i$ in the CL-1 bloc has the coefficients

$$a_{i,2n+1} = \lambda_{j+1} - \lambda_j$$

$$a_{i,3n+1} = -\wp_j$$

$$a_{i,3n+i+1} = \wp_j$$

$$b_i = -d\wp_j(d\lambda_{j+1} - d\lambda_j)$$

With $j = i$, coefficients occupy the appropriate columns. Row $n$ of the CL-1 bloc is the last, and contains the coefficients corresponding to the extreme right-hand subsystem.

The CL-1 formulation requires 'dummy' conservation laws in place of the full statements of momentum and energy conservation. These are conceptually identical to the corresponding assumptions of the Schmidt (CL-1) analysis, which constrains adjacent pressures to be equal at any given instant, and local fluid temperatures always to be equal to those of the immediately adjacent solid wall.

The 'dummy' conservation law for momentum sets $\psi_j = \psi_{j+1}$. The equality has to apply on expansion to the end of the integration step $(\phi + \mathrm{d}\phi)$, i.e., $\psi_j + \mathrm{d}\psi_j - \psi_{j+1} - \mathrm{d}\psi_{j+1} = 0$, which reduces to $\mathrm{d}\psi_j - \mathrm{d}\psi_{j+1} = 0$. The $n-1$ relationships of the CL-2 bloc occupy rows $n+1$ to $2n-1$, and the coefficients of $\mathrm{d}\psi_j$ occupy columns $1-n$, so that a typical row in the dummy CL-2 bloc has

$$a_{i,i-n} = 1$$

$$a_{i,i-n+1} = -1$$

$$b_i = 0, \qquad n+1 \leqslant i \leqslant 2n-1$$

The dummy energy conservation law states that $\tau_j = \tau_w(\lambda_j)$. The statement must be expanded to the end of the integration step, and thus becomes $\tau_j + \mathrm{d}\tau_j = \tau_w[\frac{1}{2}(\lambda_j + \mathrm{d}\lambda_j + \lambda_{j+1} + \mathrm{d}\lambda_{j+1})]$. It is not convenient to make the unknowns $\mathrm{d}\lambda_j$ and $\mathrm{d}\lambda_{j+1}$ explicit, but to do so is not essential: for reasonably small $\mathrm{d}\phi$ the $\mathrm{d}\lambda_j$ and $\mathrm{d}\lambda_{j+1}$ from the previous integration step are an excellent approximation to current values. Using these 'old' values on a first inversion of the matrix of coefficients allows a second inversion over the same integration interval to calculate accurate values.

The dummy CL-3 bloc occupies rows $2n$ to $3n-1$ (i.e., $2n \leqslant i \leqslant 3n-1$). The coefficients for the $n$ values of the $\mathrm{d}\tau_j$ occupy columns $n+1$ to $2n$. A typical row of the dummy CL-3 bloc therefore has zeros except for the following terms:

$$a_{i,i+1-2n} = 1$$

$$b_i = \tau_w[\tfrac{1}{2}(\lambda_j + \mathrm{d}\lambda_j + \lambda_{j+1} + \mathrm{d}\lambda_{j+1})] - \tau_j$$

Allocating the coefficients to the correct columns is achieved by putting $j = i + 1 - 2n$.

The ideal gas equation provides a suitable specimen gas law. In terms of absolute variables $p = \rho RT$. Normalizing, $\psi = \wp\tau$. At the end of the integration step $\psi + \mathrm{d}\psi = (\wp + \mathrm{d}\wp)(\tau + \mathrm{d}\tau)$. Subtracting the original $\psi = \wp\tau$ and retaining higher order terms (taken up in the second pass through the solution) $\mathrm{d}\psi_j - \wp_j\,\mathrm{d}\tau_j - \tau_j\,\mathrm{d}\wp_j = \mathrm{d}\wp_j\,\mathrm{d}\tau_j$.

There are $n$ such equations, one for each node. These occupy the lowermost bloc of the array – i.e., rows $3n$ to $4n-1$ if wall temperature variations are not modelled. Correct row location is achieved by $j = i + 1 - 3n$. A typical row has zeros but for the finite coefficients:

| | 1 | 2 | 3 | 4 | 5 | 6 | 7 | 8 | 9 | 10 | 11 | 12 | 13 | 14 | 15 | RHS |
|---|---|---|---|---|---|---|---|---|---|---|---|---|---|---|---|---|
| CL-1 | | | | | | | | | | | | | | | | |
| 1 | | | | | | | | | * | | | | * | | | * |
| 2 | | | | | | | | | | * | | | * | * | | |
| 3 | | | | | | | | | | | * | | | * | * | |
| 4 | | | | | | | | | | | | * | | | * | * |
| CL-2 | | | | | | | | | | | | | | | | |
| 5 | * | * | | | | | | | | | | | | | | |
| 6 | | * | * | | | | | | | | | | | | | |
| 7 | | | * | * | | | | | | | | | | | | |
| CL-3 | | | | | | | | | | | | | | | | |
| 8 | | | | | * | | | | | | | | | | | |
| 9 | | | | | | * | | | | | | | | | | |
| 10 | | | | | | | * | | | | | | | | | |
| 11 | | | | | | | | * | | | | | | | | |
| State | | | | | | | | | | | | | | | | |
| 12 | * | | | | * | | | | * | | | | | | | |
| 13 | | * | | | | * | | | | * | | | | | | |
| 14 | | | * | | | | * | | | | * | | | | | |
| 15 | | | | * | | | | * | | | | * | | | | |

Fig. 13.5 Stencil of matrix of CL-1 solution. The asterisks indicate non-zero terms and show dummy CL-2 and CL-3 laws active. The number of subdivisions (4) permits representative illustration of the pattern of array entries, but would be unrealistically small for solution of a model involving area discontinuities.

$$a_{i,i+1-3n} = 1$$

$$a_{i,i+1-2n} = -\wp_j$$

$$a_{i,i+1-n} = -\tau_j$$

$$b_i = \mathrm{d}\wp_j\,\mathrm{d}\tau_j \qquad 3n \leqslant i \leqslant 4n-1$$

Any value of $n$ equal to or greater than 3 will produce a matrix adequate for checking that the array elements are being fed into their intended locations. Fig. 13.5 was produced with $n = 4$ and by providing for the symbol $*$ to be printed for each non-zero element. The pattern of asterisks is always the same for the CL-1 and gas law blocs, but those in the CL-2 and CL-3 blocs are particular to the respective dummy laws.

Various tests of correct functioning are possible before proceeding to include momentum and energy equations. By printing out the $n$ pressures at each integration step it may be checked that these are identical. Temperatures of particles which remain in the isothermal exchangers should be constant and identical. The value of dimensionless cycle work, $\zeta = W/MRT_{\mathrm{ref}}$, obtained by integrating $\psi\alpha_x(\mathrm{d}\lambda_e + \mathrm{d}\lambda_c)$ may be compared with that for the Schmidt cycle

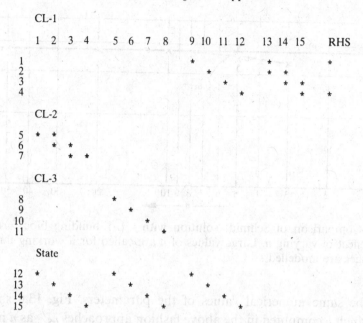

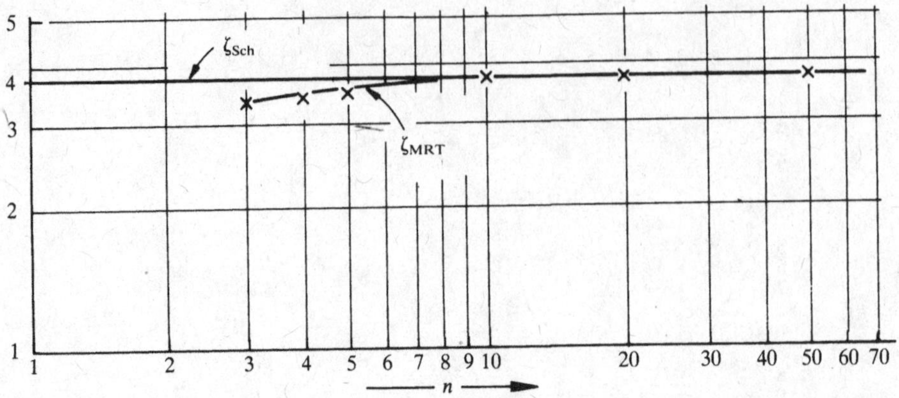


Fig. 13.6 Comparison of Schmidt solution with CL-1 building bloc formulation, showing effect of varying *n*. Large values of *n* are called for if working fluid circuit area changes are modelled.

having the same numerical values of the parameters. Fig. 13.6 shows the way in which $\zeta$ computed in the above fashion approaches $\zeta_{\mathrm{Sch}}$ as $n$ increases. Close agreement is achieved at a value of $n$ smaller than intuition might suggest.

From Fig. 13.6 it is seen that the numerical processes operate in stable fashion and return a commendable approximation to $\zeta_{\mathrm{Sch}}$ at a value of $n = 4$. The practical use of such low numbers of subdivisions is limited to program error tracing (there is full representation of the array elements, and execution time is a minimum). At the same time, the writer is not aware of any formulations in Eulerian coordinates which are viable at such coarse grid spacings. Perhaps it is the case that, when appropriate analytical and numerical tools are applied, the result is a certain simplicity and ruggedness. It remains to see whether this conjecture is borne out when the treatment is extended to include the conservation laws for momentum and energy.

## 13.7 Conservation of momentum – a CL-2MF cycle model

The law of conservation of momentum in differential form may be written

$$\frac{\mathrm{D}u}{\mathrm{d}t} + \frac{1}{\rho}\frac{\partial p}{\partial x} + F = 0 \tag{13.13}$$

The term involving $u$ is already in the required (substantial derivative) form, but that for $p$ remains untransformed as a partial differential. This presents nothing like the problem of having temperature as a loose partial differential: pressure information is not expected to propagate at particle speed. Moreover, the Lagrange approach was adopted in anticipation of application to cases where pressure wave effects are known to be slight. Any error caused by evaluating pressure gradients between moving nodes (rather than between fixed points) is thus likely to be less than the uncertainty in the data from which $F$ will be computed. On the other hand, a potential problem will arise later when flow area discontinuities are considered, and will be dealt with at that stage. Eqn (13.13) will provisionally be treated by basing the calculation of $\partial p/\partial x$ on adjacent values of $p$.

On including the expression for $F$ (see Notation) and normalizing, the equation becomes:

$$4\pi N_{MA}{}^2 \wp_j \frac{D}{d\phi} \boldsymbol{u}_{j+1} + \frac{\partial\psi}{\partial\lambda} + \tfrac{1}{2}\wp_j \boldsymbol{u}^2 C_f \frac{\boldsymbol{u}}{|\boldsymbol{u}|} \frac{N_{MA}{}^2}{\lambda_h} = 0 \qquad (13.14)$$

For reasons recently given, the contribution of the substantial derivative is likely to be small. It would, however, be reassuring to be able to take it into account to allow examination of any influence it might have. The approximation to $D\boldsymbol{u}_{j+1}$ is available in terms of $d\lambda_{j+1\,\text{current}}$ minus $d\lambda_{j+1\,\text{previous}}$ as:

$$\frac{D\boldsymbol{u}}{d\phi} = 2\pi \frac{d\lambda_{j+1\,\text{current}} - d\lambda_{j+1\,\text{previous}}}{d\phi^2}$$

This form requires that the expansion be made symmetrically about the current value of $\phi$ (Fig. 13.7), and this in turn calls for use of values of $\psi$ immediately prior to current known values (i.e. to those at $\phi - d\phi$. The expansion is carried out in terms of $d\psi$ and $d\lambda$, there being little point in expanding $\wp$, which takes its current value. The driving pressure gradient is evaluated as the mean of those at $\phi - d\phi$ and $\phi + d\phi$. The resulting CL-2 bloc locates in the matrix exactly as the corresponding dummy already

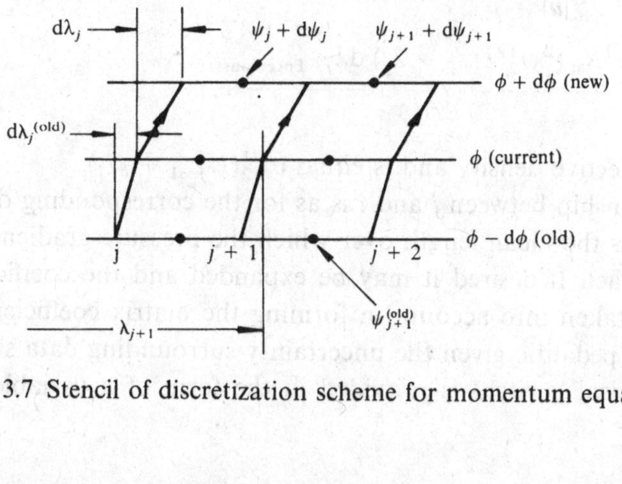


Fig. 13.7 Stencil of discretization scheme for momentum equation.

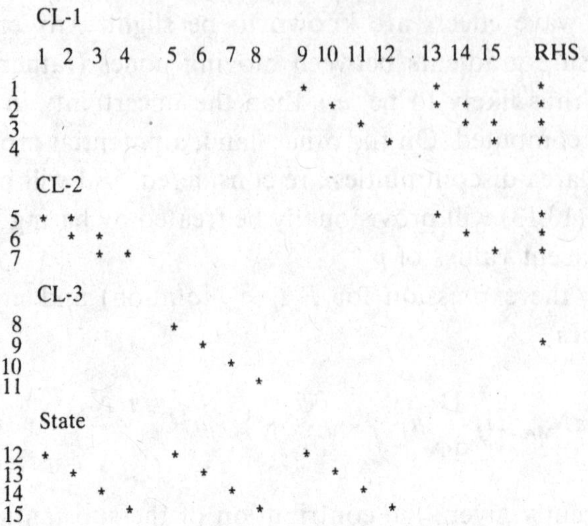


Fig. 13.8 Stencil of matrix of CL-2 MF solution. As before, asterisks indicate non-zero terms and show the dummy CL-3 law active. Again, the number of subdivisions (4) is unrealistically small, particularly for a model involving area discontinuities.

described (Fig. 13.8), and with coefficients

$$a_{i,i-n}\,\mathrm{d}\psi_j + a_{i,i-n+1}\,\mathrm{d}\psi_{j+1} + a_{i,i+2n}\,\mathrm{d}\lambda_{j+1} = b_i$$

$$a_{i,i-n} = 1 \text{ (as for the dummy)}$$

$$a_{i,i-n+1} = -1 \text{ (as for the dummy)}$$

$$a_{i,i+2n} = -\frac{8\pi^2 N_{\mathrm{MA}}{}^2 \underline{\wp}\frac{1}{2}(\lambda_{j+2}-\lambda_j)}{\mathrm{d}\phi^2}$$

$$b_i = \tfrac{1}{2}\underline{\wp}\boldsymbol{u}^2\frac{C_{\mathrm{f}}}{\lambda_{\mathrm{h}}}\frac{\boldsymbol{u}}{|\boldsymbol{u}|}N_{\mathrm{MA}}{}^2 2(\lambda_{j+2}-\lambda_j) + \psi_{j+1} - \psi_j + (\psi_{j+1}-\psi_j)_{\text{previous}}$$
$$-\frac{8\pi^2 N_{\mathrm{MA}}{}^2 \underline{\wp}\frac{1}{2}(\lambda_{j+2}-\lambda_j)\,\mathrm{d}\lambda_{j+1\,\text{previous}}}{\mathrm{d}\phi^2}$$

$\underline{\wp}$ is mean effective density and is equal to $\frac{1}{2}(\wp_{j+1}+\wp_j)$.

The relationship between $j$ and $i$ is as for the corresponding dummy case. $\frac{1}{2}(\lambda_{j+2}-\lambda_j)$ is the mean length over which the pressure gradient term $\mathrm{d}\psi$ is supposed to act. If desired it may be expanded and the coefficients of the resulting $\mathrm{d}\lambda$ taken into account in forming the matrix coefficients, $a_{i,j}$, but the option is pedantic given the uncertainty surrounding data such as $C_{\mathrm{f}}$.

If a numerical 'switch' is provided in the form of a variable taking the value 0 or 1, and if each of the terms associated with the original $D\boldsymbol{u}/d\phi$ (viz, $a_{i,i+2n}$ and the last component of $b_i$) is multiplied by this switch, the choice as to whether or not a particular run of the simulation is to include the effects of particle acceleration may be made at run time and implemented by setting the value of a single item of data. A similar switch, with a different variable name, may multiply terms associated with friction – i.e., all those containing the variable $C_f$. With both switches set to zero the remaining finite terms ($a_{i,i-n}$ and $a_{i,i-n+1}$) are equivalent to the dummy case, since the outstanding right-hand term, $\psi_{j+1} - \psi_j$ etc., is zero as a result of $\psi_{j+1}$ having been equal to $\psi_j$ at the end of the previous integration step.

The value of local hydraulic radius, $\lambda_h$, used in computing the relevant coefficients is that which applies at current, $\lambda_{j+1}$, and is found by scanning an array of data specifying $\lambda_h$ as a function of location. $C_f$ is evaluated in terms of local, instantaneous $N_{re}$, evaluated at each step from $N_{RE}$ and current $\boldsymbol{u}$, $\wp$ and $\lambda_h$.

An alternative two-conservation-law model is given by the combination of dummy momentum equation with the energy equation in full form.

## 13.8 Conservation of energy – a CL-2ME cycle model

The energy equation in the form of Eqn (13.2) is equivalent to that in the form of Eqn (13.3), and so either may be expanded for the purpose. The latter of the two has already been normalized and expanded in connection with the 'one-line' cycle model, at a stage when the variation with $\phi$ of both $\psi$ and $\boldsymbol{u}$ (and thus $\lambda$) was prescribed, leaving only $d\tau$ as unknown. It is now necessary to extract coefficients of each of $d\psi$, $d\tau$, $d\wp$ and $d\lambda$. This may be effected using the same equation if desired, or the alternative form. The former equation takes notional account of area change without the need to evaluate explicitly a term in $dA_x/dx$: it will be interesting to see whether this formulation copes directly with the abrupt changes of cross-sectional area, or whether special provision is necessary. Provisionally, therefore, Eqn (13.3) is discretized. There is the possibility of discretizing the original equation (Eqn (13.2)) also, giving an alternative set of coefficients for the CL-3 bloc. At run time the choice between one set and the other may be achieved by setting a further 'switch' via the input data, permitting direct comparison of the results of the alternatives. Since many of the terms in the coefficients are common between the two candidate formulations, the simultaneous coding of both forms of the conservation law introduces a marginal number of additional terms.

Eqn (13.3) is normalized and expanded. The multiplier $T^2/p$ of the term

$D(p/T)/dt$ cancels with terms of the differentiated form, resulting in some simplification. The unknowns are $d\psi$ and $d\tau_j$. Collecting coefficients of like terms and recalling that the CL-3 bloc extends from rows $2n$ to $3n - 1$, the discretized form is:

$$a_{i,i-2n+1}\, d\psi_j + a_{i,i-n+1}\, d\tau_j = b_i$$

$$a_{i,i-2n+1} = -\tfrac{1}{2}\text{RHS} - (\gamma - 1)\tau_j/d\phi$$

$$\text{RHS} = \tfrac{1}{4}\left[\frac{(\gamma - 1)|\underline{u}^3|C_f N_{MA}{}^2}{\pi \lambda_h}\right], \qquad \underline{u} = \frac{2\pi(d\lambda_j + d\lambda_{j+1})}{d\phi}$$

$$b_i = \frac{\gamma N_{ST}}{2\pi\lambda_h}\tau_j[\tau_w(\underline{\lambda}) - \tau_j]$$

$$+\tfrac{1}{2}\psi_j \text{RHS} + \text{terms of higher order}, \ \underline{\lambda} = \tfrac{1}{2}(\lambda_j + \lambda_{j+1})$$

The higher order terms included with $b_i$ are:

$$\tfrac{1}{2}(\gamma - 2)\, d\psi_j\, d\tau_j/d\phi - \tfrac{1}{4}\gamma N_{ST}\, d\tau_j{}^2/2\pi\lambda_h$$

For all the above expressions, the link between row number, $i$, and the column number, $j$, is $j = i - 2n + 1$. The coefficients as listed are applicable to all cells, including those having a piston face as a boundary.

Fig. 13.9(*a*) shows the specimen CL-2me matrix with coefficients of the CL-3 bloc in place. (The CL-2 bloc contains dummies.) $N_{ST}$ is defined as before as $hA_{ref}/Mc_pN_s$, so that at each evaluation of an array coefficient a figure for $h$ appropriate to local conditions has to be formed. For working fluid elements within the variable volume spaces the Annand correlation is recommended (Chap. 10). For a particle within a heat exchanger or the regenerator, the heat transfer correlation invoked will depend on which of the two similarity concepts – thermodynamic or dynamic – is embodied. The present treatment assumes thermodynamic similarity, so that friction and heat transfer correlations are those of Figs 7.3 and 7.4.

A comprehensive simulation may now be built by replacing the dummy CL-2 law with the full form (array stencil Fig. 13.9(*b*)). Numerical tests which may now be run include setting artificially high and artificially low values of $h$ and $C_f$ respectively, and checking for results close to those from the corresponding CL-1 simulation. With $h$ and $C_f$ held at zero, the initial distribution should oscillate within the exchanger system, retaining its original form but for an overall cyclic rise and fall due to adiabatic compression and expansion. Examples will be presented after the matter of flow area discontinuities has been introduced.

| | 1 | 2 | 3 | 4 | 5 | 6 | 7 | 8 | 9 | 10 | 11 | 12 | 13 | 14 | 15 | RHS |
|---|---|---|---|---|---|---|---|---|---|---|---|---|---|---|---|---|
| CL-1 | | | | | | | | | | | | | | | | |
| 1 | | | | | | | | | * | | | | * | | | * |
| 2 | | | | | | | | | | * | | | * | * | | * |
| 3 | | | | | | | | | | | * | | | * | * | * |
| 4 | | | | | | | | | | | | * | | | * | * |
| CL-2 | | | | | | | | | | | | | | | | |
| 5 | * | * | | | | | | | | | | | | | | |
| 6 | | * | * | | | | | | | | | | | | | |
| 7 | | | * | * | | | | | | | | | | | | |
| CL-3 | | | | | | | | | | | | | | | | |
| 8 | * | | | | * | | | | | | | | | | | * |
| 9 | | * | | | | * | | | | | | | | | | * |
| 10 | | | * | | | | * | | | | | | | | | * |
| 11 | | | | * | | | | * | | | | | | | | * |
| State | | | | | | | | | | | | | | | | |
| 12 | * | | | | * | | | | * | | | | | | | |
| 13 | | * | | | | * | | | | * | | | | | | |
| 14 | | | * | | | | * | | | | * | | | | | |
| 15 | | | | * | | | | * | | | | * | | | | |

(*a*)

| | 1 | 2 | 3 | 4 | 5 | 6 | 7 | 8 | 9 | 10 | 11 | 12 | 13 | 14 | 15 | RHS |
|---|---|---|---|---|---|---|---|---|---|---|---|---|---|---|---|---|
| CL-1 | | | | | | | | | | | | | | | | |
| 1 | | | | | | | | | * | | | | * | | | * |
| 2 | | | | | | | | | | * | | | * | * | | * |
| 3 | | | | | | | | | | | * | | | * | * | * |
| 4 | | | | | | | | | | | | * | | | * | * |
| CL-2 | | | | | | | | | | | | | | | | |
| 5 | * | * | | | | | | | | | | | * | | | * |
| 6 | | * | * | | | | | | | | | | | * | | * |
| 7 | | | * | * | | | | | | | | | | | * | * |
| CL-3 | | | | | | | | | | | | | | | | |
| 8 | * | | | | * | | | | | | | | | | | * |
| 9 | | * | | | | * | | | | | | | | | | * |
| 10 | | | * | | | | * | | | | | | | | | * |
| 11 | | | | * | | | | * | | | | | | | | * |
| State | | | | | | | | | | | | | | | | |
| 12 | * | | | | * | | | | * | | | | | | | |
| 13 | | * | | | | * | | | | * | | | | | | |
| 14 | | | * | | | | * | | | | * | | | | | |
| 15 | | | | * | | | | * | | | | * | | | | |

(*b*)

Fig. 13.9 Stencils of matrices for CL-2ME and CL-3 cycle models. (*a*) CL-2ME model – dummy momentum law incorporated. (*b*) CL-3 (three-conservation law) model.

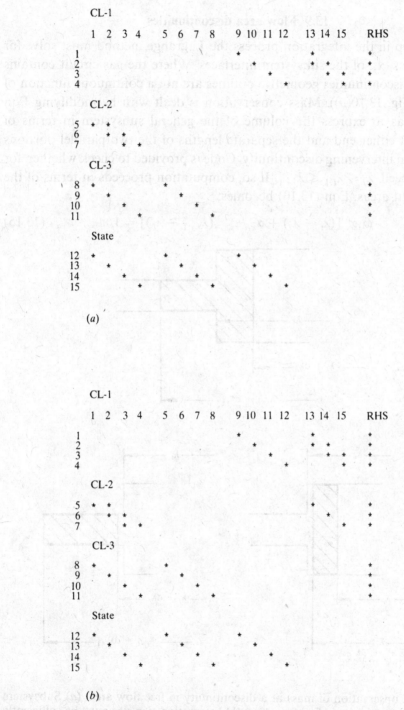

## 13.9 Flow area discontinuities

At each step in the integration process the Lagrange method must solve for the locations, $x_j$, of the subsystem interfaces. Where the gas circuit contains flow area discontinuities geometric outlines are not a continuous function of location (Fig. 13.10(*a*)). Mass conservation is dealt with by modifying Eqn (13.10) so as to express the volume of the general subsystem in terms of the areas at either end and the separate lengths of the two parallel portions assuming an intervening discontinuity. Code is provided to check whether, for a given subcell, $\lambda_j \leqslant \lambda_{f,k} \leqslant \lambda_{j+1}$. If so, computation proceeds in terms of the two different areas. Eqn (13.10) becomes:

$$\wp_j \alpha_{x_j}[(\lambda_f - \lambda_j) + \alpha_{x_j}/\alpha_{x_{j+1}}(\lambda_{j+1} - \lambda_f)] = 1/n \qquad (13.15)$$

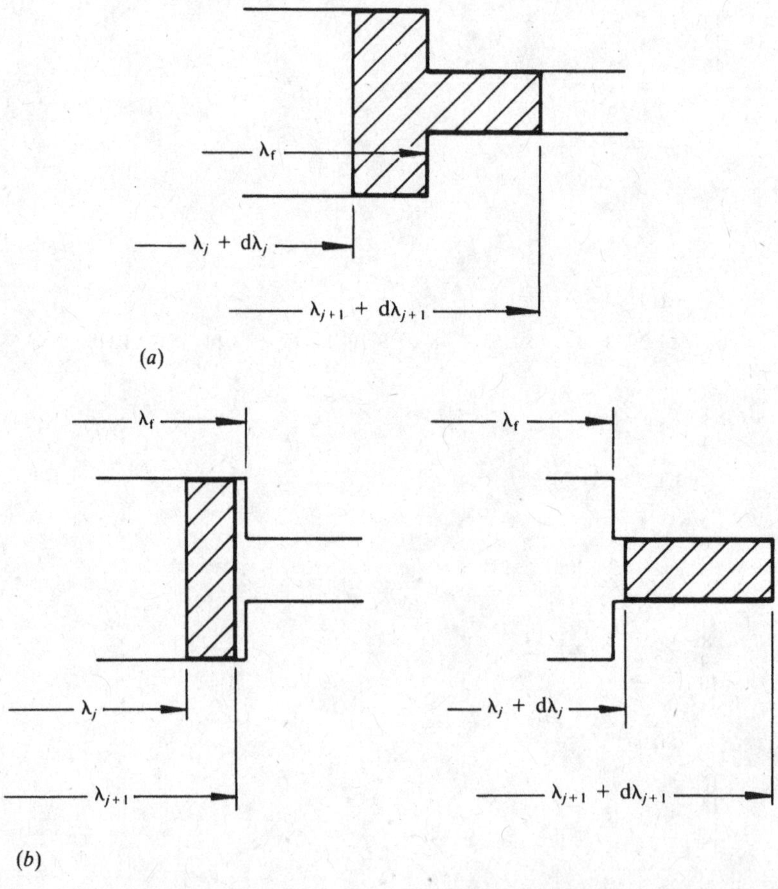


Fig. 13.10 Conservation of mass at a discontinuity in free flow area. (*a*) Subsystem straddling discontinuity in free flow area. (*b*) Integration step size must be sufficiently small that no subsystem passes completely from a duct of one cross-sectional area into a duct of different cross-sectional area during a single integration step.

On expanding Eqn (13.5) it is tempting to repeat the process which led to Eqn (10.11), i.e., to equate to zero the terms expressing mass conservation prior to expansion. However, if the subsystem had *just* entered the discontinuity since the prior integration step, the terms in question would not express total mass at that prior point, since they involve $\alpha_{x_j}/\alpha_{x_j+1}$ and $\lambda_f$, which are irrelevant to the case of the parallel duct. It is therefore necessary to retain these terms as a contribution to the term $b_i$. The coefficients for the subsystem with discontinuity are:

$$a_{i,2n+i} = \lambda_f - \lambda_j + \alpha_{x_j}/\alpha_{x_{j+1}}(\lambda_{j+1} - \lambda_f)$$

$$a_{1,3n+i} = -\wp_j$$

$$a_{1,3n+i+1} = \wp_j \alpha_{x_j}/\alpha_{x_{j+1}}$$

$$b_i = 1/(n\alpha_{x_j}) - \wp_j\{\lambda_f - \lambda_j + \alpha_{x_j}/\alpha_{x_{j+1}}(\lambda_{j+1} - \lambda_f)\}$$
$$- \mathrm{d}\wp_j[-\mathrm{d}\lambda_j + (\alpha_{x_j}/\alpha_{x_{j+1}})\,\mathrm{d}\lambda_{j+1}]$$

Evidently, the pattern of array entries is unchanged from that of Fig. 13.5.

The crank angle increment must be sufficiently small to avoid the situation represented in Fig. 13.10(*b*), where a subsystem traverses an area discontinuity completely. Moreover, *n* must be sufficient that no single subsystem straddles more than one area discontinuity at any one time. With these conditions, the foregoing formulation has been found to function infallibly over every gas circuit modelled to date.

Momentum and energy equations in the form of Eqns (13.13) and (13.14) take notional account of change in flow area between opposite faces of the subsystem. To this extent it is possible with the simulation in its present form to attempt a preliminary modelling of a realistic gas circuit. The exercise will confirm certain aspects of correct functioning, and provide guidance as to the number of subdivisions required for realistic representation of the gas processes.

Fig. 13.11 is the particle trajectory map which results when the CL-3 form of the simulation is applied to the Philips MP1002CA engine at rated operating conditions. Fig. 13.12(*a*) is a relief of gas temperature as a function of location and crank angle when the CL-1 version is run with the same data. The straight lines superimposed on the profile mark heat exchanger and regenerator boundaries. The plot confirms that conditions are isothermal in the exchangers and working spaces. The stepped surface reflects a decision to plot precisely what has been modelled, namely, a sequence of subsystems of uniform properties. Plotting temperatures interpolated between subsystem

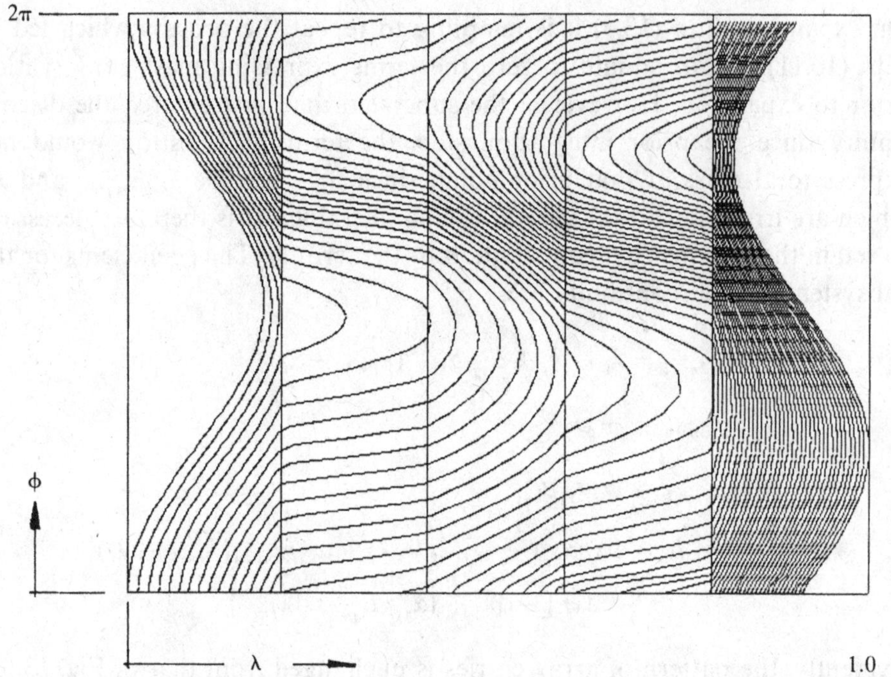


Fig. 13.11 Fluid particle trajectory map produced by CL-3 simulation in Lagrange coordinates of the Philips MP1002CA air engine (rated operating conditions).

centres gives almost perfectly smooth temperature contours, and an impression of great precision. The more 'honest' presentation provides useful information which will be returned to shortly.

If the energy conservation module operates correctly, then the CL-2ME version with exaggerated heat transfer coefficient will give a solution similar to the isothermal. Fig. 13.12(*b*) shows the results of such an exercise. It is not possible to set $N_{ST}$ infinitely high without causing problems at the matrix inversion stage, but permissible values (of the order of 80) lead to temperature profiles which convincingly approach those from the CL-1 version.

In Fig. 13.13 $N_{ST}$ has been set equal to zero to simulate the adiabatic case. The initial temperature distribution oscillates back and forth through the exchanger, demonstrating a uniform, cyclic rise and fall in sympathy with adiabatic compression and expansion. To the time of writing there are no accounts of fixed-grid schemes capable of responding appropriately to such extreme operating conditions.

Finally, Fig. 13.13(*b*) shows the development of the temperature profile over the initial cycle when the CL-3 simulation uses values of $h$ and $C_f$ calculated from local conditions.

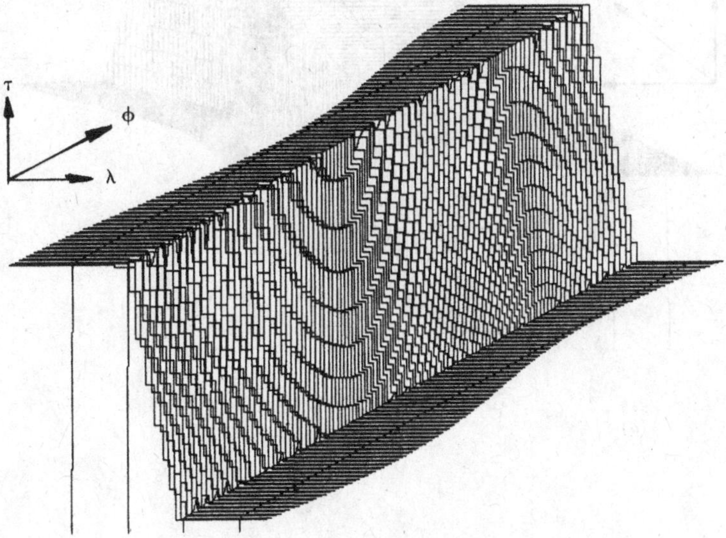


(*a*)

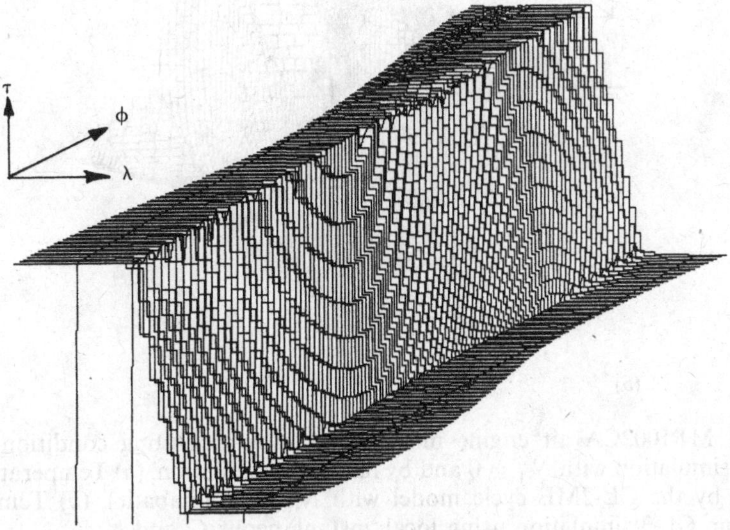


(*b*)

Fig. 13.12 MP1002CA air engine modelled at rated operating conditions by CL-1 and CL-2ME simulations. (*a*) Temperature relief produced by CL-1 cycle model. (*b*) Temperature relief produced by CL-2ME simulation with $N_{ST}$ fixed at a high value (80) simulating 'isothermal' conditions.

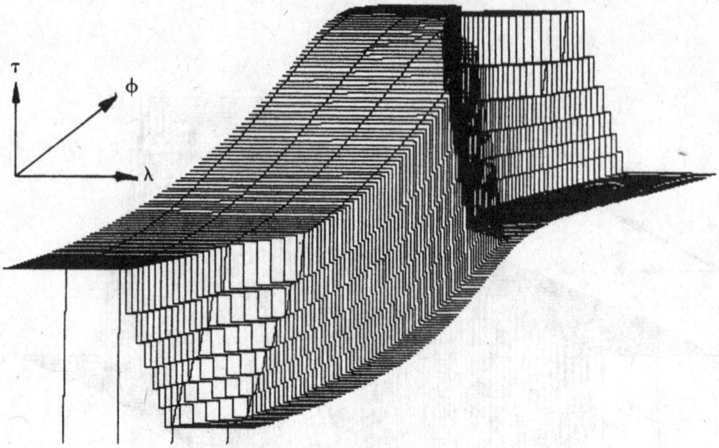


(*a*)

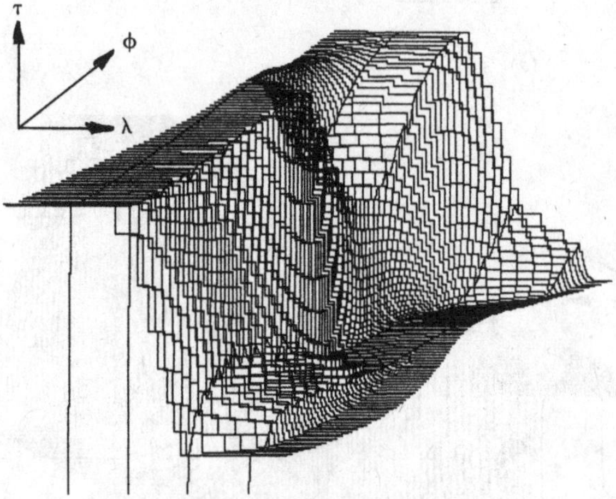


(*b*)

Fig. 13.13 MP1002CA air engine modelled at rated operating conditions by the CL-2ME simulation with $N_{ST} = 0$ and by full CL-3 simulation. (*a*) Temperature relief produced by the CL-2ME cycle model with $N_{ST} = 0$ (adiabatic). (*b*) Temperature profile from CL-3 simulation using local, instantaneous $C_f$ and $N_{ST}$.

The stepwise form of plotting imposes a harsh interpretation. Moreover, the MP1002CA machine is a demanding case, with very small heat exchanger cross-sectional area in relation to cylinder bore. The feature causes certain fluid particles to traverse the entire exchanger system from expansion to compression space and *vice versa*. Looking back to the particle trajectory plot, it is seen that there are instants when the expansion exchanger is occupied by only two separate working fluid subsystems, so that even the apparently generous number of subdivisions ($n = 49$) is probably insufficient for a realistic cycle description.

These observations raise two matters:

(*a*) increasing the number of particles in the expansion exchanger to, say, ten suggests a five-fold increase $n$ to 250 or so, and corresponding problems of extended computing time;

(*b*) within the exchanger system the approximation adopted for $\partial p/\partial x$, namely, the mean over an interval of $2d\phi$ of successive $\partial p/\partial x$ evaluated at moving points, is now suspect, since in the interval of $\phi$ concerned the later value of $\partial p/\partial x$ is washed far downstream.

The issue of computation time is clearly paramount, since further refinements to the model will only add to computational effort, and if this is already excessive then the approach has no future.

## 13.10 Computational efficiency

In an implicit integration scheme, such as that used here, the majority of cpu time goes in processing the array of coefficients. At the program development stage the appropriate way of representing and checking the array of coefficients is to work in terms of the $n \times n$ form of Figs. 13.5, 13.8 and 13.9. A routine such as the well-known SIMQX is compatible with this approach, is available on almost any computer and is simple to use. As can be seen from the figures, however, the majority of the array coefficients are zero. These occupy space and absorb time if the routine accesses all coefficients – as SIMQ and SIMQX do. At the cost of a little effort it is possible to make use of a type of routine which stores and manipulates only non-zero coefficients. The combination of Nag™ routines F01BRF (decomposition) and F04AXF functions in this way. The extra programming effort arises from the need to provide the routines not only with values of the non-zero coefficients, but also information as to the $i, j$ locations in the notional $n \times n$ array they occupy.

Keeping track of the subscripts for the building bloc has its challenges anyway, and the additional problem of addressing these correctly from a further array can be a mental strain if carried out in a single step. The task becomes manageable if broken down into two stages: the first computes the subscripts as if the coefficients were going to be assembled into the usual $n \times n$ matrix. A second, separate stage works out the location of each subscripted coefficient for the vector required by the solution routine. When it is verified that the solution is working correctly, a branch may be built around the intermediate step. (Experience shows that such intermediate code should not actually be deleted – it will inevitably be called upon again if only for reassurance!)

The possibility of further economies of computing time arises due to the fact that, once the desired combination of conservation laws has been assembled, the pattern of array coefficients is invariant – i.e., it does not change between integration steps. This means that much of the information generated by decomposition (FO4BFR) for the first integration step does not require to be generated again. The Nag library provides a routine, FO4BSF, which can be used instead of FO4BRF for second and all subsequent decompositions, and which runs several times faster than the parent routine. The program which produced the particle trajectory maps and temperature profiles of Figs 13.11–13.13 uses the Nag routines, and with $n = 49$ takes some 85 s to generate the data for a set of plots when run on a Hewlett Packard 9000/340C+ workstation. This is a worthwhile improvement over the 25 min or so required by SIMQX for the same task. There is little difference in cpu time between the CL-1, CL-2ME, CL-2MF and CL-3 forms.

Other lesser economies are possible: referring to Fig. 13.11 it is seen that at the expansion end two complete subsystems never enter the heat exchanger system. At the compression end, where both density and dead volume are greater, no less than ten subsystems never leave the variable-volume space. Allowing subsystems 1 and $n$ to be of larger than standard size promises comparable computational accuracy for a 25% (12 in 49) reduction in $n$. The $n$ estimated to give ten subsystems per exchanger accordingly falls to about 180; if a reduction to eight subsystems can be tolerated, then it falls to 150. Considering that few designs are likely to be as extreme as the MP1002CA, the value of $n$ might well reduce to 100 for most purposes, i.e., to only twice the current value – scarcely a daunting prospect.

If use of the simulation for optimization is anticipated, then all possible computational economies, however minor, must be implemented. Array elements repeatedly referred to must be set to simple variables. The subprogram which scans the gas circuit for discontinuities should set wall temperatures, $N_{re}$ etc. for the CL-2 and CL-3 blocs in the same pass. There are some attractions to carrying out the scan for one subsystem at a time, computing the values of the array coefficients for the current value of $j$ and then proceeding with the scan. If this is done, a pointer must be provided which may be set after each sweep to record how far the scan progressed before reaching the relevant subcell. The pointer tells the next sweep to enter where the previous left off.

## 13.11 Estimate of pressure gradient in the Lagrange scheme

Fig. 13.7 illustrated the centred-difference scheme for expressing a unit integration step for the momentum conservation law. Previously, the mid-interval pressure gradient was approximated as the mean of the gradients at $\phi - d\phi$ and $\phi + d\phi$ without allowance for convection of the respective gradients over the $2d\phi$ interval. The preliminary analysis of the MP1002CA has revealed that, even if this method serves, it is somewhat unsatisfying.

An alternative discretization scheme makes use of the identity:

$$\frac{D}{dt} = \frac{\partial}{\partial t} + u\frac{\partial}{\partial x}$$

from which

$$\frac{\partial}{\partial x} = \frac{1}{u}\left(\frac{D}{dt} - \frac{\partial}{\partial t}\right)$$

For the equation in question, $\boldsymbol{u}$ is evaluated as before, namely as the mean of the two values, new and old, of $2\pi\, d\lambda_{j+1}/d\phi$; $D\psi/d\phi$ may be expressed directly as $\frac{1}{2}[\frac{1}{2}(\psi_j + d\psi_j + \psi_{j+1} + d\psi_{j+1}) - \frac{1}{2}(\psi_j^{\text{old}} + \psi_{j+1}^{\text{old}})]$. The term $\partial\psi/d\phi$ calls for interpolation in terms of $\lambda_j$, $\lambda_{j+1}$ $\lambda_{j+2}$, $d\lambda_j$, $d\lambda_{j+1}$ $d\lambda_{j+2}$, $\lambda_j^{\text{old}}$, $\lambda_{j+1}^{\text{old}}$ and $\lambda_{j+2}^{\text{old}}$. The reader may care to formulate the appropriate expression, substitute together with the expression for $D\psi/d\phi$ into the previous expansion of Eqn (13.4) and collect terms. It will be found that the array pattern remains the same, and that the expressions for the coefficients are only marginally lengthier.

There are no definitive statements of momentum and energy conservation at the flow area discontinuity, so the interested reader will probably wish to formulate his own treatments. These will be more intricate than the corresponding expansions for mass conservation: time-dependent terms are involved, and it is necessary to take into account the possibility that a subsystem occupies a parallel length of duct at crank position $\phi - d\phi$ but will straddle a discontinuity at $\phi + d\phi$ (or *vice versa*). This suggests a need in the simulation for an additional pair of logical arrays, each of length $n$. The first records, for each subsystem, whether or not it straddled a discontinuity at crank angle $\phi - d\phi$, the second does the same thing for (current) crank position $\phi$.

It is not possible to carry out the elegant substitutions between conservation laws which led to the forms for continuous $dA_x/dx$ (Eqns (13.3), (13.13)) and so the array patterns for CL-*n* combinations greater than CL-1 will be different from those previously illustrated. Moreover, the pattern for a given CL-*n* formulation will generally vary from one integration step to the next depending upon which elements are involved with discontinuities, and in what manner (entering, leaving, remaining). At first sight this variability threatens to force use at each integration step of the comprehensive decomposition routine, with consequent increase in computing time. On the other hand, the routines can cope with elements having the value zero, so it is merely necessary to provide index space for each array element which *may* take finite value, and to proceed as before.

# 14
# Optimization

## 14.1 Background

There exists a variety of analytical and numerical techniques for the determination of the extreme values of a function. Application of such techniques is frequently termed 'optimization'. This text has emphasized the functional form of each of the expressions which have been developed for specific cycle work. In principle, the techniques of optimization may be applied to such expressions, revealing the numerical values of the arguments for which the expression is a maximum.

Any maximum determined in this way is meaningful only if the expression which has been maximized is a sufficiently sophisticated model of reality. If so, the process of optimization can serve to eliminate a great deal of expensive trial and error development work. In this connection it is worth noting that the 'reality' of Stirling machine performance includes bearing and seal friction (functions of inertia forces as well as of gas loads), crankcase windage and combustion chamber characteristics. It is well known that the kinematic proportions of a practical drive mechanism giving the most favourable gas circuit performance frequently give rise to high bearing loads. If a gain in thermodynamic performance is not to be offset by a loss in mechanical efficiency the function to be subjected to the optimization process must describe the whole machine. Moves in this direction have to date been only tentative.[1] Moreover, it is clear that gas circuit modelling has yet to reach a convincing level of sophistication. The *methods* of formal optimization may be developed and illustrated in anticipation of an adequate description of the whole machine, and it is this task which is undertaken here.

## 14.2 Precedents

The only attempts at formal optimization to date are those due to Finkelstein,[2] Walker[3], Kirkley,[4] Organ[5] and Takase,[6] all of whom concentrated on the cycle description due to Schmidt,[7] to Walker and Khan[8] who worked with the adiabatic analysis and to Walker and Agbi[9] who attempted to deal with a cycle using a two-phase working fluid.

The title of the Walker paper ('An optimization of the principal design parameters . . .') reflects a then current belief in the overriding importance for thermodynamic performance of phase angle and volume ratio. It is now clear, by contrast, that within wide limits, performance is little influenced by variations in these parameters, but is critically dependent on the choice of exchanger and regenerator geometry. Finkelstein's treatment was a characteristic *tour de force*. By analytical differentiation of the Schmidt expression for specific cycle work $\zeta_{MRT}$ $(W_{cyc}/MRT_c)$ he arrived at almost-explicit expressions for $\kappa_0$ and $\alpha_0$ – the values of volume ratio, $\kappa$, and piston phase angle, $\alpha$, giving maximum $\zeta_{MRT}$ for given $N_\tau$ and $\nu(N_\tau)$ in the case of the opposed-piston (alpha) machine. Finkelstein also considered optimizing cycle work per swept volume and per reference cycle pressure, but for some reason was daunted by the problem of finding a definition of swept volume consistent between the various configurations (opposed piston vs displacer).

Walker concentrated on the opposed-piston machine. He used numerical techniques to achieve the equivalent of Finkelstein's analytical optimization of $\zeta_{MRT}$. Defining swept volume of the opposed-piston configuration as $V_E(1+\kappa)$ he adapted the numerical approach to yield, as a function of $N_\tau$ and a dead-space ratio, $\mu_D$ $(=\Sigma(V_{di})/V_E)$, maxima of $\zeta_{p_{max}}$ $(W_{cyc}/V_{sw}p_{max})$, $\zeta_{p_{min}}$ $(W_{cyc}/V_{sw}p_{min})$ and $\zeta_p$ $(W_{cyc}/V_{sw}p_{mean})$. Results were presented in the form of design charts (Fig. 2.15). ($\mu_D$ is not a true parameter of the Schmidt analysis accurately formulated – Fig. 14.1.)

Kirkley also used the numerical approach, concentrated on the parameter $\zeta_{p_{max}}$ (misleadingly denoted the '*power* parameter' – it does not determine power). By defining swept volume for all three basic configurations (opposed piston, coaxial displacer and two-cylinder displacer type) design charts were produced from which could be read the $\kappa_0$ and $\alpha_0$ (or $\lambda_0$ and $\beta_0$ as appropriate) giving optimum $\zeta_{p_{max}}$ as a function of temperature ratio, $T_C/T_E$, in each case. The parameter was again dead space ratio, $\mu_D$.

Converted to the same coordinates, Kirkley's chart for the opposed-piston machine does not coincide with the corresponding chart by Walker. Moreover, use of $\mu_D$ as parameter relies on a definition of $\nu^*(N_\tau)$ of the form $\nu^*(N_\tau) = 2\mu_D/(\tau+1)$. Being derived from an approximate definition of dead space mass content, $\nu^*(N_\tau)$ is only approximately equal to $\nu(N_\tau)$. It is not possible to plot $\nu^*(N_\tau)$ against $\nu(N_\tau)$ to facilitate conversion, since the latter takes account of the *distribution* of dead space, while the former does not. A *very approximate* relationship between $\nu(N_\tau)$ and $\mu_D$ is given in Fig. 14.1 with $N_\tau$ as parameter.

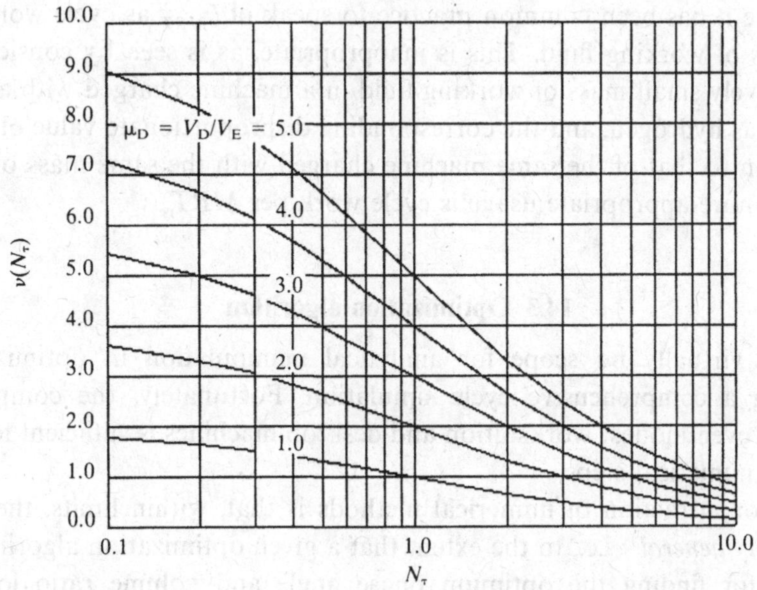


Fig. 14.1 Chart for conversion of dead space volume ratio, $\mu_D$ ($= \Sigma V_D/V_E$) to an *approximate* value of dead space parameter, $\nu(N_\tau)$.

Organ extended the Finkelstein analytical approach to the optimization of the specific cycle work per $p_{max}$ and per $V_{sw}$ for the opposed-piston (alpha configuration) machine, and derived the comparable expressions for the coaxial (beta) and gamma configurations. The optimization charts had $\mu_D$ as parameter to permit comparison with the work of the other authors. In an elegantly presented account Takase used numerical means to find the optimum parameters of a machine in which variation of piston volume ratio, $\lambda$, was achieved by stepping the coaxial bore. He went on to define the optimum kinematics of the symmetrical rhombic drive of the Schmidt machine.

Walker and Khan worked with the adiabatic analysis (Appendix VIII) and calculated in terms of $\zeta_{MRT}$. Systematic computation of $\zeta_{MRT}$ over a range of values of the independent variables $\kappa$ and $\alpha$ was preferred to the formal techniques of numerical optimization. Unfortunately, it is not obvious which value of $\gamma$ is used, but the results confirm the existence of values, $\kappa_0$ and $\alpha_0$, of $\kappa$ and $\alpha$ for which $\zeta_{MRT}$ has a maximum at specimen temperature ratio and dimensionless dead volume. The Walker and Agbi treatment compounds the sweeping assumptions of the Schmidt analysis by replacing the continuous, linear regenerator temperature gradient with a step change from $T_e$ to $T_c$ at the mid-plane. Its findings are thus unlikely to be of relevance to the task of practical design.

To date it has been common practice to speak of $\zeta_{MRT}$ as cycle work per unit *mass* of working fluid. This is inappropriate, as is seen by considering the relatively small mass of working fluid in a machine charged with a light gas such as hydrogen, and the corresponding disproportionate value of $\zeta_{MRT}$ in relation to that of the same machine charged with the same mass of, say, air. The more appropriate usage is cycle work per $MRT_{\text{ref}}$.

## 14.3 Optimization algorithm

There is virtually no scope for analytical manipulation in optimization involving a comprehensive cycle simulation. Fortunately, the computing power of even modest workstation and desktop machines is sufficient for the required numerical steps.

A major attraction of numerical methods is that, within limits, they are completely *general* – i.e., to the extent that a given optimization algorithm is suitable for finding the optimum phase angle and volume ratio for the Schmidt cycle, it is also good for finding the $N_{\text{MA}}$, $N_{\text{SG}}$ and/or other parameter values which maximize specific cycle work predicted by, say, a three-conservation-law model. A choice of algorithms is available in standard texts on numerical analysis; some computer software libraries contain precompiled versions. Differences between different algorithms amount largely to differences in computational efficiency (speed of convergence on the optimum point), the underlying principle being common, namely, repeated evaluation of the object function to determine local gradients, followed by adjustment of the argument(s) to new values indicated by the positive direction of the gradient. A typical algorithm is that of Kiefer–Wolfowitz,[10] which may be illustrated by applying it to the optimization of a generic expression for the specific work of the Schmidt cycle: consider the general form:

$$\zeta_{\text{Sch}} = \zeta_{\text{Sch}}(N_\tau, \nu(N_\tau), \kappa, \alpha) \tag{14.1}$$

In Eqn (14.1) $N_\tau$ and $\nu(N_\tau)$ are considered 'fixed', the parameters which are to be explored for optima being $\kappa$ and $\alpha$. The search process starts with 'seed' values, $\kappa_i$ and $\alpha_i$ of $\kappa$ and $\alpha$ which may have to be arbitrary. Improved values are found by computing approximations to $\partial\zeta/\partial\alpha$ and $\partial\zeta/\partial\kappa$ as:

$$\frac{\partial\zeta}{\partial\alpha} \approx \frac{\zeta[N_\tau, \nu(N_\tau), \kappa, \alpha + \Delta\alpha] - \zeta[N_\tau, \nu(N_\tau), \kappa, \alpha - \Delta\alpha]}{2\Delta\alpha} \tag{14.2a}$$

$$\frac{\partial\zeta}{\partial\kappa} \approx \frac{\zeta[N_\tau, \nu(N_\tau), \kappa + \Delta\kappa, \alpha] - \zeta[N_\tau, \nu(N_\tau), \kappa - \Delta\kappa, \alpha]}{2\Delta\kappa} \tag{14.2b}$$

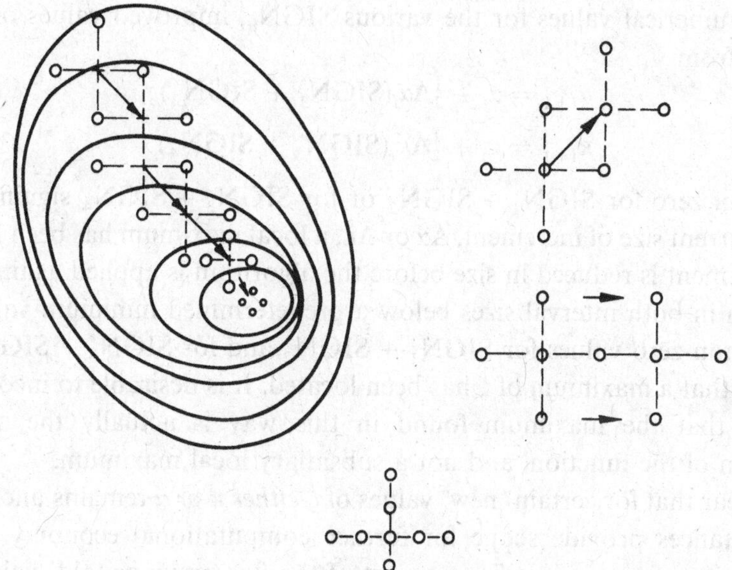

Fig. 14.2 Schematic of the Kiefer–Wolfowitz optimization algorithm.

The process identifies the directions on the surface $\zeta_{\text{Sch}} = \zeta_{\text{Sch}}(N_\tau, v(N_\tau), \kappa, \alpha)$ in which $\zeta$ is increasing locally (Fig. 14.2). (If there is more than one 'hump' on the surface, then, in general, only a local maximum will be found.)

In practice, one is less interested in the magnitude of the gradient than in its sign, and so a computer-coded version of the process would call for evaluation of the signs of four gradients in the four cardinal directions about the point $i$:

$$\text{SIGN}_1 = \frac{\zeta[N_\tau, v(N_\tau), \kappa_i, \alpha_i] - \zeta[N_\tau, v(N_\tau), \kappa_i, \alpha_i - \Delta\alpha]}{\text{ABS}\{\zeta[N_\tau, v(N_\tau), \kappa_i, \alpha_i] - \zeta[N_\tau, v(N_\tau), \kappa_i, \alpha_i - \Delta\alpha]\}} \quad (14.3a)$$

$$\text{SIGN}_2 = \frac{\zeta[N_\tau, v(N_\tau), \kappa_i, \alpha_i + \Delta\alpha] - \zeta[N_\tau, v(N_\tau), \kappa_i, \alpha_i]}{\text{ABS}\{\zeta[N_\tau, v(N_\tau), \kappa_i, \alpha_i + \Delta\alpha] - \zeta[N_\tau, v(N_\tau), \kappa_i, \alpha_i]\}} \quad (14.3b)$$

$$\text{SIGN}_3 = \frac{\zeta[N_\tau, v(N_\tau), \kappa_i, \alpha_i] - \zeta[N_\tau, v(N_\tau), \kappa_i - \Delta\kappa, \alpha_i]}{\text{ABS}\{\zeta[N_\tau, v(N_\tau), \kappa_i, \alpha_i] - \zeta[N_\tau, v(N_\tau), \kappa_i - \Delta\kappa, \alpha_i]\}} \quad (14.3c)$$

$$\text{SIGN}_4 = \frac{\zeta[N_\tau, v(N_\tau), \kappa_i + \Delta\kappa, \alpha_i] - \zeta[N_\tau, v(N_\tau), \kappa_i, \alpha_i]}{\text{ABS}\{\zeta[N_\tau, v(N_\tau), \kappa_i + \Delta\kappa, \alpha_i] - \zeta[N_\tau, v(N_\tau), \kappa_i, \alpha_i]\}} \quad (14.3d)$$

Fewer calculations of the function $\zeta$ are called for than at first appears, since (*a*) the repeated term $\zeta(N_\tau, v(N_\tau), \kappa_i, \alpha_i)$ need be calculated once only for use in all four Eqns (14.3) and (*b*) the denominator is merely the numerical value of the numerator stripped of any negative sign.

With numerical values for the various $\text{SIGN}_n$, improved values of $\kappa$ and $\alpha$ follow from:

$$\alpha_{i+1} = \alpha_i + \tfrac{1}{2}\Delta\alpha_i(\text{SIGN}_1 + \text{SIGN}_2) \tag{14.4a}$$

$$\kappa_{i+1} = \kappa_i + \tfrac{1}{2}\Delta\kappa_i(\text{SIGN}_3 + \text{SIGN}_4) \tag{14.4b}$$

A value of zero for $\text{SIGN}_1 + \text{SIGN}_2$ or for $\text{SIGN}_3 + \text{SIGN}_4$ signifies that, for the current size of increment, $\Delta\alpha$ or $\Delta\kappa$, a local maximum has been located. The increment is reduced in size before the algorithm is applied again. When reduction in both interval sizes below a predetermined minimum value fails to yield non-zero values for $\text{SIGN}_1 + \text{SIGN}_2$ and for $\text{SIGN}_3 + \text{SIGN}_4$ it is assumed that a maximum of $\zeta$ has been located. It is desirable to incorporate a check that the maximum found in this way is actually the absolute maximum of the function, and not a subsidiary local maximum.

It is clear that for certain 'new' values of $\zeta$ *either* $\kappa$ *or* $\alpha$ remains unchanged. Such instances provide scope for further computational economy: a little programming finesse identifies two out of the five previous 'old' values of $\zeta$ which may be used as 'new' values without the need for reevaluation of the object function (Fig. 14.2).

It is not difficult to see how the principle extends to a function having a larger number of arguments than the two so far supposed. Before attempting this it will be applied comprehensively to the Schmidt and adiabatic functions. There are four reasons for applying the technique to functions known in advance to be wholly inadequate representations of the real machine:

(i) 'Seed' values of the parameters are required as first approximations to the eventual optima. The optima of $\kappa$ and $\alpha$ (or equivalent parameters of drive mechanism geometry) determined from an ideal cycle are arguably the best available first approximations. Complex perturbation treatments have the isothermal or adiabatic analysis as the ideal reference. The conditions for maximum specific cycle work from the reference analysis are thus the obvious starting point for an overall optimum.

(ii) There is some evidence that general trends indicated by optimization based on the ideal analyses are confirmed in practice. For example, miniature reversed cycle Stirling coolers ($N_\tau \approx 0.25$) tend to have swept volume ratios ($\kappa$ and $\lambda$) of the order of 5–8. Reference to the ideal design charts to be presented shortly will show $\kappa_0$ and $\lambda_0$ to be in that range.

(iii) The ideal analyses provide a convenient test of the proper functioning of the optimization algorithm.

(iv) An opportunity is provided of presenting optimization charts for all configurations (opposed piston, coaxial displacer and 'gamma' types), in both modes of operation (prime mover and cooler) to a common format and with common definitions for the parameters.

## 14.4 Optimization based on the Schmidt cycle

Appendix II derived the following expression for work per cycle according to the assumptions of the Schmidt analysis:

$$\frac{W_{\text{cyc}}}{MRT_{\text{c}}} = \frac{2\pi\kappa(1 - 1/N_\tau)\sin(\alpha)}{(1/N_\tau + \kappa + 2\nu)^2} f_{\text{m}}(\xi) \tag{14.5}$$

$$f_{\text{m}}(\xi) = \frac{1}{\sqrt{(1 - \xi^2)}[1 + \sqrt{(1 - \xi^2)}]}$$

The expression applies to *all* configurations whose volume variations are described by a $\kappa$ and an $\alpha$. Therefore, the values of $\kappa$ and $\alpha$ which maximize Eqn (14.5) for given $N_\tau$ and $\nu(N_\tau)$ for the opposed-piston (alpha) machine are the $\kappa$ and $\alpha$ whose *equivalent* $\lambda$ and $\beta$ maximize the specific work of the displacer-type machines for the same $N_\tau$ and $\nu(N_\tau)$.

Moreover, since for the Schmidt cycle $W_{\text{cyc}} = Q_{\text{E}_{\text{cyc}}}(1 - 1/N_\tau)$, the condition for maximum specific cycle work is also the condition for maximum specific heat lifted in heat pump or cooler mode. One might accordingly look for a continuous variation of $\kappa_0$ and $\alpha_0$ over the range of $N_\tau$ covering operation as refrigerator and prime mover.

Fig. 14.3(*a*) shows $\kappa_0$, $\alpha_0$ and corresponding $Q_{\text{E}}/MRT_{\text{c}}$ over the range $0.1 \leqslant N_\tau \leqslant 10.0$ with $\nu(N_\tau)$ as parameter. It is seen that $Q_{\text{E}}/MRT_{\text{c}}$, as well as $\kappa_0$ and $\alpha_0$, varies continuously. Values of $\zeta$ in the range of $N_\tau$ within which the machine operates as a prime mover ($1 \leqslant N_\tau \leqslant 10$) follow by multiplication of $Q_{\text{E}}/MRT_{\text{c}}$ by $(1 - 1/N_\tau)$, but for convenience Fig. 14.3(*b*) displays $Q_{\text{E}}/MRT_{\text{c}}$ over the range $0.1 \leqslant N_\tau \leqslant 1.0$ and $\zeta$ in the range $1 \leqslant N_\tau \leqslant 10$. Subsequent charts will follow the pattern of Fig. 14.3(*b*), and so it is worth bearing in mind that the apparent discontinuity at $N_\tau = 1$ is artificial.

To determine $\lambda_0$ and $\beta_0$ for displacer-type machines, simply convert the $\kappa_0$ and $\alpha_0$ corresponding to the required temperature ratio and calculated dead space ratio using Eqns (6.9) and (6.8) respectively. For displacer machines with divided compression space (Fig. 6.1(*c*)) calculation of the dead space parameter $\nu(N_\tau)$ must take account of the additional dead space, $\mu_{\text{D}}{}^+$ identified in Section 6.2. $\mu_{\text{D}}{}^+$ may be calculated in terms of either $\kappa$ and $\alpha$ or $\lambda$ and $\beta$:

$$\begin{aligned}\mu_{\text{D}}{}^+ &= \tfrac{1}{2}(1 + \lambda - \kappa) = \tfrac{1}{2}\{1 - \kappa + \sqrt{[1 + \kappa^2 + 2\kappa\cos(\alpha)]}\} \\ &= \tfrac{1}{2}\{1 + \lambda - \sqrt{[1 + \lambda^2 - 2\lambda\cos(\beta)]}\}.\end{aligned}$$

For convenience these relationships are plotted in Figs 14.4(*a*) and (*b*).

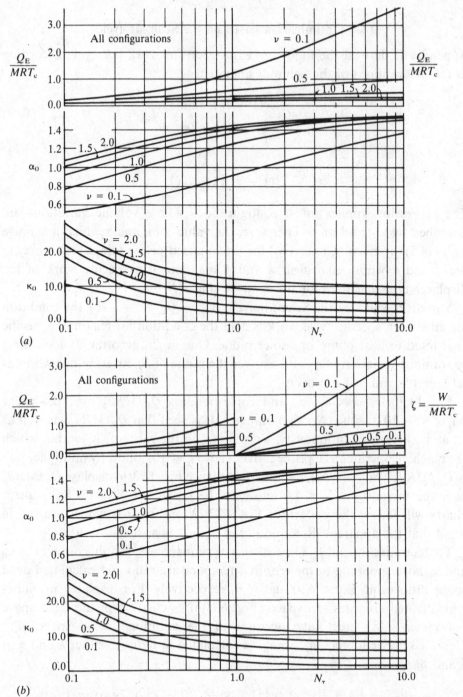

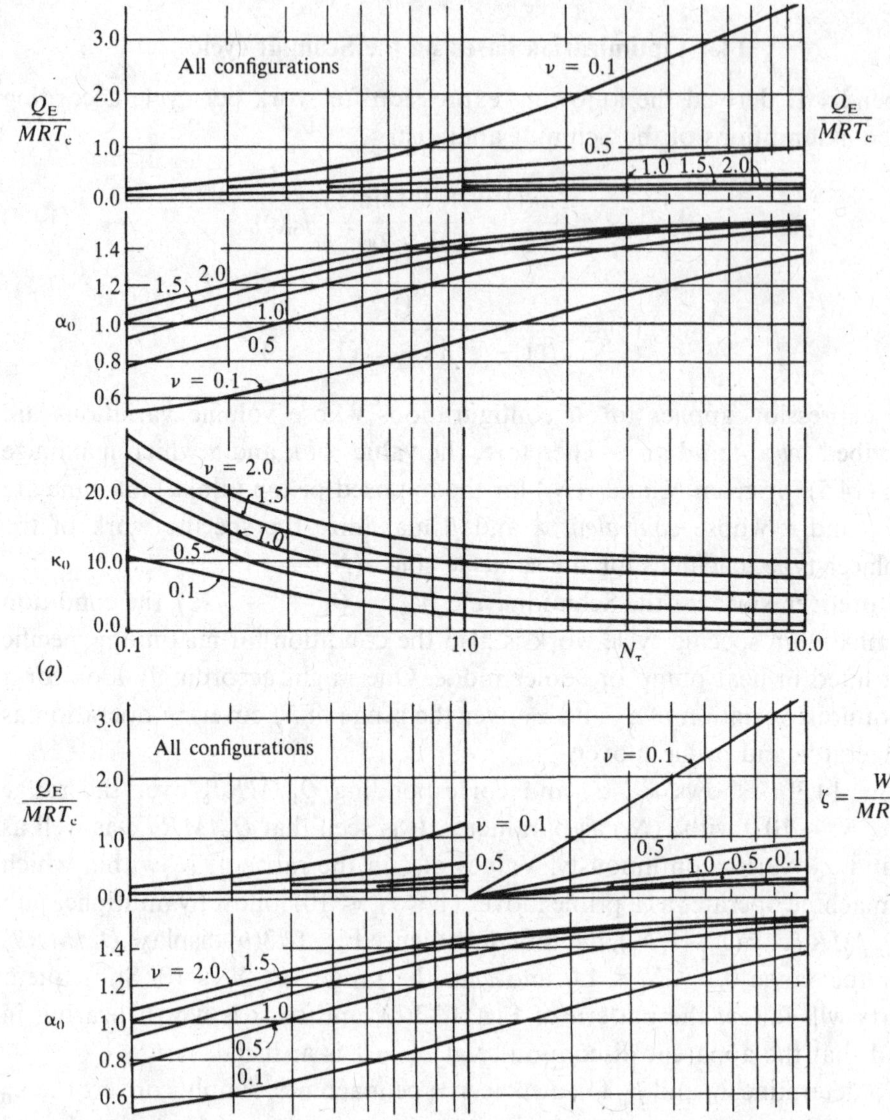


Fig. 14.3 Isothermal optimization charts. (*a*) Values of $\kappa$ and $\alpha$ giving maximum $Q_E/MRT_c$ as a function of the temperature ratio, $N_\tau$ with $\nu(N_\tau)$ as parameter. (*b*) As Fig. 14.3(*a*) but with $\zeta$ $(= W_{cyc}/MRT_c)$ displayed for values of $N_\tau > 1$ (prime mover mode).

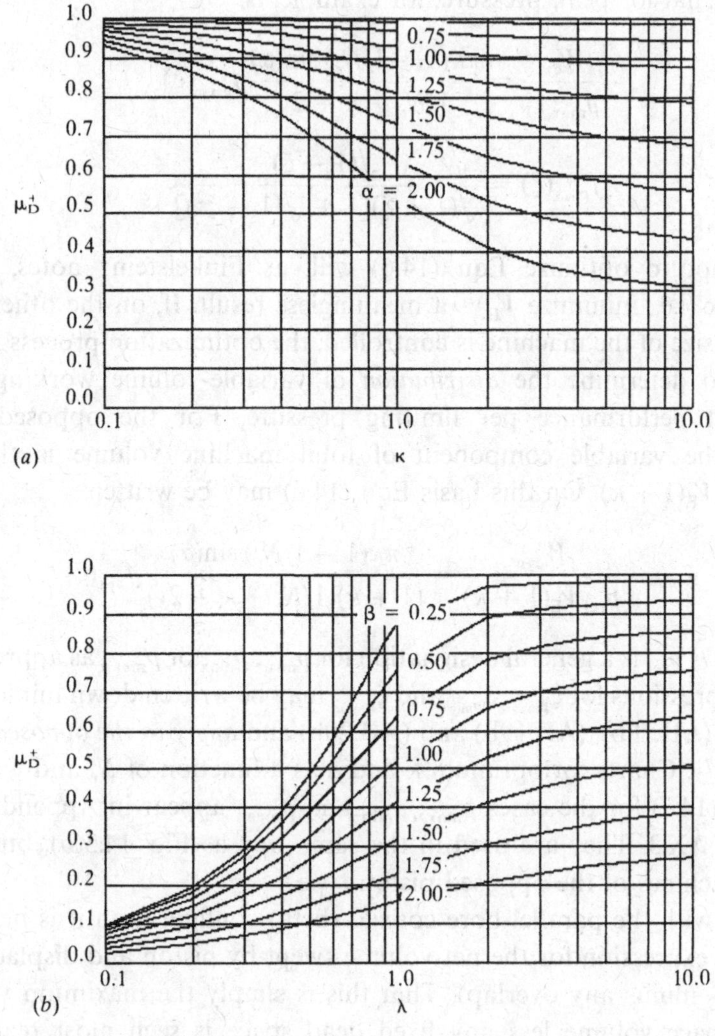


Fig. 14.4 Charts for calculating the additional amount by which normalized dead space of the displacer-type (gamma) machine exceeds that of the coaxial (beta) type. (*a*) Conversion in terms of $\kappa$ and $\alpha$. (*b*) Conversion in terms of $\lambda$ and $\beta$.

Within the (considerable) limitations of the Schmidt analysis, the design which embodies the optimum $\kappa$ and $\alpha$ (or $\lambda$ and $\beta$) from Fig 14.3 is that which makes the most effective use of given (mass $\times$ $R$) of working fluid. Alternatively, the design may be limited by some characteristic cycle pressure – by maximum pressure on account of limited high temperature strength, by minimum pressure because of limitations on charging capacity, or by mean pressure out of some consideration of power control. Appendix II has developed expressions for cycle work in terms of each of these characteristic pressures. That for peak pressure, for example, is:

$$\frac{W_{cyc}}{p_{max}V_E} = \frac{\pi\kappa(1 - 1/N_\tau)\sin(\alpha)}{1/N_\tau + \kappa + 2v} f_{p_{max}}(\xi) \tag{14.6}$$

$$f_{p_{max}}(\xi) = \frac{\sqrt{(1 - \xi)}}{\sqrt{(1 + \xi)[1 + \sqrt{(1 - \xi^2)}]}}$$

Any attempt to optimize Eqn (14.6) will, as Finkelstein[2] notes, merely maximize $\kappa$ (i.e., minimize $V_E$) – a meaningless result. If, on the other hand, the *overall* size of the machine is controlled, the optimization process may be expected to determine the *distribution* of variable-volume working space giving best performance per limiting pressure. For the opposed-piston machine, the variable component of total machine volume is the sum $V_E + V_C = V_E(1 + \kappa)$. On this basis Eqn (14.6) may be written

$$\zeta_{p_{ref}} = \frac{W_{cyc}}{p_{ref}V_E(1 + \kappa)} = \frac{\pi\kappa(1 - 1/N_\tau)\sin(\alpha)}{(1 + \kappa)(1/N_\tau + \kappa + 2v)} f_{p_{ref}}(\xi) \tag{14.7}$$

In Eqn (14.7) $p_{ref}$ is a general designation for $p_{min}$, $p_{max}$ or $p_{mean}$ as appropriate. Specific expressions for $\zeta_{p_{max}}$, $\zeta_{p_{min}}$ and $\zeta_{p_{mean}}$ may be written down immediately from Eqns (AII.18b), (AII.19b) and (AII.20b) and *apply to the opposed-piston machine only.* Charts for optimum $\kappa$ and $\alpha$ as a function of $N_\tau$ and $v$ derived from Eqn (14.7) for the cases $p_{max}$, $p_{min}$ and $p_{mean}$ appear in Appendix X as Figs AX.1–AX.3. They are used in the same way as Fig. 14.3(*b*), but apply only to machines of the opposed-piston type.

To deal with the parallel-bore coaxial (beta) configuration it is necessary to have an expression for the net volume swept by piston and displacer, viz. ($V_E$ plus $V_C$ minus any overlap). That this is simply the maximum value of working space volume less any fixed dead space is seen most readily by imagining the displacer of Fig. 6.1(*b*) to have zero axial length. Net swept volume then follows by looking for the maximum of Eqn (6.7) and discarding the dead-space term. The result is:

$$V_{sw_{beta}} = \tfrac{1}{2}(1 + \kappa + \lambda)$$
$$= \tfrac{1}{2}\{1 + \kappa + \sqrt{[1 + \kappa^2 + 2\kappa\cos(\alpha)]}\}$$

The expression for cycle work per peak cycle pressure per net swept volume in terms of $\kappa$ and $\alpha$ becomes:

$$\zeta_{p_{ref}} = \frac{\pi\kappa(1 - 1/N_\tau)\sin(\alpha)}{\tfrac{1}{2}\{1 + \kappa + \sqrt{[1 + \kappa^2 + 2\kappa\cos(\alpha)]}\}(1/N_\tau + \kappa + 2v)} f_{p_{ref}}(\xi) \tag{14.8}$$

Corresponding expressions for $\zeta_{p\max}$, $\zeta_{p\min}$ and $\zeta_{p\text{mean}}$ may be written down immediately from Eqns (AII.18b), (AII.19b) and (AII.20b). These apply *only* to machines of beta configuration.

Charts giving $\kappa_0$ and $\alpha_0$ equivalent to $\lambda_0$ and $\beta_0$ for the beta machine appear as Figs AX.4–AX.6. They are used in the same way as Fig. 14.3(*b*), conversion of $\kappa_0$ and $\alpha_0$ to $\lambda_0$ and $\beta_0$ being carried out by means of Eqns (6.9) and (6.8) respectively.

Finally, net swept volume of the gamma-configuration machine may be written down immediately as

$$V_{\text{sw}_{\text{gamma}}} = V_{\text{E}}(1 + \lambda) = V_{\text{E}}\{1 + \sqrt{[1 + \kappa^2 + 2\kappa\cos(\alpha)]}\}.$$

For this case specific cycle work is therefore

$$\zeta_{p_{\text{ref}}} = \frac{\pi\kappa(1 - 1/N_\tau)\sin(\alpha)}{\{1 + \sqrt{[1 + \kappa^2 + 2\kappa\cos(\alpha)]}\}(1/N_\tau + \kappa + 2v)} f_{p_{\text{ref}}}(\xi) \quad (14.9)$$

The corresponding expressions for $\zeta_{p\max}$, $\zeta_{p\min}$ and $\zeta_{p\text{mean}}$ derived from Eqns (AII.18b), (AII.19b) and (AII.20b) apply to gamma-configuration machines *only*. Charts in terms of $\kappa_0$ and $\alpha_0$ appear in Appendix X as Figs AX.7–AX.9. Conversion to $\lambda_0$ and $\beta_0$ is via Eqns (6.9) and (6.8) as previously.

## 14.5 Optimization based on the adiabatic cycle

The functional form for specific cycle work (all definitions) for the ideal adiabatic cycle with simple harmonic volume variations has been derived in Appendix VIII and is

$$\zeta = \zeta[N_\tau, \kappa, \alpha, v(N_\tau), \gamma] \quad (14.10)$$

For the adiabatic model there are thus two design charts for each of the corresponding charts based on the Schmidt cycle, one for $\gamma = 1.4$, another for $\gamma = 1.66$.

Each value of $\zeta$ required for the algorithm of Eqns (14.2) is obtained by the process of numerical integration specified by Eqns (AVIII.9a) and (AVIII.9b). The amount of numerical computation called for is thus orders of magnitude greater than that required to produce the Schmidt charts: whereas the data for one of the latter are generated in seconds on the Hewlett Packard 9000 series workstation used by the author, data for the equivalent adiabatic charts requires some 45 min on the same computer. It is therefore essential to take advantage of any available economies.

It is imperative to choose an appropriate step size for the basic cycle integration process: too fine a value and computation time becomes excessive, too coarse and inaccuracy in successive numerical values for cycle work make the optimization algorithm unworkable. An indication of minimum acceptable step size is had by running the adiabatic simulation in isolation from the optimization routine with $\gamma$ replaced by the value unity. For this case, pressures and individual (variable) working space temperatures around a cycle should approximate those of the Schmidt analysis, and $\zeta$ should be close to $\zeta_{Sch}$. Using the numerical scheme described in Appendix VIII an integration step size of 1° leads to working space temperature ratios, $\tau_e(\phi)$ and $\tau_c(\phi)$ which are constant and uniform to within $\pm 0.005$ over a cycle, and to values of $\psi(\phi)$ which approximate those of the Schmidt cycle to within comparable limits. With the integration step size doubled to 2°, discrepancies are an order of magnitude greater, particularly near crank angles corresponding to minimum working space volume. The results to be presented were based on 360 integration steps per crankshaft revolution.

Repeated evaluation of trigonometric functions may be avoided by noting that the normalized expansion space volume, $\mu_e(\phi) = \frac{1}{2}[1 + \cos(\phi)]$ is independent of both $\kappa$ and $\alpha$. A suitable number (in this case 360) of values of $\mu_e(\phi)$ is thus computed once only in advance of any call to the optimization algorithm and stored as an array of simple values. Normalized compression space volume $\mu_c(\phi) = \frac{1}{2}\kappa[1 + \cos(\phi - \alpha)]$. The variable part of this expression, i.e., $1 + \cos(\phi - \alpha)$, appears to require recomputation each time $\alpha$ is varied. On the other hand, the function $1 - \cos(\phi - \alpha)$ is merely the function $(1 + \cos\phi)$ offset by an angle $\alpha$ – i.e., the array of values of $\mu_e(\phi)$ displaced by $\alpha$. It is very easy to convert $\alpha$ into an integer, $n_{disp}$, corresponding to the number of array elements by which $\mu_c(\phi)$ is displaced from $\mu_e(\phi)$; $n_{disp} = INT(360 \times \alpha/2\pi)$ in this case. The precision required of the simulation calls for calculation of any fractional part of the integration interval to be taken into account. This is simply the *REAL* variable $360 \times \alpha/2\pi$ minus the *INTEGER* $n_{disp}$. In this way any required value of $\mu_c(\phi)$ is available in terms only of $\alpha$ and of the stored array of values of $\mu_e(\phi)$. An equivalent short cut is readily worked out in terms of the trigonometric identity

$$\cos(\phi - \alpha) = \cos(\phi)\cos(\alpha) + \sin(\phi)\sin(\alpha)$$

$$= (2\mu_e(\phi) - 1)\cos(\alpha) + \sin(\phi)\sin(\alpha),$$

and calls for storage of a simple array of (say, 360) values of $\sin(\phi)$.

The expressions integrated have in the denominator a term for cylinder mass content. They are therefore meaningless where the cylinder volume is zero. One remedy is to follow Finkelstein[11] and expand the solution in a series about the two singularities, but the additional algebra and resulting code promise to add substantially to computing time per cycle.

An alternative is to allocate a representative fraction of the specified dead space to each cylinder so that working space volume is never zero. The fraction should be the smallest possible, since the analysis assumes that dead space is a space in which working fluid temperature is constant at the value of the adjacent solid wall, whereas any fluid in the dead space of a variable volume has the (variable) temperature of the gas in that volume.

In the present case a fraction 0.1 of total dead space ratio

$$\mu_D \, (= (V_{de}/V_E + V_{dr}/V_E + V_{dc}/V_E) = \mu_{de} + \mu_{dr} + \mu_{dc})$$

was added to each variable volume, requiring a corresponding correction to the working value of the dead-space parameter $\nu(N_\tau)$. From the definition of $\nu(N_\tau)$, viz

$$\nu(N_\tau) = \mu_{de}/N_\tau + \mu_{dc} + \frac{\mu_{dr}\ln(N_\tau)}{N_\tau - 1}$$

it is clear that the value of $\nu(N_\tau)$ to be used in computation is 0.8 of the nominal if it is supposed that the two amounts of $0.1\mu_D$ are achieved by taking $0.2\mu_{de} + 0.2\mu_{dr} + 0.2\mu_{dc}$.

Computation of individual values of $\zeta$ remains a relatively lengthy process, and it is desirable to minimize the number of times it is carried out per search for optimum $\kappa$ and $\alpha$. In particular, it is necessary to avoid recalculation of $\zeta$ for values of $\kappa$ and $\alpha$ already used: the inset of Fig. 14.2 indicates specimen cases where arbitrary application of the algorithm would give rise to repeated evaluation. An unsophisticated but viable expedient is to store as simple variables the five values of $\zeta$ computed per application of the pattern, together with the five corresponding pairs of values of $\kappa$ and $\alpha$. Before each calculation of $\zeta$ after a pattern move, the proposed new pair of arguments is tested against each of the five previous pairs to see if it has been used before. If so, the appropriate stored value of $\zeta$ is recalled for reuse.

Once the optimum value of $\zeta$ lies within the outline of the pattern there appears to be some computational economy in making use of a series expansion to home in on the optimum: for given $N_\tau$, $\nu(N_\tau)$ and $\gamma$, $\zeta = \zeta(\kappa, \alpha)$. For $\Delta\kappa$ and $\Delta\alpha$ sufficient small:

$$\zeta(\kappa + \Delta\kappa, \alpha) \approx \zeta(\kappa, \alpha) + \Delta\kappa \frac{\partial\zeta}{\partial\kappa} + \frac{\Delta\kappa^2}{2}\frac{\partial^2\zeta}{\partial\kappa^2} \tag{14.11a}$$

$$\zeta(\kappa, \alpha + \Delta\alpha) \approx \zeta(\kappa, \alpha) + \Delta\alpha \frac{\partial\zeta}{\partial\alpha} + \frac{\Delta\alpha^2}{2}\frac{\partial^2\zeta}{\partial\alpha^2} \tag{14.11b}$$

At the optimum $\partial\zeta/\partial\kappa = \partial\zeta/\partial\alpha = 0$. Differentiating the right-hand sides with respect to $\Delta\kappa$ and $\Delta\alpha$ respectively and setting equal to zero gives the values of $\Delta\kappa$ and $\Delta\alpha$ at which the optimum is predicted to occur. Adding these to the current values of $\kappa$ and $\alpha$ gives the approximation to $\kappa_0$ and $\alpha_0$:

$$\kappa_{0_{\text{approx}}} = \kappa_{\text{current}} - \frac{\partial\zeta/\partial\kappa}{\partial^2\zeta/\partial\kappa^2} \tag{14.12a}$$

$$\alpha_{0_{\text{approx}}} = \alpha_{\text{current}} - \frac{\partial\zeta/\partial\alpha}{\partial^2\zeta/\partial\alpha^2} \tag{14.12b}$$

Numerical values of the first and second differentials are available immediately from the five values of $\zeta$ computed for the pattern.

Figs 14.5(*a*) and (*b*) are for $\gamma = 1.4$ and 1.66 respectively, and show the values of $\kappa_0$ and $\alpha_0$ giving maximum heat lifted (in the range $0.1 \leqslant N_\tau \leqslant 0.95$) and maximum cycle work ($1.5 \leqslant N_\tau \leqslant 10$) per $MRT_c$. The curves apply to all three basic configurations (provided the gamma machine does not have a

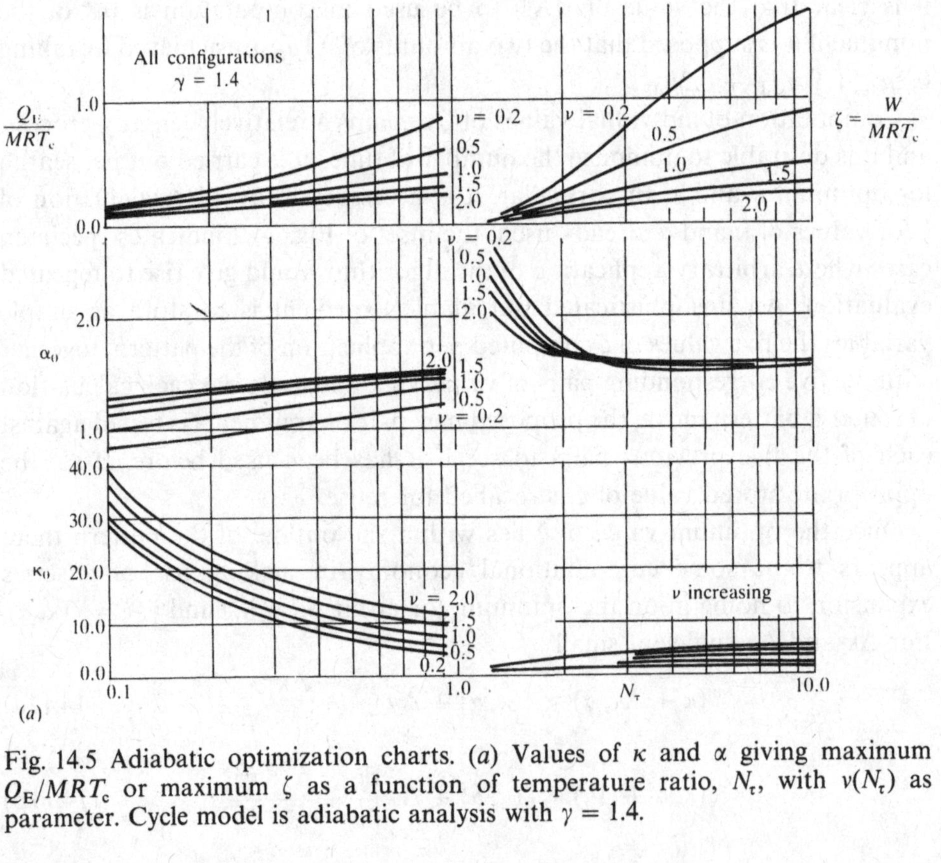


(*a*)

Fig. 14.5 Adiabatic optimization charts. (*a*) Values of $\kappa$ and $\alpha$ giving maximum $Q_E/MRT_c$ or maximum $\zeta$ as a function of temperature ratio, $N_\tau$, with $\nu(N_\tau)$ as parameter. Cycle model is adiabatic analysis with $\gamma = 1.4$.

heat exchanger dividing the two variable components of the compression space). Output between 0.95 and 1.5 has been suppressed, since in this region the optimization process leads to extreme values of $\kappa$ which, while they give favourable $Q_E/MRT$ and $\zeta$, are unlikely to be used in practice. The result may be related to a phenomenon deduced by Walker and Khan[8] whereby for small values of dead space parameter and for $\kappa = 1$ and $\alpha = \pi/2$, $\zeta$ remains negative with increasing $N_\tau$ even after $N_\tau$ has exceeded unity. The matter bears further investigation, but if verified might explain why the $\kappa_0$ and $\alpha_0$ required to make the machine behave as a heat pump just below $N_\tau = 1$ and as a prime mover at $N_\tau$ immediately above 1.0 are somewhat extreme.

The optimization of a function evaluated by complex, iterated integration is more hazardous than that based on a simple, explicit function. Possibly for this reason, the curves for optimized $\zeta_{p_{max}}$ and $\zeta_{p_{mean}}$ (Appendix X) as output by the computer are less regular than those for the corresponding Schmidt case, and a small minority have been subjected to arbitrary smoothing. At the time of writing it has not proved possible to complete a convincing set

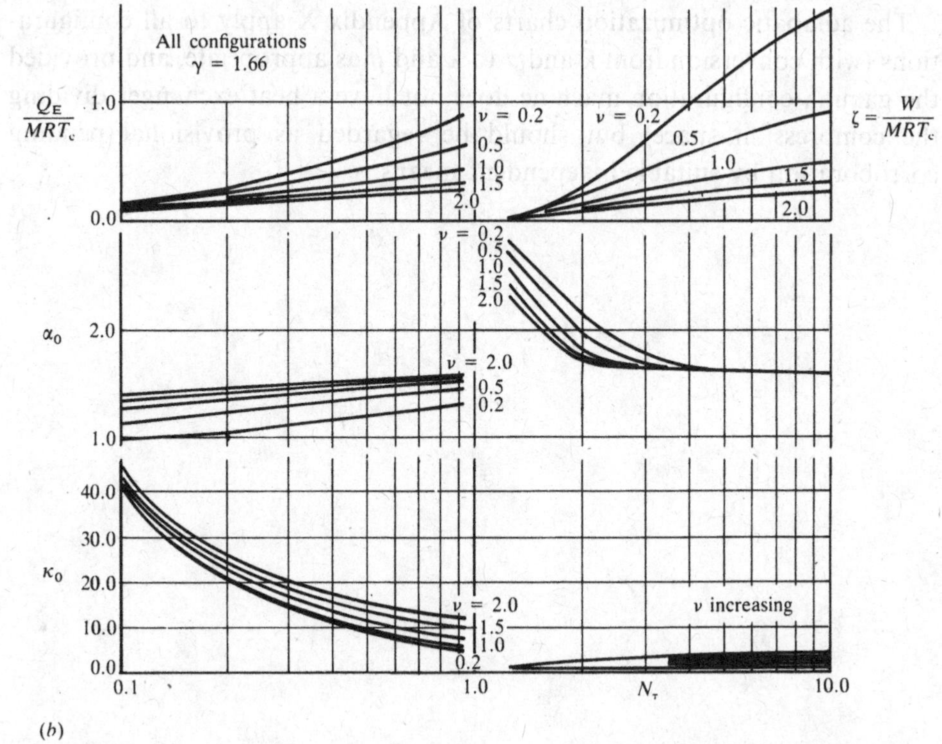


Fig. 14.5 (*cont.*) Adiabatic optimization charts. (*b*) As for Fig. 14.5(*a*) but with $\gamma = 1.66$.

of curves for $\zeta_{p_{min}}$. The difficulty is not unique to the adiabatic analysis: this was the case which demanded the greatest amount of computational 'fine tuning' when the charts for the Schmidt cycle were being prepared. In seeking the maximum of $W_{cyc}/p_{min}V_{sw}$ the process identifies the $\kappa$ and $\alpha$ which effectively minimize $\psi_{min}$. Extreme values result, and these possibly compound the previously identified problem in the vicinity of $N_\tau = 1$. At any event the optimization sequence applied to the adiabatic case tends to stall at this point.

Whereas there was some independent corroboration for the isothermal optimization charts, there is none for the adiabatic counterparts. Unfortunately, it is not possible to check the latter against the former by generating a set with the numerical value of $\gamma$ replaced by unity: although the cycle analysis responds appropriately by computing near-isothermal working space temperatures and an isothermal pressure variation, the expressions for $Q_E$ and $Q_C$ are undefined for this case (Eqns (5.3) and (5.4) of Ref 11), consistent with net heat exchange in the exchangers *per se* of the ideal isothermal cycle being zero. Therefore it is not possible with $\gamma$ replaced by 1.0 to generate values of $Q_E/MRT$ for the range $N_\tau < 1$.

The adiabatic optimization charts of Appendix X apply to all configurations (with conversion from $\kappa$ and $\alpha$ to $\lambda$ and $\beta$ as appropriate, and provided the gamma-configuration machine does not have a heat exchanger dividing the compression space) but should be regarded as provisional pending corroboration by suitable independent means.

# 15
# The hot-air engine

## 15.1 Background

The hot-air engine is a degenerate form of Stirling engine, generally distinguished from the parent machine by the lack of a proper regenerator and of extended-surface heat exchangers. It is possible that the total number of hot-air engines manufactured since the original invention of 1816 considerably exceeds the number of proper Stirling cycle machines. Lehmann, Heinrici and Kyko engines[1] all belonged in the hot-air category, as, by definition, did virtually all toy hot-air engines[1–3] which, for the decade or so after the turn of this century, were probably as common as toy steam engines. A number of so-called Stirling demonstration models available today (e.g., that described in Ref. 4) are hot-air engines.

There are three reasons for looking at the hot-air engine in some detail at this stage:

(1) Fluid flow and heat transfer in this type of engine correspond in essence to the 'shuttle transfer' phenomenon of Stirling engines proper. Shuttle heat transfer is a penalty on performance, and its exact nature is in need of clarification.
(2) Intuitively, the scope for improvement in the performance of types without a separate regenerator is not as great as that for the proper Stirling engine, but it would be interesting to assess such scope as there may be.
(3) If methods of thermodynamic analysis and design are not applicable to the hot-air engine then they will not cope with the sophisticated counterpart. Certainly it is quicker, easier and cheaper to construct and instrument a hot-air engine to test methods of thermodynamic design than to work with a proper Stirling engine.

The treatment which follows is from Ref. 5 of Chap. 1 and, far from being a simplification of earlier analysis, is an extension in three respects:

It extends the concept of exergy to the phenomena of viscous dissipation and thermal conduction internal to the working fluid.

The restrictive assumption of one-dimensional ('slab') flow is dispensed with. This eliminates the need to define the friction factor and the coefficient of convective heat transfer.

In earlier treatments use was made of values of the local, instantaneous heat addition rate obtained from an ideal, reference cycle. The sweeping assumptions of the reference cycle ('isothermal' or 'adiabatic') result in estimates of heat transfer rates in the variable volume spaces which are unusable. The present formulation proceeds independently of the need for approximate heat transfer rates in any space. For this reason it is capable of extension to the variable volume spaces.

Use of the now familiar normalization leads to the same functional form for specific cycle work, $\zeta$, as that which applies to more complex gas circuits. This is so despite the fact that the usual friction factor and heat transfer correlations have been circumvented, and confirms that use of such correlations is but one of a choice of means of describing the same physical phenomena. In full dynamic similarity form (in the case of this simple gas circuit there is little to be gained by use of the thermodynamic similarity form) the expression for $\zeta$ is:

$$\zeta = \zeta\left[ N_\tau,\ V_E/r^3,\ \begin{matrix} \lambda_{k_1}\ \lambda_{gc_1} \\ \lambda_{k_2}\ \lambda_{gc_2} \\ \lambda_{k_3}\ \lambda_{gc_3} \\ \vdots \quad \vdots \\ \lambda_{k_n}\ \lambda_{gc_n} \end{matrix},\ \gamma,\ N_{MA},\ N_{SG},\ N_H,\ k/k_w,\ (N_F) \right]$$

As before, the terms $\lambda_{k_n}$ describe the drive kinematics, and may give way to $\kappa$ and $\alpha$ if simple harmonic forms are acceptable as an approximation to actual volume variations. The ratio $k/k_w$ appears because $N_H$ is defined in terms of $k_w$, where $k_w$ is the conductivity of the solid regenerator material, whereas the conduction losses to be computed are those within the fluid. As in the case of machines with more complex gas circuits, adherence to the functional form will permit the performance envelope to be displayed in the form of maps. In turn, the maps will be shown to allow instantaneous scaling of the design.

## 15.2 Principle of operation and simplifying assumptions

The essential features of a typical hot-air engine are shown in Fig. 15.1. Other arrangements are possible, such as those involving opposed pistons or those in which the axes of piston and displacer are offset. The coaxial layout has

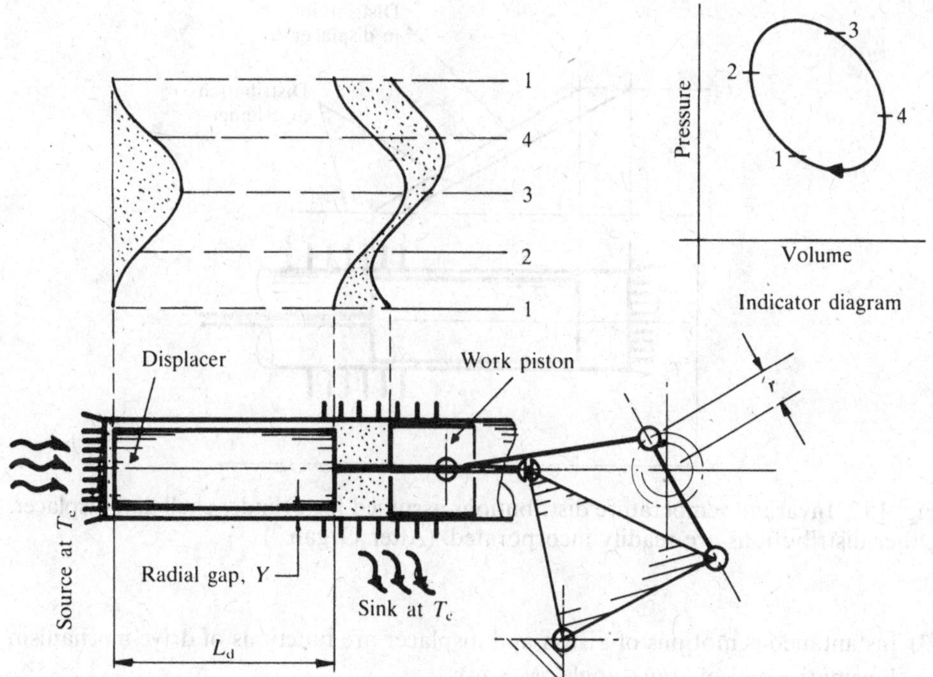


Fig. 15.1 Schematic of typical hot-air engine. (After Organ.[5])

the potential for minimum unswept volume, and thus for highest specific work. In the figure the bell-crank drive is essentially that of the Philips MP1002CA air-charged Stirling engine. Any number of alternative drive mechanisms are available to provide the required volume variations.

Compression occurs with the displacer close to its top dead-centre position and with much of the working fluid thus displaced to the compression space. Expansion occurs with the displacer and power piston moving more or less together with compression space thus at a minimum. The annular gap around the displacer serves as expansion exchanger, regenerator and compression exchanger.

The flow and heat transfer phenomena of interest are those occurring in the displacer annulus. In this connection it is assumed:

(1) that the gradient of temperature is from $T_e$ to $T_c$ in both cylinder and displacer shells, and that the gradient is linear as indicated in Fig. 15.2. The displacer shell temperature profile is fixed relative to the displacer, and thus moves with it;
(2) there is net heat addition to and rejection from the machine only at isothermal faces at $T_e$ and $T_c$, other faces being thermally insulated from the environment;

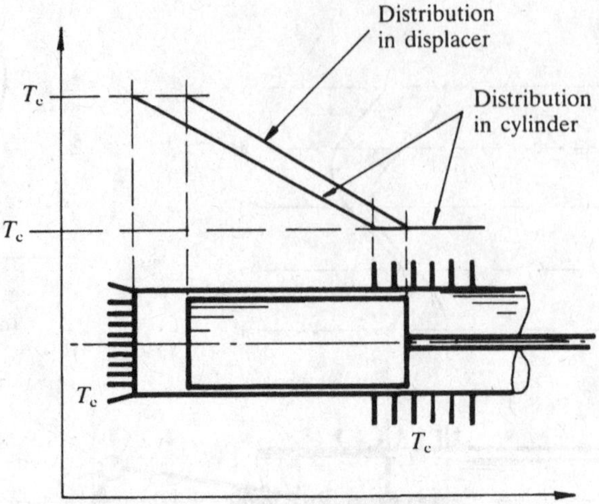


Fig. 15.2 Invariant temperature distributions assumed for cylinder shell and displacer. Other distributions are readily incorporated. (After Organ.[5])

(3) instantaneous motions of piston and displacer are functions of drive mechanism kinematics and of crank angle, $\phi(=\omega t)$;

(4) instantaneous, local pressure, density and temperature of the gas are related via the ideal gas equation, $p = \rho RT$. The local thermal conductivity, $k$, and coefficient of dynamic viscosity, $\mu$, may be functions of temperature;

(5) at both fixed and moving walls the 'no slip' condition applies. Spatial and temporal components of acceleration are ignored when formulating the momentum equation, signifying that (*a*) the velocity profile is that for fully-developed laminar flow, and (*b*) 'entry length' phenomena are not taken into account. Design or operating conditions permitting local, instantaneous $N_{re}$ associated with turbulent flow are excluded;

(6) that the problem may be treated as a linear one: local, instantaneous fluid properties are sufficiently close to values computed from an ideal, one-dimensional analysis to permit individual discrepancies to add linearly. As argued in previous chapters, the assumption is most readily substantiated in the context of analysis directed at optimization, where it becomes increasingly valid as the point of optimum performance is approached;

(7) from item (6) above, local, instantaneous velocity and density distributions integrate to give the value of local, instantaneous mass flow rate, $m'(x, t)$ computed from the ideal, reference analysis of the machine.

The treatment which follows supposes that the reference analysis has already been applied, and that $m'(x, t)$ is thus known at all locations within the regenerator annulus and for all values of $\omega t$ over a complete cycle; the reference analysis has also yielded values of $p(\omega t)$ for the complete cycle.

## 15.3 Defining equations

In Fig. 15.3 is indicated schematically a plane section through the regenerator annulus perpendicular to the direction of flow. The instantaneous, local velocity profile is computed on the basis of instantaneously prevailing pressure and a corresponding value of density based on the local mean of adjacent wall temperatures. Instantaneous, bulk mean velocity $U(x, t)$ is thus obtained in terms of $m'$:

$$U(x, t) = \frac{m'(x, t)}{\rho(x, t)A_x}$$

It would be a straightforward matter to express the corresponding instantaneous velocity distribution, $u(r, x, t)$ in cylindrical polar coordinates. However,

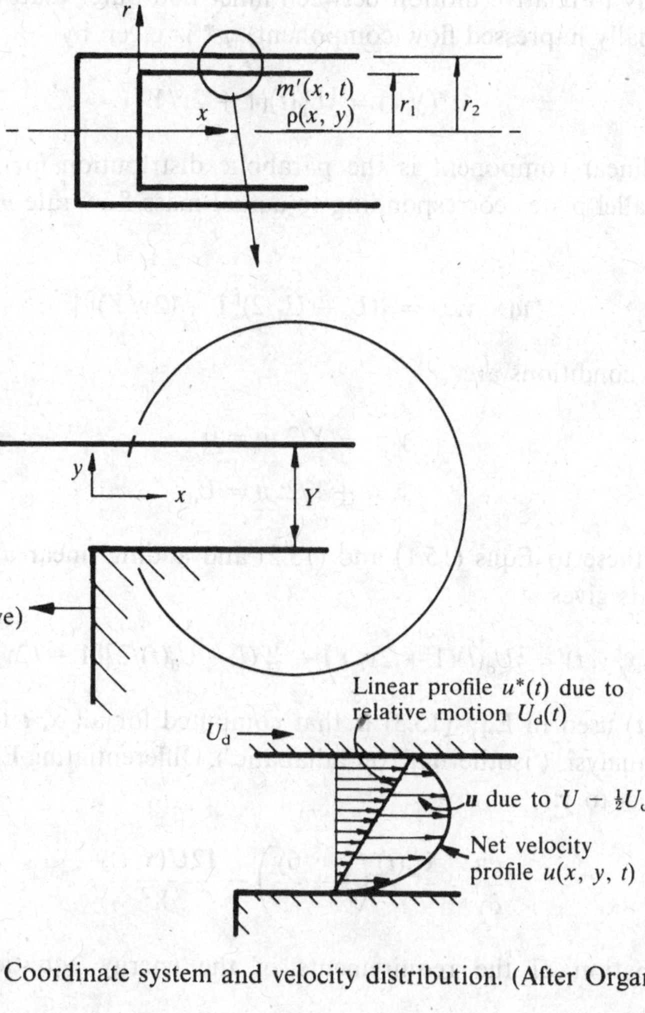


Fig. 15.3 Coordinate system and velocity distribution. (After Organ.[5]).

the resulting equation leads to difficulties[6] when $r_2/r_1 \approx 1$, as is the case here. The expedient normally employed to get around the difficulty of small differences is to work in terms of a series expansion. The effect is to convert the solution to that of flow between parallel plates. One might therefore just as well start with the parallel flat plate assumption, replacing $r$ by $y$ as the independent variable.

In Fig. 15.3 the origin for $y$ is the mid-plane, even though this is not generally the point at which $\partial u/\partial y = 0$. The $x$ origin is the moving displacer. With this convention a displacer velocity of $U_d(\phi)$ is equivalent to a velocity of $-U_d(\phi)$ at the facing wall.

Fig. 15.3 shows that instantaneous fluid velocity distribution may be thought of as a superposition of linear and parabolic components, with linear component, $u^*$, being the velocity field which would be achieved if flow were due entirely to relative motion between inner and outer plates (i.e., without any externally impressed flow component). $u^*$ is given by

$$u^*(y, t) = \tfrac{1}{2}U_d(t)(1 + 2y/Y) \tag{15.1}$$

The non-linear component is the parabolic distribution for flow between static, parallel plates corresponding to actual mass flow rate *minus* that due to $u^*$

$$u(x, y, t) = \tfrac{3}{2}(U - U_d/2)[1 - (2y/Y)^2] \tag{15.2}$$

Boundary conditions are:

$$y = -Y/2\colon u = 0$$

$$y = +Y/2\colon u = U_d$$

Applying these to Eqns (15.1) and (15.2) and adding linear and non-linear components gives

$$u(x, y, t) = \tfrac{1}{2}U_d(t)(1 + 2y/Y) + \tfrac{3}{2}[U - U_d(t)/2][1 - (2y/Y)^2] \tag{15.3}$$

The $U(x, t)$ used in Eqn (15.3) is that computed for all $x$, $t$ from the ideal reference analysis ('isothermal' or 'adiabatic'). Differentiating Eqn (15.3) once with respect to $y$:

$$\frac{\partial u}{\partial y} = \frac{U_d(t)}{Y}\left(1 + \frac{6y}{Y}\right) - \frac{12U(x, t)y}{Y^2} \tag{15.4}$$

In anticipation of the requirements of the energy equation, a second differentiation with respect to $y$ is carried out

$$\frac{\partial^2 u}{\partial y^2} = \frac{6[U_d(t) - 2U(x, t)]}{Y^2} \tag{15.5}$$

The dissipation function, $\Phi$, for the flow field may be expressed in terms of $u(x, y, t)$ and its derivatives:

$$\Phi = \tfrac{4}{3}(\partial u/\partial x)^2 + (\partial u/\partial y)^2 \tag{15.6}$$

The energy equation may be written:

$$\rho c \frac{DT}{dt} = k\nabla^2 T + \frac{Dp}{dt} + \mu\Phi \tag{15.7}$$

In Eqn (15.7) $\mu$ may be $\mu[T(x)]$. The ideal reference analysis yields the appropriate estimate of $Dp/Dt$ since

$$\frac{Dp}{dt} = \frac{\partial p}{\partial t} + u\frac{\partial p}{\partial x} = \frac{\partial p}{\partial t} + u\mu\frac{\partial^2 u}{\partial y^2} \tag{15.8}$$

To the right-hand side of Eqn (15.7) may be transferred those terms known from the reference analysis:

$$\nabla^2 T - \frac{1}{\alpha}\left(\frac{\partial T}{\partial t} + u\frac{\partial T}{\partial x}\right) = -\frac{1}{k}\left(\frac{\partial p}{\partial t} + u\mu\frac{\partial^2 u}{\partial y^2}\right) - \frac{\mu\Phi}{k} \tag{15.9}$$

Eqn (15.8) is a general heat conduction equation. Standard numerical means permit solution to give $T$ for all $x$, $y$ for an entire cycle corresponding to given boundary and initial conditions.

## 15.4 Normalization – governing dimensionless groups

Dimensionless lengths $\lambda$ and $\eta$ represent $x/L_{ref}$ $(=x/r)$ and $y/L_{ref}$ $(=y/r)$ respectively. Velocity $u$ is normalized by reference to $\omega r$, giving rise to a dimensionless dissipation function, $\Phi^*$ defined as $\Phi^* = \Phi/\omega^2$. With $\tau = T/T_{ref}$ and $\psi = p/p_{ref}$ the energy equation becomes:

$$\frac{\partial^2 \tau}{\partial \lambda^2} + \frac{\partial^2 \tau}{\partial \eta^2} - 2\pi\frac{\gamma}{\gamma - 1}N_H\left(\frac{\partial \tau}{\partial \phi} + u\frac{\partial \tau}{\partial \lambda}\right)k/k_w$$
$$= 2\pi N_H\left[\frac{\partial \psi}{\partial \phi} + \frac{2\pi}{N_{SG}}\left(u\frac{\partial^2 u}{\partial \eta^2} + \Phi^*\right)\right]k/k_w \tag{15.10}$$

Eqn (15.10) in conjunction with the ideal reference cycle determines thermodynamic performance. For a machine of fixed geometry operating with given $\gamma$ and $N_\tau$ the only operational variables of Eqn (15.8) are $N_{SG}$ and $N_H$. (A Fourier modulus, $N_F$, arises during numerical solution.) The qualitative effect of varying the numerical values of $N_{SG}$ and $N_H$ will be investigated shortly. In the meantime it is apparent that:

(1) for sufficiently small $N_H$ the solution of Eqn (15.10) is that of the Laplace (steady conduction) equation – conduction effects outweigh those of enthalpy advection;
(2) for large $N_H$ combined with large $N_{SG}$ the effects of enthalpy advection and pressure change dominate. The temperature distribution in the annulus is washed downstream;
(3) for sufficiently small $N_{SG}$ the computed temperature field is subject to the heating effect of viscous dissipation.

## 15.5 Entropy creation rate and lost available work

With the material of Chap. 5 as background the expression for entropy generation rate per unit volume due to heat transfer may be combined with that due to viscous dissipation to give the single expression:

$$S_v' = \frac{k}{T^2}(\nabla T)^2 + \frac{\mu}{T}\Phi \geqslant 0 \tag{15.11}$$

Rate of loss of available work per unit volume is $T_0 S_v'$, and the corresponding rate of lost specific work $T_0 S_v'/p_{ref} V_{ref}$. Substituting into Eqn (15.11) and taking the environment temperature $T_0$ to be equal to the heat rejection temperature, $T_c$, permits computation of net specific work loss per unit time and per unit volume. Following the methods of Chaps 5 and 8 this may be converted to an expression for $d\zeta_s$, i.e., for incremental lost available work in normalized elemental control volume, $\Delta\mu_j$, and during crankshaft increment $\Delta\phi$:

$$d\zeta_s = \left(\frac{1}{2\pi N_H}\frac{k}{k_w}\frac{(\nabla_\lambda \tau)^2}{q^2} + \frac{2\pi}{\tau N_{SG}}\Phi^*\right)\Delta\mu_j\,\Delta\phi \tag{15.12}$$

In Eqn (15.12) the terms $\nabla_\lambda \tau$ are evaluated on the basis of normalized temperatures, $T/T_{ref}$, and normalized distance coordinates, $\Delta x/r$ and $\Delta y/r$. Evaluation and summation of all increments of lost, specific work, and subtraction from ideal specific work in the manner of Section 7.3.4 leads to the required numerical value of net, specific cycle work.

## 15.6 Computational considerations

The fact that the approximate $u(x, y, t)$ are known in advance (from the reference analysis) means that the process of numerical solution of Eqn (15.10) is inherently stable. Nevertheless, two precautions are worthwhile: (*a*) to work with 'upwind' differences and (*b*) to choose an appropriate Courant number, $N_C$. The latter is $u\Delta t/\Delta x$, viz the ratio of distance travelled by a particle during the integration interval to the length of the spatial subdivision. $u$ cannot be known in advance of computation, but the peak value is closely related to peak rate of volume change which, in turn, is linked to $\omega r$. The ratio of the cross-sectional area of the displacer of diameter $d$ to that of the annular gap is $D/4Y$, giving $\omega rD/4Y$ as the notional peak particle speed. With $n_x$ for the number of spatial subdivisions in the $x$ direction, the Courant number is estimated as $n_x\Delta\phi D/(4YL_d/r)$. For suitable values of Courant number the numerical solution of Eqn (15.10) may be expected to approximate the (hypothetical) analytical solution.

A solution for net specific cycle work, $\zeta$, is obtained in three stages: the 'isothermal' or 'adiabatic' reference cycle is simulated first, using only the data $N_\tau$, $\lambda_1$, . . . , $\lambda_n$ and $\gamma$. The arrays of dimensionless pressure $\psi(\phi)$ and velocity, $(u/\omega r)$ are stored corresponding to each angular increment, $\Delta\phi$, of crankshaft position and for a line of points within the regenerator annulus. These data are picked up by a second program which computes the two-dimensional array of normalized velocities and their derivatives (Eqns (15.3)–(15.5)) together with a corresponding array of values of the dissipation function. Solution of the heat conduction equation to give (normalized) temperature distribution calls for working from a matrix of 'old' temperatures at crank angle $\phi - \Delta\phi$ and taking account of the motion of the boundary temperature profile during the interval. As in the computation of the temperature field internal to the regenerator wire (Section 8.5) it is necessary to carry the integration process through more than a single crankshaft revolution in order to establish cyclic equilibrium (two full revolutions have been found satisfactory).

The final computational stage is evaluation of individual contributions to lost specific work and their summation over a revolution of the crankshaft. Total loss for the whole cycle is a linear sum of the losses from the individual increments, and is subtracted from ideal, normalized cycle work to yield the net, indicated value.

## 15.7 Specimen velocity and temperature profiles

The velocity field derives from the mass flow rates computed from the ideal reference cycle. For a machine of given geometry operating at specified $N_\tau$ there are only two velocity fields, one for $\gamma = 1.4$, the other for $\gamma = 1.66$. The resulting slight difference is indicated by the profiles in Fig. 15.4, which exaggerates the $\eta$ $(y/r)$ scale for clarity. Although the profiles would generally be different at a different value of crank angle, $\phi$, the profiles at any given $\phi$ are independent of rotational speed, $N_s$.

Specimen profiles of dimensionless temperature are shown in Fig. 15.5 for $\gamma = 1.4$. As in the case of Fig. 15.4 the scale for $\eta$ $(y/r)$ is exaggerated to facilitate display. In Fig. 15.5(*a*) $N_{SG}$ has a large value, and viscous dissipation effects are therefore negligible. A small value of $N_H$ implies

low $p_{ref}$ and/or
low $N_s$ and/or
high $k$

For any or all of these conditions there results a Laplace temperature field in which the dominant effect is steady conduction. With $N_H$ sufficiently small, entropy creation due to thermal conduction causes losses per cycle to exceed useful work.

A high value of $N_H$ (Fig. 15.5(*b*)) results from

high $p_{ref}$ and/or
high $N_s$ and/or
low $k$.

For this case the profiles of temperature in the $\lambda$ direction are more the

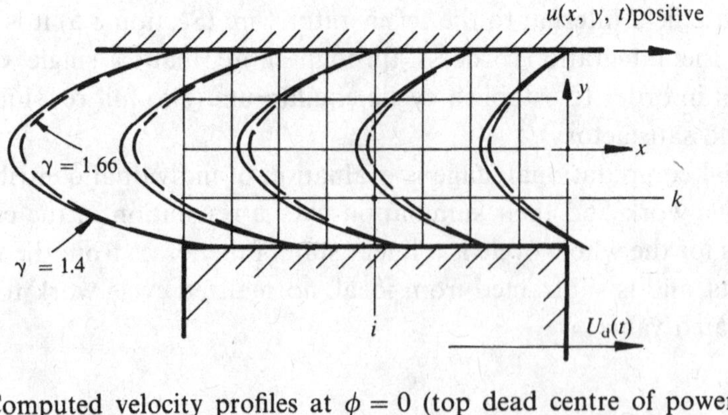


Fig. 15.4 Computed velocity profiles at $\phi = 0$ (top dead centre of power piston). Geometry and operating conditions as in Table 15.1. Computational grid point, *i*, *k*, identified. (After Organ.[5])

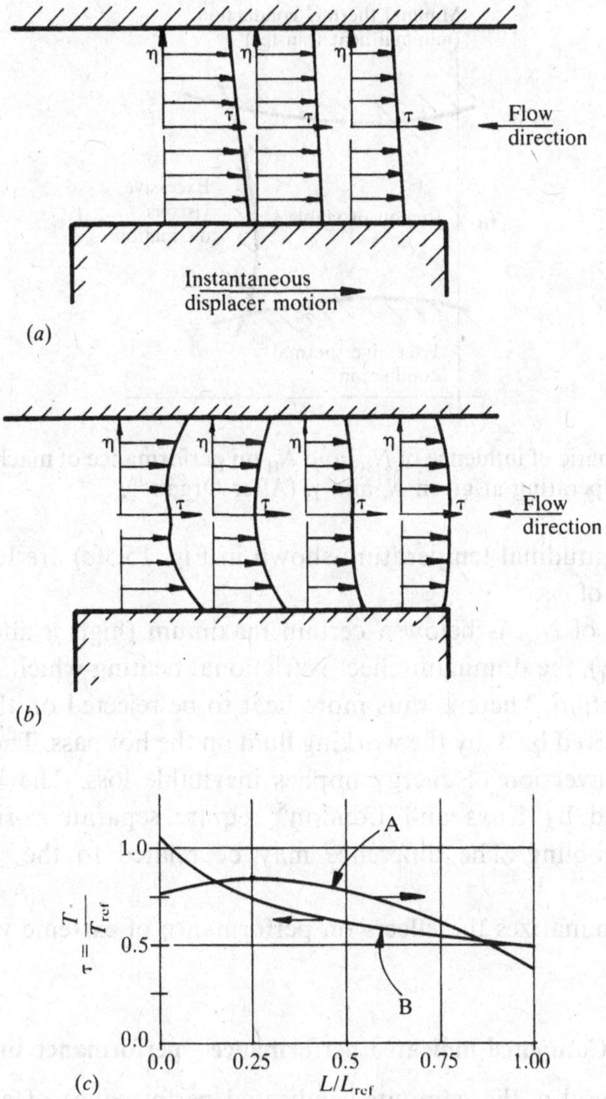


Fig. 15.5 Specimen temperature distributions. (After Organ.[5]) (*a*) Laplacian temperature distribution for $N_H = 10^2$. (*b*) Temperature distribution dominated by enthalpy flux for $N_H = 10^5$. (*c*) Temperature distribution at mid-annular surface $\eta = 0$, at different points in the cycle: A at top dead-centre position of power piston with $dp/dt$ positive; B at 294° after top dead-centre with $dp/dt$ negative and low expansion space outlet temperature.

result of fluid motion (enthalpy advection) and rate of change of pressure with time than of conduction effects. By plotting lengthwise temperature distribution at the mid-surface ($y = Y/2$) against $\lambda$ at different crank positions the temperature waves predicted by Organ[7] are confirmed. The specimen

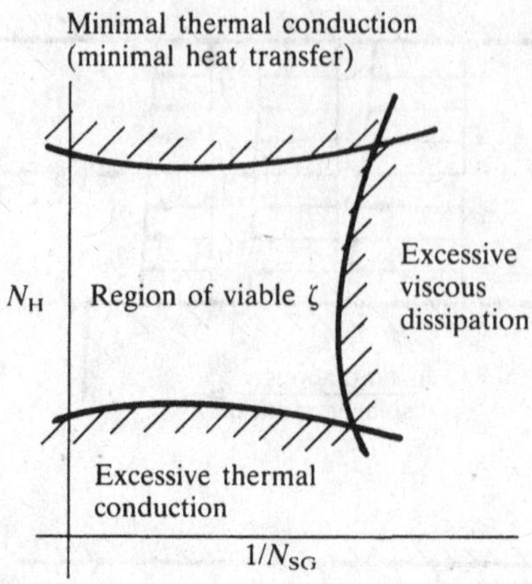


Fig. 15.6 Schematic of influence of $N_{SG}$ and $N_H$ on performance of machine with fixed geometry and operating at given $N_\tau$ and $\gamma$. (After Organ.[1])

profiles of longitudinal temperature shown in Fig. 15.5(*c*) are for arbitrarily chosen values of $\phi$.

If the value of $N_{SG}$ is below a certain maximum (high $\mu$ and/or high $N_s$ and/or low $p_{ref}$), the dominant effect is frictional heating which accompanies viscous dissipation. There is thus more heat to be rejected on the cold blow than to be received back by the working fluid on the hot pass. The continuous irreversible conversion of energy implies inevitable loss. The heat transfer data assembled by Kays and London[8] require separate correlations for heating and cooling. The difference may be related to the phenomenon identified here.

Fig. 15.6 summarizes the effects on performance of extreme values of $N_{SG}$ and of $N_H$.

## 15.8 Computed indicated performance – performance maps

For given $N_\tau$ and $\gamma$, the computed indicated performance of a machine of fixed geometry is a function only of $N_{SG}$ and of $N_H$ and may thus be represented in the now familiar map form. (If transient conduction losses in the metal parts are modelled, or if temperature coefficients of $\mu$ or $k$ are taken into account, then a given map applies only for given values of the corresponding additional dimensionless groups.)

Table 15.1 summarizes the data used to produce the performance map of Fig. 15.7. Data on the geometry of the drive mechanism of the MP1002CA engine have been given in full in Appendix VI and are therefore not repeated in the table. The values of $Y/r$, $L_d/r$ and $D/r$ are arbitrary, but representative of hot-air engine practice.

Table 15.1. *Data for performance map of Fig. 15.7*

| crank geometry | As for Philips MP1002CA air engine (Appendix VI) |
|---|---|
| $\lambda_j$: | |
| $Y/r$ | 0.054 |
| $L_d/r$ | 4.0 |
| $D/r$ | 1.8 |
| $N_\tau$ | 2.6 |
| $\gamma$ | 1.4 |
| $N_C$ | 0.5 |
| $N_{SG}$, $N_H$ | variable |

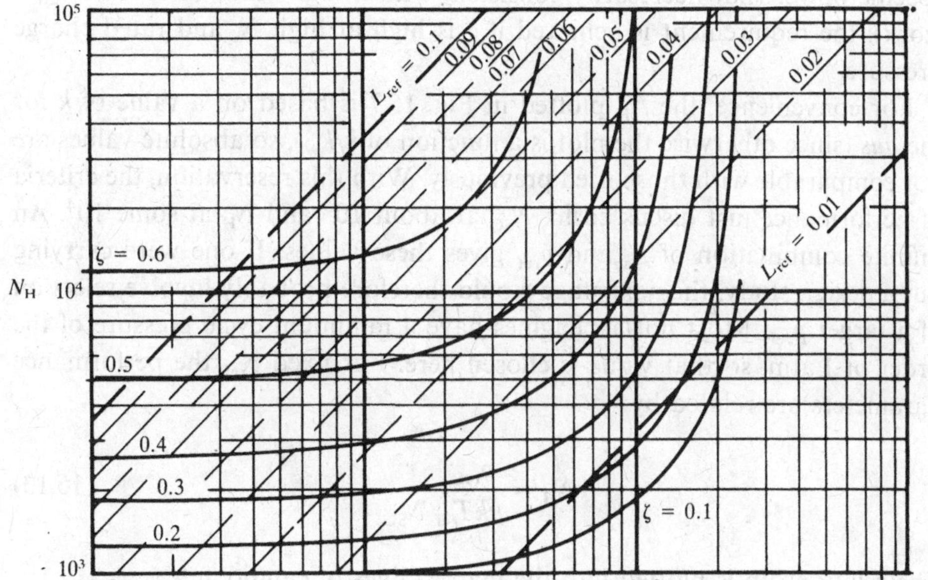


Fig. 15.7 Map of Eqn (15.12) for a hot-air engine defined by the data of Table 15.1. (After Organ.[5])

For the machine specified by the data of Table 15.1 the map applies to all $N_{SG}$, $N_H$ (i.e., to all $p_{ref}$, $N_s$, $\mu$ etc.) for which the assumptions of the analysis are valid. By contrast with the findings of the earlier, one-dimensional analyses, a high computed value of $\zeta$ no longer necessarily signifies that conditions are close to those of the ideal cycle: transient conduction losses may have a small computed value for one of two reasons:

(*a*) temperature gradients are small;
(*b*) $k$ is small, even though temperature gradients are severe.

At extreme values of $N_H$ high rates of pressure change and high enthalpy flux rates swamp conduction effects. Conditions in the annular gap are thus virtually adiabatic. Large temperature gradients are possible in both radial and axial directions, with associated departures from the ideal cycle model. If viscous losses happen to be low, a meaninglessly high computed value of $\zeta$ may result.

For good specific performance $N_H$ may be neither too low (excessive conduction) nor too high (excessive departure from reference cycle conditions). On the other hand, $N_{SG}$ should always be as high as possible (low $\mu$, high $N$ and/or high $p_{ref}$). This is consistent with the requirement for high specific power (not necessarily consistent with a high value of specific cycle work): the requirement is achieved if $\zeta$ is high at high $N_s$ and rated charge pressure.

For convenience, the $N_H$ plotted in Fig. 15.7 is based on a value of $k$ for the *gas* (since otherwise the plot is a function of $k/k_w$), so absolute values are not comparable with those used previously. With this reservation, the criteria of performance just discussed fix $N_{SG}$ at about $10^6$ and $N_H$ at some $10^4$. An infinite combination of $N_s$ and $p_{ref}$ gives these values. If one were carrying out a design study, the next stage would therefore probably involve selection of a target $p_{ref}$. Most hot-air engines have a minimum cycle pressure of the order of 1 atm, so that value is chosen here. For fixed $p_{ref}$ the performance parameters are related by

$$N_H = \frac{p_{ref}^2}{\mu k T_{ref}} \frac{r^2}{N_{SG}} \qquad (15.13)$$

The relationship is plotted into the map as lines of candidate $r$.

## 15.9 Experimental corroboration

Meaningful data on the performance of the hot-air engine have never been assembled. With a view to putting this matter right, apparatus has been designed under the terms of a grant from the UK Science and Engineering Research Council, and is currently under construction. The analysis will be tested, modified if necessary and reported upon in the light of experimental findings.

# 16
# The way forward

## 16.1 Résumé

The cost of modification during the development of a Stirling cycle engine greatly exceeds that of modifying a novel internal combustion engine of comparable specification. The unfavourable comparison promises to hold for the foreseeable future. Therefore, as long as Stirling cycle machines are being developed there will be a demand for improved understanding of the thermodynamic cycle, and for design tools incorporating that understanding.

Any contribution made by this text towards fulfilling that demand will be capable of assessment only at some future date. Assessment will be in terms of eventual acceptance (or otherwise) of the views which have been advanced, and of the success (or otherwise) of the analytical and numerical methods in achieving improved thermodynamic performance or reduced development cost. The meantime will be used to comment on the state of Stirling machine analysis as it now stands, and to speculate on the scope for its future development.

## 16.2 Stirling cycle analysis as an academic discipline

It was Finkelstein[1] who pointed out in the early days of cycle modelling that the gas cycle of the Stirling machine is devoid of the effects of combustion, dissociation, shocks, and the intermittent operation of valves, and is therefore straightforward as problems in unsteady, viscous flow go. Notwithstanding, simulation of the Stirling cycle lags behind that of, say, two-dimensional viscid flow around airfoil sections and of the gas processes of the internal combustion engine. The deficiency is in terms not only of intricacy, but of utility for design also. The internal combustion analysts have had no difficulty in embodying the Method of Characteristics in induction and exhaust blow-down modelling. Has there been something fundamentally wrong with the way the Stirling cycle modelling problem has been approached to date?

There is little question that Stirling cycle thermodynamics currently enjoys (and deserves) a lower level of respectability than do the majority of other areas of engineering study: if each stress analyst had a different approach to the finite-element modelling of two-dimensional loading cases to the extent that Stirling specialists disagree over their models of the gas circuit, and particularly if they were as reluctant to submit their numerical models to comparison with cases having known solution, then someone would quite rightly want to know the reason why. By contrast, the indifference shown by Stirling analysts to the distinction between partial, total and substantial differential coefficients provokes no comment. The principles of dynamic similarity enjoy almost total neglect with consequent high cost in terms of unnecessary computational effort and lost insight – and no protest is voiced. The recent text by Wurm, Kinast, Roose and Staats[2] on the design and applications of Stirling and Vuilleumier heat pumps continues the tradition of uncritical review and subscription to first-, second- and third-order classification of cycle models. The definition proffered for second-order is that it is the level of analysis which falls between first and third! The celebrated Schmidt treatment is misrepresented (p. 77 of Ref. 2) as depending on the assumption of zero dead volume.

## 16.3 Multi-dimensional as opposed to one-dimensional modelling

One conclusion above all others which surely emerges from the present treatment, then, is that the modelling of the Stirling gas processes is as yet in its infancy – even judged by the modest standards of one-dimensional or 'slab' flow. With luck, this fact may yet be taken into account in deliberations, in train at the time of writing, as to the future of the American programme of Stirling engine research and development coordinated[3] by the Lewis Research Center of NASA. The coordinating organization is not happy about the margin of performance which has to be built into engine designs to ensure the meeting of specifications. It sees identification and quantification of individual cycle losses as a key to closing the gap between thermodynamic performance designed for and realized. Two- and three-dimensional modelling of selected features of the gas processes is proposed for gaining additional insight into the loss mechanisms. As paymaster for the programmes of theoretical and experimental work subcontracted to American universities and research laboratories, NASA is in a strong position to influence academic research policy.

The problems in the path of multi-dimensional modelling are not insurmountable, but by comparison with the slab flow case, are substantial. The

Mach *line* of the one-dimensional Method of Characteristics gives way to the (three-dimensional) Mach *cone* in the two-dimensional flow-field. Illustrative solutions have been prepared for simple two-dimensional problems,[4] but successful application to a representative Stirling gas circuit threatens a monumental research effort. If the Characteristics method or an equivalent is *not* used, pressure information will again propagate at arbitrary speed, and to this extent nothing will have been gained over the one-dimensional treatment.

In two dimensions the Lagrange scheme completely loses the edge of simplicity and elegance which it holds over the Euler scheme in one dimension, and in the presence of high shear deformations becomes effectively unworkable. *Any* two-dimensional model (other than perturbation schemes such as that of Chap. 15) can be expected to call for an order of magnitude more cpu time than the slab-flow case. Embodiment of such models into optimization programmes for general purpose use may therefore have to wait for computer technology to evolve through two further generations.

Thus there appear to be three compelling reasons for not committing resources to multi-dimensional modelling at this stage:

(*a*) there is not yet sufficient appreciation of either the potential of the one-dimensional case, or of its limitations, to allow judgement of its adequacy or otherwise for the intended purpose. Multi-dimensional modelling can therefore not be conclusively shown to be necessary;

(*b*) it promises to demand many times the analytical and computing capacity of the one-dimensional case for returns which cannot be guaranteed; and

(*c*) unless it can be coordinated with better effect than one-dimensional work to date, it will be decades before it assumes an intelligible order.

The Stirling gas circuit is a set of conventional heat exchangers in series, and functions as any heat exchanger but for two points of difference: (i) gas flow is variable in direction and intensity and (ii) work is extracted simultaneously with heat exchange. It ought surely to be possible to modify conventional (one-dimensional) methods of heat exchange and flow analysis to cope with the combination of features.

## 16.4 Verification of numerical solutions

This isolates the major outstanding problem as that of *verification*. With the exception of pressure, the gas circuit variables effectively defy direct measurement. Only by coincidence is agreement between measured and computed net cycle quantities (work, efficiency) an indication of appropriate analysis and error-free programming. In the circumstances the logical priority is the highest achievable fidelity between numerical output and the underlying analytical model: after all, one sets up an analytical model first and foremost to see what the *model* predicts. If the numerical method does not solve the equations accurately, then this basic purpose is defeated. Nothing of any generality (and therefore nothing of value to design) is learned simply because a numerical method can be made to produce an 'answer' which coincides with experimentally measured performance.

In this connection there is an urgent requirement for a comprehensive specification of a representative Stirling machine citing free-flow areas, hydraulic radii, wetted areas, flow passage lengths etc. In principle, an extended version of the specification of the Philips MP1002CA (Chap. 4) would serve, but official manufacturer's data on something more modern – say, of the United Stirling V-160 – would have more authority. It should not take long to establish a recognized value for work per cycle under rated conditions for a correctly implemented CL-1 analysis, with separate figures for actual crank kinematics and for the equivalent sinusoidal case. Corresponding particle trajectory maps would complete the ideal cycle description, although it might be helpful to extend this by mapping local, instantaneous, notional $N_{re}$ over the cycle corresponding to isothermal, lossless conditions.

CL-2MF treatments could be compared next, since they are free of propagating temperature discontinuities. No candidate CL-2ME solution is worth applying unless it can be shown to pass the 'adiabatic' test: the solution is started with a gas temperature distribution identical to that of the adjacent solid wall with heat transfer coefficients set equal to zero. The original temperature profile must shuttle back and forth within the exchanger system without distortion but for a cyclic rise and fall in sympathy with adiabatic compression and expansion, and corresponding (reversible) cyclic shortening and extension in the axial direction.

If the simulation is being constructed in modular 'building bloc' form, then the combination of individual blocs from proven CL-2MF and CL-2ME forms should make for a correctly functioning CL-3 combination. Only when it can be shown that the three-conservation laws can be correctly solved for is it worthwhile to introduce the laws for heat diffusion in the solid parts of the regenerator, heat exchangers and containing walls.

Once correct numerical answers are being obtained from the analytical description it becomes worthwhile to look for ways of comparing actual and computed gas processes. Few analysts have access to a comprehensively instrumented modern Stirling engine, but it could be that a power-producing engine *per se* is not the ideal test bed: the author's experience with the MP1002CA is that thermodynamic performance is more consistent and reproducible when the machine is motored using a speed-controlled electric drive. In this mode, the heat exchangers may be operated 'back-to-back', i.e., the heat rejected from the compression exchanger may be recirculated to the expansion exchanger. In combination with suitable mechanical modifications, the expedient allows heat leak between hot and cold ends to be all but eliminated, and permits cyclic pumping losses to be inferred.

Fig. 16.1 is a section through the working spaces of an MP1002CA engine modified for 'back-to-back' operation: section-lining with paired parallel lines shows where metallic components have been substituted by replacements in high-density polythene for low conductivity. Overall height of the machine is increased by insertion of an insulating phenolic resin spacer between expansion-end cylinder shell and compression-end exchanger body. The changed axial length is compensated by making the (plastic) displacer mounting flange of greater thickness than the original metal component.

Individual volumes of cavities in the external coolant (water) circuit are kept to a minimum to reduce thermal capacity, and all coolant jackets are of low conductivity plastic. A phenolic resin flywheel (Fig. 16.2) substitutes the original cast-iron component to minimize dissipation of heat from the crankshaft. The crankcase is totally immersed in a close-fitting water bath of high-density polythene (not illustrated), and a plastic water jacket replaces the original aluminium alloy housing around the lower part of the cylinder liner (Fig. 16.1). Coolant connections to inlet and outlet of all four jackets are of phenolic resin, as are the thermocouple tappings at each of these points. External plumbing makes use of the shortest possible lengths of insulated plastic tubing.

With a view to checking on errors which might be due to the small $\Delta T$ between expansion and compression ends during back-to-back running, a changeover switch (Fig. 16.3) provides more or less instantaneous reversal of flow between compression and expansion exchangers. At any given setting of the operating parameters ($N_{SG}$, $N_{MA}$) the machine may be run with the expansion end temperature marginally higher than that of the compression end or *vice versa.* Switching of the coolant path is achieved by constricting the low-conductivity plastic tubing and components of the switch in contact with the tubing are of phenolic resin – again to minimize conduction.

Early experiments[5] underestimated the experimental finesse called for when looking for a small quantity measured as the difference between two relatively large heat fluxes, and gave misleading results. Subsequent, painstaking experiment by a visitor to the author's laboratory, Dr P. S. Jung from Korea,

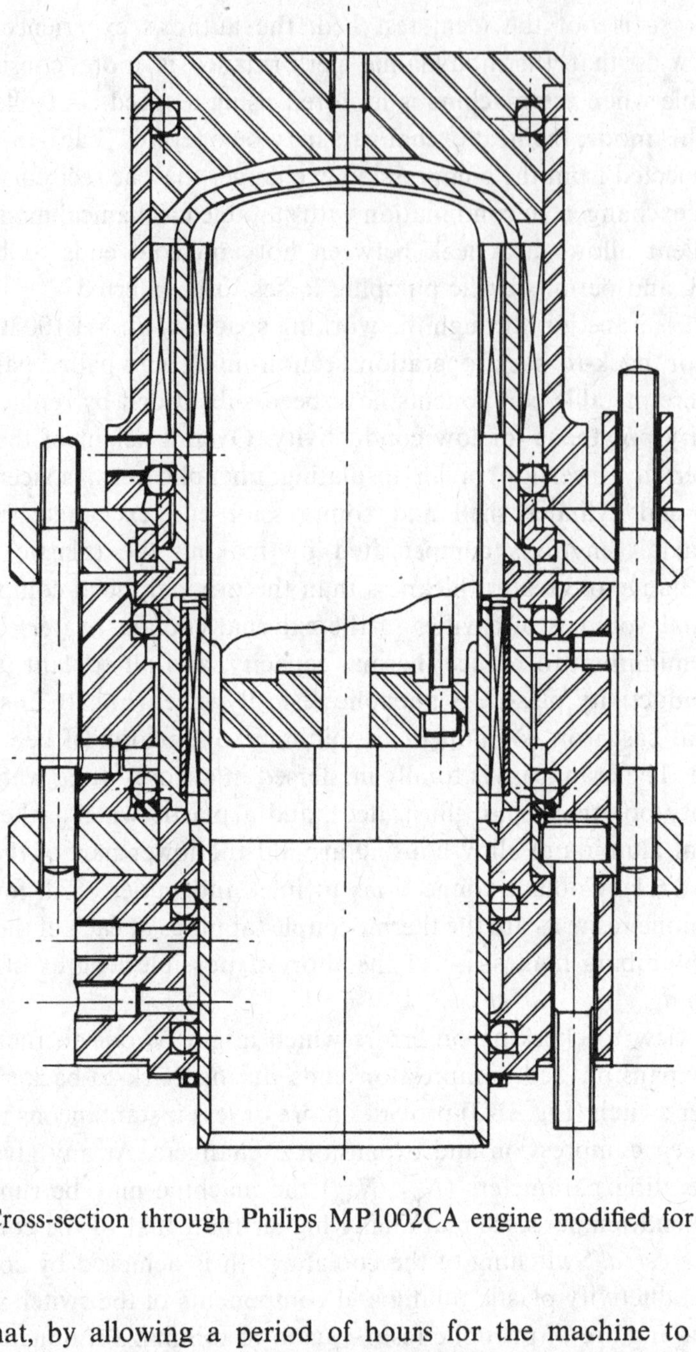

Fig. 16.1 Cross-section through Philips MP1002CA engine modified for 'back-to-back' test.

showed that, by allowing a period of hours for the machine to settle to equilibrium, total mechanical input power could be balanced reproducibly to about $\pm 2\%$. Continuation of the work awaits resources to cover the substantial running time required for tests to cover a representative range of the operating parameters.

Fig. 16.2 Philips MP1992CA air engine modified for back-to-back test, showing low-conductivity jackets for the coolant circuit and phenolic resin flywheel. (Crank case coolant jacket and external plumbing not shown.)

Eventually, a cycle model must reflect all aspects of thermal conduction, radiation and leakage, combined, if desired, under umbrellas such as 'reheat loss' and 'shuttle loss'.[6] Means are required for verifying at each intermediate step in building up the cycle description that correct numerical solutions are being achieved to the symbolic descriptions proposed. The final, numerically correct simulation is unlikely to predict measured performance exactly, and it is suggested that nothing is to be gained by arbitrary adjustment of correlations and empirical factors to make it do so. Its greatest value is displaying the operating envelope defined by the underlying mathematical model and mapping performance *trends* within that envelope: making an analogy with turbo-machinery performance, a numerical model which displays a region of the proposed operating envelope in which surge or stall is

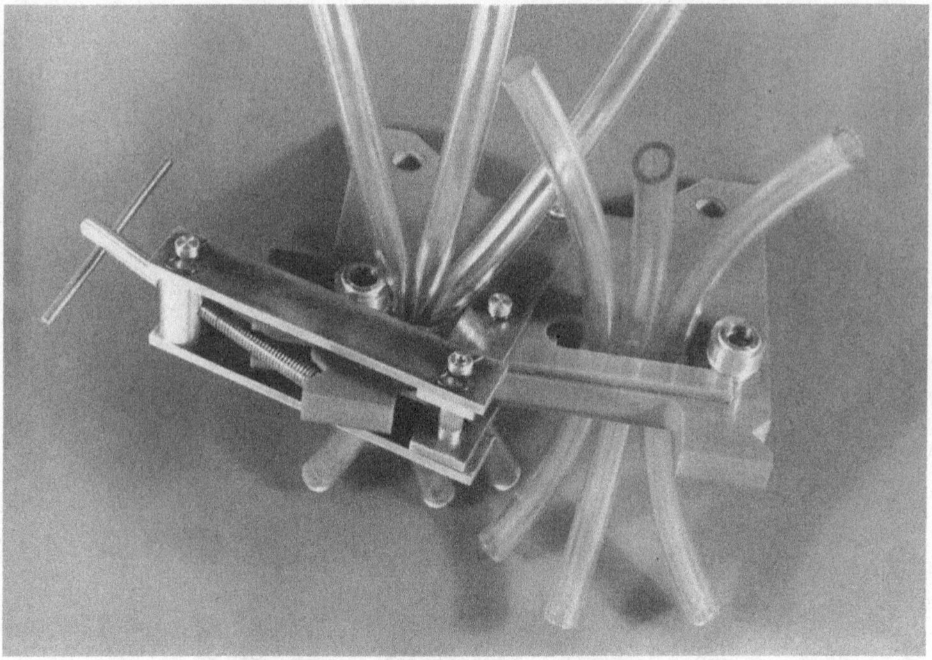

Fig. 16.3 Switch for rapid changeover of direction of fluid flow between expansion and compression exchangers.

possible is of greater value than one returning an accurate prediction of operation at a single point. In the case of the Stirling cycle machine an intriguing challenge is to see whether this aim can be realized in terms of one-dimensional flow.

## 16.5 An overall strategy for thermodynamic design

It is appropriate to end with a summary of guidelines for thermodynamic design. The following suggested outline has the merit of being quite general, i.e., of applying to coolers as well as prime movers. Note that steps (1)–(9) may be completed *without use of the computer* (but with the aid, if desired, of design charts – see Section 16.6). A design manual dedicated to the method is available,[7] and the reader may find an interesting comparison between this and, e.g., the Martini design recipe.[6]

(1) Locate the existing design which, regardless of absolute size, comes closest to the desired geometric configuration and for which data are available on performance, geometry and materials. (In the unlikely event of there being no suitable precedent, proceed to step (10).)

(2) Determine individual values of free-flow area, wetted area, dead volume and length for all flow passages. Establish the dimensions determining the kinematics of the drive mechanism. Make an inventory of the materials of construction of the gas circuit, and look up and tabulate relevant respective properties ($k$, $\rho$ etc.).

### A Thermodynamic similarity phase

(3) For the existing design, work out the *dimensionless* gas circuit and operational parameters listed in Table 4.1. (Depending on the amount of performance data available, there may be a *range* of $N_{SG}$, $N_{MA}$ etc. rather than single values.) Additionally determine equivalent phase angle and swept volume ratio.

(4) Provided the proposed mechanical construction does not introduce disproportionate thermal resistance (where heat flow is to be maximized) and/or thermal shorting (where heat flow is to be minimal) *any* derivative design having the same values of the dimensionless characteristics will have indicated thermal efficiency and specific cycle work similar to that of the case study prototype. Choose tentative rpm, $N$, (or reference pressure) and working fluid, and work back via $N_{SG}$ and $N_{MA}$ to a value of characteristic hydraulic radius, $R_H$.

(5) Calculate absolute values of the individual hydraulic radii of the new design.

(6) From $\zeta$ at chosen $N$ and $p_{ref}$ obtain required $V_E$. From $\mu_v$ obtain $V_{ref}$. From $V_{ref}$ and $R_H$ acquire $A_{ref}$, and hence absolute values of cross-sectional areas. Absolute dead volumes follow, and then passage lengths from respective dead volumes and cross-sectional areas.

(7) For a wire gauze regenerator choose a weave giving the required combination of porosity and $r_h$ (Eqns (5.4), (5.6) and Fig. 5.6).

(8) Make a first attempt at selecting constructional materials to satisfy the target numerical values of $N_H$ and $N_F$ for the regenerator. If the outstanding dimensionless parameters for the new design are markedly different from the starting values, the design *may* still function satisfactorily, but with the loss of thermodynamic similarity there are no grounds for confidence. It may be prudent to return to step (4), trying alternative $N$, $p_{ref}$.

(9) Commence the mechanical, metallurgical and tribological design, repeating steps (4)–(8) as necessary to reconcile the needs of mechanical and thermodynamic aspects.

If no existing design was available on which to base the scaling process, then simulation by computer is the only option. In any case, the computer model will provide insight which the simple scaling process cannot give, so step (10) is virtually essential:

(10) Model the design, as defined by the parameters specifying its thermodynamic similarity to the prototype.

The reader may wish to contact the author with regard to availability of simulations described in this text, or alternatively may prefer to build up a suite of programmes to personal requirement. If the latter, a simple 'isothermal' or 'adiabatic' treatment can be useful if it provides for particle tracking. Assuming the scaling processes of steps (4)–(8) have been carried out successfully, the particle trajectory map (see e.g., Fig. 13.11) for the proposed design has little option but to correspond to that for the prototype machine. A facility for producing further trajectory maps reflecting slight modifications to the gas circuit is desirable, particularly if there is any possibility of the modification's introducing 'inactive' elements of working fluid which oscillate within an exchanger without emerging at either end.

The isothermal or adiabatic treatment is an essential basis for simulations of the perturbation type (see Chap. 7), so an Availability analysis of losses might be the next logical step. If, on the other hand, the preference is for working in terms of absolute variables, the choice is between proceeding via a lumped-parameter model or the (Lagrange) finite-cell alternative. (Unless the reader is confident that he or she has identified a numerical discretization scheme which deals satisfactorily with propagation phenomena, a finite-cell scheme in Eulerian coordinates will be no more reliable than the far simpler lumped-parameter formulation. The latter is therefore the preferred Eulerian scheme.)

If the building-bloc approach in Lagrange coordinates is too daunting a programming proposition, it is possible to fall back on the 'thumb-nail sketch' of heat transfer and friction processes afforded by the treatment of Section 13.3: this is readily extended from the case study presented (regenerator only) to the complete gas circuit. The reader's isothermal or adiabatic model will provide the required estimates of pressure excursion and particle trajectory amplitude, and of the phase angle between the two. Eqn (13.3) (which the 'thumb-nail' analysis essentially solves) is readily converted into an expression for instantaneous entropy creation rate following Section 7.3. There is accordingly a means of probing variations in *net* available work lost when alterations to heat exchanger geometry cause interdependent changes in the effects of both heat transfer and friction.

## B Dynamic similarity phase

When the designer is satisfied that the indicated performance of the new design is thermodynamically similar to that of the case study, it is time for refinement of the performance prediction. For this there is no alternative to simulation by computer, and to the dynamic similarity approach. By accepting $R_H$ at reference length, the thermodynamic similarity formulations may be readily converted to the dynamic similarity form in three steps:

(i) Exact drive kinematics replace the simple harmonic assumption. Phase angle and volume ratio give way to (dimensionless) kinematics of the actual drive mechanism. (Use crank radius, $r$, to normalize, and use the ratio $r/R_H$ where necessary to interface thermodynamic calculations with crank kinematics.)
(ii) Embody heat transfer and flow friction correlations appropriate to chosen heat exchanger geometry.
(iii) Provide for the effects of conduction within the heat exchanger walls.

Step (11) of the outline design process is evidently:

(11) Exercise the simulations using slight variations of the (dimensionless) parameters of the evolving design. Attempt to confirm any apparent improvement in performance by an approach embodying independent assumptions.

If the code has been formulated in functional form (see e.g., Eqn (7.1)) it is a straightforward matter to apply the techniques of optimization (Chap. 14). It may be wise to confine attention to two parameters at any one time. If, as an outcome of the optimization process, it is decided to alter significantly the value of a parameter such as $N_{MA}$, it will be necessary to recalculate the dependent *absolute* design variables and modify the physical layout accordingly.

Analysts who are confident of having exhausted the possibilities of the lumped-parameter and finite-cell schemes (thermodynamic and dynamic similarity versions) may wish to apply the Characteristics and linear waves treatments.

## 16.6 Case study using chart form of conditions for thermodynamic similarity

The thermodynamic similarity phase of Section 16.5 may be facilitated by the use of nomograms to represent the individual similarity conditions. The advantage of the graphical form is that it can embody specific values of physical constants (e.g., coefficient of dynamic viscosity of air, He, $H_2$ etc.), thereby obviating the need to key these into the calculator.

To provide a specific numerical example it is supposed that the Philips MP1002CA air engine is to be scaled from its rated operating point on air (Table 4.1) so as to run on $H_2$ (the new gas must be diatomic for similarity unless an operating point is available for a monatomic gas) at 4000 rpm and to produce a five-fold power increase at the new rpm. Temperature ratio, $N_\tau$, and equivalent phase angle and volume ratio must retain the original values.

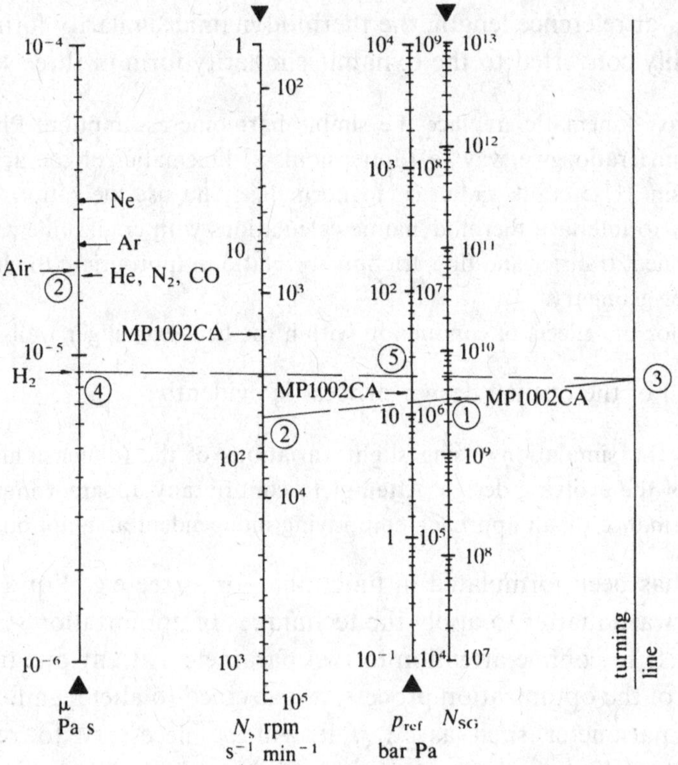


Fig. 16.4 Nomogram for evaluation of $N_{SG} = p_{ref}/N_s\mu$.

The new rated pressure, $p_{ref}$ is found by substituting new rpm (converted to $N_s$) and $\mu$ (at $T_{ref}$) into the expression for $N_{SG}$ and equating to the rated value for the MP1002CA, viz, to 3.3E+09, giving $p_{ref}$ = 19.8 bar or, say, 20. Fig. 16.4 is a nomogram with the rated MP1002CA operating point and reference viscosity values labelled. This and all subsequent nomograms are entered with two, generally separate straight-line elements. It will be noted that two of the graduated scales of each nomogram are marked by triangles at a lower extremity, the other two being so marked at the upper extremity. An individual straight-line segment connects values on one pair of scales only, the intersection of that line with the other graduated scales having no significance. The link between the two pairs of scales is the (ungraduated) turning axis.

In the present case, connect the rated value of $N_{SG}$ (marked 1) to required $N_s$ (marked 2) and extrapolate to the turning line at 3. A line joining dynamic viscosity of $H_2$ (point 4) to the turning point gives the required new $p_{ref}$ at point 5.

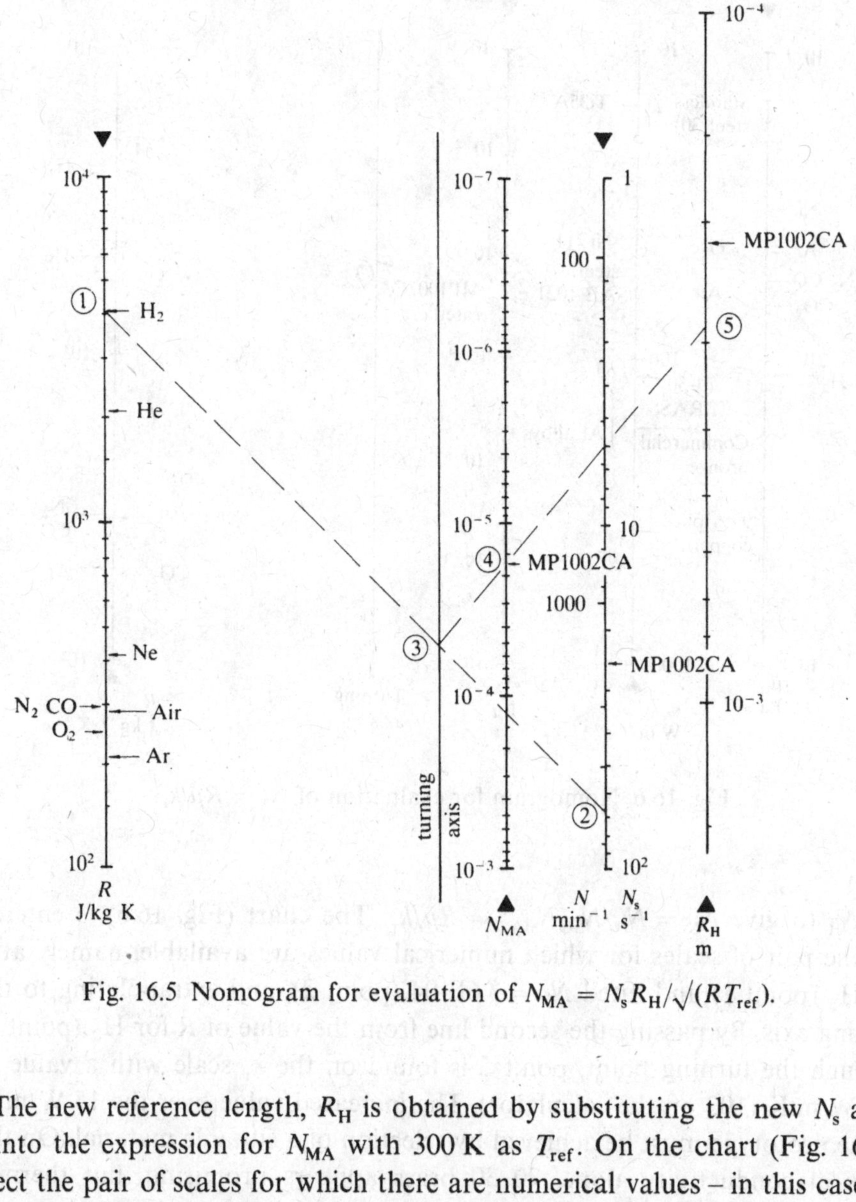


Fig. 16.5 Nomogram for evaluation of $N_{MA} = N_s R_H / \sqrt{(RT_{ref})}$.

The new reference length, $R_H$ is obtained by substituting the new $N_s$ and $R$ into the expression for $N_{MA}$ with 300 K as $T_{ref}$. On the chart (Fig. 16.5) select the pair of scales for which there are numerical values – in this case $R$ (for $H_2$) and $N_s$, join the values and note the point (3) at which the turning line is cut. Finally extend a second line from the turning point through rated $N_{MA}$ (point 4) to the $R_H$ scale and read off the new value of about 3.0E-04 (2.92E-04 using the calculator) at point 5. This new value is seen to be some 50% greater than that of the prototype.

There is now sufficient information to permit selection of a suitable value of nominal regenerator wire conductivity, $k_r$. Presentation of the process in chart form is facilitated by substituting expressions for $N_{SG}$, $N_{MA}$ into that

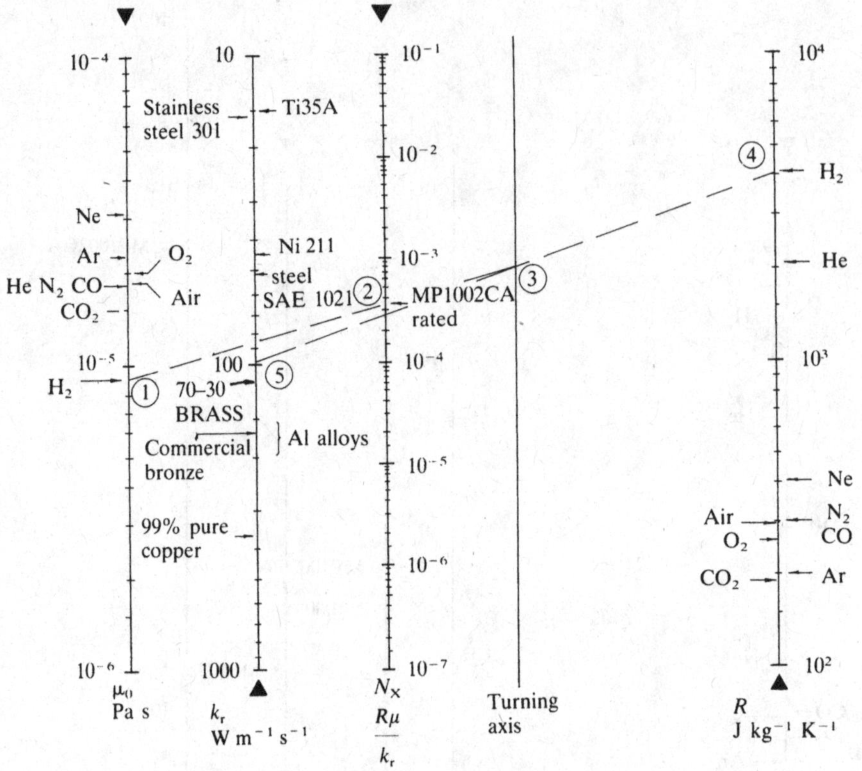


Fig. 16.6 Nomogram for evaluation of $N_X = R\mu/k_r$.

for $N_H$ to give $N_X = N_H/N_{SG}N_{MA}{}^2 = R\mu/k_r$. The chart (Fig. 16.6) is entered via the pair of scales for which numerical values are available, namely at $\mu$ for $H_2$ (point 1) and rated $N_X = 3.6E\text{-}04$ (point 2), and extrapolating to the turning axis. By passing the second line from the value of $R$ for $H_2$ (point 4) through the turning point, point 5 is found on the $k_r$ scale with a value of 100 W/m K (103 on the calculator). The increased value over the 15 W/m K of the prototype must be achieved by selection of a suitable material. On the basis of conductivity alone, 70–30 brass appears promising, but thermal diffusivity has yet to be checked.

This may be achieved using the nomogram of Fig. 16.7 which represents the range of values of $N_F$. (Note that $N_F$ and $R_H$ scales coincide. Read $R_H$ from the right-hand graduations, $N_F$ from the left.) Again, the pair of scales is selected for which values are already to hand, namely $N_F$ (the rated value of 4.26) and $N_s$. Extrapolating from point 1 via point 2 to the turning axis gives point 3, from which a second line through the new value of $R_H$ at point 4 gives $\alpha_r = 25E\text{-}06$ at point 5. Aluminium bronze (Al 7, Fe 2.5, Cu bal) has

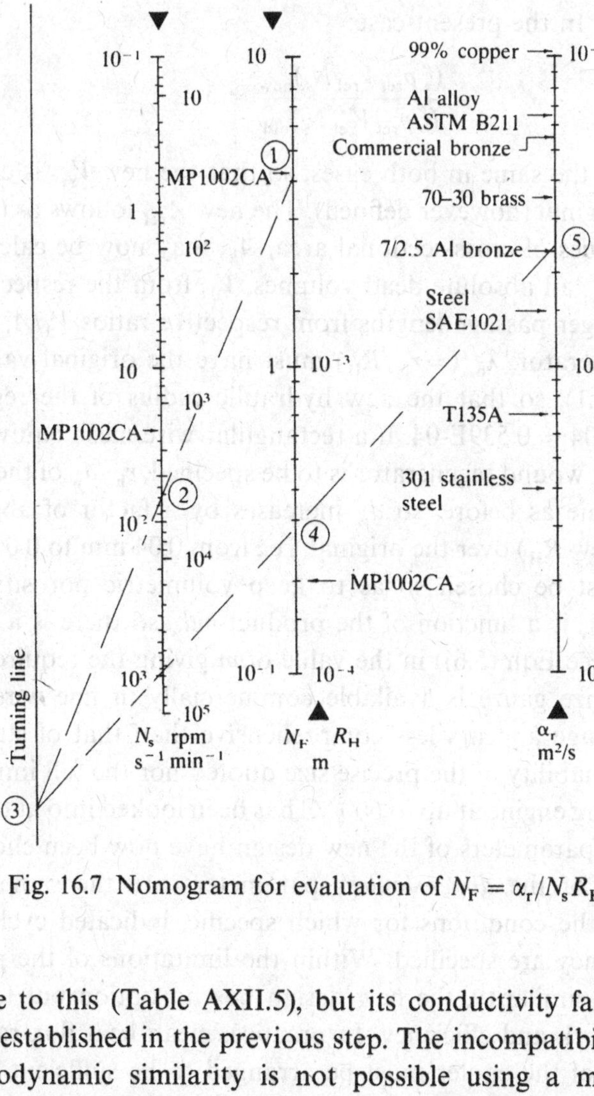


Fig. 16.7 Nomogram for evaluation of $N_F = \alpha_r / N_s R_H{}^2$

a value close to this (Table AXII.5), but its conductivity falls short of the requirement established in the previous step. The incompatibility means that exact thermodynamic similarity is not possible using a material for the regenerator from the selection embodied in the chart. At this point, therefore, we note that diffusivity, $\alpha$, is defined as $k/\rho C$. Provided $k$ is equal to or in excess of the value calculated, thermal performance of the regenerator may be expected to be satisfactory if a material is used whose $\rho C$ (thermal capacity/unit mass) gives a value of $\alpha$ at *or below* the computed requirement, since this means adquate or excess thermal capacity. On this basis the candidate material is phosphor bronze, mechanical suitability and commercial availability remaining to be checked when wire diameter and mesh have been calculated.

The power requirement is $\zeta p_{ref} V_{ref} N_s$ – regardless of respective definitions of $\zeta$, $p_{ref}$, and $V_{ref}$, provided these are consistent between case study prototype and new design. In the present case:

$$\frac{(\zeta p_{\text{ref}} V_{\text{ref}} N_s)_{\text{new}}}{(\zeta p_{\text{ref}} V_{\text{ref}} N_s)_{\text{old}}} = 5$$

$\zeta$ is numerically the same in both cases, so that the new $V_{\text{ref}}$ is calculated as 1.5 times the original (however defined). The new $A_{\text{ref}}$ follows as $(V_{\text{ref}}/R_{\text{H}})_{\text{new}}$. All absolute values of cross-sectional area, $A_x$, may now be calculated from the respective $\alpha_x$, all absolute dead volumes, $V_d$, from the respective $\mu_d$, and absolute exchanger passage lengths from respective ratios $V_d/A_x$.

For the regenerator, $\lambda_{h_r}$ $(=r_{h_r}/R_{\text{H}})$ must have the original value, namely, 0.1847 (Table 4.1), so that the new hydraulic radius of the regenerator is 0.1847 × 2.92E-04 = 0.539E-04. If a rectangular wire mesh equivalent to the original filament wound regenerator is to be specified, $r_{h_r}/d_w$ of the new design must be the same as before, so $d_w$ increases by a factor of about 1.5 (see calculation of new $R_{\text{H}}$) over the original, i.e., from 0.04 mm to 0.06 mm. Mesh number, $m$, must be chosen so as to keep volumetric porosity, $¶_v$, at its original value. $¶_v$ is a function of the product $md_w$ so there is a decrease by a factor of 1.5 (see Eqn (5.6)) in the value of $m$ giving the required (original) $¶_v$. Woven bronze gauze is available commercially in fine wire diameters, although the range appears less comprehensive than that of stainless steel. Neither the availability of the precise size quoted nor the full implications of its use in a Stirling engine at up to 600 °C has been looked into at this stage.

The absolute parameters of the new design have now been chosen in such a way that all of the *dimensionless* parameters are the same as for the MP1002CA at the conditions for which specific, indicated cycle work and indicated efficiency are specified. Within the limitations of the principles of thermodynamic similarity, the new design has no option but to afford the same specific work and efficiency. In practice, mean heat flux rates into the heater and out of the cooler must be arranged to be sufficient to maintain inner surface temperatures at the values achieved in the MP1002CA. With this reservation, preliminary thermodynamic design is complete.

Scaling of an air-charged machine to give a thermodynamically similar performance when using $H_2$ is perhaps the reverse of the most probable design requirement. The study was pursued partly on account of the interesting outcome, whereby realization of required thermodynamic behaviour is made difficult by the limited range of *thermal* properties of available regenerator materials. The reader may care to reverse the scaling process, starting with an $H_2$ charged machine and determining the specification of a thermodynamically similar machine using air (or $N_2$) and operating at different $p_{\text{ref}}$ and $N_s$. The exercise may give rise to numerical values which fall outside the scope of the design charts. If so, and if the designer prefers the chart method to the use of a calculator, it is a simple matter to extrapolate the scales simply by repeating the graduations over one or more decades.

## 16.7 Closure

This may be the point at which to recall the stance taken in the Preface in favour of enhanced research into the practical Stirling cycle by analysis and by computer simulation. In the words of Hamming[8] as quoted by Roache[9] '*The purpose of computing is insight – not numbers*'. It is inconceivable that the development of the Stirling cycle machine for eventual commercial use can fail to be speeded by improved insight into its intimate inner workings. Achievement of such insight is a stimulating and rewarding aim in its own right; the eventual replacement of certain types of conventional energy converters by Stirling machines working to the limits of their efficiency, quietness and freedom from pollution is nothing short of a pressing imperative.

# Appendix I
# Stirling's patent No. 4081 of 1816

**(Reproduced with the permission of the Comptroller HMSO)**

## AI.1 Background

Although the title of Stirling's original patent had been duly entered in the 'Docquet Book of the Great Seal' in which patents granted up to 1852 were registered, the actual specification was, for some reason, not enrolled. Finkelstein[1] explains that enrolment involved copying the specification and drawings onto parchment and attaching to the preceding patent by stitching to make one, continuous roll. It is possible that the £5 enrolment fee – a substantial outlay in the early nineteenth century – may have deterred Stirling from completing the registration of his patent. At any event, the patent appears to have lapsed, and the detailed documents to have disappeared until 100 years later, when their reappearance and acquisition intact by the Library of the Patent Office was commemorated by the printing of a complete transcript in *The Engineer* for Dec. 14th. 1917.

## AI.2 Patent No. 4081 'Improvements for Diminishing the Consumption of Fuel, and in particular an Engine capable of being Applied to the Moving of Machinery on a Principle Entirely New'

All my improvements for diminishing the consumption of fuel consist of different forms or modifications of a New Method, Contrivance, or Mechanical arrangement for heating and cooling Liquids, Airs and Gases, and other Bodies by the use of which Contrivance Heat is abstracted from one portion of such fluids, airs, and other bodies and communicated to another portion with very little loss; so that in all cases where a constant succession of heated liquids or other bodies is required, the quantity of fuel necessary to maintain or supply it is by this contrivance greatly diminished. The First Modification of said Contrivance or arrangement is described as follows: A B fig. 1st is a pipe, channel, or passage, formed of Metal, Stone, Bricks or other Materials according to circumstances, i.e. according to the Chemical agencies of the bodies to be heated or cooled and the degree of heat in said bodies. The Hot liquid, gas or body to be cooled is by any means made to enter the passage at A and to pass along to its other extremity B. In its progress it gives out its heat to the sides of the passage or to any bodies contained in it and issues at B at nearly the original temperature of the passage.

In this manner the extremity at A and a considerable portion of the passage is heated to nearly the temperature of the Hot liquid, while the extremity B still retains its original temperature nearly. When the temperature of the passage at B has been raised a few degrees by the motion of the fluid from A to B it is stopped and a portion of the fluid which is required to be Heated and which is supposed to be a few degrees colder than the extremity of the passage at B is made to traverse the same passage in a contrary Direction i.e. from B to A; by which means it receives heat from the sides of the passage or other bodies contained in it and issues at A at nearly the same temperature with the fluid to be cooled. When the temperature of the passage at A has thus been lowered a few degrees, the process is again stopped and a portion of the fluid to be cooled is made to pass from A to B and so on alternatively.

The Second Modification of my said Contrivance or arrangement consists in interposing a thin plate of Metal or other materials, according to circumstances, between two currents of liquid gas or vapour which are made to run in opposite directions.

A A B fig. 2nd. represents such a plate and *c*, *d*, two passages between which it is interposed. The fluid to be cooled is made to traverse the passage *d* from A to B and the fluid to be heated is made to traverse the passage *d* from B to A. The extremity of the plate at A is kept Hot by the current in *d* and the extremity B is kept cold by the current in *c*. The plate A B abstracts the heat from the fluid in the passage *d* and communicates it to that in *c* with very little loss. The waste or escape of heat from the passages is prevented by their being surrounded with Charcoal powder, wood, bricks or any substance that does not easily permit heat to pass through it, as is represented at *c*, *d*, *e*, *f* fig. 1st and 2nd. In the construction of the passages in both Modifications of my contrivance for heating and cooling liquids and other bodies I observe the following rules:

1. When the passages are made of metal or any other substance that conducts or transmits heat easily I make the metal or other substance as thin as possible to prevent the heat from being transmitted in this manner from the hot to the cold extremity of the passages.
2. Liquids and airs being very imperfect conductors of heat I make the passages very narrow (at least in one direction) in proportion to their lengths, for the purpose of heating and cooling more completely the liquids and airs that pass through them. A transverse section of the passages is given at A and B C fig 3rd.
3. When the passages cannot be made sufficiently narrow I make their sides jagged or rough by bodies projecting from them as represented at fig. 4 or I adopt any similar method for promoting the internal motions of the fluids and the ready communication of heat to them or to the passage.
4. When the width of the passage cannot be sufficiently diminished I encrease its length in order to attain the same end.

The form and construction of the tubes, passages and plates in both the modifications of my general Contrivance or Arrangement may be varied according to circumstances; but the benefit to be derived from this contrivance arises from the fluids and other bodies to be heated and those to be cooled being made to move in opposite directions and it is for the invention or improvement of this arrangement that I have applied for and obtained His Majesty's Letters Patent.

Having thus described and ascertained the nature of my Invention I shall now describe several of its numerous and useful applications. The First form or Modification of my general contrivance or arrangement may be applied to diminishing the consumption of fuel in Glass house and other furnaces wherever a high degree of heat is required, and the nature of the processes carried on in these furnaces admits of their being accurately closed, and this application is also capable of producing a much more intense heat than can be raised by the ordinary methods.

A fig. 5 is a furnace filled with fuel A, E G and A D E (*sic*) are two flues or passages (a transverse section of which is represented at A fig. 6). B is a passage

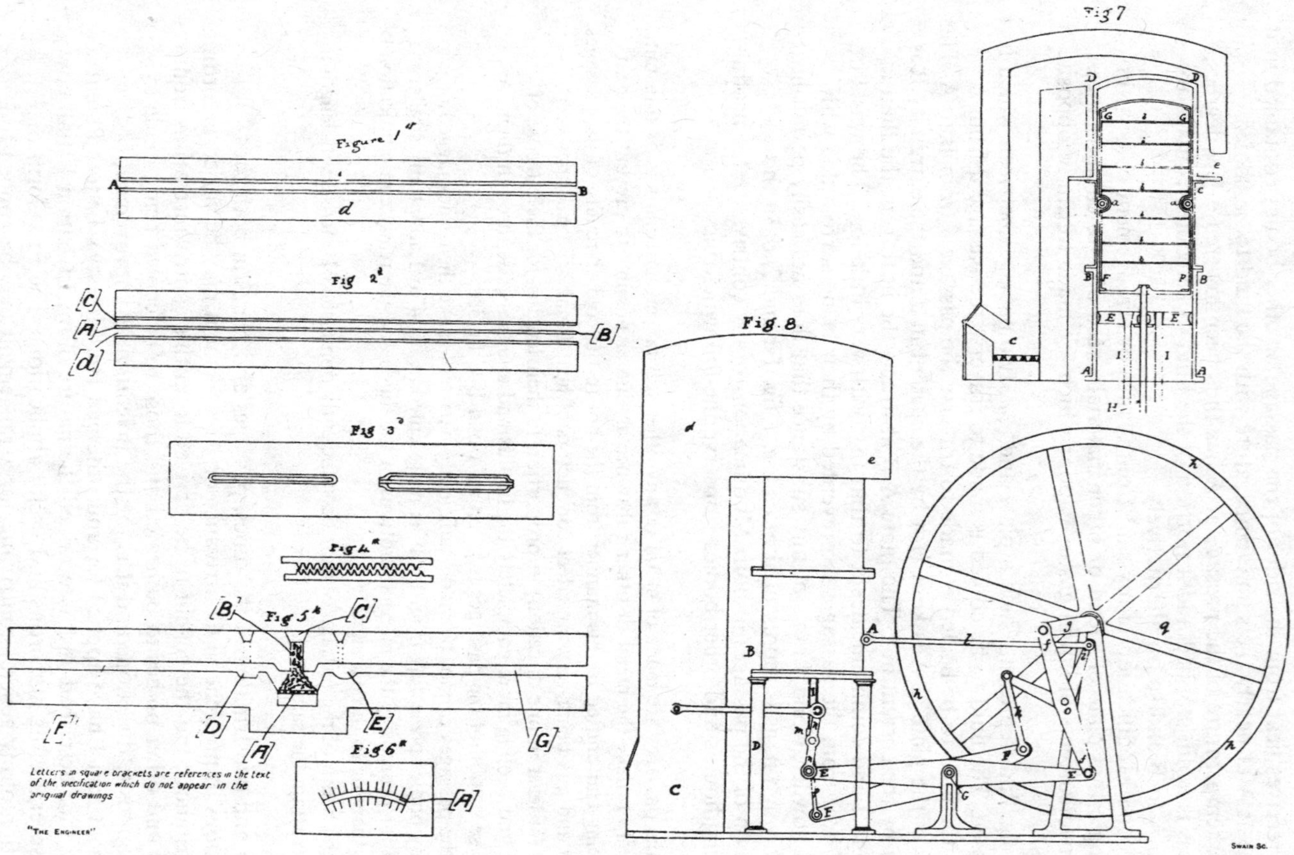


Fig. AI.1 The drawings which accompanied the patent description. (Characters in square brackets are references in the text of the specification which do not appear in the original drawings.) (Courtesy of the Comptroller, HMSO.)

for introducing the fuel accurately closed at C. – All other parts are likewise closed so that no air can pass through or have access to the fire but through said flues. – The air which supports the combustion is alternatively introduced at F and G by blowing engines, bellows or any other means and the gases and vapours which arise from the fire pass off in an opposite direction leaving the greater part of their heat in the flues in the manner above described. – This heat is taken up by the air which supports the fire and is again left in the opposite flue along with the additional quantity produced by combustion. In this manner the heat is accumulated in the furnace and the parts of the flues adjacent to it. The bodies to be heated are placed at D and E and the openings by which they are introduced are accurately closed. The furnace and flues are constructed of the best firebricks or any material that may better resist the intense heat, and the Mason work, as in other furnaces, is made sufficiently thick to prevent the waste of heat. – The form and management of the furnace and flues may be varied according to circumstances; but it must be remembered that the narrower and longer the flues are and the more frequently the direction of the air is changed the greater is the saving of fuel and intensity of the heat produced.

The Second form or modification of my invention may be applied to the saving of fuel in Breweries, Distilleries, Dye Works and other Manufactures, by transferring heat from one portion of liquid, air, or vapour to another. – No further directions seem to be necessary for these applications but this that the quantity of fluids which is allowed to run through the respective passages must be inversely proportional to the specific heat of the respective fluids which may be learned from books or from observing the degree to which they are respectively heated or cooled in their passage. By either of the above described Modifications of my said Arrangement I construct my Engine, for moving Machinery and the following is a general description of the manner in which this is performed. – I employ the Expansion and Contraction (or either) of atmospheric air or any of the permanent gases by heat and cold to communicate motion to a piston or other similar contrivance. – In order to produce this expansion and contraction I cause the air to pass from a cold to a hot part of the engine and the contrary alternatively either in the same passage in the manner described at the explanation of fig. 1st, or in different passages as described at fig. 2nd. – I apply fire to the warmest part of the engine, in order to supply the waste of heat occasioned by its transmission from the hot to the cold parts by the radiating and conducting power of the materials of which they are formed, by the change of capacity for heat which the air suffers from condensation and rarefaction, and by the impossibility of transferring the whole of the heat from the air to the passages and the contrary I apply a stream of cold air or water to the coldest part of the engine to carry off said waste heat. – The passages are of course Hot at the one extremity and Cold at the other, and in passing through them the air is alternatively heated and cooled or expanded and contracted. – The following is a particular description of that form of my engine which I consider the best. – A A D D fig. 7 is a cylinder composed of three parts accurately joined together by rivets or screws and rendered airtight by hammering or soldering the joints. – The part A B is formed of cast iron and accurately bored, the part B C is made of sheet or cast iron as thin as possible (as one tenth of an inch), and the part C D is of Sheet or Cast iron. – To this cylinder is fitted a piston E E which is made airtight in the usual way, and provided with rods I I for communicating its motion. – F F G G is a hollow cylinder also as thin as possible made of sheet iron covered with thin plates of polished brass or silver to prevent the waste from radiation and divided into compartments by plates *b*, *b*, for the same purpose. It is shut on all sides and airtight, kept at a small distance from the outer cylinder by wheels *a*, *a*, or any similar contrivance and furnished with a rod H working through a stuffing in the centre of the piston, by means of which it is moved up and down. This inner cylinder I call the Plunger. The part B D of the outer cylinder is kept hot by the flame or heated airs of a fire C applied at D D made to descend on all sides of the cylinder at C and allowed to escape at *e*. – The part A B is kept cool by a stream of air or of water directed upon it, and the part B C encreases in temperature from B to C. – The temperature of the plunger encreases from F to G and the interval between the cylinder and plunger is partially filled with wires wound round the latter and kept at a small distance from it and from one another by wires laid along it at right angles to the former, in order to heat and cool the air more completely. – This interval is on each side about one fiftieth of the whole diameter of the cylinder. – Figures 7 and 8 are drawn to a scale of one half inches to a foot except the thickness of the metal which is to no scale.

The space contained by the cylinder and piston is filled with atmospheric air. – Figure 8 is an elevation of the engine. – A B C the cylinder D D pillars supporting it, E E a beam centred at G to which the rods I I of the piston are connected by a parallel joint N N; *f f* an arm which connects the said beam to the crank *g*. *i o i* is a bent lever joined to said arm at *o* and connected to the fixed joint A by the rod *l*, and to the other extremity of the beam F F by the rod *k*. – This beam is also centred at G and moves between two plates or similar beams of which the beam E E is composed. *p a* rod by which F F is connected to the rod of the plunger, M a slider upon the rods of the piston fixed to the rod of the plunger to render its motion steady and parallel. – *h h h* the fly upon the same axle with the crank. *c d e* the furnace and flues.

The Operation of the engine is explained as follows. – The part of the cylinder surrounded by the flues is heated to a temperature of 480° higher than part A B. – In the position represented at fig. 8 the plunger is in contact with the piston, by which means the included air is brought to the warm part of the cylinder has its elasticity increased and presses upon the piston with a force greater than that of the atmosphere. – The piston is thus forced downwards and the rod *f f* and crank *g* upwards till the pressure of the included air and that of the atmosphere becomes equal. – The impulse communicated to the fly carries the end of the crank towards *q*, and the arm *f f* and bent lever *i i* are brought to such a position as to depress the rod *k* and thus to raise the plunger from the piston. – The included air is thus made to descend between the plunger and cylinder and brought to the cold part; it is cooled in its descent, has its elasticity diminished, and its pressure becomes less than that of the atmosphere, the piston is forced upwards, and the crank downwards. – The revolution of the fly and crank again bring the plunger towards the piston, the air ascends through the same passage by which it descended, is heated in its ascent and forces the piston downwards and the crank upwards, and so on alternately. – In this manner a rotatory motion is produced which may be applied to the moving of machinery. – The force of the engine is regulated by allowing a portion of air to escape outwards and inwards by a small cock which is opened and shut by a governor as in Steam engines. and placed in the cold part of the cylinder immediately above the highest ascent of the piston. – The distance which the rod of the plunger H fig 7 moves through the piston I call the Stroke of the plunger and I make it equal to that of the piston when the difference of temperature in the hot and cold parts of the cylinder is 480°. – When the difference is less than 480° I make the stroke of the plunger proportionally greater than that of the piston, and the contrary when the difference is greater. – The length of the arms of the bent lever *i o i* and of the rod *k* which is necessary to make the plunger just touch the piston on the one hand and the upper end of the cylinder on the other I determine by experiment as being the most convenient method known to me. – I do not answer for the absolute correctness of those in the plan. – The cylinder is inverted to prevent the oil used to make the piston airtight from getting to the hot part and wasting the heat. In the foregoing description, wherever I have specified more than one Material or more than one Method of performing the same thing, I have placed that first to which I give the preference; But I reserve to myself the power of using any materials and applying my new contrivance for heating and cooling bodies, to the purpose to which it is applicable in any form or manner which further experience may prove to be advantageous and which is not inconsistent with the terms of his Majesty's Letters Patent.

In witness whereof I the said Robert Stirling have hereunto set my hand and seal this Twentieth day of January in the year of our Lord one thousand eight hundred and seventeen.

Signed sealed and delivered (Seal of
in the presence of R. S.) Robt. Stirling

(Rob. Cameron) Accountant of Edinburgh
(Francis F Cameron) Preacher of the Gospel in Edinburgh

# Appendix II

# Ideal Stirling cycles

## AII.1 Discontinuous piston motion – mass balance as a basis for a statement of instantaneous pressure

The algebra which follows refers to the schematics Fig. 3.3 and Fig. AII.1 of an opposed-piston machine. The independent variable which marks instantaneous values of volumes $V_e(\phi)$ and $V_c(\phi)$ is an angle, $\phi$ $(0 < \phi < 2\pi)$. To make possible the condition for constant-volume transfer, $V_C = V_E$.

The basis of the analysis is a mass balance expressed in terms of the ideal gas equation. Taking account of the expression for regenerator mass content (Appendix III) total mass enclosed between the two pistons may be written in terms of instantaneous pressure and local values of variable volume, $V_i(\phi)$, dead volume, $V_{di}$ and corresponding temperatures.

$$M = \frac{p}{R}\left[\frac{V_e(\phi)}{T_e} + \frac{V_{de}}{T_e} + \frac{V_{dr}\ln(1/N_\tau)}{T_c(1 - N_\tau)} + \frac{V_{dc}}{T_c} + \frac{V_c(\phi)}{T_c}\right]$$

$$= \frac{pV_E}{RT_c}[\mu_e(\phi)/N_\tau + \nu(N_\tau) + \mu_c(\phi)], \qquad (\mu_j = V_j/V_E)$$

$$\nu(N_\tau) = \mu_{de}/N_\tau + \mu_{dc} + \frac{\mu_{dr}\ln(N_\tau)}{N_\tau - 1}$$

If $\phi_{ref}$ is the value of $\phi$ at which pressure, $p$, has some reference value, $p_{ref}$ (e.g., min, max, mean) then, abbreviating $\nu(N_\tau)$ to $\nu$:

$$\frac{p}{p_{ref}} = \frac{\mu_e(\phi_{ref}) + N_\tau\nu + N_\tau\mu_c(\phi_{ref})}{\mu_e(\phi) + N_\tau\nu + N_\tau\mu_c(\phi)} \qquad \text{(AII.1a)}$$

$$= \frac{C}{\mu_e(\phi) + N_\tau\nu + N_\tau\mu_c(\phi)} \qquad \text{(AII.1b)}$$

For the case of the prime mover ($N_\tau > 1$) $C_{min}$ for $p_{ref} = p_{min} = N_\tau(\nu + 1)$ because for this case $\mu_e(\phi) = 0$ and $\mu_c(\phi) = 1$ at $p = p_{min}$. By the same token, $C_{max} = \mathbf{r}^* + N_\tau\nu$.

## AII.2 Work per cycle – discontinuous piston motion

Work per cycle may be found by integrating $p\,dV$ over the individual intervals for which the $V$ are continuous functions of $\phi$. Correspondingly, work per reference

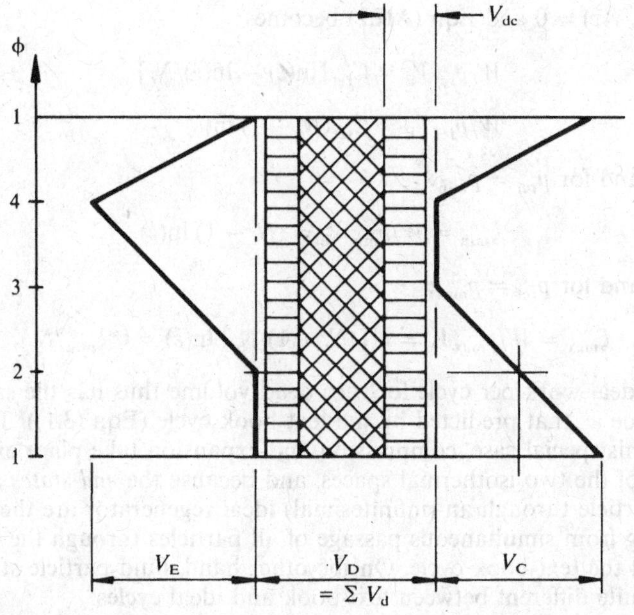


Fig. AII.1 Compression ratio, **r**, of a machine with finite dead volume may be expressed in terms of **r*** – the ratio of $V_c(\phi)$ at which constant volume transfer begins to maximum compression space volume, $V_C$, which may be written $\mathbf{r}^* = V_c(\phi_2)/V_C$.

cycle pressure and reference volume, $\zeta$, is the integral $(p/p_{ref})\,d\mu$. The integration is carried out noting that, for example, $p\,dV$ is equivalent to $p(dV/d\phi)\,d\phi$. In this connection:

$$\mu_c(\phi)_1{}^3 = 1 - \phi/\phi_3, \qquad \mu_c'(\phi) = -1/\phi_3$$

Likewise:

$$\mu_e(\phi)_1{}^2 = 0$$

$$\mu_e(\phi)_2{}^4 = (\phi - \phi_2)/(\phi_4 - \phi_2), \qquad \mu_e'(\phi) = -\mu_c'(\phi)_1{}^3$$

with similar expressions for other phases. Carrying out the two integrations over which there are variations of volume gives:

$$\frac{W}{p_{ref}\,V_E} = C_{ref}[\ln(A_1) - \ln(A_2)/N_\tau] \qquad \text{(AII.2)}$$

$$A_1 = \frac{1 + N_\tau \nu}{\mathbf{r}^* + N_\tau \nu}$$

$$A_2 = \frac{1 + \nu}{\mathbf{r}^* + \nu}$$

$$\mathbf{r}^* = \mu_c(\phi_2)/\mu_c(\phi_1) = \mu_c(\phi_2) = 1 - \phi_2/\phi_3$$

$$= \mathbf{r} \text{ for } \sum \mu_d = \mu_{de} + \mu_{dr} + \mu_{dc} = 0$$

For $\sum \mu_d = 0$, $v(\tau) = 0$ and Eqn (AII.2) becomes

$$W/p_{ref} V_E = C_{ref}[\ln(\varepsilon) - \ln(\varepsilon)/N_\tau] \qquad \text{(AII.2a)}$$

$$W/p_{ref} V_E = C_{ref}(N_\tau - 1) \ln(\varepsilon) \qquad \text{(AII.3)}$$

For $v(\tau) = 0$ and for $p_{ref} = p_{min}$:

$$\zeta_{min} = W/p_{min} V_E = (N_\tau - 1) \ln(\varepsilon) \qquad \text{(AII.4a)}$$

For $v(\tau) = 0$ and for $p_{ref} = p_{max}$:

$$\zeta_{max} = W/p_{max} V_E = \mathbf{r}^*[(N_\tau - 1)/N_\tau] \ln(\varepsilon) = \mathbf{r}^* \zeta_{min}/N_\tau \qquad \text{(AII.4b)}$$

Normalized, ideal work per cycle for zero dead volume thus has the same numerical value as that predicted by the text-book cycle (Eqn (3.1)). This is because, for this special case, compression and expansion take place exclusively in one or other of the two isothermal spaces, and because the *end states* after transfer particle-by-particle through an (infinitesimal) ideal regenerator are the same as those resulting from simultaneous passage of all particles through the hypothetical regenerator of the text-book cycle. On the other hand, fluid particle states *during* transfer are quite different between text-book and ideal cycles.

### AII.3 Compression and expansion ratios, r and ε – discontinuous piston motion

Compression ratio, **r**, is given (Fig. AII.1) by:

$$\mathbf{r} = \frac{\mathbf{r}^* + \sum \mu_d}{1 + \sum \mu_d} \qquad \text{(AII.5)}$$

for $\sum \mu_d = 0$

$$\mathbf{r} = \mathbf{r}^* = V_{c2}/V_{c1} = V_{c2}/V_C = \mu_c(\phi_2)$$

$$\varepsilon^* = \frac{\varepsilon}{\sum \mu_d + 1 - \varepsilon \sum \mu_d} \qquad \text{(AII.6)}$$

$$= \varepsilon$$

for $\sum \mu = 0$.

### AII.4 General isothermal cycle

The material of Sections AII.1 and AII.2 may be generalized to allow volume variations which are arbitrary functions of crank angle. Fig. AII.2 shows a one-dimensional machine. $V_e(\phi)$ and $V_c(\phi)$ represent the variations with $\phi$ of those respective parts, $V_E$ and $V_C$ of expansion and compression space completely swept by the pistons. Corresponding working space dead volumes are represented by $V_{de}$ and $V_{dc}$ respectively. If $V_e(\phi)/V_E$ is denoted by $fv_e(\phi)$ and $V_c(\phi)/V_C$ by $fv_c(\phi)$, then $fv_e(\phi)$ and $fv_c(\phi)$ are functions which both vary between 0 and 1.

The choice of reference volume, $V_{ref}$, is immaterial, since it can be changed at any stage to an alternative, $V_{alt}$, through multiplication by $V_{alt}/V_{ref}$. The algebra will be most compact if $V_E$ is used as reference volume during the derivation. The

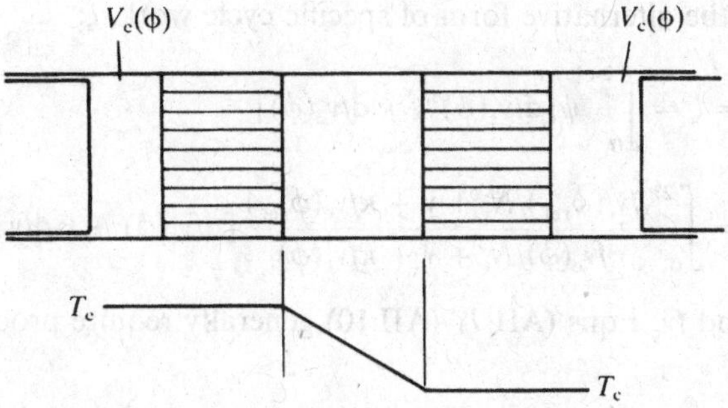


Fig. AII.2 One-dimensional machine with arbitrary piston motion.

expression for working space mass content then becomes:

$$M = \frac{pV_E}{RT_c}[fv_e(\phi)/N_\tau + v(N_\tau) + \kappa fv_c(\phi)]$$

The definition of $v(N_\tau)$ remains based on $V_E$ as reference volume and is as before, namely:

$$v(N_\tau) = \mu_{de}/N_\tau + \mu_{dc} + \frac{\mu_{dr}\ln(N_\tau)}{N_\tau - 1}$$

*and will be seen to be one half of the value of $v(N_\tau)$ as defined by Finkelstein.*[1]

The formulation in terms of $fv_e$ etc. throws up a volume ratio, $\kappa$, regardless of the fact that volume variations are not simple harmonic. Instantaneous pressure, $p$, normalized with respect to $MRT_c$, becomes:

$$\psi = \frac{pV_E}{MRT_c} = \frac{1}{fv_e(\phi)/N_\tau + v + \kappa fv_c(\phi)} \qquad \text{(AII.7)}$$

Work per cycle $= \int p[\mathrm{d}v_e(\phi) + \mathrm{d}v_c(\phi)]$, where the integral is taken between 0 and $2\pi$, so that specific cycle work, based on the current definition of $\psi$ is

$$W_{cyc}/MRT_c = \int \psi[\mathrm{d}fv_e(\phi) + \kappa\,\mathrm{d}fv_c(\phi)]$$

$$= \int_0^{2\pi} \frac{\mathrm{d}fv_e(\phi) + \kappa\,\mathrm{d}fv_c(\phi)}{fv_e(\phi)/N_\tau + v + \kappa fv_c(\phi)} \qquad \text{(AII.8)}$$

If $\phi_{ref}$ is the value of the independent variable, $\phi$, at which $p = p_{ref}$, then from Eqn (AII.7):

$$\frac{p}{p_{ref}} = \frac{fv_e(\phi_{ref})/N_\tau + v + \kappa fv_c(\phi_{ref})}{fv_e(\phi)/N_\tau + v + \kappa fv_c(\phi)} \qquad \text{(AII.9)}$$

Eqn (AII.9) introduces an alternative definition of $\psi$ as $p/p_{ref}$. This alternative definition leads to the alternative form of specific cycle work, $\zeta$:

$$W_{cyc}/p_{ref}V_{ref} = \zeta = \int_0^{2\pi} \psi[dfv_e(\phi) + \kappa\, dfv_c(\phi)]$$

$$= \int_0^{2\pi} \frac{fv_e(\phi_{ref})/N_\tau + \nu + \kappa fv_c(\phi_{ref})}{fv_e(\phi)/N_\tau + \nu + \kappa fv_c(\phi)}[dfv_e(\phi) + \kappa\, dfv_c(\phi)] \quad \text{(AII.10)}$$

For arbitrary $fv_e$ and $fv_c$ Eqns (AII.7)–(AII.10) generally require processing numerically.

## AII.5 The Schmidt analysis

### *AII.5.1 Pressure as a function of crank angle*

This is the most celebrated analysis of the practical Stirling cycle. The account by Finkelstein[1] is the most general, and will be followed here. In his formulation (Fig. AII.3) the harmonic volume variations are represented as $V_e(\phi) = \frac{1}{2}V_E[1 + \cos(\phi)]$, $V_c(\phi) = \frac{1}{2}V_C[1 + \cos(\phi - \alpha)]$, in which $\alpha$ is the constant phase angle between volume variations. $fv_e(\phi)$ and $fv_c(\phi)$ become $\frac{1}{2}[1 + \cos(\phi)]$ and $\frac{1}{2}[1 + \cos(\phi - \alpha)]$ respectively. With these definitions:

$$M = \frac{pV_E}{2RT_c}\{[1 + \cos(\phi)] + 2\nu + \kappa[1 + \cos(\phi - \alpha)]\}$$

After expansion of the term $\cos(\phi - \alpha)$ and consolidation of the trigonometric quantities:

$$\frac{1}{2}\frac{pV_E}{MRT_c} = \frac{1}{(1/N_\tau + \kappa + 2\nu)[1 + \xi\cos(\phi - \theta)]} \quad \text{(AII.11)}$$

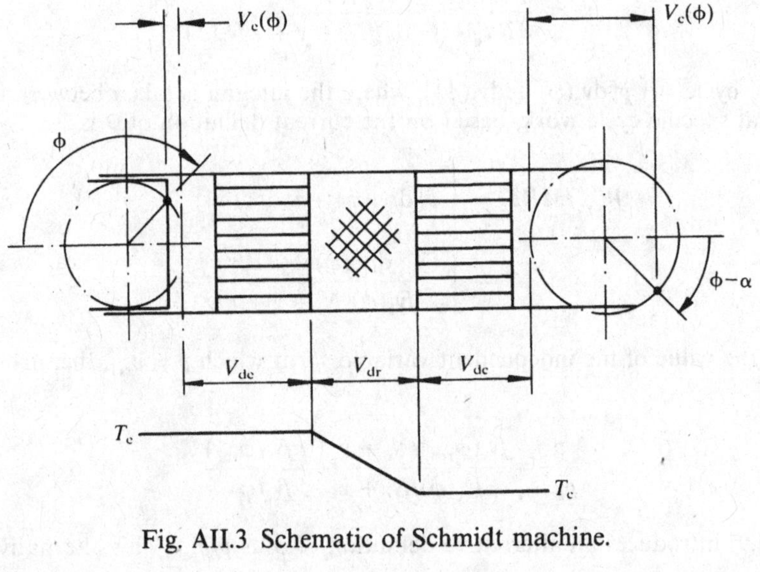


Fig. AII.3 Schematic of Schmidt machine.

$$\xi = \frac{\sqrt{[(1/N_\tau)^2 + \kappa^2 + 2\kappa\cos(\alpha)/N_\tau]}}{1/N_\tau + \kappa + 2\nu} \tag{AII.12}$$

$$\theta = \arctan\frac{\kappa\sin(\alpha)}{[1/N_\tau + \kappa\cos(\alpha)]} \tag{AII.13}$$

From inspection of Eqn (AII.11)

$$\frac{p_{\max}V_E}{MRT_c} = \frac{1}{\frac{1}{2}(1/N_\tau + \kappa + 2\nu)(1-\xi)} \tag{AII.14a}$$

$$\frac{p_{\min}V_E}{MRT_c} = \frac{1}{\frac{1}{2}(1/N_\tau + \kappa + 2\nu)(1+\xi)} \tag{AII.14b}$$

Ideal cycle pressure ratio, $p_{\max}/p_{\min}$, $\Pi$, follows by division of Eqn (AII.14a) by (AII.14b):

$$\Pi = \frac{1+\xi}{1-\xi} \tag{AII.15}$$

$$\frac{p_{\text{mean}}V_E}{MRT_c} = \frac{1}{\frac{1}{2}(1/N_\tau + \kappa + 2\nu)}\frac{1}{2\pi}\int_0^{2\pi}\frac{d(\phi-\theta)}{1+\xi\cos(\phi-\theta)}$$

$$= \frac{1}{\frac{1}{2}(1/N_\tau + \kappa + 2\nu)}\frac{1}{\sqrt{(1-\xi^2)}} \tag{AII.14c}$$

By taking ratios of Eqn (AII.11) to each of Eqns (AII.14a)–(AII.14c), in turn it is possible to express $p(\phi)$ in terms of $p_{\max}$, $p_{\min}$ and $p_{\text{mean}}$ respectively:

$$\frac{p}{p_{\max}} = \frac{\psi}{\psi_{\max}} = \frac{1-\xi}{1+\xi\cos(\phi-\theta)} \tag{AII.16a}$$

$$\frac{p}{p_{\min}} = \frac{\psi}{\psi_{\min}} = \frac{1+\xi}{1+\xi\cos(\phi-\theta)} \tag{AII.16b}$$

$$\frac{p}{p_{\text{mean}}} = \frac{\psi}{\psi_{\text{mean}}} = \frac{\sqrt{(1-\xi^2)}}{1+\xi\cos(\phi-\theta)} \tag{AII.16c}$$

Eqns (AII.16) assume $\psi$ defined as $pV_{\text{ref}}/MRT_{\text{ref}} = pV_E/MRT_c$. They are valid for the alternative definition of $\psi$ as $p/p_{\text{ref}}$.

### AII.5.2 Work per cycle, $W_{cyc}$

$W_{\text{cyc}}$ is computed as

$$\int p[dv_e(\phi) + dv_c(\phi)] = \int p\left[\frac{dv_e(\phi)}{d\phi}d\phi + \frac{dv_c(\phi)}{d\phi}d\phi\right]$$

From Eqn (AII.11) this may be expressed:

$$\frac{W_{\text{cyc}}}{MRT_c} = -\frac{1}{1/N_\tau + \kappa + 2\nu}\int_0^{2\pi}\frac{\sin(\phi) + \kappa\sin(\phi-\alpha)}{1+\xi\cos(\phi-\theta)}d\phi$$

This is a standard integral[2] and yields

$$\frac{W_{cyc}}{MRT_c} = \frac{2\pi\kappa(1 - 1/N_\tau)\sin(\alpha)}{(1/N_\tau + \kappa + 2\nu)^2} f_m(\xi) \tag{AII.17}$$

$$f_m(\xi) = \frac{1}{\sqrt{(1 - \xi^2)}[1 + \sqrt{(1 - \xi^2)}]} \tag{AII.18}$$

Expressions for specific cycle work in terms of the various characteristic reference pressures, $p_{max}$, $p_{min}$ and $p_{mean}$, follow by substitution into Eqn (AII.17) of Eqns (AII.14) for characteristic pressure in terms of $MRT_c$. The formulation is that due to Finkelstein:[1]

$$\frac{W_{cyc}}{p_{max} V_E} = \frac{\pi\kappa(1 - 1/N_\tau)\sin(\alpha)}{1/N_\tau + \kappa + 2\nu} f_{p_{max}}(\xi) \tag{AII.19a}$$

$$f_{p_{max}}(\xi) = \frac{\sqrt{(1 - \xi)}}{\sqrt{(1 + \xi)}[1 + \sqrt{(1 - \xi^2)}]} \tag{AII.20a}$$

$$\frac{W_{cyc}}{p_{min} V_E} = \frac{\pi\kappa(1 - 1/N_\tau)\sin(\alpha)}{1/N_\tau + \kappa + 2\nu} f_{p_{min}}(\xi) \tag{AII.19b}$$

$$f_{p_{min}}(\xi) = \frac{\sqrt{(1 + \xi)}}{\sqrt{(1 - \xi)}[1 + \sqrt{(1 - \xi^2)}]} \tag{AII.20b}$$

$$\frac{W_{cyc}}{p_{mean} V_E} = \frac{\pi\kappa(1 - 1/N_\tau)\sin(\alpha)}{1/N_\tau + \kappa + 2\nu} f_{p_{mean}}(\xi) \tag{AII.19c}$$

$$f_{p_{mean}}(\xi) = \frac{1}{1 + \sqrt{(1 - \xi^2)}} \tag{AII.20c}$$

### *AII.5.3 Mass flow rates*

Fig. AII.4 represents a space of variable volume with its inlet/outlet port. From the ideal gas equation, instantaneous mass content, $m$, is given by:

$$m = pV/RT$$

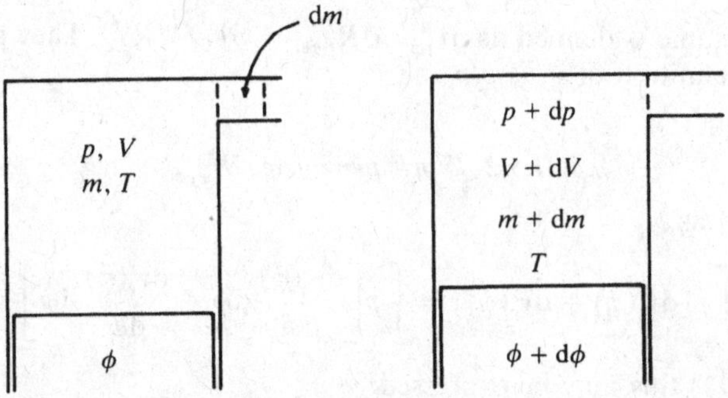


Fig. AII.4 Variable volume space with contents at wall temperature and provision for inflow/outflow of mass at instantaneous rate $m'$.

The rate of change of $m$ with $\phi$ is had by differentiating:

$$\frac{dm}{d\phi} = \frac{1}{RT}\left(p\frac{dV}{d\phi} + V\frac{dp}{d\phi}\right)$$

*It is important to note that this is the rate of* increase *of working space mass contents, and so is positive for inflow, negative for outflow.* In terms of normalized variables, and with $\psi$ defined as $pV_{ref}/MRT_c$, flow rate *into* the expansion space at $T_e$ is:

$$\frac{d\sigma_e}{d\phi} = \frac{1}{N_\tau}\left(\psi\frac{dfv_e}{d\phi} + fv_e\frac{d\psi}{d\phi}\right) \tag{AII.21a}$$

$$\frac{d\sigma_c}{d\phi} = \kappa\left(\psi\frac{dfv_c}{d\phi} + fv_c\frac{d\psi}{d\phi}\right) \tag{AII.21b}$$

If the definition $\psi = p/p_{ref}$ is preferred:

$$\frac{d\sigma_e}{d\phi} = \frac{\psi_{ref}}{N_\tau}\left(\psi\frac{dfv_e}{d\phi} + fv_e\frac{d\psi}{d\phi}\right) \tag{AII.22a}$$

$$\frac{d\sigma_c}{d\phi} = \kappa\psi_{ref}\left(\psi\frac{dfv_c}{d\phi} + fv_c\frac{d\psi}{d\phi}\right) \tag{AII.22b}$$

$$\psi_{ref} = \frac{p_{ref}V_{ref}}{MRT_{ref}} = \frac{p_{ref}V_E}{MRT_c}$$

in the present case.

### *AII.5.4 Application*

The Schmidt analysis (and the adiabatic) are proposed as an integral part of the thermodynamic similarity approach described in Chap. 4. But for the definition of $\nu(N_\tau)$ the foregoing development of the Schmidt analysis has been identical with that of Finkelstein.[1] It is therefore necessary to investigate how an analysis normalized on one basis is made compatible with a similarity principle normalized according to another.

The rôle of the ideal equations in connection with thermodynamic similarity is to give values for specific work and for $\sigma_e'$ and for $\sigma_c'$. (The prime indicates differentiation with respect to the independent variable crank angle.) Reference volume, $V_{ref}$, suggested for thermodynamic similarity is the working fluid circuit volume at some appropriate $p_{ref}$ and corresponding crank angle, $\phi_{ref}$. The various definitions of specific work, $\zeta$, convert from a base of $V_E$ to a base of $V_{ref}$ simply by multiplying $\zeta$ by the ratio $\mu_v$, where $\mu_v = V_E/V_{ref}$.

Computation of $\sigma_e'$ and of $\sigma_c'$ involves the use of dimensionless pressure, $\psi$. This is converted in the same way as $\zeta$. If volume variations are always handled in terms of the $fv_e$, etc., the only other quantity subject to different numerical value according to choice of normalizing variable is $\nu(N_\tau)$. In a typical application of the thermodynamic similarity principle it would be evaluated using the equation immediately preceding Eqn (AII.1a), in which the various $\mu_{d_j}$ will have been calculated as $V_{d_j}/V_{ref}$. The $\nu(N_\tau)$ of the present treatment is thus converted to the form required by thermodynamic similarity as for $\zeta$ – i.e., by multiplying by $\mu_v$, where $\mu_v = V_E/V_{ref}$.

# Appendix III

## Ideal cycle with thermal resistance in series with source and sink

The treatment follows Bejan (Ref. 5 of Chap. 3). Referring to Fig. AIII,1 it is seen that power output, $W'$, is given by

$$W' = Q_e'(1 - 1/N_\tau^*) \qquad \text{(AIII.1)}$$

$$N_\tau^* = T_e^*/T_c^*$$

In Eqn (AIII.1) $Q_e' = (hA)_e(T_e - T_e^*)$. Also $Q_c' = (hA)_c(T_c^* - T_c)$.

Power output thus depends on $(hA)_e$, $(hA)_c$, $T_e^*$ and $T_c^*$. If the machine already exists then $(hA)_e$ and $(hA)_c$ are fixed and known and power is a function only of $N_\tau^*$ for which an optimum is sought. Eliminating $T_e^*$ and $T_c^*$ using the expressions for $Q_e'$ and $Q_c'$:

$$\frac{W'}{(hA)_e T_e} = \frac{(1 - N_\tau^*/N_\tau)(1 - 1/N_\tau^*)}{1 + (hA)_e/(hA)_c} \qquad \text{(AIII.2)}$$

$$N_\tau = T_e/T_c$$

The expression confirms that the optimum ideal cycle temperature ratio, $N_\tau^*$, depends only on source-to-sink temperature ratio, $N_\tau$. Solving for $\partial W'/\partial N_\tau^* = 0$ gives:

$$N_{\tau\,opt}^* = \sqrt{N_\tau} \qquad \text{(AIII.3)}$$

The corresponding peak cycle power, $W_{max}'$ is given by

$$\frac{W_{max}'}{(hA)_e T_e} = \frac{[1 - \sqrt{(1/N_\tau)}]^2}{1 + (hA)_e/(hA)_c} \qquad \text{(AIII.4)}$$

Thus, power per unit heat transfer capacity has a maximum at $N_\tau^* = N_{\tau opt}^*$.

Power, $W' = \zeta p_{ref} V_{ref} N/60$. For a machine having equal heat transfer capacity at expansion and compression ends the denominator of Eqn (AIII.4) has the value 2. Noting that $A_e/V_{ref} = (A_e/V_{de})(V_{de}/V_{ref}) = \mu_{de}/r_{he}$:

$$\zeta_{max} = 30 N_\tau [1 - \sqrt{(1/N_\tau)}]^2 \mu_{de} \frac{hT_e}{N p_{ref} r_{he}} \qquad \text{(AIII.5)}$$

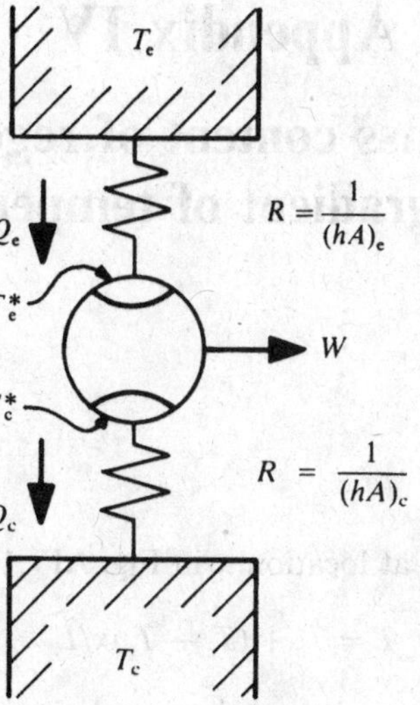


Fig. AIII.1 Thermal resistance between source and reversible machine and between reversible machine and sink.

# Appendix IV

## Instantaneous mass content of regenerator having linear gradient of temperature

The absolute temperature, $T$, at location $x$ in Fig. AIV.1 is:

$$T = T_e + (T_c - T_e)x/L$$

From the ideal gas equation the elemental mass, d$m$, contained between plane faces at $x$ and $x + \mathrm{d}x$ is

$$\mathrm{d}m = \frac{pA_x\,\mathrm{d}x}{R[T_e + (T_c - T_e)x/L]} \tag{AIV.1}$$

$$= \frac{pA_x L}{(1 - N_\tau)RT_c}\,\frac{\mathrm{d}x}{[LN_\tau/(N_\tau - 1) + x]}$$

The integral between limits of 0 and $L$ of Eqn (AIV.1) with respect to $x$ gives the total mass content. This is a standard integral involving the logarithm of the denominator. Noting that $A_x L$ is equal to $V_{dr}$ (regenerator dead space):

$$m_r = \frac{pV_{dr}\ln(N_\tau)}{RT_c(N_\tau - 1)} \tag{AIV.2}$$

$$= \frac{pV_{dr}}{RT_c} \quad \text{for } N_\tau = 1 \tag{AIV.3}$$

In terms of normalized variables:

$$\sigma_r = \psi\mu_{dr}\frac{\ln(N_\tau)}{N_\tau - 1} \tag{AIV.2a}$$

$$= \psi\mu_{dr} \quad \text{for } N_\tau = 1 \tag{AIV.3a}$$

The ostensible discrepancy between Eqns (AIV.2) and (AIV.2a) and corresponding expressions published elsewhere is due to the different definitions of characteristic temperature ratio, $N_\tau$. (N.B. $N_\tau = T_e/T_c$) here.)

In a computer program designed to make use of Eqn (AIV.2) a trap for $N_\tau \approx 1.0$ is desirable. A suitable approach is:

XMR = (Eqn (AIV.3) or (AIV.3a))
IF(ABS(XNT – 1.0) > 1.0E-06)XMR = (Eqn (AIV.2) or (AIV.2a))

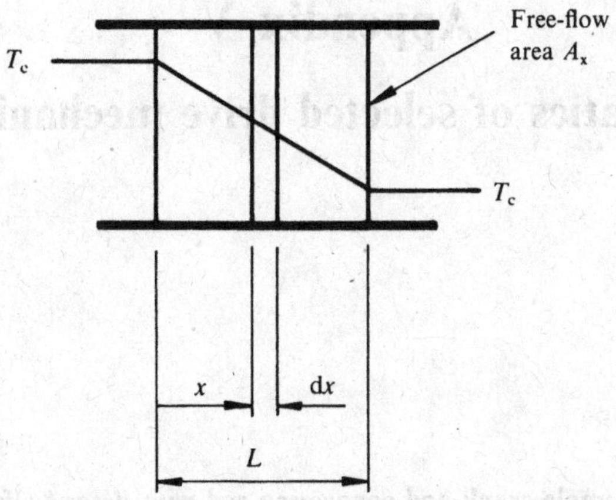


**Fig. AIV.1 Schematic of regenerator with linear gradient of temperature.**

# Appendix V

## Kinematics of selected drive mechanisms

### AV.1 Simple crank and connecting rod with *désaxé* offset

Fig. AV.1(*a*) shows a simple system comprising crank and connecting rod. To cover the general case, the axis of the cylinder is offset from that of the crankshaft by an amount $e$ – the *désaxé* offset. The problem of specifying cylinder volume is essentially that of expressing vertical position, $y$, of the gudgeon pin as a function of crank angle, $\phi$.

In the terminology of Fig. AV.1:

$$y = r\cos(\phi) + l\cos(\theta) \tag{AV.1}$$

But the instantaneous horizontal offset of the crank pin may be written

$$l\sin(\theta) + e = r\sin(\phi) \tag{AV.2}$$

To eliminate $\theta$ in favour of $\phi$ Eqn (AV.2) is squared:

$$l^2\sin^2(\theta) = [r\sin(\phi) - e]^2$$

But from the basic trigonometric relationship

$$l^2\sin^2(\theta) + l^2\cos^2(\theta) = l^2$$

so that

$$l\cos(\theta) = \surd[-l^2\sin^2(\theta) + l^2]$$
$$= \surd\{-[r\sin(\theta) - e]^2 + l^2\}$$

Substituting into Eqn (AV.1)

$$y = r\cos(\phi) + \surd\{-[r\sin(\theta) - e]^2 + l^2\}$$

Normalizing with respect to $r$:

$$y/r = \cos(\phi) + \surd\{(l/r)^2 - [\sin(\phi) - e/r]^2\} \tag{AV.3}$$

The stroke of the arrangement is equal to $2r$ only for $e = 0$. For finite $e$ stroke is determined by subtracting $y_{bdc}$ from $y_{tdc}$. Figs. AV.1(*b*) and (*c*) represent these cases, showing that determination of $y_{bdc}$ and $y_{tdc}$ amounts to applying Pythagoras' theorem to the triangles *o*–*p*–*g* for the two cases. With $s$ for stroke there results:

$$s/r = \surd[(l/r + 1)^2 - (e/r)^2] - \surd[(l/r - 1)^2 - (e/r)^2] \tag{AV.4}$$

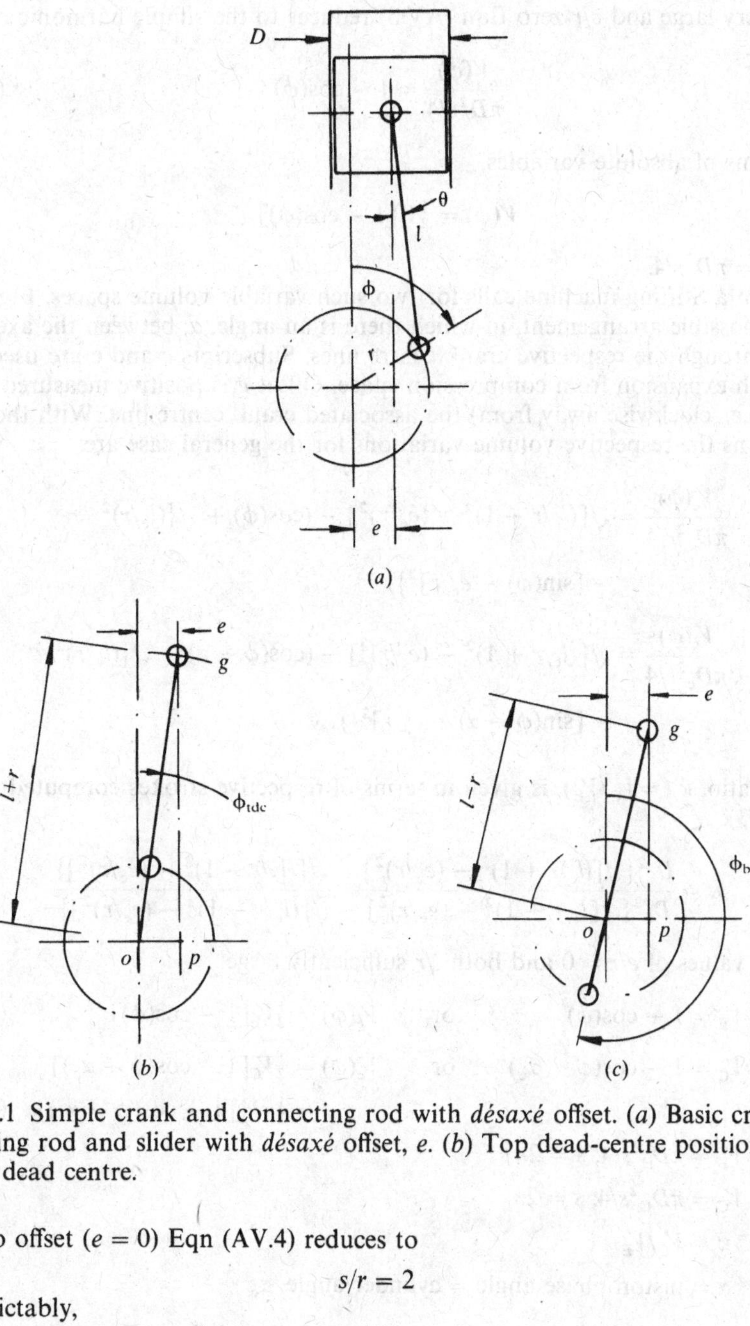


Fig. AV.1 Simple crank and connecting rod with *désaxé* offset. (*a*) Basic crank, connecting rod and slider with *désaxé* offset, *e*. (*b*) Top dead-centre position. (*c*) Bottom dead centre.

For zero offset ($e = 0$) Eqn (AV.4) reduces to

$$s/r = 2$$

or, predictably,

$$s = 2r$$

The volume swept out by the piston face between top dead-centre position and current location may now be written down:

$$\frac{V(\phi)}{\pi D^2 r/4} = \sqrt{[(l/r + 1)^2 - (e/r)^2]} - (\cos(\phi) + \sqrt{\{(l/r)^2 - [\sin(\phi) - e/r]^2\}}) \quad \text{(AV.5)}$$

For $l/r$ very large and $e/r$ zero Eqn (AV.5) reduces to the simple harmonic case:

$$\frac{V(\phi)}{\pi D^2 r/4} = 1 - \cos(\phi) \qquad \text{(AV.5a)}$$

or, in terms of absolute variables,

$$V(\phi) = \tfrac{1}{2}V[1 - \cos(\phi)]$$

where $V = \pi D^2 s/4$.

To form a Stirling machine calls for two such variable volume spaces. Fig. AV.2 shows a possible arrangement, in which there is an angle, $\alpha$, between the axes passing through the respective crank centre lines. Subscripts e and c are used to distinguish expansion from compression space. Offset $e$ is positive measured to the right of (i.e., clockwise away from) the associated crank centre line. With these conventions the respective volume variations for the general case are:

$$\frac{V_e(\phi)}{\pi D_e{}^2 r/4} = \sqrt{[(l_e/r + 1)^2 - (e_e/r)^2]} - (\cos(\phi) + \sqrt{\{(l_e/r)^2} - [\sin(\phi) - e_e/r]^2\}) \qquad \text{(AV.6)}$$

$$\frac{V_c(\phi)}{\pi D_c{}^2 r/4} = \sqrt{[(l_c/r + 1)^2 - (e_c/r)^2]} - (\cos(\phi - \alpha) + \sqrt{\{(l_c/r)^2} - [\sin(\phi - \alpha) - e_c/r]^2\}) \qquad \text{(AV.7)}$$

Volume ratio, $\kappa$ $(= V_C/V_E)$, is given in terms of respective strokes computed as per Eqn (AV.3):

$$\kappa = \frac{D_c{}^2\{\sqrt{[(l_c/r + 1)^2 - (e_c/r)^2]} - \sqrt{[(l_c/r - 1)^2 - (e_c/r)^2]}\}}{D_e{}^2\{\sqrt{[l_e/r + 1)^2 - (e_e/r)^2]} - \sqrt{[(l_e/r - 1)^2 - (e_e/r)^2]}\}} \qquad \text{(AV.8)}$$

For both values of $e/r = 0$ and both $l/r$ sufficiently large:

$$V_e(\phi)/V_E = 1 - \cos(\phi) \quad \text{or} \quad V_e(\phi) = \tfrac{1}{2}V_E[1 - \cos(\phi)] \qquad \text{(AV.6a)}$$

$$V_c(\phi)/V_C = 1 - \cos(\phi - \alpha_c) \quad \text{or} \quad V_c(\phi) = \tfrac{1}{2}V_C[1 - \cos(\phi - \alpha_c)] = \tfrac{1}{2}\kappa V_E[1 - \cos(\phi - \alpha_c)] \qquad \text{(AV.7a)}$$

$$V_E = \pi D_E{}^2 s/4;\; s = 2r$$

$$V_C = \pi D_C{}^2 s/4;\; s = 2r$$

$$\kappa = V_C/V_E$$

$\alpha$ = piston phase angle = cylinder angle, $\alpha_c$

A *désaxé* offset large relative to $r$ (i.e., a large $e/r$), especially when combined with a small $l/r$, introduces harmonics which can be shown to have a substantial influence on the thermodynamic processes (see, e.g., the chapter on the Stirling machine as a linear wave phenomenon). The same conditions also lead to an effective volume phase angle which is not quite equal to piston (or cylinder) phase angle, $\alpha_c$, and which therefore requires to be evaluated.

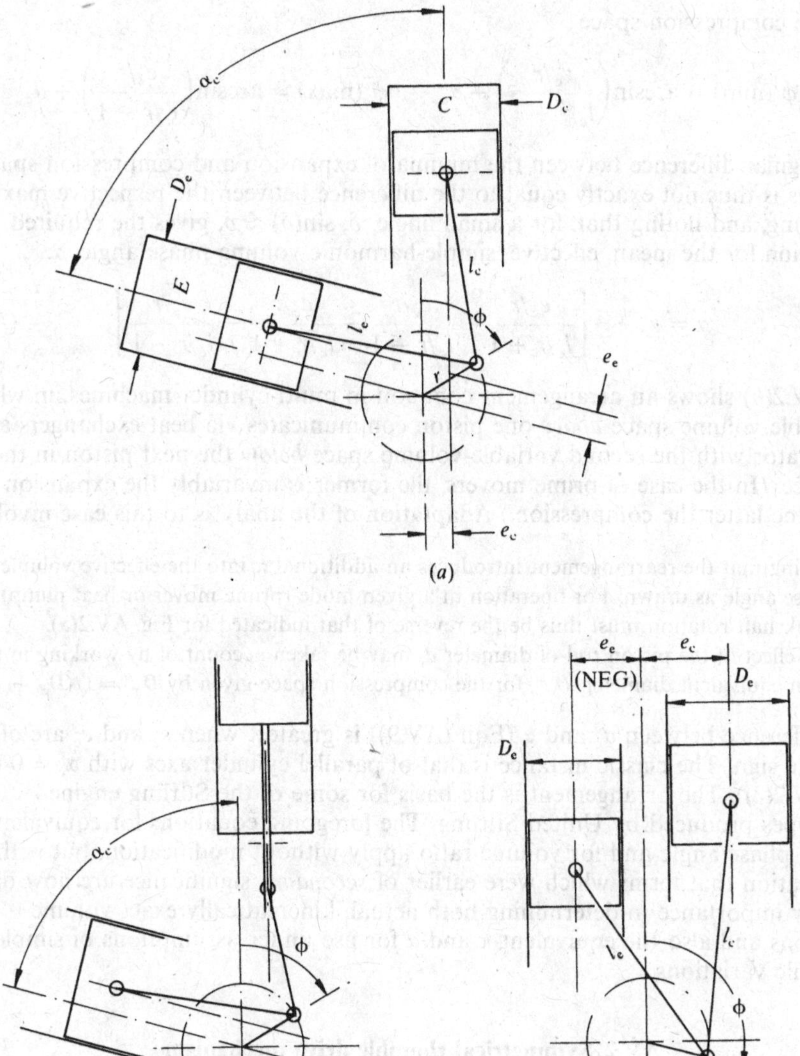


Fig. AV.2 Kinematics of V-configuration drive. (*a*) Compression and expansion spaces both above piston faces. (*b*) Compression space between piston and bdc position. (*c*) Parallel cylinder machine – $\alpha_c = 0$.

Referring again to Figs. AV.1(*b*) and (*c*), the minimum-volume (top dead-centre) and maximum-volume values of $\phi$ are respectively:

$$\phi_e(\text{min}) = \arcsin\left(\frac{e_e/r}{l_e/r + 1}\right) \qquad \phi_e(\text{max}) = \arcsin\left(\frac{e_e/r}{l_e/r - 1}\right)$$

For the compression space

$$\phi_c(\min) = \arcsin\left(\frac{e_c/r}{l_c/r + 1}\right) + \alpha_c \qquad \phi_c(\max) = \arcsin\left(\frac{e_c/r}{l_c/r - 1}\right) + \alpha_c$$

The angular difference between the minima of expansion and compression space volumes is thus not exactly equal to the difference between the respective maxima. Averaging, and noting that, for a small angle, $\delta$, $\sin(\delta) \approx \delta$, gives the required expression for the mean, effective, simple-harmonic volume phase angle, $\alpha$:

$$\alpha = \alpha_c + \tfrac{1}{2}\left[\frac{e_c/r}{l_c/r + 1} + \frac{e_c/r}{l_c/r - 1} - \frac{e_e/r}{l_e/r + 1} - \frac{e_e/r}{l_e/r - 1}\right] \qquad \text{(AV.9)}$$

Fig. AV.2(*b*) shows an arrangement common in multi-cylinder machines, in which a variable-volume space *above* one piston communicates via heat exchangers and regenerator with the second variable-volume space *below* the next piston in the sequence. (In the case of prime movers, the former is invariably the expansion space, the latter the compression.) Adaptation of the analysis to this case involves:

(*a*) Noting that the rearrangement introduces an additional $\pi$ into the effective volume phase angle as drawn. For operation in a given mode (prime mover or heat pump) crankshaft rotation must thus be the reverse of that indicated for Fig. AV.2(*a*).
(*b*) The effect of the piston rod of diameter $d_r$ may be taken account of by working in terms of an equivalent diameter, $D_c^*$, for the compression space given by $D_c^* = \surd(D_c^2 - d_r^2)$.

The difference between $\alpha_c$ and $\alpha$ (Eqn (AV.9)) is greatest when $e_e$ and $e_c$ are of opposite sign. The classic instance is that of parallel cylinder axes with $\alpha_c = 0$ (see Fig. AV.2(*c*)). The arrangement is the basis for some of the Stirling engine prototypes produced by United Stirling. The foregoing equations for equivalent volume phase angle and for volume ratio apply without modification, but with the qualification that terms which were earlier of *secondary* significance are now of primary importance in determining both actual, kinematically exact volume variations and also the equivalent $\kappa$ and $\alpha$ for use under assumptions of simple harmonic variations.

## AV.2 Symmetrical rhombic drive mechanism

Fig. AV.3(*a*) illustrates the case. A more general form is possible, and may have thermodynamic advantages in the sense of permitting increased specific cycle work based on net swept volume. However, net swept volume accounting for a small fraction of the bulk of coaxial machine embodying the mechanism, the symmetrical case is the usual choice on account of the relative ease with which the conditions for dynamic balance can be established.

Attention focusses on the *left-hand* crank, which rotates clockwise for the prime-mover mode, for which $e$ is positive as drawn, and which will be seen to correspond directly with Fig. AV.1 for the offset single-cylinder case.

By analogy with the arrangement of Fig. AV.1, the height, $y_c$, of the power piston little-end above the crankshaft axis is given by:

$$y_c/r = \cos(\phi) + \surd\{(l/r)^2 - [\sin(\phi) - e/r]^2\} \qquad \text{(AV.3a)}$$

where crank angle $\phi$ is measured from the vertical as indicated. Similarly, the height, $y_e$, from crankshaft centre line to displacer gudgeon pin is given:

$$y_e/r = \cos(\phi) - \sqrt{\{(l/r)^2 - [\sin(\phi) - e/r]^2\}} \qquad \text{(AV.10)}$$

The expression for dimensionless stroke of piston and displacer individually, $s/r$, is the same as that for the simple crank/connecting rod mechanism with *désaxé* Figs. AV.1($b$) and ($c$)):

$$s/r = \sqrt{[(l/r + 1)^2 - (e/r)^2]} - \sqrt{[(l/r - 1)^2 - (e/r)^2]} \qquad \text{(AV.4)}$$

For the machine with symmetrical drive and $D_e = D_c$, piston swept volume ratio, $\lambda$, = 1 always. The volume amplitude ratio, $\kappa$, is calculated by noting that, for $D_e = D_c$, volume ratios are equal to corresponding linear distance ratios. Thus the ratio $V_C/V_E$ is equal to the ratio (maximum–minimum separation of the little-end eyes)/$s$.

Minimum separation (corresponding to maximum compression-space volume) occurs at $\phi = 3\pi/2$ (Fig. AV.3($b$)) and is equal to $2r\sqrt{[(l/r)^2 - (1 + e/r)^2]}$. The maximum separation case (minimum compression-space volume) has two instances, depending upon whether $e \geqslant r$, or $e < r$. For $e \geqslant r$ (Fig. AV.3($c$)) the maximum occurs at $\phi = \pi/2$ (innermost crank-pin position) and has the value $2r\sqrt{[(l/r)^2 - (1 - e/r)^2]}$. For $e < r$ (Fig. AV.3($d$)) the maximum occurs twice per revolution at the instants when the connecting rods are aligned and vertical, in which cases the distance in question is $2l$. The two cases lead to:

$$\kappa_{e \geqslant r} = \frac{\sqrt{[(l/r)^2 - (1 - e/r)^2]} - \sqrt{[(l/r)^2 - (1 + e/r)^2]}}{s/2r} \qquad \text{(AV.11)}$$

$$\kappa_{e < r} = \frac{l/r - \sqrt{[(l/r)^2 - (1 + e/r)^2]}}{s/2r} \qquad \text{(AV.12)}$$

Piston top dead-centre position occurs at $\phi = \arcsin[(e/r)/(l/r + 1)]$, and bottom dead centre at $\phi = \pi + \arcsin[(e/r)/(l/r - 1)]$. Displacer top dead centre occurs at $\phi = -\arcsin[(e/r)/(l/r - 1)]$ – see Fig. AV.3($e$) – and bottom dead centre at $\phi = \pi - \arcsin[(e/r)/(l/r + 1)]$. The dead-centre positions thus both differ by piston phase angle, $\beta$:

$$\beta = \arcsin\left(\frac{e/r}{l/r + 1}\right) + \arcsin\left(\frac{e/r}{l/r - 1}\right) \qquad \text{(AV.13)}$$

Volume phase angle, $\alpha$, may either be estimated from $\lambda$ and $\beta$ using the methods of Chap. 6, or determined as the difference in $\phi$ between the respective minima of expansion space and compression space volumes.

Expansion space volume is a minimum at displacer top dead centre, i.e., at $\phi = -\arcsin[(e/r)/(l/r - 1)]$ – Fig. AV.3($e$). The absolute minimum of compression space volume depends on whether $e \geqslant r$, or $e < r$, but in either case, the value of $\phi$ about which the minimum is disposed is $\pi/2$. Expansion space maximum is at displacer bottom dead centre, i.e., at $\phi = \pi - \arcsin[(e/r)/(l/r + 1)]$, and compression space is maximum, regardless of the relationship between $r$ and $e$, at $\phi = 3\pi/2$. The angular differences between respective minima and maxima thus differ, and so an average is taken:

$$\alpha = \frac{\pi}{2} + \tfrac{1}{2}\left[\arcsin\left(\frac{e/r}{l/r + 1}\right) + \arcsin\left(\frac{e/r}{l/r - 1}\right)\right] \qquad \text{(AV.14)}$$

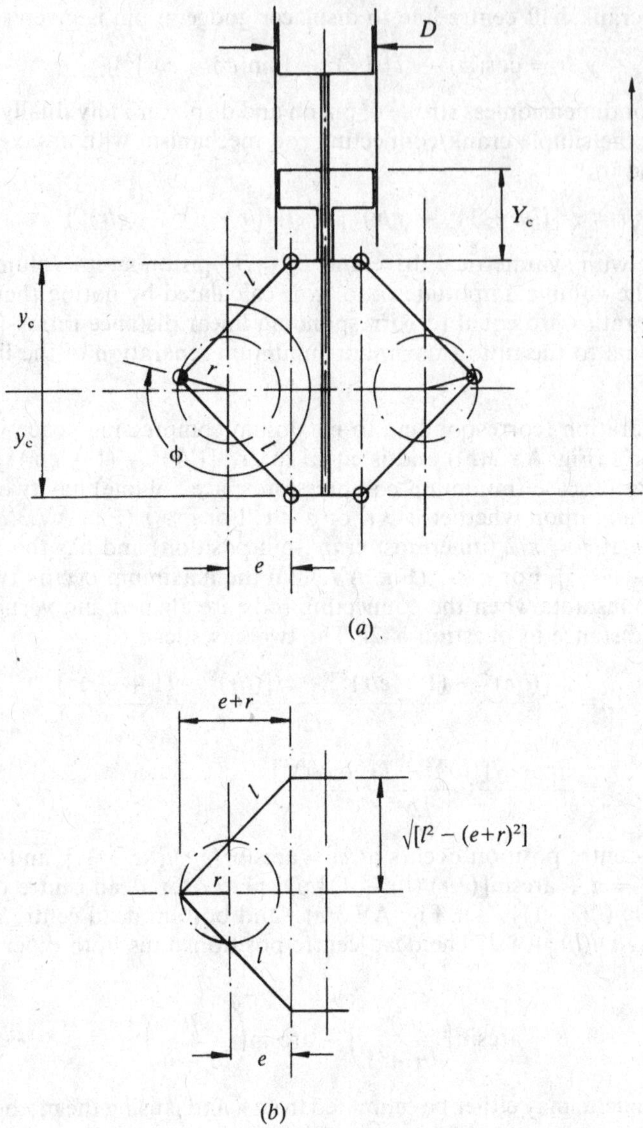


Fig. AV.3 Terminology for symmetrical rhombic drive. (*a*) Schematic of symmetrical rhombic drive. (*b*) Crank position at maximum compression space volume.

Certain classes of simulation work call for exact values of volume as a function of crank angle. The variable part of the expansion space volume may be written directly in terms of the linear distance between top dead-centre position and current position of the displacer cross-head yoke:

$$\frac{V_e(\phi)}{\pi D^2 r/4} = \sqrt{[(l/r - 1)^2 - (e/r)^2]} - (\cos(\phi) + \sqrt{\{(l/r)^2 - [\sin(\phi) - e/r]^2\}}) \quad \text{(AV.15)}$$

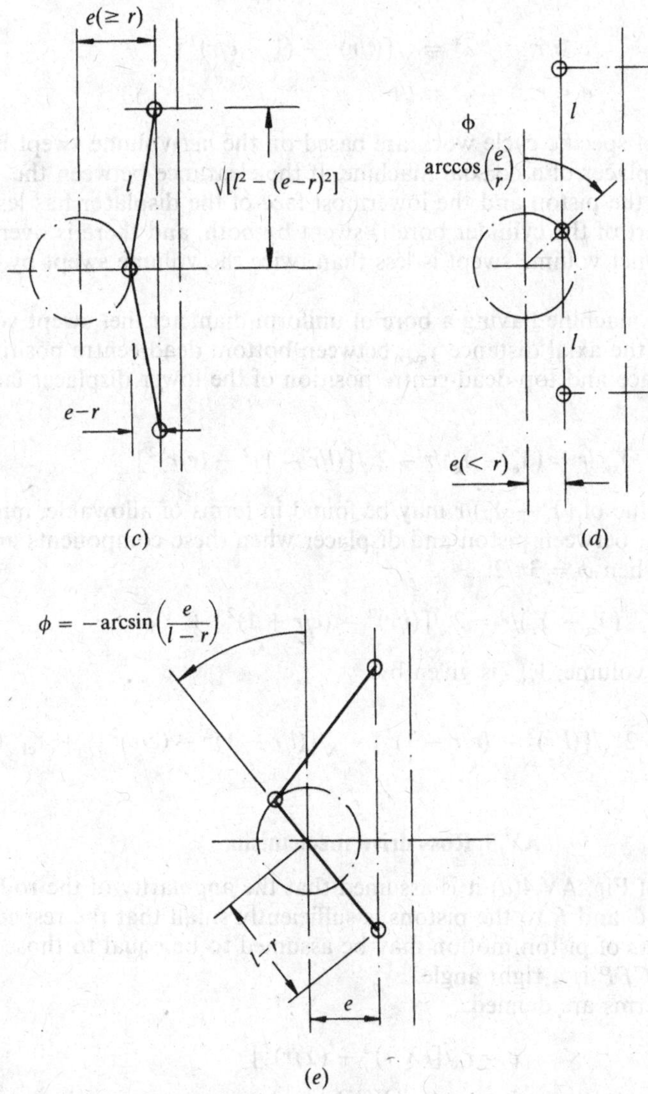


Fig. AV.3 (*cont.*) Terminology for symmetrical rhombic drive. (*c*) Minimum compression-space position ($e \geqslant r$). (*d*) One of two possible positions for minimum compression volume ($e < r$). (*e*) Upper dead-centre position of displacer.

Compression space is at a minimum when displacer and piston cross-heads are at their furthest distance apart, and increases in proportion to reduction in this distance. There are thus two cases, corresponding again to the alternatives $e \geqslant r$ and $e < r$,

$$\frac{V_c(\phi)}{\pi D^2 r/4} = 2(\lambda^* - \sqrt{\{(l/r)^2 - [\sin(\phi) - e/r]^2\}}) \qquad \text{(AV.16)}$$

where

$$e \geqslant r: \quad \lambda^* = \sqrt{[(l/r)^2 - (1 - e/r)^2]}$$
$$e < r: \quad \lambda^* = l/r$$

Some definitions of specific cycle work are based on the *net* volume swept by the piston and the displacer of a coaxial machine. If the clearance between the uppermost face of the piston and the lowermost face of the displacer has less than a certain value, part of the cylinder bore is swept by both, and there is overlap. In general, therefore, net volume swept is less than twice the volume swept by either component.

For any coaxial machine having a bore of uniform diameter, net swept volume is proportional to the axial distance $y_{sw}$, between bottom dead-centre position of the upper piston face and top dead-centre position of the lower displacer face. From Fig. AV.3:

$$Y_{sw}/r = (Y_e - Y_c)/r - 2\sqrt{[(l/r - 1)^2 - (e/r)^2]}$$

The 'unknown' value of $(Y_e - Y_c)/r$ may be found in terms of allowable, minimum axial clearance, $Y_{cl}$, between piston and displacer when these components are at their closest, i.e., when $\phi = 3\pi/2$:

$$(Y_e - Y_c)/r = 2\sqrt{[(l/r)^2 - (e/r + 1)^2]} + Y_{cl}$$

Finally, net swept volume, $V_{sw}$, is given by:

$$\frac{V_{sw}}{\pi D^2 r/4} = 2\{\sqrt{[(l/r)^2 - (e/r + 1)^2]} - \sqrt{[(l/r - 1)^2 - (e/r)^2]}\} + Y_{cl} \quad \text{(AV.17)}$$

## AV.3 Ross drive mechanism

In the schematic of Fig. AV.4(*a*) it is assumed that the angularity of the rods connecting points *C* and *E* to the pistons is sufficiently small that the respective vertical components of piston motion may be assumed to be equal to those of *C* and *E*. The angle *CFP* is a right angle.

Two constant terms are defined:

$$c = r\sqrt{[(X/r)^2 + (Y/r)^2]}$$
$$\xi = \arctan(Y/X)$$

Expressions for the following intermediate variables are deduced from the figure:

$$\theta = \pi/2 - \phi - \xi$$
$$d^2/r^2 = [c/r - \cos(\theta)]^2 + \sin^2(\theta)$$
$$= (c/r)^2 + 1 - 2(c/r)\cos(\theta)$$
$$\delta_1 = \arctan\{\sin(\theta)/[c/r - \cos(\theta)]\}$$
$$\delta_2 = \arccos\frac{(d/r)^2 + (R/r)^2 - (h/r)^2}{2DR/r^2}$$
$$\psi = \delta_1 + \delta_2$$

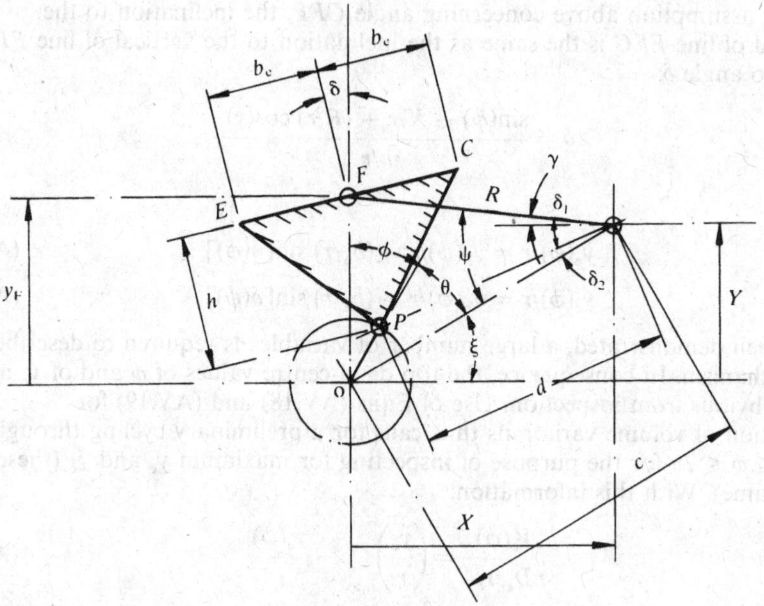


(*a*)

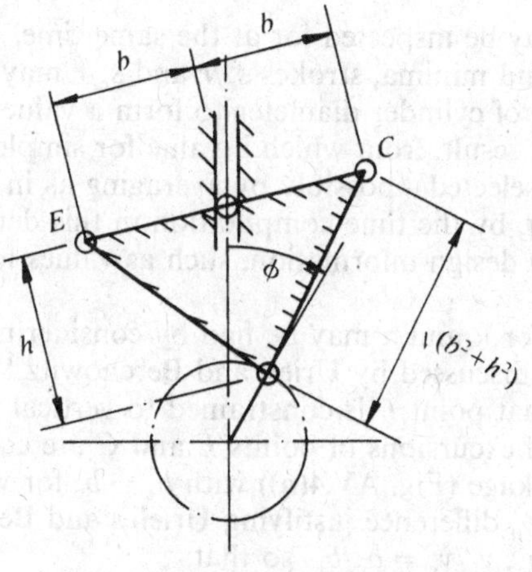


(*b*)

Fig. AV.4 Terminology for the symmetrical Ross drive mechanism. (*a*) The full symmetrical Ross drive mechanism. (*b*) Ross mechanism as simplified by Urieli and Berchowitz.[1]

The angle $\gamma$ may now be computed as $\gamma = \psi - \xi$. The $x$ and $y$ coordinate positions of point $F$ may now be stated in terms of intermediate variables:

$$y_F(\phi) = Y/r + R/r \sin(\gamma)$$

$$x_F(\phi) = R/r \cos(\gamma)$$

From the assumption above concerning angle *CFP*, the inclination to the horizontal of line *EFC* is the same as the inclination to the vertical of line *FP*, i.e., is equal to angle $\delta$:

$$\delta = \frac{\sin(\phi) - X/r + (R/r)\cos(\gamma)}{h/r}$$

Finally:

$$y_e(\phi)/r = y_F(\phi)/r - (b_e/r)\sin[\delta(\phi)] \tag{AV.18}$$

$$y_c(\phi)/r = y_F(\phi)/r + (b_c/r)\sin[\delta(\phi)] \tag{AV.19}$$

As has been demonstrated, a large number of variables is required to describe the Ross mechanism. In consequence, the top dead-centre values of $\phi$ and of $y_e$ and $y_c$ are not obvious from inspection. Use of Eqns (AV.18) and (AV.19) for computation of volume variations thus calls for a preliminary cycling through the range $0 \leqslant \phi \leqslant 2\pi$ for the purpose of inspecting for maximum $y_e$ and $y_c$ (these are not the same). With this information:

$$\frac{V_e(\phi)}{\pi D_e^2 r/4} = \left(\frac{y_e}{r}\right)_{\max} - \frac{y_e(\phi)}{r} \tag{AV.20}$$

$$\frac{V_c(\phi)}{\pi D_c^2 r/4} = \left(\frac{y_c}{r}\right)_{\max} - \frac{y_c(\phi)}{r} \tag{AV.21}$$

Minima of $y_e$ and $y_c$ may be inspected for at the same time. From respective differences of maxima and minima, strokes $s_e/r$ and $s_c/r$ may be found and used in conjunction with values of cylinder diameter to form a value of $\kappa$. Two values of volume phase difference result, from which a value for simple harmonic volume phase angle, $\alpha$, may be selected – possibly by averaging as in the case of the rhombic drive. However, by the time computation in this detail has been undertaken, preliminary design information, such as values for $\kappa$ and $\alpha$, is of academic interest only.

Approximate values for $\kappa$ and $\alpha$ may be had by considering the simplified form of the Ross mechanism discussed by Urieli and Berchowitz.[1] This is shown in Fig. AV.4(*b*), and assumes that point *F* is constrained to vertical motion. Under these assumptions the vertical excursions of points *E* and *C* are equal for $b_e = b_c$. (Rix[2] describes a full Ross linkage (Fig. AV.4(*a*)) with $b_e = b_c$ for which $y_e = 25.4$ mm and $y_e = 25.9$ mm – a 2% difference justifying Urieli's and Berchowitz' simplification.) If $b_e \neq b_c$, $y_c/y_e = b_c/b_e$, so that:

$$\kappa \approx \frac{D_c^2 b_c}{D_e^2 b_e} \tag{AV.22}$$

For an estimate of volume phase angle, note the similarity between Fig. AV.2(*c*) and Fig. AV.4(*a*) when $b_e$ is used for $e_e$, $b_c$ for $e_c$, $\sqrt{(b_e^2 + h^2)}$ for $l_e$ and $\sqrt{(b_c^2 + h^2)}$ for $l_c$. In the approximate expression for $\alpha$ which follows, the quantities labelled with an asterisk are the equivalents just identified:

$$\alpha \approx \tfrac{1}{2}\left(\frac{e_c^*/r}{l_c^*/r + 1} + \frac{e_c^*/r}{l_c^*/r - 1} - \frac{e_e^*/r}{l_e^*/r + 1} - \frac{e_e^*/r}{l_e^*/r - 1}\right) \tag{AV.23}$$

(NB: $e_e^*$ *is negative and* $\alpha_c = 0$ *for this case – see Fig. AV.2.*)

## AV.4 Philips' bell-crank mechanism

With the caution that not all symbols serve the same function in both instances, comparing the schematic of the bell-crank mechanism (Fig. AV.5) with that of the Ross drive (Fig. AV.4) will show that there are more features in common than differences.

Intermediate variables are again defined:

$$c = r\sqrt{[(X/r)^2 + (Y/r)^2]}$$
$$\xi = \arctan(Y/X)$$

Expressions for intermediate variables are deduced from the figure:

$$\theta = \pi/2 - \phi - \xi$$
$$d^2/r^2 = [c/r - \cos(\theta)]^2 + \sin^2(\theta)$$
$$= (c/r)^2 + 1 - 2(c/r)\cos(\theta)$$

$$\delta_1 = \arctan\{\sin(\theta)/[c/r - \cos(\theta)]\}$$

$$\delta_2 = \arccos\frac{(d/r)^2 + (R/r)^2 - (h/r)^2}{2dR/r^2}$$

$$\psi = \delta_2 - \delta_1$$

The foregoing set of equations establishes the motion of bar $R$ $(Q\text{–}F)$ as a function of $\phi$. Referring to Fig. AV.5(*b*):

$$\gamma + \chi = \pi/2 + [\psi - (\pi/2 - \xi)]$$
$$\gamma = \psi - \chi + \xi$$

The position, $y_e(\phi)$ relative to a horizontal line through point $Q$ may now be written down as the basis for specifying displacer motion:

$$y_e(\phi)/r = -(QE/r)\sin[\gamma(\phi)] \qquad \text{(AV.24)}$$

The motion of the compression space piston is as per Eqn (AV.3), viz

$$y_c/r = \cos(\phi) + \sqrt{\{(l_c/r)^2 - [\sin(\phi) - e/r]^2\}} \qquad \text{(AV.3a)}$$

As in the case of the Ross mechanism, $\kappa$, $\alpha$, $\lambda$ and $\beta$ do not follow by simple inspection. However, top and bottom dead-centre positions of the displacer evidently occur at the two crank positions shown in Figs. AV.6(*a*) and (*b*) respectively. For these two cases, values of the angle $A$ may be written down immediately:

$$A_{y\max} = \arccos\frac{(c/r)^2 + (h/r - 1)^2 - (R/r)^2}{2(h/r - 1)c/r}$$

$$A_{y\min} = \arccos\frac{(c/r)^2 + (h/r + 1)^2 - (R/r)^2}{2(h/r + 1)c/r}$$

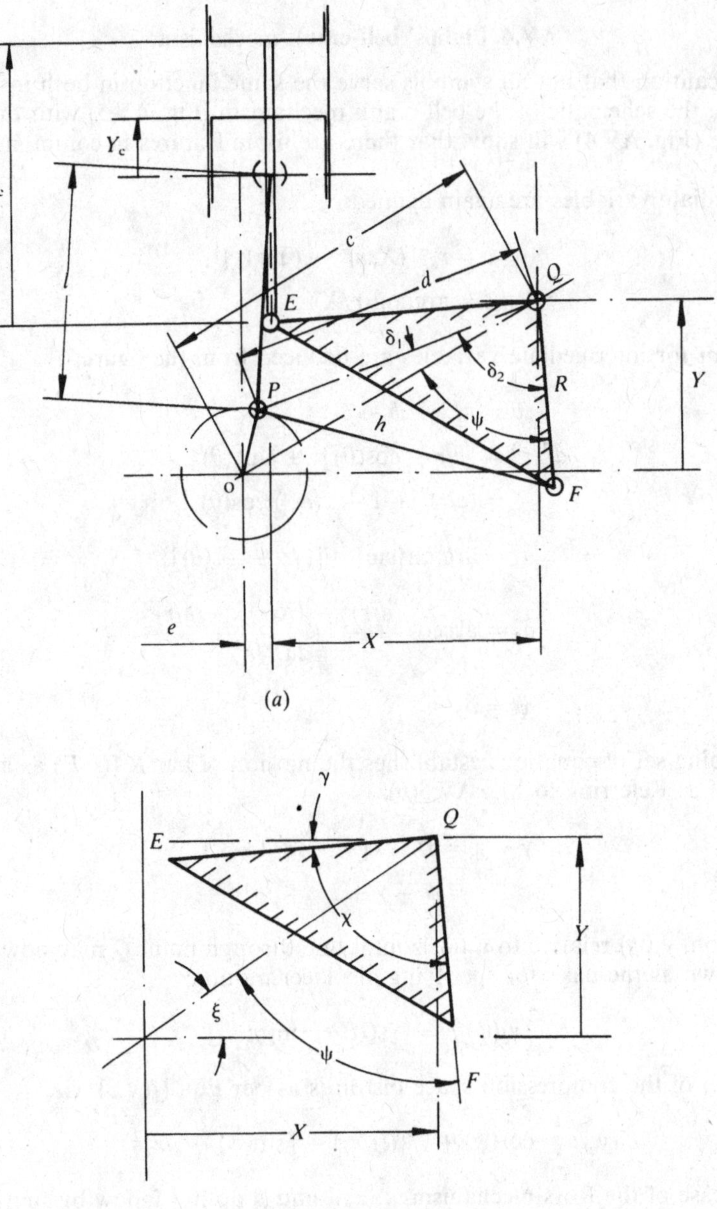


Fig. AV.5 Terminology for the Philips bell-crank mechanism. (*a*) The complete bell-crank mechanism. (*b*) Detail of bell crank.

From the figure the crank angles giving these maximum and minimum values:

$$\phi_{y\max} = 3\pi/2 - (\xi - A_{y\max}) \tag{AV.25}$$

$$\phi_{y\min} = \pi/2 - (\xi - A_{y\min}) \tag{AV.26}$$

Use of these values of $\phi$ in Eqn (AV.24) gives $y_{e\max}/r$ and $y_{e\min}/r$ respectively.

Compression piston stroke is given by an expression of the form of Eqn (AV.4), so that $\lambda$ for the coaxial machine with uniform bore becomes

$$\lambda = \frac{\sqrt{[(l/r + 1)^2 - (e/r)^2]} - \sqrt{[(l/r - 1)^2 - (e/r)^2]}}{(QE/r)[\sin\gamma(\phi_{y\min}) - \sin\gamma(\phi_{y\max})]} \tag{AV.27}$$

Piston phase angle, $\beta$, is defined as the angle between dead centre positions (top or bottom). Eqn (AV.25) provides the top dead-centre position of the displacer. For the compression-space piston $\phi = \arcsin[(e_c/r)/(l_c/r + 1)]$, so that, based on top dead-centre values:

$$\beta_{tdc} = 2\pi + \arcsin\left(\frac{e_c/r}{l_c/r + 1}\right) - \phi_{y\max}$$

Based on conditions at bottom dead centre:

$$\beta_{bdc} = \phi_{y\min} - \arcsin\left(\frac{e_c/r}{l_c/r - 1}\right)$$

A numerical value would logically be based on an average:

$$\beta = \tfrac{1}{2}(\beta_{tdc} + \beta_{bdc})$$

Values of $\kappa$ and $\alpha$ for use in the equivalent machine are probably best obtained from $\lambda$ and $\beta$ by use of Eqs (6.10) and (6.11).

A value of net swept volume is required for use in certain definitions of specific cycle work, and the definition is that already given in connection with the rhombic drive. By analogy with this earlier case:

$$y_{sw}/r = (Y_e - Y_c)/r - (QE/r)\sin[\gamma(\phi_{y\max})] - \sqrt{[(l_c/r - 1)^2 - (e/r)^2]} \tag{AV.28}$$

$(Y_e - Y_c)/r$ is found from the condition that the distance between the lowermost displacer face and piston top face has a minimum value equal to $Y_{cl}$, and that this clearance is achieved when the two faces are at their closest:

$$(Y_e - Y_c)/r = Y_{cl}/r + \cos(\phi^*) + \sqrt{\{(l_c/r)^2 - [\sin(\phi^*) - e/r]^2\}} - Y/r - (QE/r)\sin[\gamma(\phi^*)] \tag{AV.29}$$

$\phi^*$ is the value of $\phi$ for which piston face separation is a minimum. It cannot be found by visual examination of the equations for $y_e$ and $y_c$, and is best inspected for during a run of Eqns (AV.24) and (AV.3a) programmed for computer.

The various geometric variables determining piston and displacer motion are not all open to free choice. Certain combinations of values would cause clashing, or would prevent full 360° motion of the crank. Some instances of constraint are

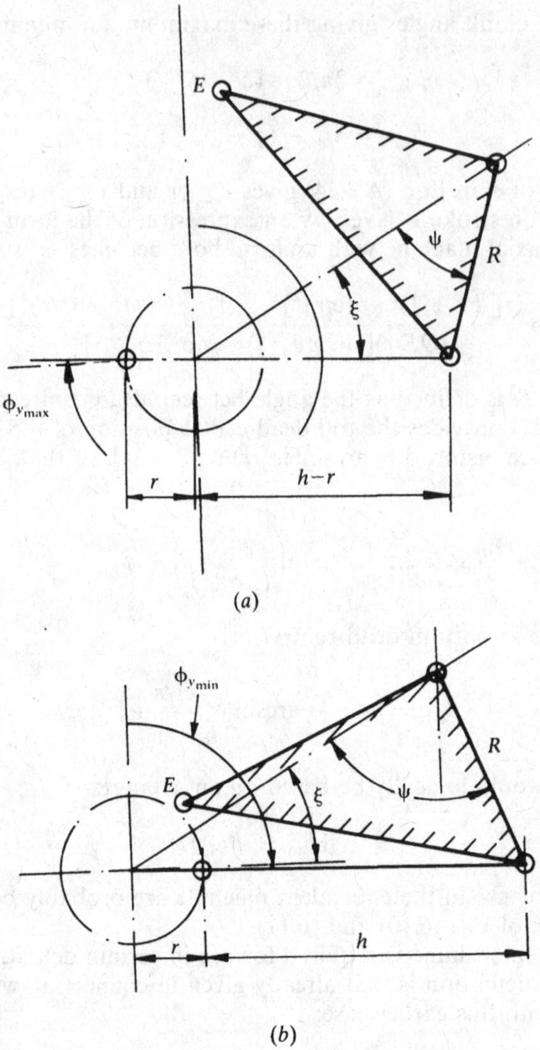


Fig. AV.6 Crank positions corresponding to top and bottom dead-centre positions of displacer. (*a*) Top dead-centre position of displacer. (*b*) Bottom dead-centre position.

obvious on first inspection. Others become so when a particular embodiment of the mechanism is laid out on the draughting machine. An example of the way in which the number of variables would probably be reduced during the course of preparation of computer code is given by consideration of the motion of point *E*: the required articulation between *E* and the point of attachment to the displacer is almost certainly best achieved by a means which minimizes the horizontal component of motion of *E*. This condition obtains when the motion of *E* takes it as far *above* a horizontal line through *q* as below. The condition may be developed algebraically with reference to Figs. AV.5 and AV.6:

$$\xi + \psi_{tdc} - \chi = -\xi - \psi_{bdc} + \chi$$
$$2(\xi - \chi) = -(\psi_{bdc} + \psi_{tdc})$$

Finally:

$$\chi = \frac{2\xi + \psi_{bdc} + \psi_{tdc}}{2} \tag{AV.30}$$

# Appendix VI

## Specifications of selected Stirling cycle machines

### AVI.1 General Motors GPU-3

(Source of data – principally Refs 1, 3 and 4. The machine is illustrated at Fig. AVI.1.)

*General*

| | |
|---|---|
| Configuration | Single-cylinder, coaxial, uniform-diameter bore. Rhombic drive crank mechanism |
| Working fluid(s) | He, $H_2$ |
| Rated maximum output | 7.5 kW with hydrogen at 69 bar |
| Specimen operating point: | (Ref. 4, Appendix F) 2 500 rpm with He at 41.3 bar. Expansion temperature 977 K, compression temperature 288 K |
| Inferred indicated power: | 3.96 kW at 12.7% *brake* thermal efficiency |
| Bore | 69.9 mm |
| Stroke (piston and displacer) | 31.2 mm |

*Working fluid circuit dimensions*

| | |
|---|---|
| Heater | |
| mean tube length | 245.3 mm |
| length exposed to heat source | 77.7 mm |
| tube length (cylinder side) | 116.4 mm |
| tube length (regenerator side) | 128.9 mm |
| tube inside diameter | 3.02 mm |
| tube outside diameter | 4.83 mm |
| no. complete tubes per cylinder | 40 |
| no. of tubes per regenerator | 5 |
| Cooler | |
| tube length | 46.1 mm |
| length exposed to coolant | 35.5 mm |
| tube inside diameter | 1.08 mm |
| tube outside diameter | 1.59 mm |
| no. of tubes per cylinder | 312 |
| no. of tubes per regenerator | 39 |

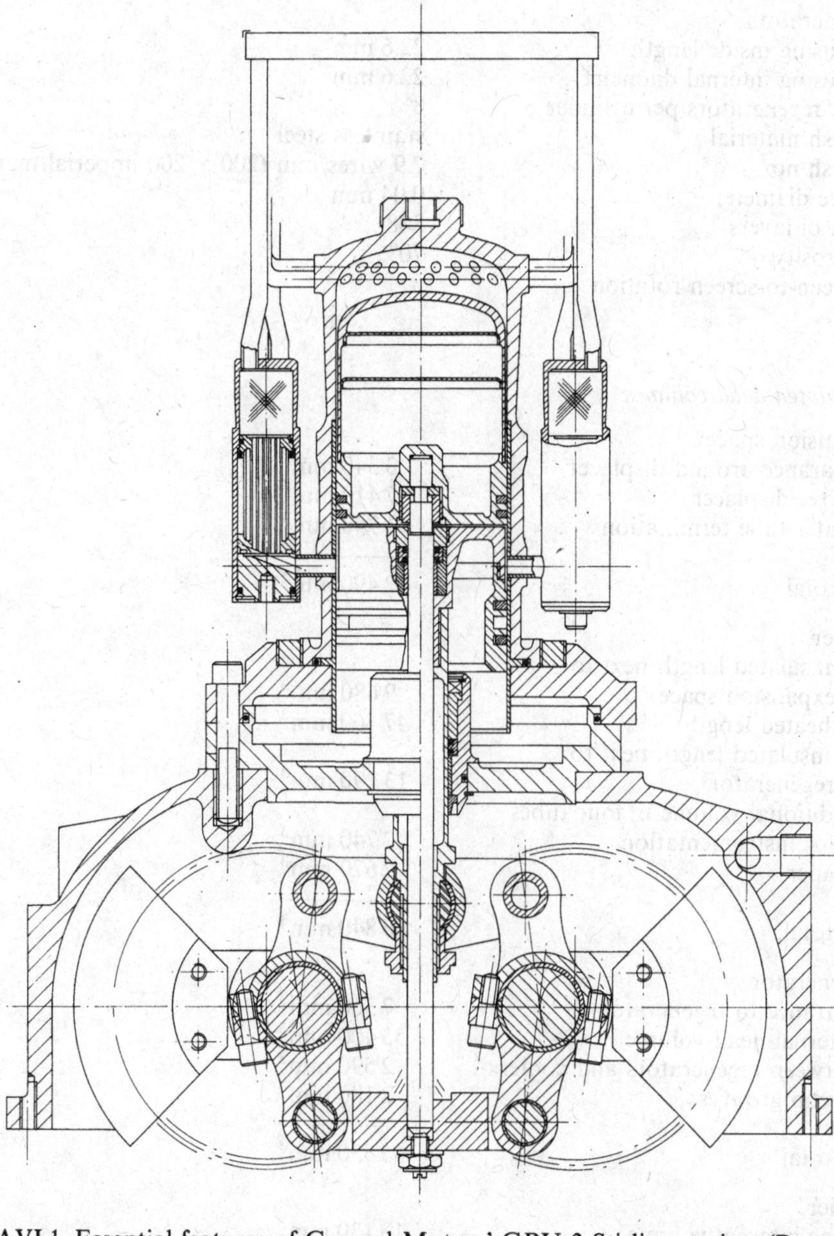

Fig. AVI.1 Essential features of General Motors' GPU-3 Stirling engine. (Based on drawings and data in Ref. 3.)

| | |
|---|---|
| Compression-end connecting ducts | |
| length | 15.9 mm |
| duct inside diameter | 5.97 mm |
| no. of ducts per cylinder | 8 |
| cooler end cap | 279 mm$^3$ |

| | |
|---|---|
| Regenerators | |
| housing inside length | 22.6 mm |
| housing internal diameter | 22.6 mm |
| no. regenerators per cylinder | 8 |
| mesh material | stainless steel |
| mesh no. | 7.9 wires/mm (200 × 200 imperial mesh) |
| wire diameter | 0.04 mm |
| no. of layers | 308 |
| porosity | 70% |
| screen-to-screen rotation | 5° |

*Calculated dead volumes*

| | |
|---|---|
| Expansion space | |
| clearance around displacer | 3 340 $mm^3$ |
| above displacer | 7 410 $mm^3$ |
| heater tube terminations | 1 740 $mm^3$ |
| total | 12 490 $mm^3$ |
| Heater | |
| in insulated length next to expansion space | 9 680 $mm^3$ |
| in heated length | 47 460 $mm^3$ |
| in insulated length next to regenerator | 13 290 $mm^3$ |
| additional volume in four tubes for instrumentation | 2 740 $mm^3$ |
| header | 7 670 $mm^3$ |
| total | 80 840 $mm^3$ |
| Regenerator | |
| entrance to regenerators | 7 360 $mm^3$ |
| internal dead volume | 53 400 $mm^3$ |
| between regenerators and coolers | 2 590 $mm^3$ |
| circlip grooves | 2 180 $mm^3$ |
| total | 65 530 $mm^3$ |
| Cooler | |
| tube internal volume | 13 130 $mm^3$ |
| Compression space | |
| exit from cooler | 3 920 $mm^3$ |
| cooler end caps | 2 770 $mm^3$ |
| compression-end connecting ducts | 3 560 $mm^3$ |
| clearance around compression piston | 7 290 $mm^3$ |
| between displacer and compression piston | 1 140 $mm^3$ |
| connections to cooler end caps | 2 330 $mm^3$ |

| | |
|---|---|
| piston cut-outs | 60 mm$^3$ |
| between displacer and rod | 110 mm$^3$ |
| total | 21 180 mm$^3$ |
| **Total dead volume** | 193 170 mm$^3$ |
| Minimum live volume | 39 180 mm$^3$ |
| Minimum total working space volume | 232 350 mm$^3$ |
| Measured minimum total working working volume | 233 500 mm$^3$ |

*Drive mechanism* (*Fig.* 6.4)

| | |
|---|---|
| crank eccentricity, $r$ | 13.8 mm |
| connecting rod length, $l$ | 46.0 mm |
| *désaxé* offset, $e$ (i.e., $r < e$) | 20.8 mm |
| displacer rod diameter | 9.52 mm |
| piston rod diameter | 22.2 mm |
| displacer shell diameter | 69.6 mm |
| displacer wall thickness | 1.59 mm |
| linear expansion space clearance | 1.63 mm |
| linear compression space clearance | 0.3 mm |
| max. buffer space volume | 521 000 mm$^3$ |
| min. working space volume | 232 350 mm$^3$ |

## AVI.2 Philips MP1002CA air engine

(Source: direct measurement by the author. The machine is illustrated at Fig. AVI.2.)

*General*

| | |
|---|---|
| Configuration | Single-cylinder, coaxial, uniform-diameter bore. Bell-crank drive mechanism |
| Rated working fluid | Air |
| Rated output | 200W at output of belt-driven generator. The engine drives its own cooling fan. Manufacturer's rated gross shaft output not stated |
| Rated minimum cycle pressure | 15 atm (14 atm gauge) |
| Bore | 55.0 mm |
| Stroke | |
| power piston | 27.0 mm |
| displacer | 26.0 mm |
| Expansion-end temperature | 873 K |
| Compression-end temperature | 333 K |
| Heating | Combustion of liquid hydrocarbon fuel |
| Cooling | Air – forced convection from fan driven at generator shaft speed |

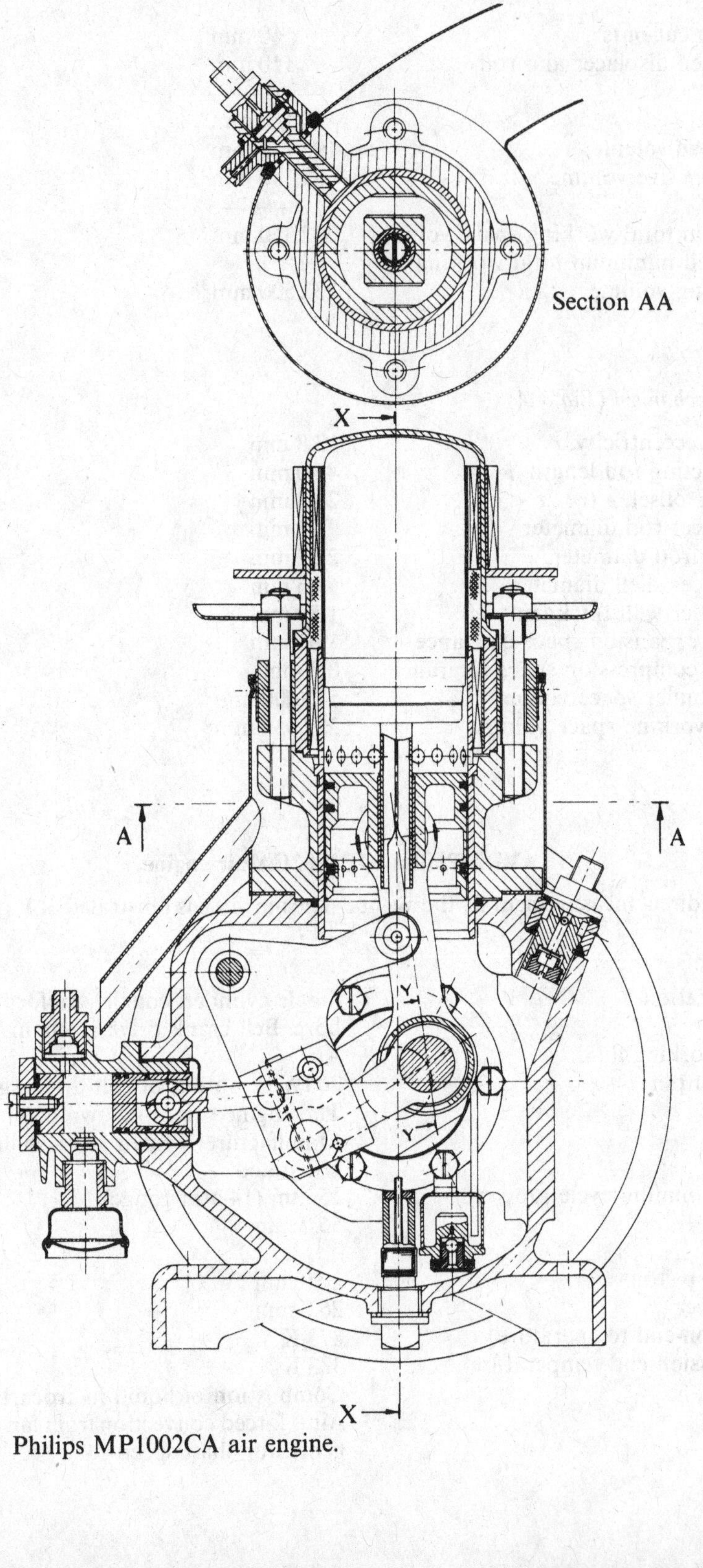


Fig. AVI.2 Philips MP1002CA air engine.

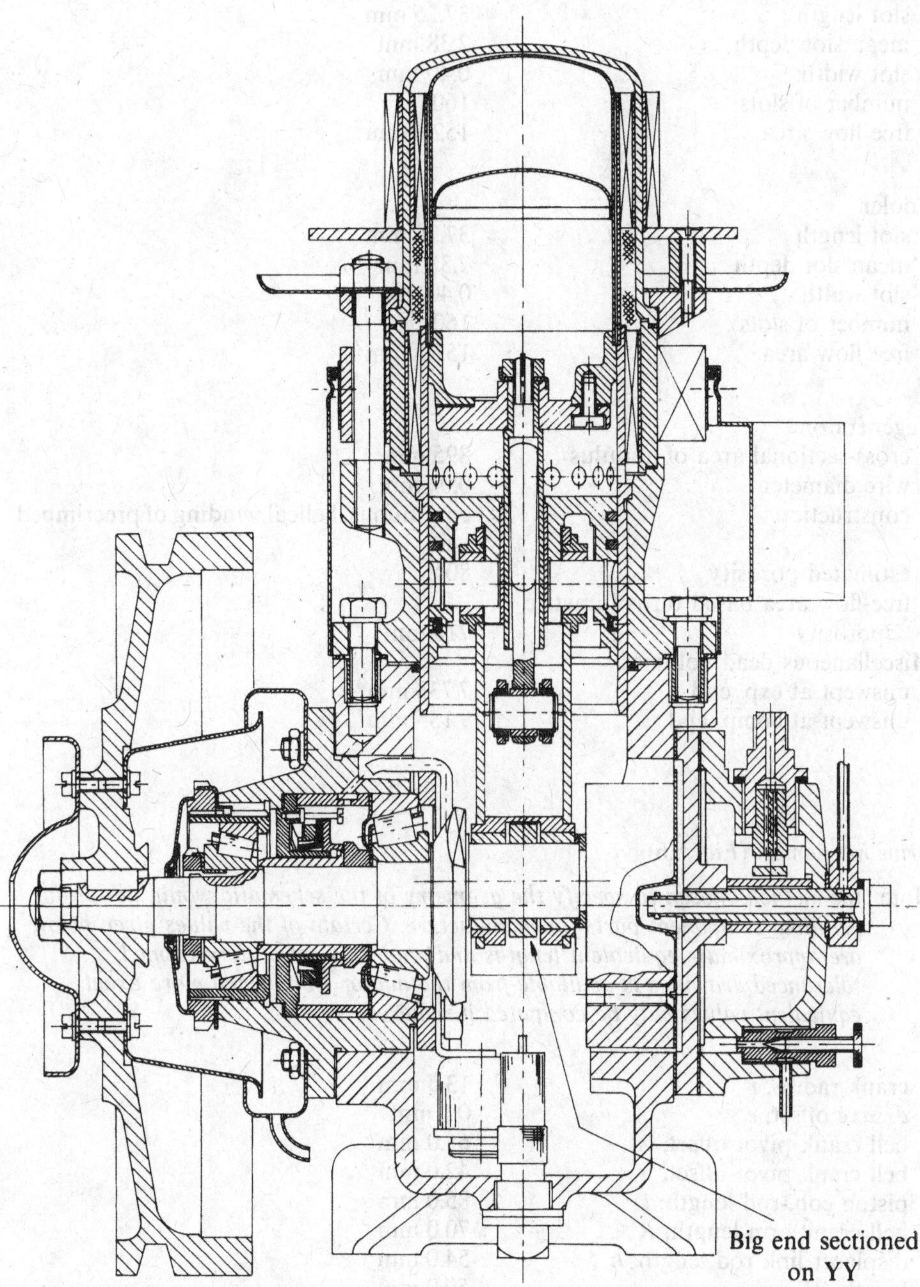


Fig. AVI.2 *Continued.*

*Working fluid circuit dimensions*

| | |
|---|---|
| Heater | |
| slot length | 37.25 mm |
| mean slot depth | 2.38 mm |
| slot width | 0.40 mm |
| number of slots | 160 |
| free flow area | 152.6 mm$^2$ |
| Cooler | |
| slot length | 37.25 mm |
| mean slot depth | 2.38 mm |
| slot width | 0.40 mm |
| number of slots | 160 |
| free flow area | 152.6 mm$^2$ |
| Regenerator | |
| cross-sectional area of annulus | 895 mm$^2$ |
| wire diameter | 0.04 mm |
| construction | continuous, helical winding of precrimped wire |
| estimated porosity | 80% |
| free-flow area based on volumetric porosity | 716 mm$^2$ |
| Miscellaneous dead volumes | |
| unswept at exp. end | 7 733 mm$^3$ |
| unswept at comp. end | 9 135 mm$^3$ |

*Drive mechanism (Fig. 6.6)*

Note: *the method chosen to specify the geometry of the schematic would not do for detailing component parts for manufacture. Certain of the values given below are approximate equivalent lengths and angles. A set of dimensioned, toleranced drawings is available from the author from which more exact equivalent values may be computed if required.*

| | |
|---|---|
| crank radius, $r$ | 13.5 mm |
| *désaxé* offset, $e$ | 0.0 mm |
| bell crank pivot offset, $X$ | 65.0 mm |
| bell crank pivot offset, $Y$ | 42.0 mm |
| piston con.-rod length, $l$ | 85.0 mm |
| bell crank arm length, $R$ | 70.0 mm |
| displacer link rod length, $h$ | 54.0 mm |
| length $QE$ | 50.0 mm |
| fixed angle $\chi$ | 75° |
| $Y_e$ | 80.0 mm |
| $Y_c$ | 24.0 mm |
| displacer rod diameter | 12.0 mm |

## AVI.3 SM1 research machine

(Source: Original design data and Ref. AVI.2. The machine is illustrated at Fig. AVI.3.)

*General*

| | |
|---|---|
| Configuration | Twin inverted parallel-cylinder Ross crank mechanism |
| Working fluid(s) | Air, He, Ne, Ar, $CO_2$ |
| Operating conditions at design life of 100 hours | 2 000 rpm at 70 bar max. Heater tube temperature 932 K |
| Bore (both) | 54.1 mm |
| Strokes | |
| expansion piston | 25.4 mm |
| compression piston | 25.9 mm |

*Working fluid circuit*

Note: *The machine was designed expressly for investigation of pressure wave effects. Expansion and compression exchanger tube lengths reflect the fact, and are in excess of values which would have been chosen if a high value of specific cycle work had been a priority.*

| | |
|---|---|
| Expansion-end exchanger | |
| mean effective tube length | 471 mm |
| internal diameter | 3.64 mm |
| external diameter | 4.74 mm |
| number of tubes | 16 |
| Compression-end exchanger | |
| mean effective tube length | 471 mm |
| internal diameter | 3.64 mm |
| external diameter | 4.74 mm |
| number of tubes | 16 |
| Regenerator | |
| housing internal diameter | 53.2 mm |
| no. of regenerators per cylinder set | 1 |
| mesh material | stainless steel |
| mesh no. | 7.9 wires/mm ('200 mesh') |
| wire diameter | 0.05 mm |
| no. of discs in stack | 330 |

*Calculated dead volumes*

| | |
|---|---|
| unswept in expansion space | 22 750 $mm^3$ |
| expansion side of regenerator | 31 490 $mm^3$ |
| comp. side of regenerator | 82 980 $mm^3$ |
| unswept in comp. space | 10 110 $mm^3$ |

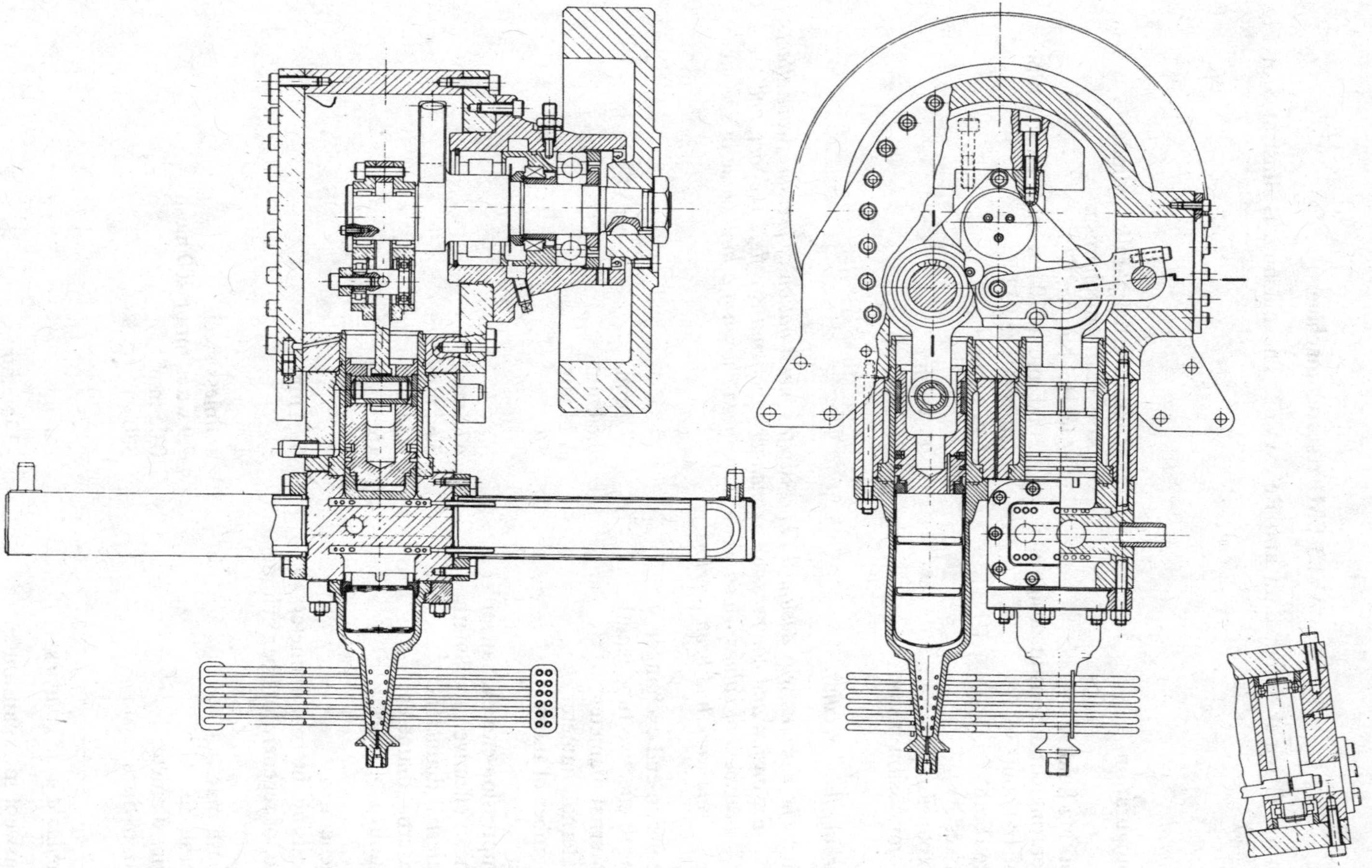

Fig. AVI.3 SM1 research machine.

*Drive mechanism* (*Fig.* 6.5)

| | |
|---|---|
| crank radius, $r$ | 9.0 mm |
| $b_e$, $b_c$ | 50.0 mm |
| $h$ | 50.0 mm |
| swinging link centre distance, $R$ | 112.0 mm |
| connecting rod centres | 85.0 mm |
| pivot offset, $X$ | 112.0 mm |
| pivot offset, $Y$ | 50.0 mm |

# Appendix VII

## Entropy creation rate

### AVII.1 Entropy creation

The treatment which follows is adapted from the work of Bejan.[1] The adaptation originally appeared in Ref. 2.

The Second Law of thermodynamics may be written:

$$\sum_{\text{in}} \mathrm{d}ms - \sum_{\text{out}} \mathrm{d}ms + \frac{\mathrm{d}Q}{T} \leqslant \mathrm{d}S$$

The amount by which d$S$ exceeds the algebraic sum of the terms on the left-hand side measures the degree of irreversibility accompanying the transfer process. The excess rate of entropy generation over that involved in the reversible process may be called the *entropy generation rate*, $S'$. This is not a thermodynamic property. $S'$ may be expressed:

$$S' = \frac{\partial s}{\partial t} - \frac{Q'}{T} + \sum_{\text{out}} m's - \sum_{\text{in}} m's \tag{AVII.1}$$

In Fig. 7.1 the elemental length of duct, d$x$, has cross-sectional area $A_x$ and hydraulic radius $r_h$. Fluid flows at local, instantaneous rate $m'$ kg/s. There is heat transfer between wall and fluid at instantaneous local rate $q'$ d$x$.

### AVII.2 Application to flow with friction and heat transfer in a duct

Applying Eqn (AVII.1) to the control volume of Fig. 7.1 gives:

$$S' = A_x\left(s + \frac{\partial s}{\partial x}\mathrm{d}x\right)\left(u + \frac{\partial u}{\partial x}\mathrm{d}x\right)\left(\rho + \frac{\partial \rho}{\partial x}\mathrm{d}x\right) - A_x su\rho + \frac{\partial}{\partial t}(\rho s)A_x\,\mathrm{d}x - \frac{q'}{T+\Delta T}\mathrm{d}x$$

For the one-dimensional case under consideration the mass conservation law is

$$\frac{\partial \rho}{\partial t} + \frac{\partial}{\partial x}\rho u = 0$$

Expanding the preceding expression and taking account of mass conservation leads to

$$\frac{S'}{A_x\,\mathrm{d}x} = \rho\frac{\mathrm{D}s}{\mathrm{d}t} - \frac{q'}{T+\Delta T}\frac{1}{A_x} \tag{AVII.1a}$$

The energy equation (First Law) for the one-dimensional, unsteady flow case is

$$\frac{\mathrm{D}}{\mathrm{d}t}(c_v T) = uF + \frac{q'}{\rho A_x} - \frac{p}{\rho}\frac{\partial u}{\partial x}$$

where $F$ is the friction parameter defined under Notation.

The energy equation may be rewritten:

$$\frac{\rho}{T}\frac{\mathrm{D}}{\mathrm{d}t}(c_v T) = \frac{q'}{TA_x} + \frac{\rho u F}{T} - \frac{p}{T}\frac{\partial u}{\partial x} \qquad \text{(AVII.2)}$$

The Second Law states that

$$T\,\mathrm{d}s = \mathrm{d}u + p\,\mathrm{d}(1/\rho)$$

The equivalent statement as a function of time is

$$\rho\frac{\mathrm{D}s}{\mathrm{d}t} = \frac{\rho}{T}\frac{\mathrm{D}u}{\mathrm{d}t} - \frac{p}{\rho T}\frac{\mathrm{D}\rho}{\mathrm{d}t}$$

The law of mass conservation is invoked to give:

$$\frac{\mathrm{D}\rho}{\mathrm{d}t} = \frac{\partial\rho}{\partial t} + u\frac{\partial\rho}{\partial x} = -\rho\frac{\partial u}{\partial x}$$

The Second Law equation becomes

$$\rho\frac{\mathrm{D}s}{\mathrm{d}t} = \frac{\rho}{T}\frac{\mathrm{D}u}{\mathrm{d}t} + \frac{p}{T}\frac{\partial u}{\partial x} \qquad \text{(AVII.3)}$$

Eqn (AVII.3) is substituted into Eqn (AVII.2) giving

$$\rho\frac{\mathrm{D}s}{\mathrm{d}t} = \frac{q'}{TA_x} + \frac{\rho u F}{T}$$

The left-hand term is replaced by the corresponding term from Eqn (AVII.1) giving, on the assumption that $\Delta T \ll T$

$$S' = \left(q'\frac{\Delta T}{T^2} + \rho u\frac{FA_x}{T}\right)\mathrm{d}x$$

# Appendix VIII

## The adiabatic cycle analysis

### AVIII.1 The variable-volume working space

Fig. AVIII.1 is a schematic of a control volume. Mass entering or leaving does so via the single port shown. Provisionally this restricts applicability to machines having just two variable-volume spaces. In fact, the treatment may be extended to devices having three variable volumes, as is demonstrated shortly.

It is assumed that, when an element of mass, $\mathrm{d}m$, enters, it has the temperature, $T_w$, of the wall. When $\mathrm{d}m$ leaves it has the bulk temperature of the contents. In what follows, the independent variable is crank angle, $\phi$.

The First Law states:

$$\mathrm{d}H - p\,\mathrm{d}V = \mathrm{d}U \quad (\mathrm{d}Q = 0)$$

The inflow case is described:

$$\mathrm{d}mc_pT_w - p\,\mathrm{d}V = c_v(T\,\mathrm{d}m + m\,\mathrm{d}T)$$

From the definition of $R$ in terms of $c_v$ and $\gamma$:

$$RT_w\,\mathrm{d}m = \frac{V}{\gamma}\mathrm{d}p + p\,\mathrm{d}V \qquad \text{(inflow)} \qquad \text{(AVIII.1a)}$$

$$RT\,\mathrm{d}m = \frac{V}{\gamma}\mathrm{d}p + p\,\mathrm{d}V \qquad \text{(outflow)} \qquad \text{(AVIII.1b)}$$

Eqns (AVIII.1a) and (AVIII.1b) differ only in the second symbol of the left-hand sides and may be combined by defining a step function, $U$

$$U(\mathrm{d}m) = \begin{cases} 1 & \text{for } \mathrm{d}m \geqslant 0 \quad \text{(inflow)} \\ 0 & \text{otherwise} \quad \text{(outflow)} \end{cases}$$

Then

$$R\,\mathrm{d}m\{T[1 - U(\mathrm{d}m)] + U(\mathrm{d}m)T_w\} = \frac{V}{\gamma}\mathrm{d}p + p\,\mathrm{d}V \qquad \text{(AVIII.2)}$$

Replacing $\gamma$ by unity and setting $T_w = T$ gives the well-known 'isothermal' case:

$$RT\,\mathrm{d}m = V\,\mathrm{d}p + p\,\mathrm{d}V \qquad \text{(AVIII.2a)}$$

The number of cycle analyses which have made use of Eqns (AVIII.1a) and (AVIII.1b) is too large to list. Finkelstein[1] appears to have been alone in

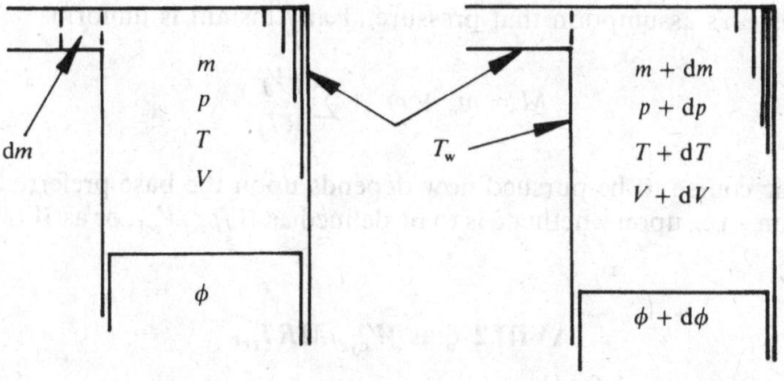


Fig. AVIII.1 Control volume for 'adiabatic' analysis.

exploiting the more compact form involving the step function. His pioneer work was carried out in the early days of electronic computing, when storage was at a premium, and when it therefore made sense to carry out the maximum of algebraic reduction and elimination before coding.

Finkelstein followed this policy and, in one of his several algebraic *tours de force* reduced an algebraic description of a complete machine to a pair of differential equations in two unknowns, $m_e$ and $m_c$, namely, instantaneous expansion- and compression-space mass content. He then solved these numerically with the aid of an electromechanical desk computer. Nowadays one would code separately the working space expressions and those for the heat exchangers. This leaves the reduction and elimination to take place numerically, and allows unlimited flexibility in the analytical description of the heat exchanger gas processes.

The Finkelstein algebraic approach nevertheless offers a substantial attraction: it yields a comprehensive description (mass flow rates, heat transfer rates etc.) of a plausible cycle in a fraction of the time required for any comparable alternative. Thus it may be called an indefinite number of times in code which requires approximate values for more sophisticated routines, or as an integral part of a perturbation approach. Under the heading which follows the algebraic treatment is therefore pursued as far as the final, differential equation form. There are some departures from Finkelstein's original treatment:

the restriction of built-in simple-harmonic volume variations is removed. Aside from permitting incorporation of the volume variations produced by, say, the rhombic drive, the modification eliminates the numerical problem caused by singularities due to zero dead volume at the top dead-centre positions;

the resulting description of specific cycle work may now be in terms of a characteristic cycle pressure (e.g., $p_{min}$, $p_{max}$) as an optional alternative to being in terms of total working fluid mass, $M$;

the solution algorithm proposed for the eventual differential equations is explicit, avoiding the usual problems of simultaneous solution.

With $m_e$, $m_c$ and $m_d$ to represent the instantaneous mass content of expansion, compression and dead spaces respectively, total mass content, $M$ may be expressed:

$$M = m_e + m_c + m_d$$

With Finkelstein's assumption that pressure at any instant is uniform:

$$M = m_e + m_c + \sum \frac{pV_j}{RT_j} \tag{AVIII.3}$$

The algebraic course to be pursued now depends upon the base preferred for normalization – i.e., upon whether $\zeta$ is to be defined as $W/p_{ref}V_{ref}$, or as $W/MRT_{ref}$.

## AVIII.2 $\zeta$ as $W_{cyc}/MRT_{ref}$

Normalized pressure is defined

$$\psi = pV_{ref}/MRT_{ref}$$

Dividing Eqn (AVIII.3) through by $M$ gives

$$\sigma_e + \sigma_c + \frac{pV_{ref}}{MRT_{ref}} \sum \frac{\mu_j}{\tau_j} = 1 \tag{AVIII.4}$$

$$\psi = \frac{1 - \sigma_e - \sigma_c}{\nu} \tag{AVIII.5}$$

$$d\psi = \frac{-d\sigma_e - d\sigma_c}{\nu} \tag{AVIII.6}$$

Substituting Eqns (AVIII.5) and (AVIII.6) into Eqn (AVIII.2) yields, for a given working space:

$$\left\{\mu \frac{d\sigma}{\sigma}[1 - U(d\sigma)] - d\mu\right\}\left(\frac{1 - \sigma_e - \sigma_c}{\nu}\right) + U(d\sigma)\, d\sigma\tau_w + \frac{\mu}{\gamma}\left(\frac{d\sigma_e + d\sigma_c}{\nu}\right) = 0 \tag{AVIII.7}$$

There is one such expression for each of the two variable-volume spaces. These give a pair of simultaneous difference equations in $d\sigma_e$ and $d\sigma_c$, susceptible to solution by standard numerical techniques, and sufficient to determine $\sigma_e$ and $\sigma_c$ for all crank angles, $\phi$. Finkelstein applied a second-order Runge–Kutta scheme, which required alternately stepping forward from $\phi$ to $\phi + d\phi$ in one equation of the pair, substituting the intermediate new values into the other equation and repeating. On the other hand, it is possible to formulate a scheme which has second-order accuracy, which yields an explicit value for each of $\sigma_e$ and $\sigma_c$ at each time step, and which is thus more efficient of computing time.

Eqn (AVIII.7) for both expansion and compression spaces may now be expanded about $\phi = \phi_0$. This involves replacing $\sigma_e$ by $\sigma_e + \frac{1}{2}\, d\sigma_e$, with a comparable expression for $\sigma_c$. Carrying out the substitution and collecting terms (it is not necessary to try to make explicit the argument of $U$) results in two simultaneous, linear expressions in $d\sigma_e$ and $d\sigma_c$ having the form:

$$\left.\begin{aligned} a_{11}\, d\sigma_e + a_{12}\, d\sigma_c &= b_1 \\ a_{21}\, d\sigma_e + a_{22}\, d\sigma_c &= b_2 \end{aligned}\right\} \tag{AVIII.8}$$

The $d\sigma_e$, $d\sigma_c$ etc. are *differences* over a crank angle increment, $d\phi$; there is no need to work in terms of differential coefficients, $d/d\phi$. In Eqn (AVIII.8) the coefficients are:

$$a_{11} = \mu_e \frac{1 - U(d\sigma_e)}{\sigma_e}(1 - \sigma_e - \sigma_c) + U(d\sigma_e)\nu\tau_{we} + \tfrac{1}{2}\,d\mu_e + \frac{\mu_e}{\gamma}$$

$$a_{12} = \frac{\mu_e}{\gamma} + \tfrac{1}{2}\,d\mu_e$$

$$a_{21} = \frac{\mu_c}{\gamma} + \tfrac{1}{2}\,d\mu_c$$

$$a_{22} = \mu_c \frac{1 - U(d\sigma_c)}{\sigma_c}(1 - \sigma_e - \sigma_c) + U(d\sigma_c)\nu\tau_{wc} + \tfrac{1}{2}\,d\mu_c + \frac{\mu_c}{\gamma}$$

The process of algebraic expansion yields higher order (non-linear) terms in $d\sigma_e$ and $d\sigma_c$ which cannot be incorporated explicitly. For highest accuracy of solution, these terms are included with the 'known' right-hand sides (the $b_1$, $b_2$ – see below). Iterating each integration step at least once incorporates appropriate values into the non-linear components of the right-hand sides.

$$b_1 = d\mu_e(1 - \sigma_e - \sigma_c) + \text{terms of higher order}$$
$$b_2 = d\mu_c(1 - \sigma_e - \sigma_c) + \text{terms of higher order}$$

In the foregoing expressions $\sigma_e$ and $\sigma_c$ are the 'known' values at the start of the integration step. The $\mu_e$ and $\mu_c$ and $d\mu_e$ and $d\mu_c$ (known) are given their respective mid-interval ($\phi + \tfrac{1}{2}\,d\phi$) values. The solutions, i.e., the values of $\sigma_e$ and $\sigma_c$ at the end of the interval $d\phi$ are:

$$\sigma_{e\phi + d\phi} = \frac{b_1 a_{22} - b_2 a_{12}}{a_{11}a_{22} - a_{21}a_{12}} + \sigma_{e\phi} \qquad \text{(AVIII.9a)}$$

$$\sigma_{c\phi + d\phi} = \frac{b_2 a_{11} - b_1 a_{21}}{a_{11}a_{22} - a_{21}a_{12}} + \sigma_{c\phi} \qquad \text{(AVIII.9b)}$$

A comment on implementing the solution follows the next section.

## AVIII.3 ζ as $W_{cyc}/p_{ref}V_{ref}$

If $\zeta$ defined as $W_{cyc}/p_{ref}V_{ref}$ is preferred over $\zeta$ defined as $W_{cyc}/MRT_{ref}$ there are two ways of arriving at the appropriate numerical value: in the first, numerical integration on Eqns (AVIII.9a) proceeds as already explained, and the maximum, minimum and/or mean value of $\psi$ is stored. Values of the former two quantities are obtained by simple inspection, that of the latter by noting:

$$\psi_{mean} = \frac{1}{2\pi}\int_0^{2\pi} \psi\, d\phi \qquad \text{(AVIII.10)}$$

Bringing the integration process to cyclic equilibrium involves repeated determination of $\psi_{ref}$ ($\psi_{max}$, $\psi_{min}$ or $\psi_{mean}$) but after the final iteration it will be possible to form the definitive ratio $\psi/\psi_{ref}$ for all the values of $\phi$ for which $\psi$ has been computed. From the definition of $\psi$, $\psi/\psi_{ref} = p/p_{ref}$. The required value of $\zeta$ thus follows immediately by carrying out numerically the integration:

$$\zeta = W_{cyc}/p_{ref}V_{ref} = \int_0^{2\pi} (\psi/\psi_{ref})(d\mu_e/d\phi + d\mu_c/d\phi)\,d\phi \qquad \text{(AVIII.11)}$$

The alternative method[2] applies when the normalizing reference volume is $V_{wfc}$ (working fluid circuit volume at reference conditions). For this particular case $p_{ref}V_{ref}/(MRT_{ref}) = p_{ref}V_{wfc}/(MRT_{ref}) = 1$ by definition. Eqn (AVIII.4) still applies:

$$\sigma_e + \sigma_c + \psi\nu = 1 \qquad \text{(AVIII.4a)}$$

It therefore applies in particular at the crank angle, $\phi^*$ at which $p = p_{ref}$ and at which $\psi$ has the value of unity:

$$\sigma_e{}^* + \sigma_c{}^* + \nu = 1 \qquad \text{(AVIII.4b)}$$

Dividing Eqn (AVIII.4a) by (AVIII.4b) gives

$$\psi = 1 - \frac{\sigma_e - \sigma_e{}^* + \sigma_c - \sigma_c{}^*}{\nu} \qquad \text{(AVIII.5a)}$$

$d\psi$ is defined by Eqn (AVIII.6) as before. Substituting Eqns (AVIII.5a) and (AVIII.6) into Eqn (AVIII.2) yields, for a given working space:

$$\left\{\mu\frac{d\sigma}{\sigma}[1 - U(d\sigma)] - d\mu\right\}\left(1 - \frac{\sigma_e - \sigma_e{}^* + \sigma_c - \sigma_c{}^*}{\nu}\right)$$

$$+ U(d\sigma)\,d\sigma\tau_w + \frac{\mu}{\gamma}\left(\frac{d\sigma_e - d\sigma_c}{\nu}\right) = 0 \qquad \text{(AVIII.7a)}$$

Again, there is one such expression for each of the two variable-volume spaces. Expanding these as before about $\phi = \phi_0$ leads to two simultaneous difference equations of the form of Eqns (AVIII.8), in which the coefficients are now:

$$a_{11} = \mu_e\frac{1 - U(d\sigma_e)}{\sigma_e}[\nu - (\sigma_e - \sigma_e{}^* + \sigma_c - \sigma_c{}^*)] + U(d\sigma_e)\nu\tau_{we} + \tfrac{1}{2}\,d\mu_e + \frac{\mu_e}{\gamma}$$

$a_{12}$ and $a_{21}$ are as before.

$$a_{22} = \mu_c\frac{1 - U(d\sigma_c)}{\sigma_c}[\nu - (\sigma_e - \sigma_e{}^* + \sigma_c - \sigma_c{}^*)] + U(d\sigma_c)\nu\tau_{wc} + \tfrac{1}{2}\,d\mu_c + \frac{\mu_c}{\gamma}$$

$$b_1 = d\mu_e[\nu - (\sigma_e - \sigma_e{}^* + \sigma_c - \sigma_c{}^*)]$$

$$b_2 = d\mu_c[\nu - (\sigma_e - \sigma_e{}^* + \sigma_c - \sigma_c{}^*)]$$

The same integration process as that already described may be used. There is no way of knowing in advance the values of $\phi^*$ and corresponding $\mu_e{}^*$, $\mu_c{}^*$, $\sigma_e{}^*$ etc., so values are initially guessed. During the course of a cycle, the values of $\mu_e{}^*$, $\mu_c{}^*$,

$\sigma_e^*$ etc. at which $p$ attains $p_{ref}$ are noted and stored for a second cycle. The process repeats until cyclic stability is achieved. An indication of correct programming may be had by running with $\gamma$ replaced by 1 and checking that $\tau_e$ $(=\psi\mu_e/\sigma_e) \approx N_\tau$ and that $\tau_c$ $(=\psi\mu_c/\sigma_c) \approx 1$ for all $\phi$, and that the $\psi(\phi)$ are those given by the corresponding 'isothermal' analysis. The inspection is most convincing if the higher order terms (see p. 337) are taken into account in evaluating $b_1$ and $b_2$.

## AVIII.4 Implementation

The cyclic nature of the gas processes requires that, eventually, $\sigma_{e_{\phi=2(n+1)\pi}} = \sigma_{e_{\phi=2n\pi}}$ simultaneously with $\sigma_{c_{\phi=2(n+1)\pi}} = \sigma_{c_{\phi=2n\pi}}$. The process of step-by-step integration must be continued until this condition is satisfied. In Finkelstein's original treatment, cyclic steady state was reached after both pistons had passed their respective top dead-centre positions. He attributed the rapid achievement of steady state to the fact that the pistons swept the entire cylinder (i.e., there was no unswept volume in the live spaces) and to the fact that the effects of the arbitrary starting conditions had therefore been eliminated after the contents of both cylinders has been completely expelled once.

The two-fold price for the rapid convergence is the need for supplementary algebraic work of series expansion to prevent numerical singularity at top dead-centre, coupled with the fact that zero unswept volume is, in any case, unrepresentative of reality. The more realistic case of finite swept volume results in relatively slow convergence to steady state.

If computation is in terms of a $\psi$ defined by reference to $MRT_{ref}$ and $V_{ref}$, then forming the integrals $\int \psi \, d\mu_e + \int \psi \, d\mu_c$ immediately gives specific cycle work $\zeta$, where $\zeta = W/MRT_{ref}$.

$$\zeta = \zeta[N_\tau, \kappa, \alpha, \nu(\tau), \gamma]$$

In view of the definition of $\nu(\tau)$ this may be written:

$$\zeta = \zeta\begin{pmatrix} \mu_{dce} \\ \mu_{de} \\ N_\tau, \kappa, \alpha, \mu_{dr}, \gamma \\ \mu_{dc} \\ \mu_{dcc} \end{pmatrix} \qquad \text{(AVIII.12a)}$$

In Eqn (AVIII.12a) the arguments are seen to be a subset of those in Eqn (4.9). The alternative $\zeta$ $(=W/p_{ref}V_{sw}, V_{sw} \neq V_{ref})$ may be formed by taking account of the fact that the integrals $\int \psi \, d\mu_e + \int \psi \, d\mu_c$ for the case of $\psi/\psi_{ref} = p/p_{ref}$, $\mu = V/V_{ref}$ yield $W/p_{ref}V_{ref}$ rather than $W/p_{ref}V_{sw}$. This is overcome either by using $V_{sw}$ as $V_{ref}$ throughout, or merely by retaining $V_{ref}$ as the normalizing quantity and adjusting the final value of $\zeta$ in the ratio $V_{ref}/V_{sw}$.

Either way, $\zeta$ for this case becomes:

$$\zeta = \zeta\begin{pmatrix} \mu_{dce} \\ \mu_{de} \\ N_\tau, \kappa, \alpha, V_{ref}/V_{sw}, \mu_{dr}, \gamma \\ \mu_{dc} \\ \mu_{dcc} \end{pmatrix} \qquad \text{(AVIII.12b)}$$

which is the same as Eqn (AVIII.12a) but for inclusion of the additional dimensionless group $V_{ref}/V_{sw}$.

## AVIII.5 The three-space machine

The adiabatic analysis is capable of extension to the case of machines having the compression space divided by a heat exchanger, i.e., to the so-called three-space machine. Fig. AVIII.2 is a schematic, indicating an intermediate variable-volume space with properties identified by subscript *i*. Miniature, reversed cycle coolers frequently have this configuration, but the author is aware of only one previously published treatment, namely, that of Halpern and Shtrikman.[3] The authors claim to have generalized Finkelstein's analysis to cover systems comprising an arbitrary number, *n*, of intermediate spaces. The presentation of their algebra is extremely terse, but gives the impression that an equation of the type Eqn (AVIII.2) is applied to each of the *n* spaces, giving a set of *n* non-linear equations to be solved simultaneously for the *n* values of d*m*. If this is a correct interpretation, then the approach is flawed: as Fig. AVIII.2(*a*) shows, all variable volumes intermediate to the two end spaces are subject to *two* possibilities for inflow and outflow, one from the right, as drawn, one from the left. Flow may be inward from one port (and therefore at $T_w$) while being outward from the other (and therefore at $T$). A three-space machine thus gives rise to *four* values of d*m*, a four-space machine to *six* etc. That the results of Halpern and Shtrikman should agree with those of Rule and Qvale[4] when $\gamma$ is replaced by unity does not amount to substantiation, since for this unique case (equivalent to 'isothermal' operation) the two, separate d*m* of the intermediate space enter and leave at common, constant temperature, and may be treated (exceptionally) as one. For this case only, the *n* space machine may be modelled in terms of the proposed *n* simultaneous equations – i.e., as per the claim for the general case. Halpern and Shtrikman report satisfactory computation with 24 steps per cycle (i.e., with intervals of 15°). The interval is exceptionally coarse, and adds to misgivings about the account.

Referring to the three-space machine of Fig. AVIII.2(*a*), the energy equation for the extreme left- and right-hand spaces is Eqn (AVIII.2), as before. However, the statement of total mass must now be

$$\sigma_e + \sigma_i + \sigma_c + \frac{pV_{ref}}{MRT_{ref}} \sum \frac{\mu_j}{\tau_j} = 1 \qquad \text{(AVIII.13)}$$

$$\psi = \frac{1 - \sigma_e - \sigma_i - \sigma_c}{\nu} \qquad \text{(AVIII.14)}$$

$$\mathrm{d}\psi = \frac{-\mathrm{d}\sigma_e - \mathrm{d}\sigma_i - \mathrm{d}\sigma_c}{\nu} \qquad \text{(AVIII.15)}$$

In Eqns (AVIII.14) and (AVIII.15), $\sigma_i = \sigma_{ia} + \sigma_{ib}$, $\mathrm{d}\sigma_i = \mathrm{d}\sigma_{ia} + \mathrm{d}\sigma_{ib}$.

Temporarily dropping the subscript i, the energy balance for the intermediate space is:

$$\mathrm{d}\sigma_a\{\tau[1 - U(\mathrm{d}\sigma_a)] + U(\mathrm{d}\sigma_a)\tau_w\} + \mathrm{d}\sigma_b\{\tau[1 - U(\mathrm{d}\sigma_b)] + U(\mathrm{d}\sigma_b)\tau_w\} = \frac{\mu}{\gamma}\mathrm{d}\psi + \psi\,\mathrm{d}\mu \qquad \text{(AVIII.16)}$$

Substituting from the gas law $\tau = \psi\mu/\sigma$ as for the treatment of the two-space machine, expanding the defining equations, including Eqn (AVIII.16), to $\phi + \mathrm{d}\phi$ and collecting terms as before gives four simultaneous relationships between the

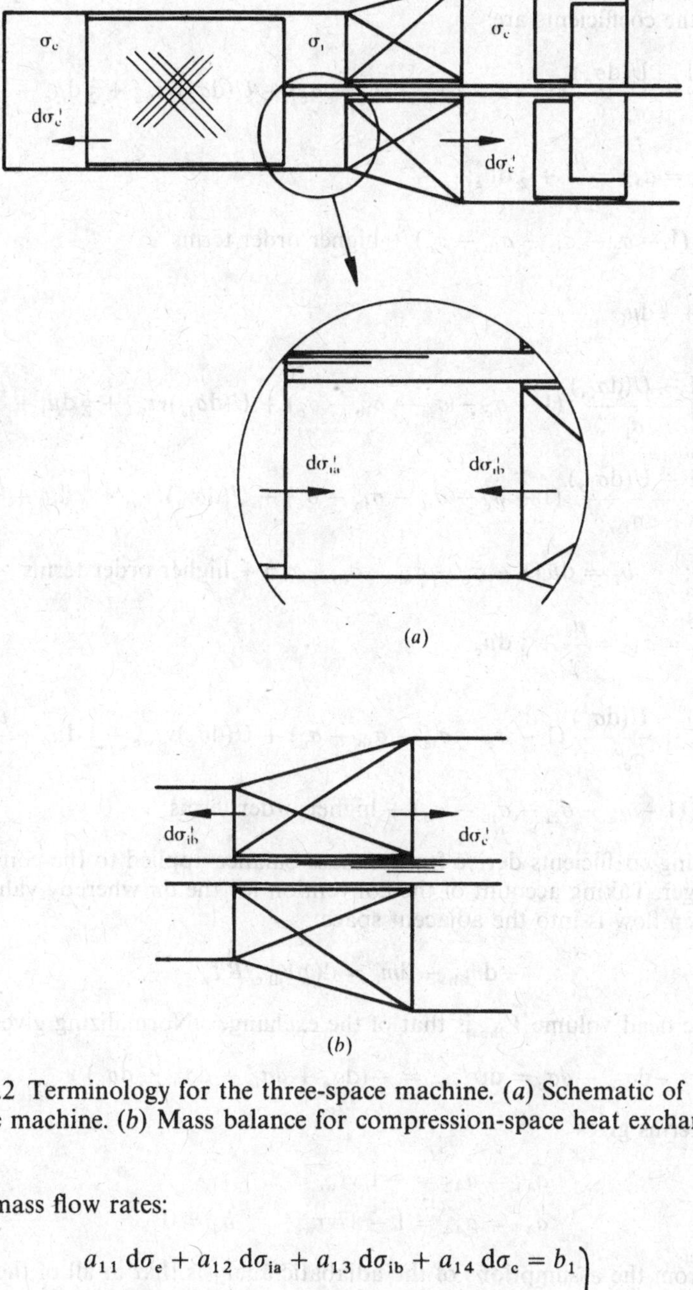


Fig. AVIII.2 Terminology for the three-space machine. (*a*) Schematic of three-space machine. (*b*) Mass balance for compression-space heat exchanger.

unknown mass flow rates:

$$\left.\begin{aligned} a_{11}\,\mathrm{d}\sigma_{\mathrm{e}} + a_{12}\,\mathrm{d}\sigma_{\mathrm{ia}} + a_{13}\,\mathrm{d}\sigma_{\mathrm{ib}} + a_{14}\,\mathrm{d}\sigma_{\mathrm{c}} &= b_1 \\ a_{21}\,\mathrm{d}\sigma_{\mathrm{e}} + a_{22}\,\mathrm{d}\sigma_{\mathrm{ia}} + a_{23}\,\mathrm{d}\sigma_{\mathrm{ib}} + a_{24}\,\mathrm{d}\sigma_{\mathrm{c}} &= b_2 \\ a_{31}\,\mathrm{d}\sigma_{\mathrm{e}} + a_{32}\,\mathrm{d}\sigma_{\mathrm{ia}} + a_{33}\,\mathrm{d}\sigma_{\mathrm{ib}} + a_{34}\,\mathrm{d}\sigma_{\mathrm{c}} &= b_3 \\ a_{41}\,\mathrm{d}\sigma_{\mathrm{e}} + a_{42}\,\mathrm{d}\sigma_{\mathrm{ia}} + a_{43}\,\mathrm{d}\sigma_{\mathrm{ib}} + a_{44}\,\mathrm{d}\sigma_{\mathrm{c}} &= b_4 \end{aligned}\right\} \qquad \text{(AVIII.17)}$$

As before, the $\mathrm{d}\sigma_{\mathrm{e}}$, $\mathrm{d}\sigma_{\mathrm{ia}}$, etc. are *differences* over a crank angle increment, $\mathrm{d}\phi$;

there is no need to work in terms of differential coefficients, $d/d\phi$. In Eqn (AVIII.17) the coefficients are:

$$a_{11} = \mu_e \frac{1 - U(d\sigma_e)}{\sigma_e}(1 - \sigma_e - \sigma_{ia} - \sigma_{ib} - \sigma_c) + U(d\sigma_e)\nu\tau_{we} + \tfrac{1}{2}\,d\mu_e + \frac{\mu_e}{\gamma}$$

$$a_{12} = a_{13} = a_{14} = \frac{\mu_e}{\gamma} + \tfrac{1}{2}\,d\mu_e;$$

$$b_1 = d\mu_e(1 - \sigma_e - \sigma_{ia} - \sigma_{ib} - \sigma_c) + \text{higher order terms}$$

$$a_{21} = \frac{\mu_i}{\gamma} + \tfrac{1}{2}\,d\mu_i$$

$$a_{22} = \mu_i \frac{1 - U(d\sigma_{ia})}{\sigma_i}(1 - \sigma_e - \sigma_{ia} - \sigma_{ib} - \sigma_c) + U(d\sigma_{ia})\nu\tau_{wi} + \tfrac{1}{2}\,d\mu_i + \frac{\mu_i}{\gamma}$$

$$a_{23} = \mu_i \frac{1 - U(d\sigma_{ib})}{\sigma_i}(1 - \sigma_e - \sigma_{ia} - \sigma_{ib} - \sigma_c) + U(d\sigma_{ib})\nu\tau_{wi} + \tfrac{1}{2}\,d\mu_i + \frac{\mu_i}{\gamma}$$

$$a_{24} = a_{21}; \qquad b_2 = d\mu_i(1 - \sigma_e - \sigma_{ia} - \sigma_{ib} - \sigma_c) + \text{higher order terms}$$

$$a_{31} = a_{32} = a_{33} = \frac{\mu_c}{\gamma} + \tfrac{1}{2}\,d\mu_c$$

$$a_{34} = \mu_c \frac{1 - U(d\sigma_c)}{\sigma_c}(1 - \sigma_e - \sigma_{ia} - \sigma_{ib} - \sigma_c) + U(d\sigma_c)\nu\tau_{wc} + \tfrac{1}{2}\,d\mu_c + \frac{\mu_c}{\gamma}$$

$$b_3 = d\mu_c(1 - \sigma_e - \sigma_{ia} - \sigma_{ib} - \sigma_c) + \text{higher order terms}$$

The remaining coefficients derive from a mass balance applied to the compression-end exchanger. Taking account of the convention for the $d\sigma$ whereby values are positive when flow is into the adjacent space:

$$-dm_{ib} - dm_c = d(pV_{dhc}/RT_w)$$

in which the dead volume $V_{dhc}$ is that of the exchanger. Normalizing gives

$$-d\sigma_{ib} - d\sigma_c = d\psi/\tau_{wc} = -(d\sigma_e + d\sigma_{ia} + d\sigma_{ib} + d\sigma_c)/\nu\tau_{wc}$$

Collecting terms gives

$$a_{41} = a_{42} = -1/\nu\tau_{wc}$$

$$a_{43} = a_{44} = 1 - 1/\nu\tau_{wc}; \qquad b_4 = 0$$

It follows from the assumptions of the adiabatic analysis that in all of the foregoing expressions, $\tau_{we} = N_\tau$ and that $\tau_{wi} = \tau_{wc} = 1$.

Since the solution matrix has a known, maximum size of 4 × 4 it may well be worthwhile to multiply it out by the rules for determinants, thereby obtaining explicit expressions for the unknowns. If the prospect of the algebraic work does not appeal, then a simple matrix inversion routine such as SIMQ or SIMQX used in single precision will be found suitable for solving the equation set.

# Appendix IX

## Mass rate and heat transfer in frictionless flow

### AIX.1 Assumptions

It is assumed that:

pressure is variable, but that there are no pressure gradients, so that at any instant, pressure is uniform throughout the system of heat exchangers and variable-volume spaces;

the working fluid behaves as an ideal gas;

local fluid temperature is always equal to that of the immediately adjacent solid wall;

an expression is available in function of crank angle for mass flow rate into and out of one of the variable-volume working spaces.

The treatment is that of Ref. 1.

### AIX.2 Mass flow rate, $m'$

From Fig. 7.2

$$m_{j+1}' = m_j' - \partial m/\partial t$$
$$= m_j' - \frac{1}{RT_{\mathrm{w}_{j+1/2}}}\left(A_{x_j}\,\mathrm{d}x\,\frac{\mathrm{d}p}{\mathrm{d}t}\right) \qquad \text{(AIX.1)}$$

According to Eqn (AIX.1), $m_{j+1}'$ may be calculated if $m_j'$ is known. The latter quantity depends in turn on $m_{j-1}'$.

In the standard schematic representation of a Stirling cycle machine the left-most control volume is the expansion space. Denoting by $-m_\mathrm{e}'$ mass flow rate *out* of this extreme left-hand control volume:

$$m_{j+1}' = -m_\mathrm{e}' - \frac{1}{R}\sum_{2}^{n}\frac{1}{T_{\mathrm{w}_{j+1}}}A_j\,\mathrm{d}x\,\frac{\mathrm{d}p}{\mathrm{d}t} \qquad \text{for } n = j+1 \qquad \text{(AIX.2)}$$

### AIX.3 Heat transfer rate per unit length, $q_\mathrm{L}'$

Referring again to Fig. 7.2 of the main text and invoking the First Law:

$$q_\mathrm{L}'\,\mathrm{d}x = \frac{\partial}{\partial t}(mc_\mathrm{v}T) + \frac{\partial}{\partial x}(m'c_\mathrm{p}T_\mathrm{w})\,\mathrm{d}x \qquad \text{(AIX.3)}$$

From mass continuity

$$\frac{\partial m}{\partial t} = -\frac{\partial}{\partial x} m' \, \mathrm{d}x \qquad \text{(AIX.4)}$$

Substituting Eqn (AIX.4) into (AIX.3) and making the replacement $c_p - c_v = R$:

$$q_L' = -A_x \frac{\mathrm{d}p}{\mathrm{d}t} + R \frac{\gamma}{\gamma - 1} m' \frac{\mathrm{d}T_w}{\mathrm{d}x} \qquad \text{(AIX.5)}$$

(The negative sign omitted from the corresponding equation in the source[1] has been reinstated.) The finite-difference form of Eqn (AIX.5) required for computational purposes follows immediately.

When the control volume connects directly to a variable-volume space (as in Fig. 7.2) the standard assumption is that inflow is at the instantaneous temperature, $T$, of the contents of the variable-volume space. Outflow is at local wall temperature. In the terminology of the 'adiabatic' cycle analysis (Appendix VIII) the extra heat rate per unit length to the fluid from the exchanger connected to the expansion end is:

$$c_p U(-m_e) m_e' \frac{T - T_w}{\mathrm{d}x}$$

Coding for numerical computation may be made more compact by incorporating this term into Eqn (AIX.5):

$$q_L' = -A_{x_j} \frac{\mathrm{d}p}{\mathrm{d}t} + R \frac{\gamma}{\gamma - 1} \left[ m' \left.\frac{\mathrm{d}T_w}{\mathrm{d}x}\right|_{j+1/2} - U(-m_e) m_e' \frac{T - T_w}{\mathrm{d}x} \right] \qquad \text{(AIX.6)}$$

(and not as misprinted in the appendix of Ref. 1.) The final term in Eqn (AIX.6) is evaluated only when the equation is applied to a control volume immediately adjoining a variable-volume space.

# Appendix X
## Selected design charts

### AX.1 Scope

This appendix contains a set of design charts to common format for selection of the phase angle and volume ratio which maximize the specific cycle work (or specific heat lifted) of ideal machines operating on the 'isothermal' and 'adiabatic' cycles.

All configurations – opposed-piston (alpha), parallel-bore coaxial (beta) and separate-displacer (gamma) – are treated in terms of their *equivalent* optimum volume ratio, $\kappa_0$, and volume phase angle $\alpha_0$. The result is the same as calculating in terms of $\lambda$ and $\beta$. To convert to the corresponding optima, $\lambda_0$ and $\beta_0$, simply apply Eqns (6.9) or (6.8) as appropriate.

In the case of machines of gamma configuration, the 'adiabatic' curves apply only to those variants which do not have a heat exchanger between the two variable-volume components of the compression space. (Such a heat exchanger makes for a three-space machine, which is not covered by the charts of this section.)

One or more curves of a minority of the charts (namely those of Figs AX.12a, AX.13b and AX.15b) have been subjected to arbitrary smoothing.

## AX.2 Charts based on Schmidt analysis

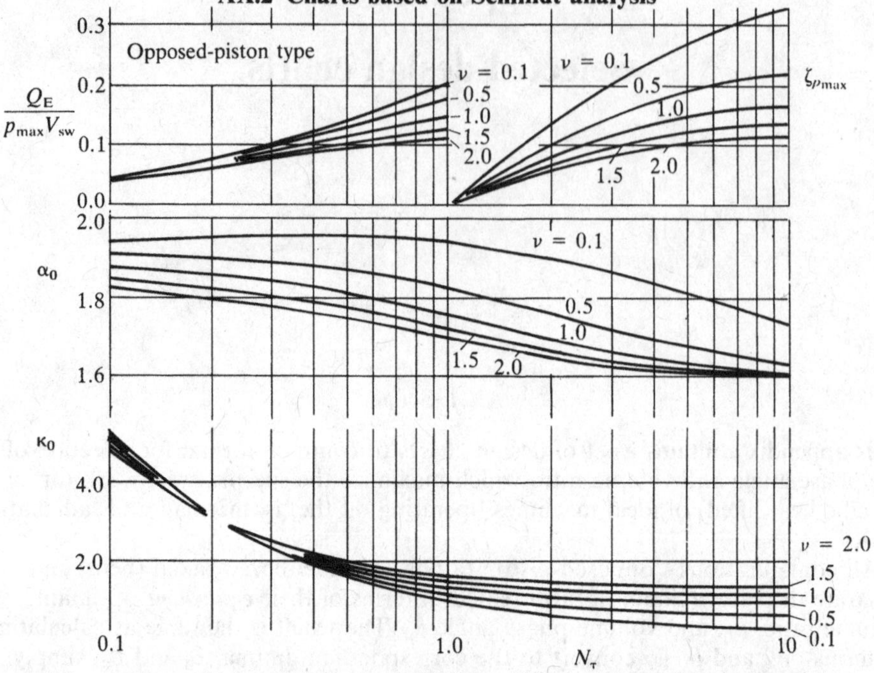


Fig. AX.1 $\kappa$ and $\alpha$ which maximize specific heat lifted and specific cycle work, $W_{cyc}/p_{max}V_{sw}$ of the opposed piston machine operating on the Schmidt cycle.

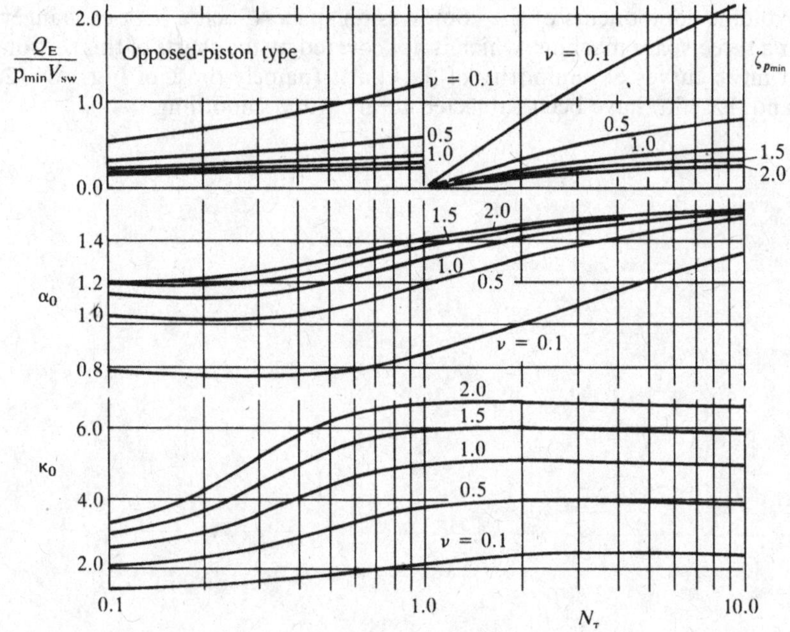


Fig. AX.2 As for Fig. AX.1, but for specific cycle work based on $p_{min}$, viz $\zeta = W_{cyc}/p_{min}V_{sw}$.

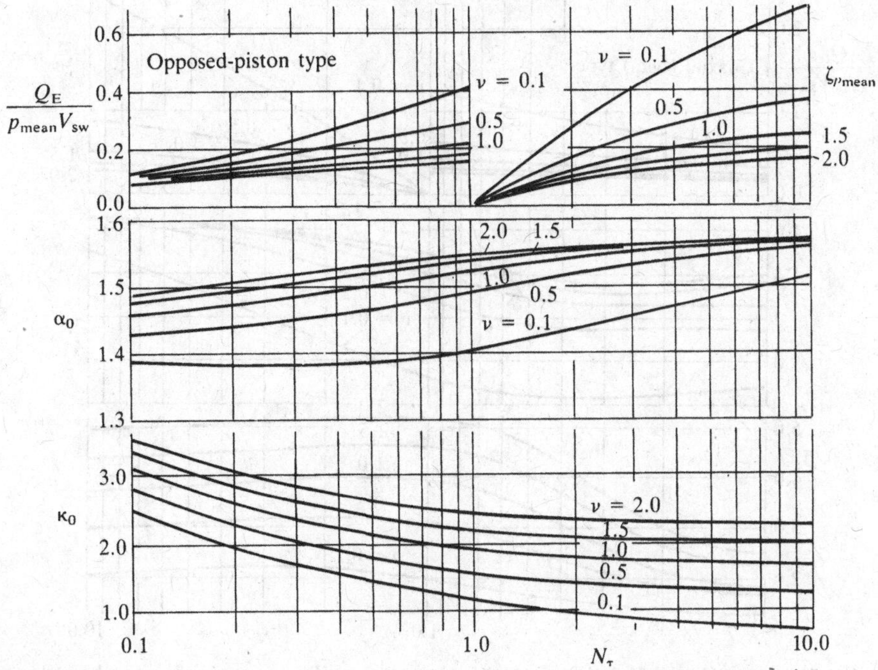


Fig. AX.3 As for Fig. AX.1, but for specific cycle work based on $p_{mean}$, viz $\zeta = W_{cyc}/p_{mean}V_{sw}$.

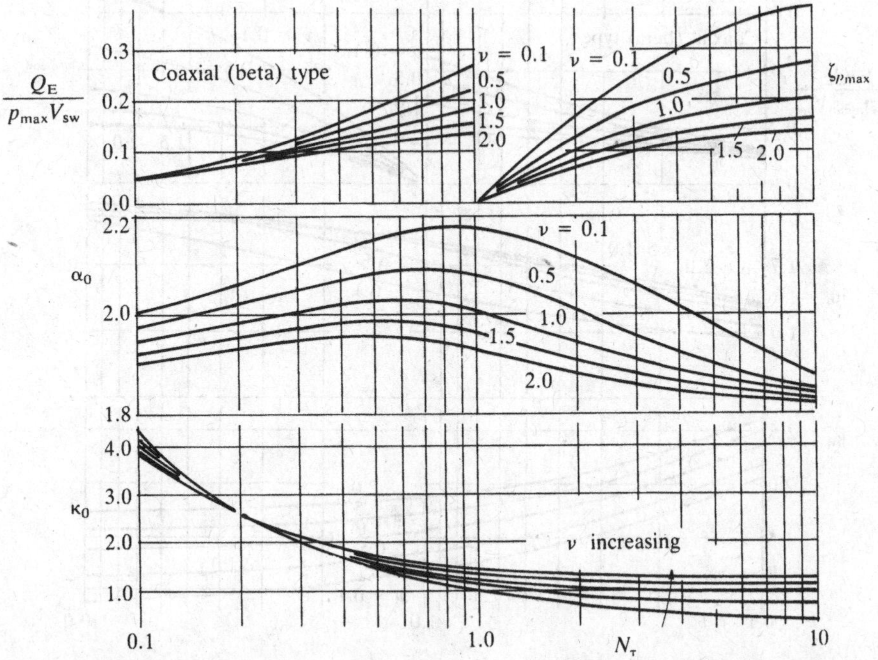


Fig. AX.4 Equivalent $\kappa$ and $\alpha$ which maximize specific heat lifted and specific cycle work, $W_{cyc}/p_{max}V_{sw}$, of the displacer (beta) machine operating on the Schmidt cycle.

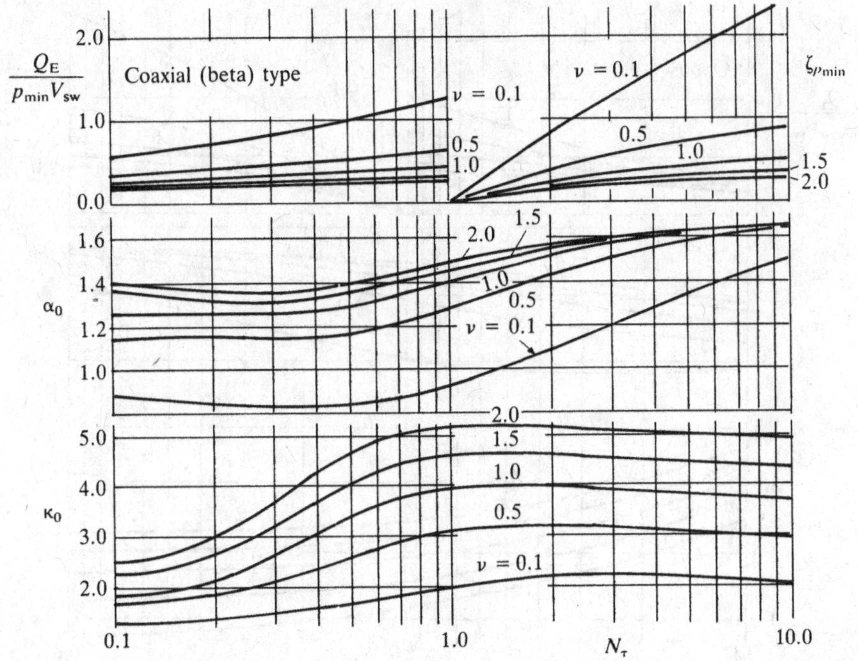


Fig. AX.5 As for Fig. AX.4 (beta machine), but specific cycle work now based on $p_{min}$, viz $\zeta = W_{cyc}/p_{min}V_{sw}$.

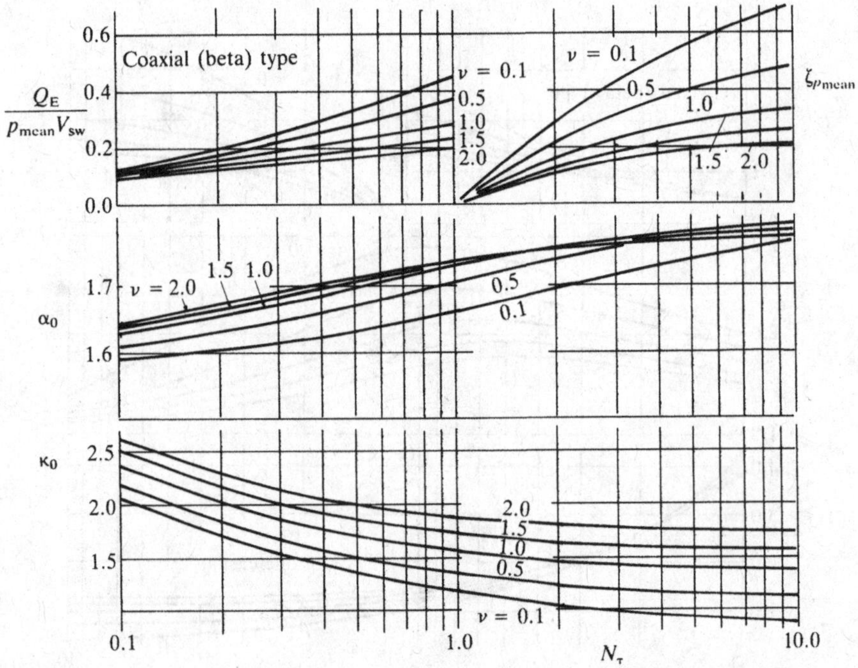


Fig. AX.6 As for Fig. AX.4 (beta machine), but specific cycle work based on $p_{mean}$, viz $\zeta = W_{cyc}/p_{mean}V_{sw}$.

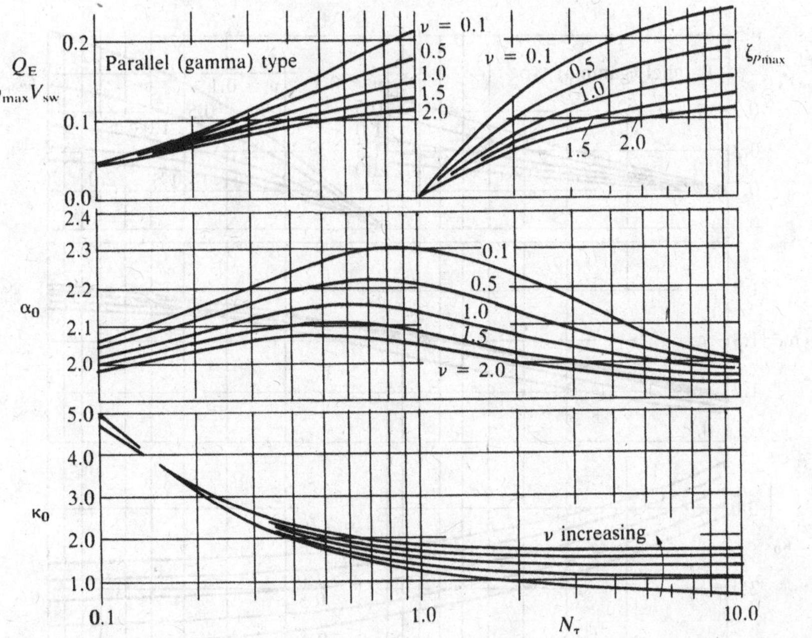


Fig. AX.7 Equivalent $\kappa$ and $\alpha$ which maximize specific heat lifted and specific cycle work, $W_{cyc}/p_{max}V_{sw}$, of the separate displacer (gamma) machine operating on the Schmidt cycle.

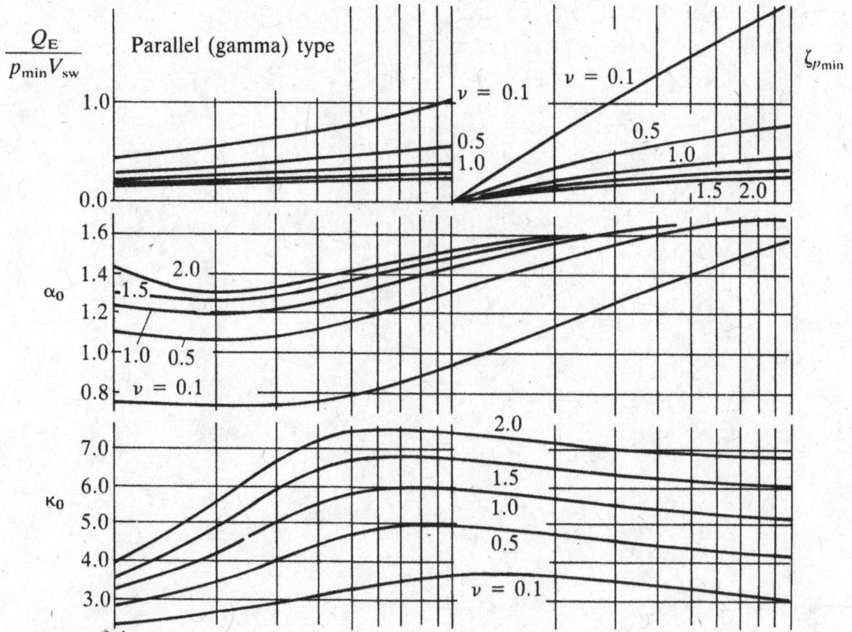


Fig. AX.8 As for Fig. AX.7 (gamma machine), but specific cycle work based on $p_{min}$, viz $\zeta = W_{cyc}/p_{min}V_{sw}$.

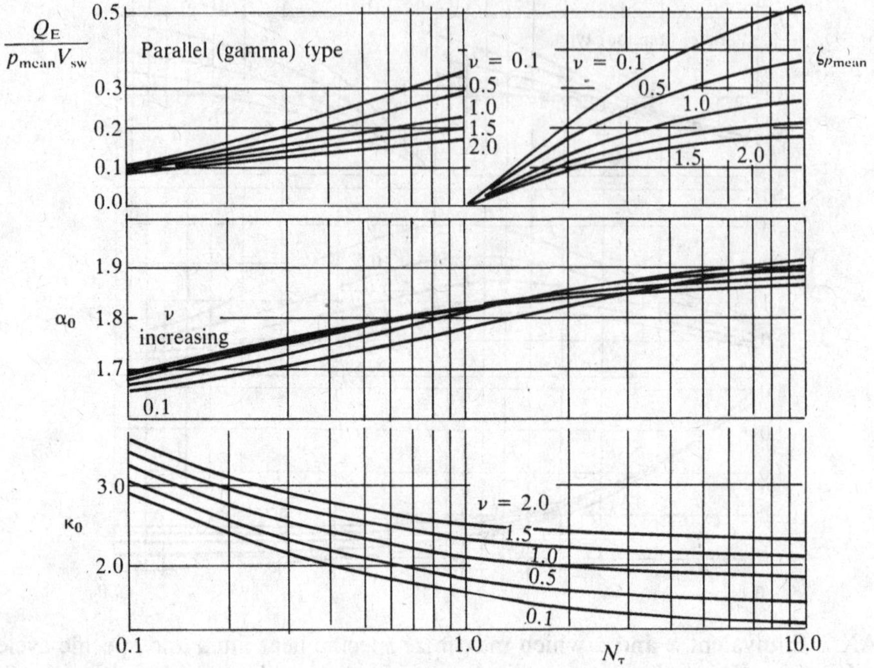


Fig. AX.9 As for Fig. AX.7 (gamma machine), but specific cycle work based on $p_{mean}$, viz $\zeta = W_{cyc}/p_{mean}V_{sw}$.

## AX.3 Charts based on adiabatic analysis

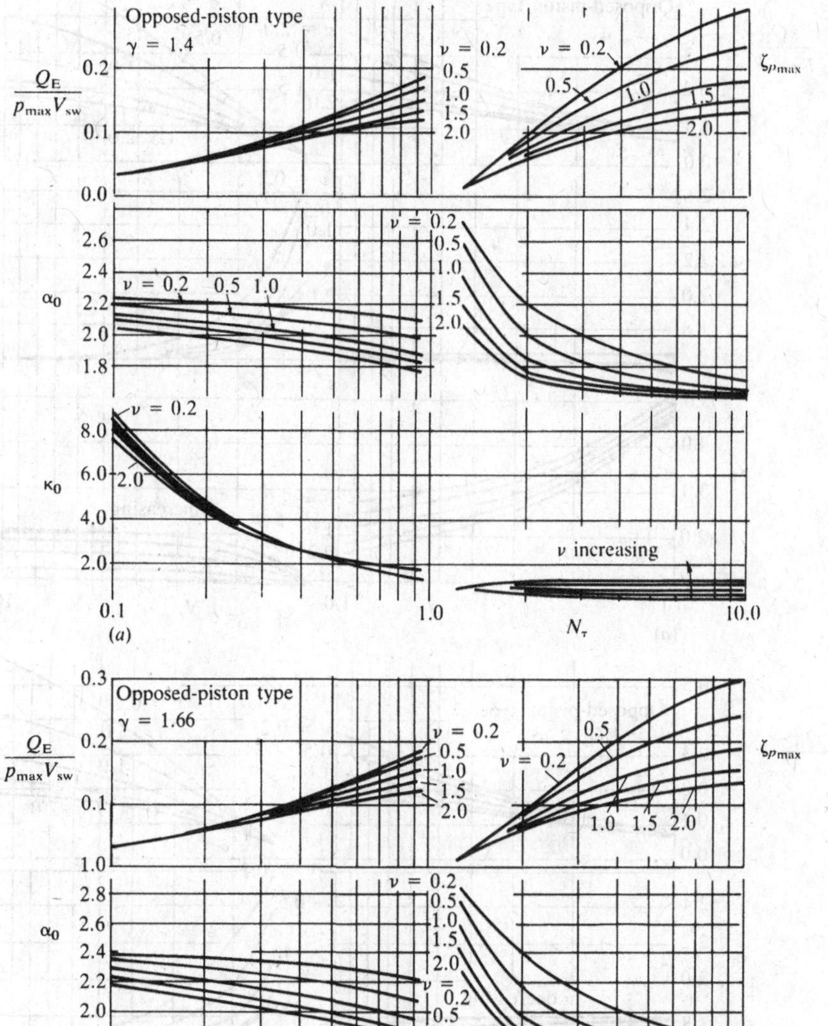

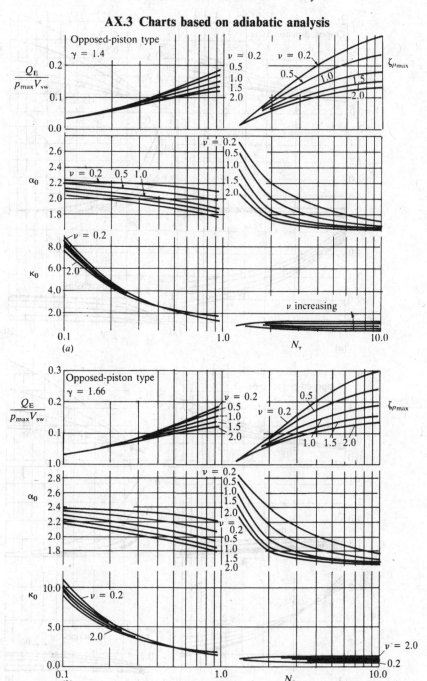

Fig. AX.10(*a*) $\kappa$ and $\alpha$ which maximize specific heat lifted and specific cycle work, $W_{cyc}/p_{max}V_{sw}$ of the opposed-piston machine operating on the adiabatic cycle. $\gamma = 1.4$. (*b*) As for (*a*), but for $\gamma = 1.66$.

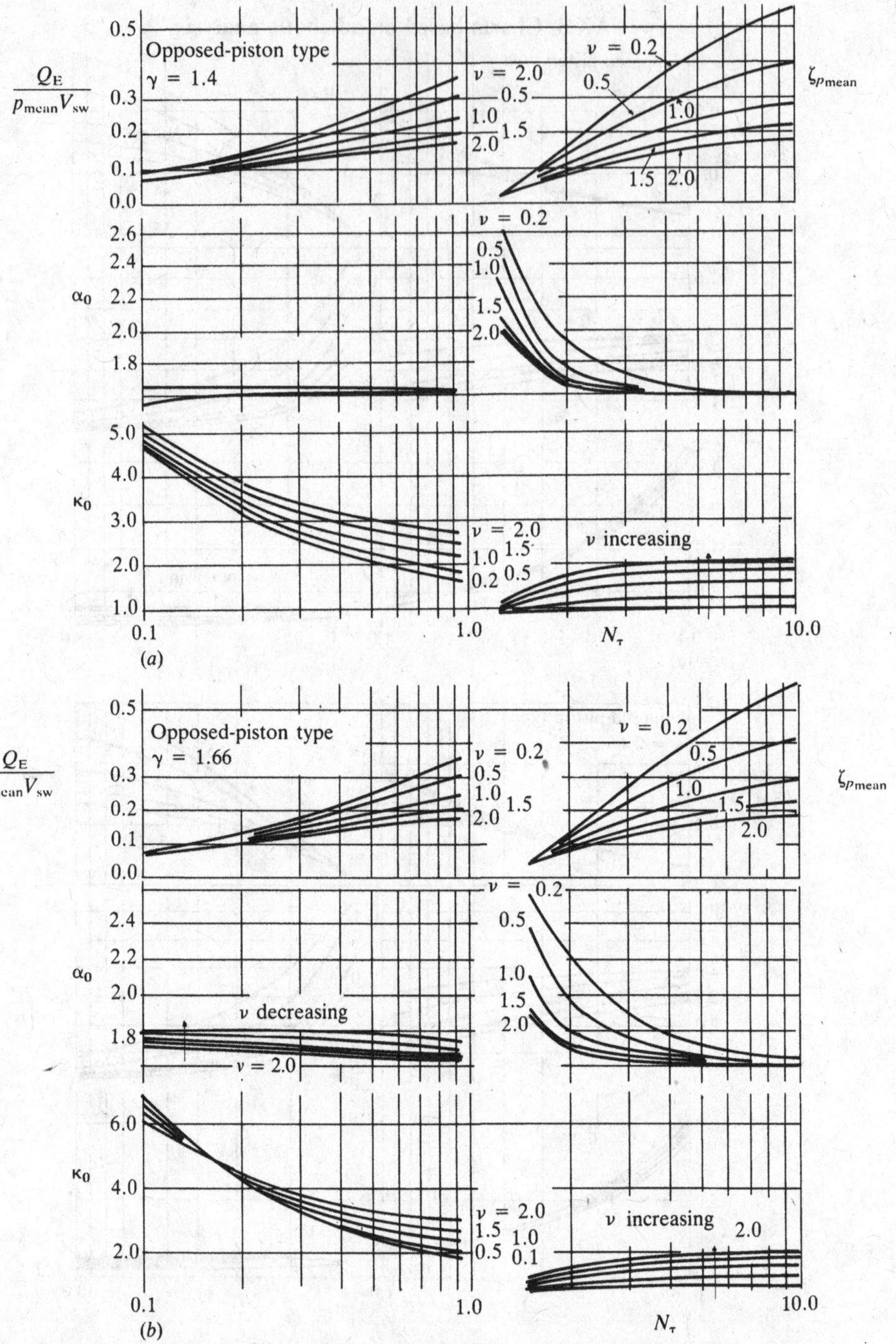


Fig. AX.11(*a*) $\kappa$ and $\alpha$ which maximize specific heat lifted and specific cycle work, $W_{cyc}/p_{mean} V_{sw}$ of the opposed-piston machine operating on the adiabatic cycle. $\gamma = 1.4$. (*b*) As for Fig. (*a*), but for $\gamma = 1.66$.

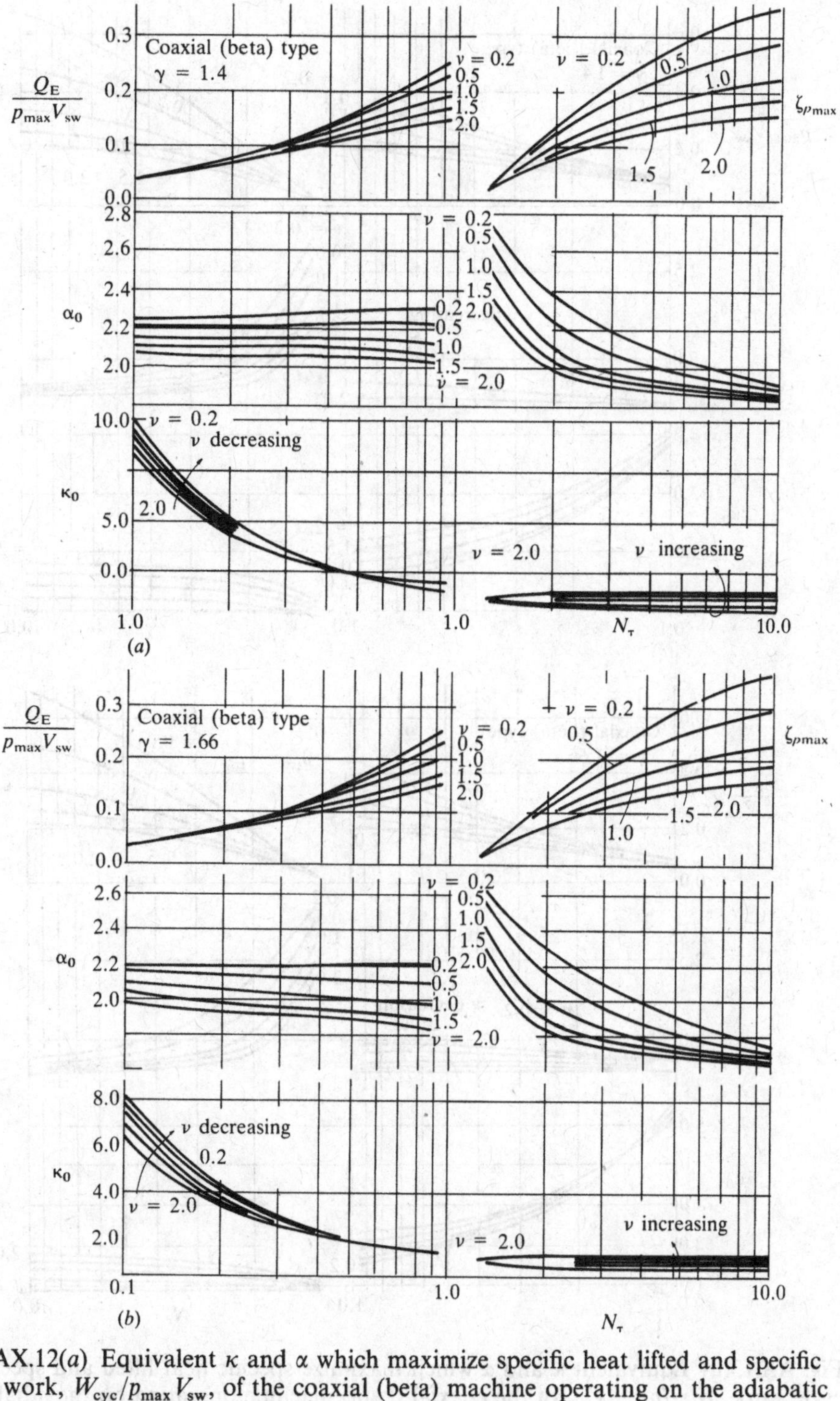


Fig. AX.12(*a*) Equivalent $\kappa$ and $\alpha$ which maximize specific heat lifted and specific cycle work, $W_{cyc}/p_{max}V_{sw}$, of the coaxial (beta) machine operating on the adiabatic cycle. $\gamma = 1.4$. (*b*) As for (*a*), but for $\gamma = 1.66$.

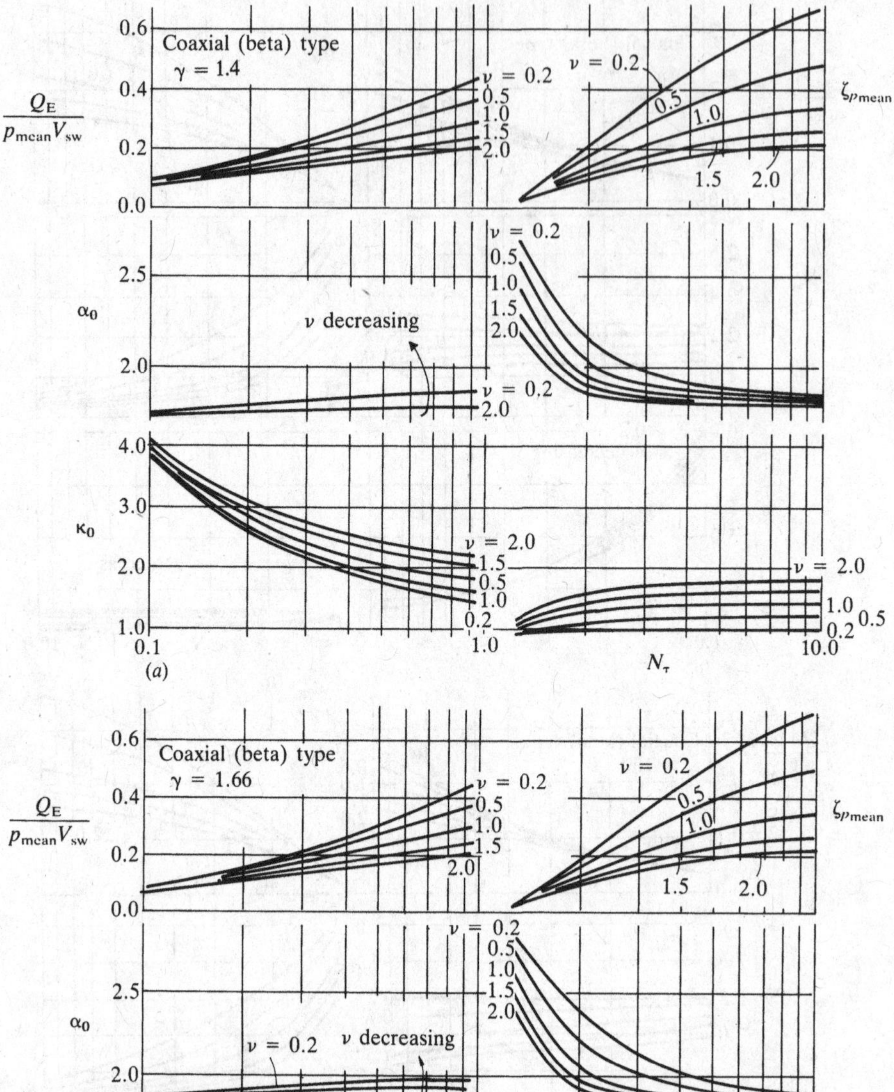

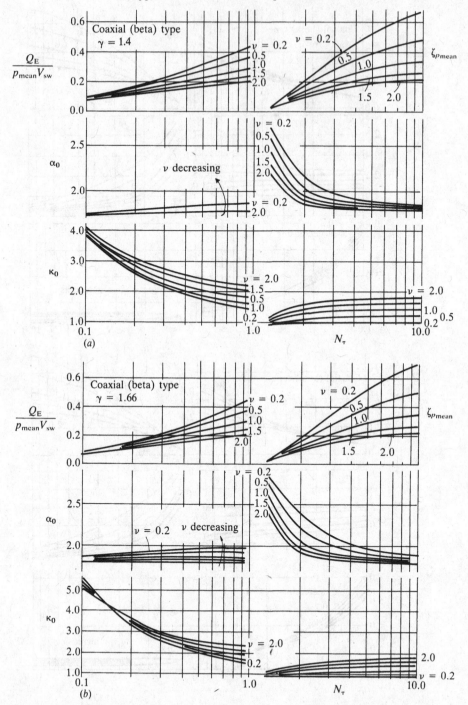

Fig. AX.13(*a*) Equivalent $\kappa$ and $\alpha$ which maximize specific heat lifted and specific cycle work, $W_{cyc}/p_{mean} V_{sw}$, of the coaxial (beta) machine operating on the adiabatic cycle. $\gamma = 1.4$. (*b*) As for (*a*), but for $\gamma = 1.66$.

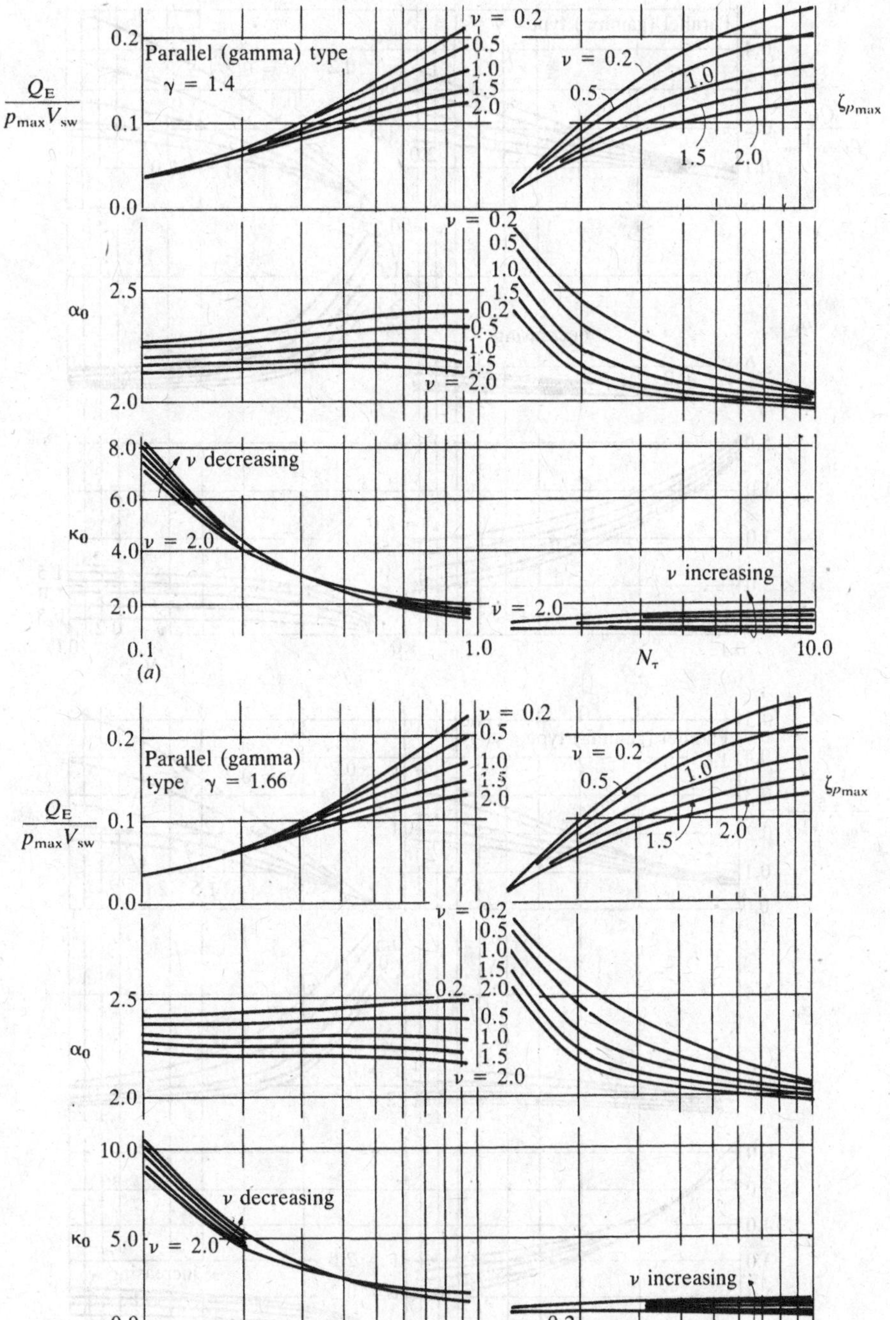


Fig. AX.14(*a*) Equivalent $\kappa$ and $\alpha$ which maximize specific heat lifted and specific cycle work, $W_{cyc}/p_{max}V_{sw}$, of the gamma machine operating on the adiabatic cycle. $\gamma = 1.4$. (*b*) As for (*a*), but for $\gamma = 1.66$.

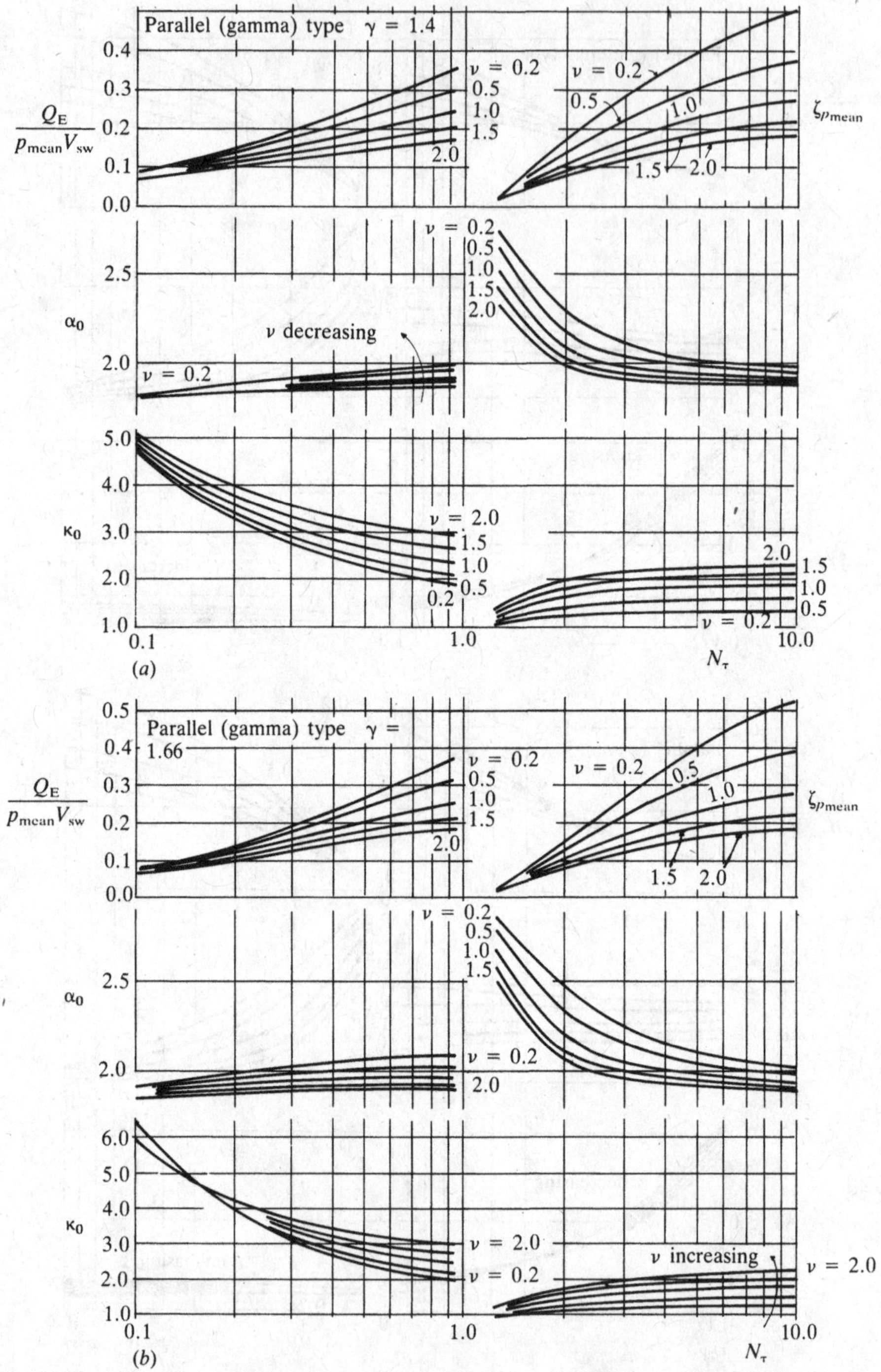


Fig. AX.15(*a*) Equivalent $\kappa$ and $\alpha$ which maximize specific heat lifted and specific cycle work, $W_{cyc}/p_{mean}V_{sw}$, of the gamma machine operating on the adiabatic cycle. $\gamma = 1.4$. (*b*) As for (*a*), but for $\gamma = 1.66$.

# Appendix XI

## Introduction to the Method of Characteristics

### AXI.1 Analytical models and solutions

No-one intending to embark upon the numerical modelling of the gas processes of the Stirling machine should proceed without first carrying out a simple and salutary experiment, here described by reference to Fig. AXI.1(*a*). There is shown a schematic of steady, one-dimensional, unidirectional flow at uniform velocity, $u$, in a simple, cylindrical heat exchanger of length $L$ and diameter $D$ having constant, uniform wall temperature $T_w$. The heat transfer coefficient, $h$ is constant, as are specific heats, $c_p$ and $c_v$. Initially, the fluid entry temperature is uniformly $T_w$. At some instant it changes to and remains at a new value $T_i$. Neglecting kinetic energy, a differential expression may be written for a typical control volume:

$$h\pi D(T_w - T)\,dx = \rho c_v A_x(\partial T/\partial t)\,dx + \rho c_p u A_x(\partial T/dx)\,dx \qquad \text{(AXI.1)}$$

With the simplification that $u$ behind the temperature front retains the original value, $\rho$ adjusting to maintain the assumption of constant pressure, there is an analytical solution for the temperature distribution behind the front. With $t$ for time since entry of a particle at $x$:

$$\frac{T_w - T}{T_w - T_i} = e^{-hu\pi Dt/(m'c_p)} = e^{-xN_{st}/r_h} \qquad \text{(AXI.2)}$$

Fig. AXI.1(*b*) shows the same exchanger divided into control volumes in anticipation of numerical solution by a fixed grid scheme. The reader should put Eqn (AXI.1) into finite difference form and embody it in a calculator sequence or simple computer program which processes all $n$ control volumes systematically, and attempt to achieve the numerical equivalent of Eqn (AXI.2).

Having confirmed the conceptual and programming finesse required to achieve anything approximating the analytical result, the reader may care to reflect on the implications for numerical solution of the considerably more complex case of flow in the exchanger of the Stirling cycle machine. By contrast with the case of the simple, steady-state heat exchanger, there is no verified analytical solution with which to compare numerical results. Consequently it is impossible to say how good or how bad a result has been achieved. (Perturbation solutions of the type described in Chaps. 7 and 8, while subject to their own, particular approximations, do not involve the solution of the unsteady terms of the conservation equations, and so do not give rise to the difficulties currently in question.)

Flow in the gas circuit of a Stirling cycle machine is, of course, not one-dimensional. This fact is sometimes used as part of an argument that two wrongs

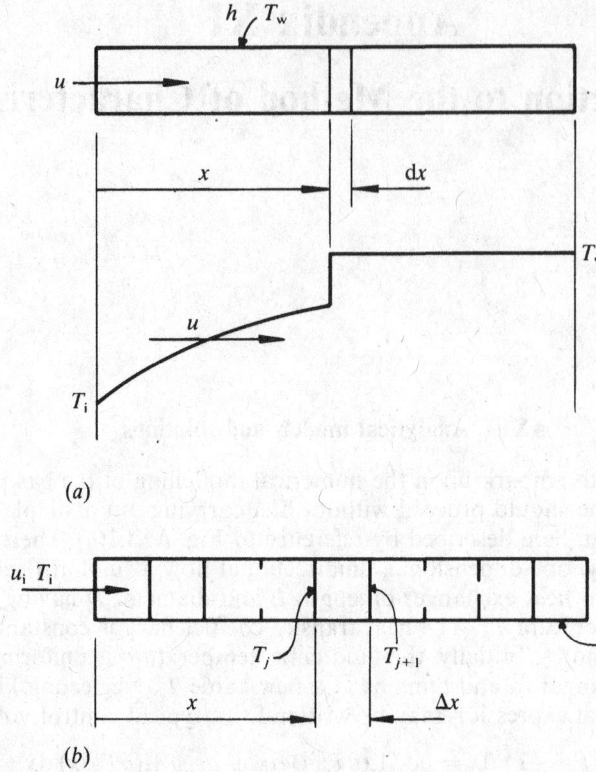


Fig. AXI.1 Scheme for assessing the effectiveness of numerical methods for heat transfer computations. (*a*) Steady, uniform flow in a parallel duct with heat transfer and uniform wall temperature. Fluid entry temperature undergoes step change from $T_w$ to $T_i$ at time $t = 0$. The analytical solution for the temperature distribution from entry to wave front is sketched, and is identical to the solution for the steady state case (entry temperature $T_i$) over that region. (*b*) The same duct subdivided for numerical solution of Eqn (AXI.1).

make a right, that the numerical discrepancies introduced by the fixed-grid scheme offset the discrepancies between actual and ideal, one-dimensional flow, and that somehow the net result is an acceptable numerical model of the real physical process. The argument has a certain appeal, but it is not necessary to accept it as though there were no alternative: there exists a method of numerical solution of the differential equations of one-dimensional, unsteady, compressible flow which, for practical purposes, is exact and correct. There is thus a means of examining in detail the flow field phenomena which *would* arise in the Stirling machine gas circuit *if* flow were truly one-dimensional. The numerical scheme, well established outside the field of Stirling cycle modelling, but hardly acknowledged within that field, is known as the Method of Characteristics.

Briefly, the Method of Characteristics involves a transformation of the partial differential form of the conservation laws to axes along which the changes in thermodynamic properties are in the form of substantial derivatives. This form has a legitimate, finite difference counterpart, so that numerical or graphical solution may proceed without the problems (numerical dispersion, etc.) which caused the difficulties with the introductory problem above. An alleged disadvantage of the Method of Characteristics, namely, that integration proceeds along a variable lattice rather than the usual cartesian grid, will be shown later to be imagined rather than real.

## AXI.2 The conservation equations in Characteristic form

The treatment to be presented follows that of Shapiro[1] and Abbott[2]. It is adapted from a summary by Organ[3].

Fig. AXI.2 shows an element of duct having length d$x$ along which there are variations in cross-sectional area, $A_x$, and in fluid properties, $p$, $u$, $\rho$ and $T$. In general, fluid properties inside the control volume vary with time. The criterion suggested by Shapiro for permissible rate of area change with $x$ is that d(ln $A_x$)/d$x$ should be small compared with unity.

### *AXI.2.1 Conservation of mass*

The differential expression for conservation of mass must now take account of area change:

$$\frac{\partial}{\partial x}(\rho u A_x)\,\mathrm{d}x = -\frac{\partial}{\partial t}(\rho A_x\,\mathrm{d}x)$$

Multiplying out:

$$\frac{\partial \rho}{\partial t} + \rho\frac{\partial u}{\partial x} + u\frac{\partial \rho}{\partial x} + \frac{\rho u}{A_x}\frac{\mathrm{d}A_x}{\mathrm{d}x} = 0 \qquad \text{(AXI.3)}$$

### *AXI.2.2 Conservation of momentum*

Following Shapiro, the effects of friction are considered to be confined to the wall and expressed in terms of a friction factor, $C_f$, where

$$C_f = \frac{\tau_w}{\rho u^2/2}$$

and where $\tau_w$ is the local, instantaneous wall shearing stress. The law of conservation of momentum must include a term for the effect of wall friction, and for the axial component of pressure force on the inclined wall. The following form

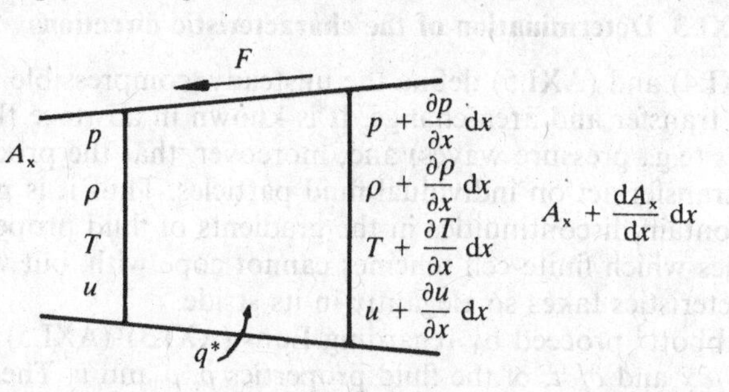


Fig. AXI.2 Control volume for flow with friction, heat transfer and area change.

is that used by Shapiro, with a small modification to account for the possibility of non-cylindrical cross-section. The symbol $P$ represents the periphery of the section at a point mid-way between the planes at $x$ and $x + dx$.

$$-\frac{\partial}{\partial x}(pA_x)\,dx + p\frac{dA_x}{dx}dx - C_f\frac{\rho u^2}{2}P\,dx = \frac{\partial}{\partial t}(\rho A_x u\,dx) + \frac{\partial}{\partial x}(\rho A_x u^2)\,dx$$

This equation may be expanded and the mass conservation equation subtracted as before:

$$\frac{\partial u}{\partial t} + u\frac{\partial u}{\partial x} + \frac{1}{\rho}\frac{\partial p}{\partial x} + F = 0 \qquad \text{(AXI.4)}$$

In Eqn (AXI.4) $F$ is a wall friction term defined by Shapiro as

$$F = \frac{C_f}{r_h}\frac{u^2}{2}\frac{u}{|u|}$$

$r_h$ is hydraulic radius, $A_x/P$. For a cylindrical or conical duct it has the value $D/4$. The term $u/|u|$ ensures that the force due to friction always opposes motion.

### *AXI.2.3 Conservation of energy*

Again, it is the formulation by Shapiro which is presented here. The law equates heat transfer rate per unit time and per unit mass of fluid, $q^*$, to time rate of increase of energy (internal and kinetic) of the fluid within the control volume and the *difference* between enthalpy and kinetic energy convected in and out of the control volume by flow in the $x$ direction:

$$q^*\rho A_x\,dx = (\partial/\partial t)[(\rho A_x\,dx)[c_v T + u^2/2)] + (\partial/dx)[(\rho u A_x)(c_v T + p/\rho + u^2/2)]\,dx$$

The equation amounts to the First Law of thermodynamics for the control volume of Fig. AXI.2 Shapiro expands the equation, simplifies by subtracting the mass conservation equation (Eqn (AXI.3)), eliminates terms in $\partial p/\partial x$ with the aid of the momentum equation (Eqn (AXI.4)), and expresses $T$ in terms of $p$ and $\rho$ with the aid of the ideal gas equation. With $a$ (acoustic speed) for $\sqrt{(\gamma RT)}$ the working form of the energy conservation law is

$$(\gamma - 1)\rho(q^* + uF) = \frac{\partial p}{\partial t} + u\frac{\partial p}{\partial x} - a^2\left(\frac{\partial \rho}{\partial t} + u\frac{\partial \rho}{\partial x}\right) \qquad \text{(AXI.5)}$$

## AXI.3 Determination of the characteristic directions

Eqns (AXI.3), (AXI.4) and (AXI.5) define the unsteady, compressible flow-field with friction, heat transfer and area change. It is known in advance that such fields may contain *waves* (e.g., pressure waves) and, moreover, that the processes of friction and heat transfer act on individual fluid particles. Thus it is possible for the flow-field to contain discontinuities in the gradients of fluid properties. It is these discontinuities which finite-cell schemes cannot cope with, but which the Method of Characteristics takes so elegantly in its stride.

Shapiro[1] and Abbott[2] proceed by regarding Eqns (AXI.3)–(AXI.5) as equations in the gradients, $\partial/\partial x$ and $\partial/\partial t$, of the fluid properties $p$, $\rho$ and $u$. There is clearly a total of six such gradients, $\partial u/\partial x$, $\partial u/\partial t$, $\partial p/\partial x$, $\partial p/\partial t$, $\partial \rho/\partial x$, $\partial \rho/\partial t$, but so far only three equations. If, however, the assumption is made that the fluid properties themselves, $p$, $u$, $\rho$ are continuous over the $x$–$t$ region to be considered, then use can be made of the three further differential expressions which define this continuity. The expression for pressure typifies those for $u$ and for $\rho$:

$$dp = \frac{\partial p}{\partial x}\,dx + \frac{\partial p}{\partial t}\,dt \tag{AXI.6}$$

The set of six, simultaneous equations for the property gradients are set out in systematic form:

$$\begin{array}{l}
dx\dfrac{\partial u}{\partial x} + dt\dfrac{\partial u}{\partial t} = du \\[2ex]
dx\dfrac{\partial p}{\partial x} + dt\dfrac{\partial p}{\partial t} = dp \\[2ex]
dx\dfrac{\partial \rho}{\partial x} + dt\dfrac{\partial \rho}{\partial t} = d\rho \\[2ex]
\rho\dfrac{\partial u}{\partial x} + u\dfrac{\partial \rho}{\partial x} + \dfrac{\partial \rho}{\partial t} = \dfrac{\rho u\, dA_x}{A_x\, dx} \\[2ex]
u\dfrac{\partial u}{\partial x} + \dfrac{\partial u}{\partial t} + \dfrac{1}{\rho}\dfrac{\partial p}{\partial x} = -F \\[2ex]
u\dfrac{\partial p}{\partial x} + \dfrac{\partial p}{\partial t} - ua^2\dfrac{\partial \rho}{\partial x} - a^2\dfrac{\partial \rho}{\partial t} = (\gamma - 1)\rho(uF + q^*)
\end{array}$$

The solution for any chosen partial differential coefficient may be expressed in determinant form:

$$\frac{\begin{array}{cccccc}
du & dt & 0 & 0 & 0 & 0 \\
dp & 0 & dx & dt & 0 & 0 \\
d\rho & 0 & 0 & 0 & dx & dt \\
\dfrac{\rho u\, dA_x}{A_x\, dx} & 0 & 0 & 0 & u & 1 \\
-F & 1 & 1/\rho & 0 & 0 & 0 \\
(\gamma - 1)\rho(uF + q^*) & 0 & u & 1 & -ua^2 & -a^2
\end{array}}{\begin{array}{cccccc}
dx & dt & 0 & 0 & 0 & 0 \\
0 & 0 & dx & dt & 0 & 0 \\
0 & 0 & 0 & 0 & dx & dt \\
\rho & 0 & 0 & 0 & u & 1 \\
u & 1 & 1/\rho & 0 & 0 & 0 \\
0 & 0 & u & 1 & -ua^2 & -a^2
\end{array}} = \frac{\partial u}{\partial x} \tag{AXI.7}$$

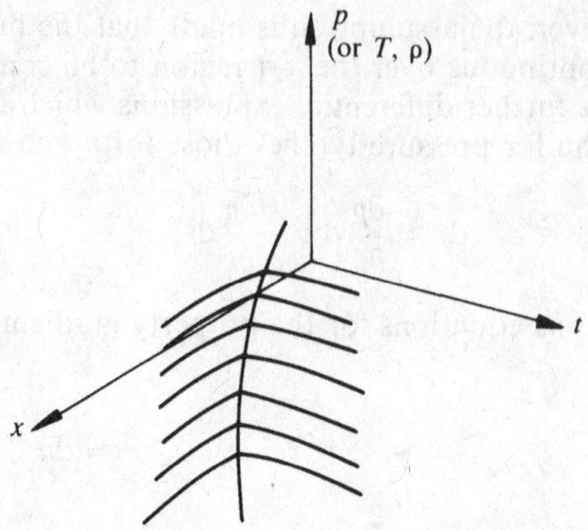


Fig. AXI.3 Discontinuity in the gradient of a continuous property.

Now suppose that a one-dimensional, compressible flow-field contains a feature such as that indicated in Fig. AXI.3. The feature is a discontinuity in the *gradient* of a field property – e.g., in the gradient of velocity, $u$, or of pressure, $p$. The mathematical condition that the gradient should be discontinuous is that the determinant expressing the gradient should be indeterminate – i.e., should be 0/0. Multiplying out the denominator of *any* of the six solutions of the form of Eqn (AXI.7) leads to

$$\frac{\mathrm{d}x}{\mathrm{d}t_{\mathrm{I}}} = u + a \qquad \text{(AXI.8a)}$$

$$\frac{\mathrm{d}x}{\mathrm{d}t_{\mathrm{II}}} = u - a \qquad \text{(AXI.8b)}$$

Eqns (AXI.8a) and (AXI.8b) confirm the possibility of discontinuities in the gradients of fluid properties in the $x$–$t$ plane along the loci of pressure waves. In a one-dimensional flow-field (cf. the simple, parallel duct) pressure waves may run right at $u + a$ and left at $u - a$. Subscripts **I** and **II** refer to right- and left-running waves respectively.

$$\mathrm{d}x/\mathrm{d}t_{\mathrm{path}} = u \qquad \text{(AXI.8c)}$$

Eqn (AXI.8c) reveals that, where heat transfer and friction effects are present there is the possibility of discontinuities in fluid property gradients *along the trajectories in the t–x plane of individual fluid particles.* The physical meaning is that if a discontinuity in temperature gradient exists in the flow-field at some stage, then that discontinuity propagates at local particle speed.

The denominator of Eqn (AXI.7) is set equal to zero to complete the statement of indeterminacy. The roots of the resulting equation are:

$$\mathrm{d}u_{\mathrm{I}} = -\frac{a}{\gamma}\frac{\mathrm{d}p_{\mathrm{I}}}{p} - \frac{au}{A_x}\frac{\mathrm{d}A_x}{\mathrm{d}x}\mathrm{d}t_{\mathrm{I}} - (\gamma - 1)\frac{q^*}{a}\mathrm{d}t_{\mathrm{I}} - F[1 - (\gamma - 1)u/a]\,\mathrm{d}t_{\mathrm{I}} \qquad \text{(AXI.9a)}$$

$$\mathrm{d}u_{\mathrm{II}} = +\frac{a}{\gamma}\frac{\mathrm{d}p_{\mathrm{II}}}{p} + \frac{au}{A_x}\frac{\mathrm{d}A_x}{\mathrm{d}x}\mathrm{d}t_{\mathrm{II}} + (\gamma - 1)\frac{q^*}{a}\mathrm{d}t_{\mathrm{II}} - F[1 + (\gamma - 1)u/a]\,\mathrm{d}t_{\mathrm{II}} \qquad \text{(AXI.9b)}$$

There are four differences in algebraic sign between Eqn (AXI.9a) for the **I** direction and Eqn (AXI.9b) for the **II** direction.

Shapiro demonstrates how Eqns (AXI.9) may be converted to the form:

$$du_{\mathrm{I/II}} = \mp\frac{2}{\gamma-1}\,da_{\mathrm{I/II}} + \left\{\mp\frac{au}{A_x}\frac{dA_x}{dx} \pm \gamma\frac{q^*}{a} - F\left(1 \mp \frac{u}{a}\right) + \frac{a^2}{\gamma}\frac{\partial}{\partial x}\left[\ln\left(\frac{a^{2\gamma/\gamma-1}}{p}\right)\right]\right\} dt_{\mathrm{I/II}} \tag{AXI.10}$$

In Eqn (AXI.10) the first of the two alternative algebraic signs holds when the equation is applied along the **I** direction, the second when it is applied along the **II** direction.

That certain terms are multiplied by $dt_{\mathrm{I}}$ for calculations along the **I** direction, and by $dt_{\mathrm{II}}$ for calculations along the **II** direction, corresponds to the fact that the effects of friction, heat transfer and area change over an interval $dt$ are proportional to the time for which they act, i.e., to respective $dt$. These are the $dt$ indicated on the physical plane of Fig. AXI.4, which also indicates path lines.

Evaluation of the final term in Eqns (AXI.10) calls for knowledge of pressure distribution in the $x$ direction. The term in $p$ may not be eliminated in favour of $a$ without introducing another unknown, via $\rho$. A further relationship is therefore required between $p$ and the other variables of the problem. A suitable, independent relationship is that which defines the entropy change in terms of changes in $T$ and $\rho$ along the path lines between intersections with Mach lines.

A standard thermodynamic expression for change in entropy corresponding to changes in $T$ and $\rho$ may be written:

$$T\,ds = c_v\,dT + p\,d(1/\rho)$$

For a process involving heat transfer and friction over a time interval $dt$ the change in entropy, $ds$ is

$$ds_{\mathrm{path}} = \frac{uF + q}{T}\,dt_{\mathrm{path}}$$

Reference to Shapiro's treatment will show how these equations are manipulated to give the following relationship for the change in states occurring in time $dt_{\mathrm{path}}$ along a fluid particle track such as that marked in Fig. AXI.4:

$$dp_{\mathrm{path}} = \frac{p}{a}\left[\frac{2\gamma}{\gamma-1}\,da_{\mathrm{path}} - \frac{\gamma}{a}(uF + q^*)\,dt_{\mathrm{path}}\right]$$

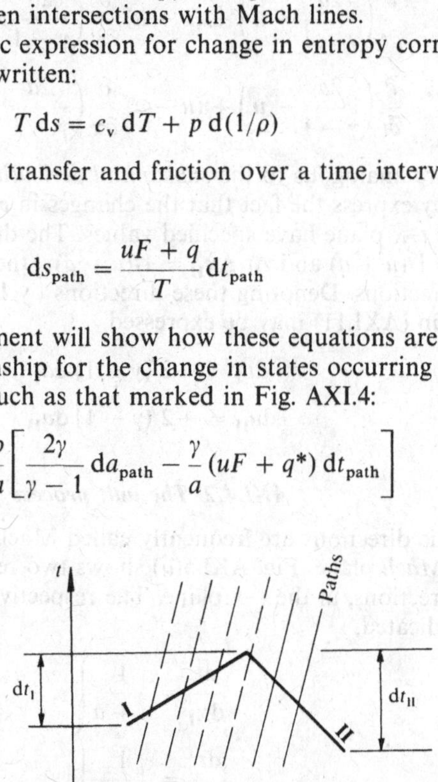


Fig. AXI.4 Characteristic directions – Mach and path lines.

In this last equation the $p$, $a$, $u$, $F$ and $q^*$ are evaluated at the mid-point of the path in question.

### AXI.4 Computational strategy

Graphical and numerical solution involves a parallel sequence of computational steps. A graphical illustration thus typifies both, and is the standard form of presentation for the Method of Characteristics. Moreover, all of the computational nuances may be appreciated from specimen constructions involving the Characteristic equations in a form simplified by omission of heat transfer and friction effects. It is true that applying the resulting (homentropic) form to the Stirling machine will yield a value of zero for computed indicated work. However, reinstatement of the terms for the irreversibilities will give rise to no difficulty once the basic methods of manipulation have been presented.

#### *AXI.4.1 Characteristics in the homentropic flow field*

What the foregoing algebra has achieved is a statement of the directions in the time–distance ($t$–$x$) plane along which a specified combination of fluid properties remains unchanged. In alternative mathematical form:

$$\frac{\partial}{\partial t}\left(\frac{2a}{\gamma-1}+u\right)+(u+a)\frac{\partial}{\partial x}\left(\frac{2a}{\gamma-1}+u\right)=0 \qquad \text{(AXI.11a)}$$

$$\frac{\partial}{\partial t}\left(\frac{2a}{\gamma-1}-u\right)+(u-a)\frac{\partial}{\partial x}\left(\frac{2a}{\gamma-1}-u\right)=0 \qquad \text{(AXI.11b)}$$

Eqns (AXI.11) are analogous to the *substantial derivative*, $\mathrm{D}/\mathrm{d}t = \partial/\partial t +$ (speed)$\partial/\partial x$. They express the fact that the changes in $u$ and $a$ along specific directions in the $t$–$x$ plane have specified values. The directions in question are locally $\Delta t/\Delta x_{\mathrm{I}} = 1/(u + a)$ and $\Delta t/\Delta x_{\mathrm{II}} = 1/(u - a)$ – the right- and left-running characteristic directions. Denoting these directions by **I** and **II** respectively, the arguments of Eqn (AXI.11) may be expressed:

$$\mathrm{d}u_{\mathrm{I}} = -2/(\gamma - 1)\,\mathrm{d}a_{\mathrm{I}} \qquad \text{(AXI.12a)}$$

$$\mathrm{d}u_{\mathrm{II}} = +2/(\gamma - 1)\,\mathrm{d}a_{\mathrm{II}} \qquad \text{(AXI.12b)}$$

#### *AXI.4.2 The unit process*

The characteristic directions are frequently called Mach lines, and the $t$–$x$ plane the *physical* or *Mach* plane. Fig. AXI.5($a$) shows two representative Mach lines, or characteristic directions, in the $t$–$x$ plane. The respective gradients of **I** and **II** waves are, as indicated,

$$\left.\begin{aligned}\frac{\mathrm{d}t}{\mathrm{d}x_{\mathrm{I}}} &= \frac{1}{u+a}\\ \frac{\mathrm{d}t}{\mathrm{d}x_{\mathrm{II}}} &= \frac{1}{u-a}\end{aligned}\right\} \qquad \text{(AXI.13)}$$

Eqns (AXI.12) and (AXI.13) are the celebrated Riemann relationships.

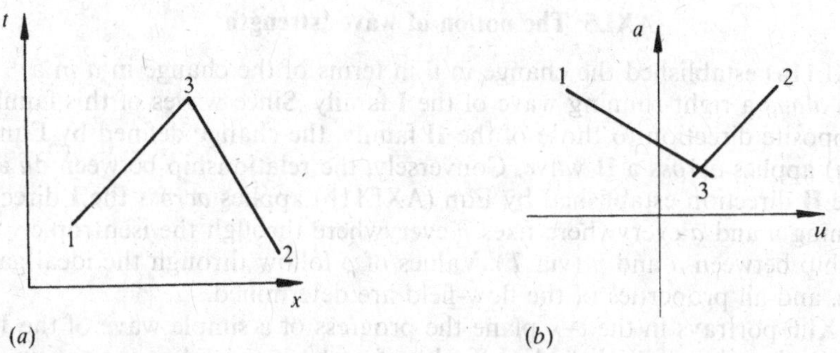


Fig. AXI.5 Characteristics in physical and state diagrams for homentropic flow. (*a*) Physical or Mach plane. (*b*) State plane.

The Method of Characteristics involves integrating in the local coordinates of the Mach plane. As with any process of numerical integration, a computational step starts with known values. In this case, the information required to initiate a step is the values of $u$ and $a$ at two points (e.g., points 1 and 2 in Fig. AXI.5(*a*)) suitably close together in the Mach plane. Along a right-running (I) characteristic through 1 the *change* in $u$ is known in terms of the change in $a$ from Eqn (AXI.12a). Correspondingly, along a left-running (II) wave through point 2 the change in $u$ is again known in terms of the change in $a$, this time from Eqn (AXI.12b). The gradients of the wave loci in the $x$–$t$ plane are known (from Eqns (AXI.13)), the values of $u$ and $a$ must coincide at the new point 3, and so the solution has advanced to that point. It may now proceed to a further point in the $t$–$x$ plane by stepping forward on the basis of point 3 and a further known point suitably close to 3.

The simultaneous solution of Eqns (AXI.12) is conveniently displayed in a *state diagram* (Fig. AXI.5(*b*)). The gradient $da/du_{\mathrm{I}}$ is everywhere equal to $-(\gamma - 1)/2$. This gives the relative change of properties whilst travelling *along* a right-running (I) Mach wave (or *across* a left-running (II) wave). *Along* a left-running (II) wave (i.e., *across* a right-running (I) wave) the relative change in $u$ and $a$ is given by $da/du_{\mathrm{I}} = (\gamma - 1)/2$ (always). Fluid properties at a point 3 are found by constructing lines of the appropriate gradient through the known point and reading off $u$ and $a$ at the intersection. The graphical construction has an obvious numerical counterpart.

The *unit process* of the Characteristics Method of integration may be summarized (see Fig. AXI.5):

Given $u_1$, $a_1$, $u_2$ and $a_2$, solve Eqns (AXI.12) simultaneously to give $u_3$ and $a_3$. By using appropriate values of $u$ and $a$ in Eqns (AXI.13), evaluate the slopes of the **I** and **II** line elements in the Mach plane and thus locate point 3 there.

Although there is nothing inherently difficult about applying the Method of Characteristics, there is no doubt that it calls for a different view of the problem from the more 'obvious' numerical techniques tied to a cartesian $x$–$t$ frame. Moreover, the user must usually make a choice between the so-called 'field' and 'lattice' methods of implementation. The sections which follow will therefore attempt a graded introduction for those to whom the way of working may be novel. Homentropic flow will be assumed for the rest of the appendix.

## AXI.5 The notion of wave 'strength'

Eqn (AXI.11a) established the change in $u$ in terms of the change in $a$ in a direction *along* a right-running wave of the **I** family. Since waves of this family run in the opposite direction to those of the **II** family, the change defined by Eqn (AXI.11a) applies *across* a **II** wave. Conversely, the relationship between d$u$ and d$a$ *along* the **II** direction established by Eqn (AXI.11b) applies *across* the **I** direction. Determining $u$ and $a$ everywhere fixes $p$ everywhere through the isentropic relationship between $a$ and $p$ (via $T$). Values of $\rho$ follow through the ideal gas equation, and all properties of the flow-field are determined.

Fig. AXI.6 portrays in the $t$–$x$ plane the progress of a simple wave of the **II** family into a previously undisturbed fluid uniformly at $u_0$ and $a_0$ (i.e., at known $p_0$, $T_0$, and $\rho_0$). If either $u$ or $a$ behind the wave is known, the other property may be found by applying Eqn (AXI.12a) along the right-running wave which crosses from state 0 into state 1. So if the simple wave has been initiated by imparting a small, initial velocity $u_1$ (positive, thus making the simple wave an expansion wave) then by Eqn (AXI.12) the change in $a$ along the **I** wave which crosses the expansion wave is $-(\gamma - 1)u_1/2$. But there is no right-running wave so far, so how is Eqn (AXI.12a) to be applied?

The answer lies in appreciating that the Riemann relationships (Eqns (AXI.12) and (AXI.13)) apply to waves of zero as well as of finite strength. The

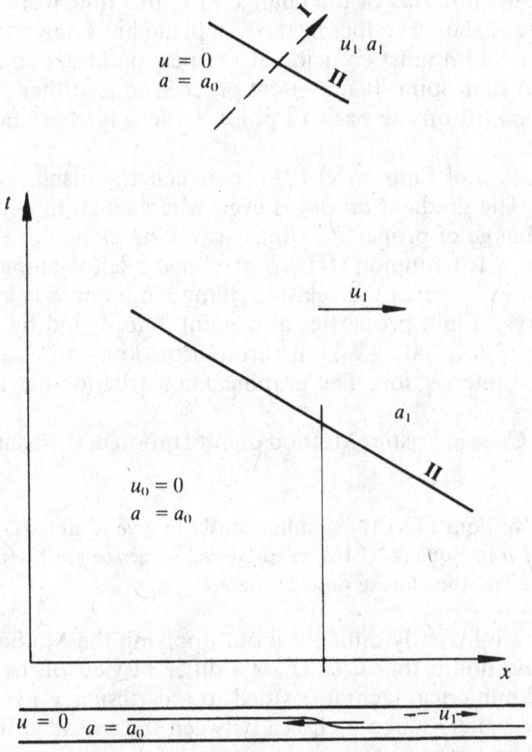


Fig. AXI.6 Determination of conditions behind a simple, left-running (**II**) expansion wave.

characteristic directions are thus best regarded as those directions along which a disturbance *would* propagate *if* initiated. There is an infinity of such latent waves in both directions, and one takes one's pick according to the needs of the problem. In the inset diagram the dashed wave in the **I** direction crosses the expansion wave from state 0 to state 1, and Eqns (AXI.12a) and (AXI.13) are applied *along* it, i.e., *across* the expansion wave.

In the general case an unsteady, compressible flow-field contains an infinity of criss-crossing Mach lines. If these are tracks in the $t$–$x$ plane of pressure disturbances, then this same $x$–$t$ plane must be a plan view of a surface whose local elevation must be pressure, $p$ (or acoustic speed, $a$, or $T$ or $\rho$). It would be helpful to have a perspective view of that surface.

## AXI.6 Non-simple waves

There is a choice of three representations, two of which coincide with slightly different methods of applying the characteristics solution process (the so-called 'lattice' and 'field' methods). The first interpretation is that of a continuous, three-dimensional surface with selected Mach lines having continuous curvature ruled on it (Fig. AXI.7(*a*)). But for the fact that the choice of Mach lines is arbitrary, (and with the reservation that one cannot actually *see* Mach lines anyway) this picture is that which most closely represents reality.

The way in which Eqns (AXI.12) and (AXI.13) are applied is step-by-step – either graphically or numerically. The corresponding form of Eqns (AXI.12) is:

$$\Delta u_{\mathrm{I}} = -[2/(\gamma - 1)]\Delta a_{\mathrm{I}} \qquad \text{(AXI.14a)}$$

$$\Delta u_{\mathrm{II}} = +[2/(\gamma - 1)]\Delta a_{\mathrm{II}} \qquad \text{(AXI.14b)}$$

An implementation of Eqns (AXI.14) for computer solution would have the form

$$(u_{n+1} - u_n)_{\mathrm{I}} = -[2/(\gamma - 1)](a_{n+1} - a_n)_{\mathrm{I}} \qquad \text{(AXI.15a)}$$

$$(u_{m+1} - u_m)_{\mathrm{II}} = +[2/(\gamma - 1)](a_{m+1} - a_m)_{\mathrm{II}} \qquad \text{(AXI.15b)}$$

In implementation of a solution step, values at point $n$ and $m$ are known, points $n + 1$ and $m + 1$ coincide, and simultaneous numerical or graphical solution gives the new $u$ and $a$.

Corresponding to the linear steps between the various $u$ and $a$, the $t$–$x$ (physical) plane (i.e., the plan view of the pressure surface) consists of a mesh made up of straight-line segments of Mach line running either right or left. There are two geometric interpretations corresponding to this plan view, and either one may be implied in published treatments of the Method of Characteristics:

Between intersections the Mach line segments of Fig. AXI.7(*b*) are straight lines. The lozenge-shaped pressure (or acoustic speed) facets between individual quadrilaterals are not necessarily plane. The facets are inclined to the horizontal ($t$–$x$) plane at various angles, but adjacent facets have a common edge. There are thus discontinuities in *gradient* across each Mach line, but pressure (or acoustic speed) is continuous. This is the interpretation of the *lattice* methods of graphical and numerical solution.

Finally, as indicated in Fig. AXI.7(*c*), the Mach lines may be considered to be plan views of the horizontal faces of individual, infinitesimal 'steps' in the surface of

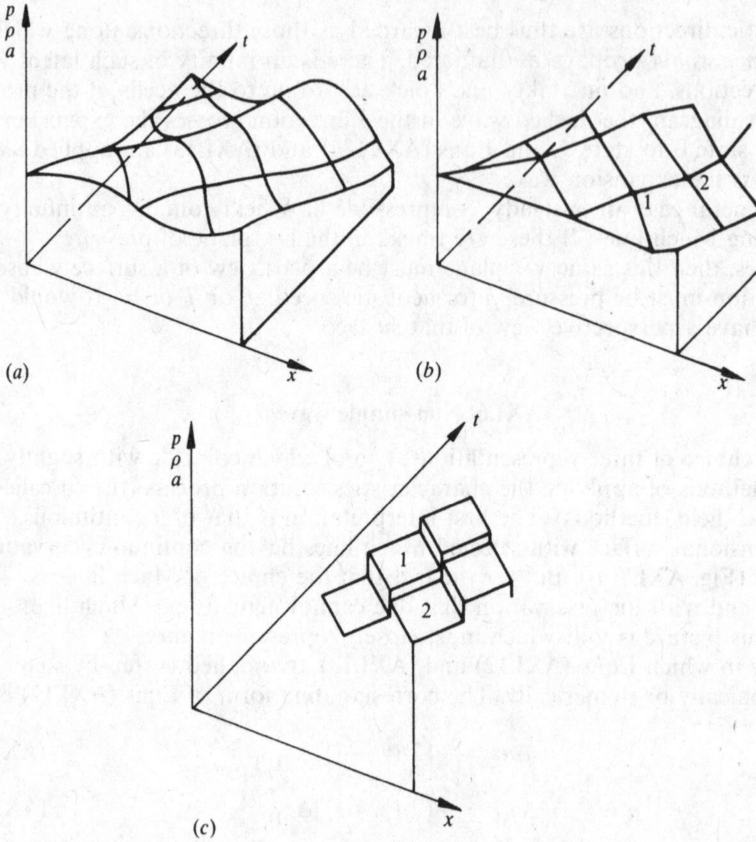


Fig. AXI.7 Interpretations of the surface of thermodynamic properties in unsteady, one-dimensional flow. (*a*) Real (continuous) surface of *p*, *a*, *T*. (*b*) Lattice representation. (*c*) Field interpretation.

fluid properties. Consistent with this interpretation, and by contrast with the lattice view, individual quadrilateral surfaces all lie parallel with the *x*–*t* plane – i.e., are all horizontal. The *field* methods of graphical and numerical solution see the property surface in this way.

The choice of interpretation of the pressure surface determines the detail of how numerical solution will proceed. If the *lattice* picture is accepted, then a wave between points 1 and 2 in Fig. AXI.7(*b*) proceeds in the *t*–*x* plane with gradient $1/(u + a)$ where *u* and *a* are the values at the point mid-way between 1 and 2. If the *field* picture is preferred, then a wave between *fields* 1 and 2 in Fig. AXI.7(c) proceeds at a gradient calculated in terms of the mean of the *u* and *a* in the two fields *at either side*.

## AXI.7 Cases which may be dealt with as simple waves

### *AXI.7.1 Left- and right-running simple waves*

Fig. AXI.8 is an attempt to portray the most elementary form of simple wave. In Fig. AXI.8(*a*) a simple compression wave of the **I** family proceeds from left to

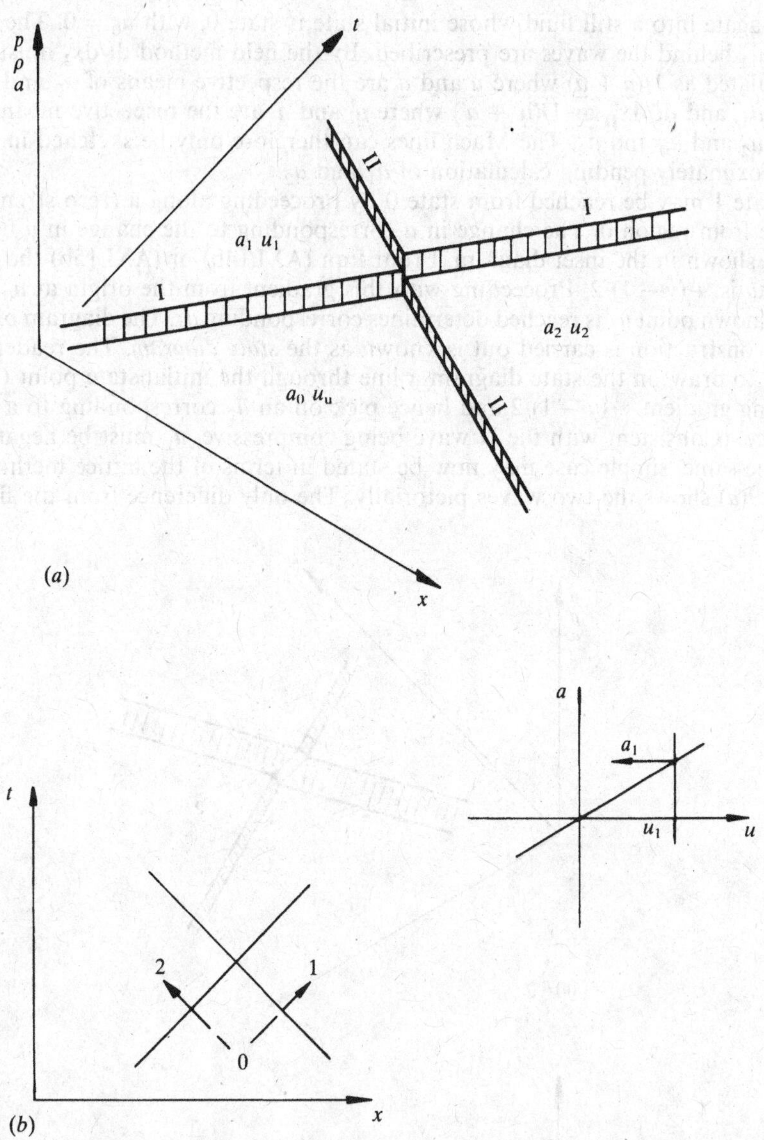


Fig. AXI.8 Simple wave of **I** and **II** families approaching. Interpreted in terms of the field concept. (*a*) Perspective of physical plane. (*b*) Physical or Mach plane.

right. The field representation is used. It retains its sense (compression) after crossing a second compression wave of the **II** family, which also continues as a compression wave after the crossing. At the crossing point conditions are indeterminate, consistent with the need in the field method to calculate $dt/dx_{\mathbf{I\,and\,II}}$ in terms of the mean of states on either side of the Mach line.

Fig. AXI.8(*b*) is the corresponding Mach plane – i.e. the 'plan view' of Fig. AXI.8(*a*). The two simple compression 'steps' are seen in plan view as single lines and approach each other from opposite ends of the parallel duct. The waves propagate into a still fluid whose initial state is state 0, with $u_0 = 0$. The values $u_1$ and $u_2$ behind the waves are prescribed. By the field method $dt/dx_{I}$ must be calculated as $1/(\underline{u} + \underline{a})$ where $\underline{u}$ and $\underline{a}$ are the respective means of $u_0$ and $u_1$ and $a_0$ and $a_1$, and $dt/dx_{II}$ as $1/(\underline{u}' + \underline{a}')$ where $\underline{u}'$ and $\underline{a}'$ are the respective means of $u_0$ and $u_2$ and $a_0$ and $a_2$. The Mach lines can therefore only be sketched in approximately pending calculation of $a_1$ and $a_2$.

State 1 may be reached from state 0 by proceeding along a (zero strength) **II** wave from region 0. The change in $a$ corresponding to the change in $u$ from 0 to $u_1$ is shown in the inset diagram. From Eqn (AXI.14b) or (AXI.15b) the gradient $\Delta a/\Delta u$ is $+(\gamma - 1)/2$. Proceeding with this gradient from the origin at $u_0$, $a_0$ until the known point $u_1$ is reached determines corresponding $a_1$. The diagram on which this construction is carried out is known as the *state diagram*. The reader may wish to draw on the state diagram a line through the initial state point ($u_0$, $a_0$) having gradient $-(\gamma - 1)/2$ and hence pick off an $a_2$ corresponding to a $u_2$ of his choice. (Consistent with the **II** wave being compressive, $u_2$ must be negative.)

The same, simple case may now be stated in terms of the lattice method. Fig. AXI.9(*a*) shows the two waves pictorially. The only difference from the field

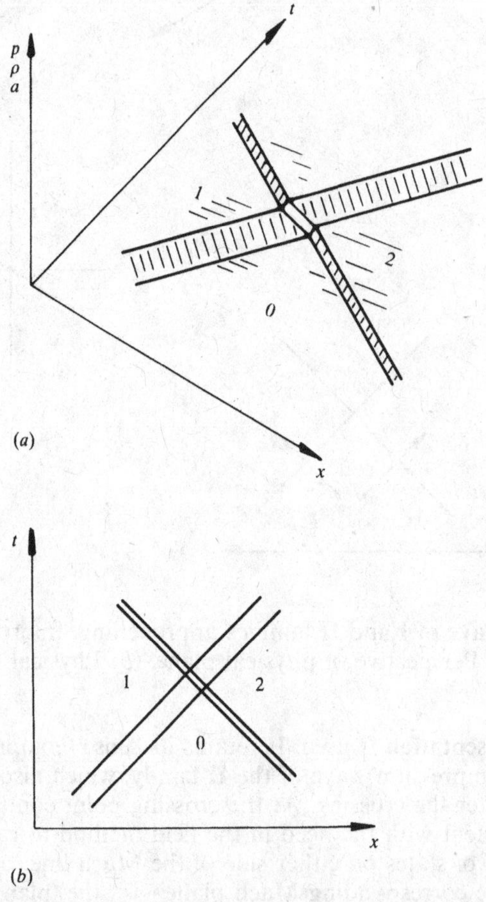


Fig. AXI.9 Simple wave of **I** and **II** families approaching. Interpreted in terms of the lattice method. (*a*) Perspective of physical plane. (*b*) Physical or Mach plane.

representation is that property discontinuities have been replaced by continuous properties with discontinuous gradients. In the plan view (the Mach plane) each wave must be represented by two lines (not necessarily parallel), one for the bottom of the incline, one for the top. It is worth noting from this that a construction carried out by the lattice method will contain one more **I** line and one more **II** line than the corresponding field construction carried out to the same numerical accuracy (Fig. AXI.9(*b*)).

Because of the unrepresentative simplicity of the simple wave case, the states at the extremities of the Mach lines shown are the same as those at the beginning. Indeed, outside the bounds of the respective pairs of lines, states are uniform, and the case is as for the field method. But the right-most line of the **I** pair must clearly proceed at a gradient calculated in terms of $u_0$ and $a_0$, while the left-most of the pair has a generally different gradient calculated from $u_1$ and $a_1$. The latter states are calculated by proceeding across the **I** waves along *any* **II** wave, regardless of strength. These and states 2 may thus be found by the means given for the field method.

Further examples will be in terms of the lattice method only. A return will be made to separate lattice and field treatments when non-simple waves arise and when the distinction is significant.

### *AXI.7.2 Simple waves crossing*

Fig. AXI.10 is essentially Fig. AXI.9 extended in the $t$ direction to incorporate state 3 which results from the two waves crossing. The uniform state 3 may be reached by crossing the **II** wave from state 1 and by crossing the **I** wave from state 2. This process is indicated on the state diagram, and amounts to the simultaneous solution of Eqns (AXI.14) (or (AXI.15)) for $u_3$ and $a_3$ in terms of the now known $u_1$, $a_1$, $u_2$ and $a_2$.

### *AXI.7.3 Simple wave reflected at a solid wall*

In Fig. AXI.11 a simple compressive wave of the **I** family propagates into a still fluid at $a_0$. Conditions behind the wave are $u_1$ and $a_1$. The wave is eventually incident on a solid wall. That the wave will be reflected (as a wave of the **II** family) is certain. To find out at what speed, and whether the reflected wave is a compression or expansion wave it is necessary to determine states $u_2$ and $a_2$.

$u_2$ is effectively known: region 2 is of uniform state, since it contains no other waves. At its right-hand extremity $u_2 = 0$ because there can be no flow perpendicular to the wall. It remains only to find $a_2$. Region 2 is accessed by travelling along a **I**-type wave from region 1. The starting point on the state diagram is $u_1$, $a_1$, and the direction is along a line of gradient $-(\gamma - 1)/2$. The required value of $a_2$ is that indicated when the line intersects the axis at a $u = 0$, and the right-most of the set of reflected **II** characteristics can now be constructed with gradient $1/(u_2 - a_2)$, since this is the lattice method.

### *AXI.7.4 Reflection from an open end (subsonic flow)*

In certain designs of Stirling machine, tubular heat exchangers connect with spaces of cross-sectional area considerably larger than the combined area of the tubes. In these circumstances, flow into and out of an individual tube is similar to flow into

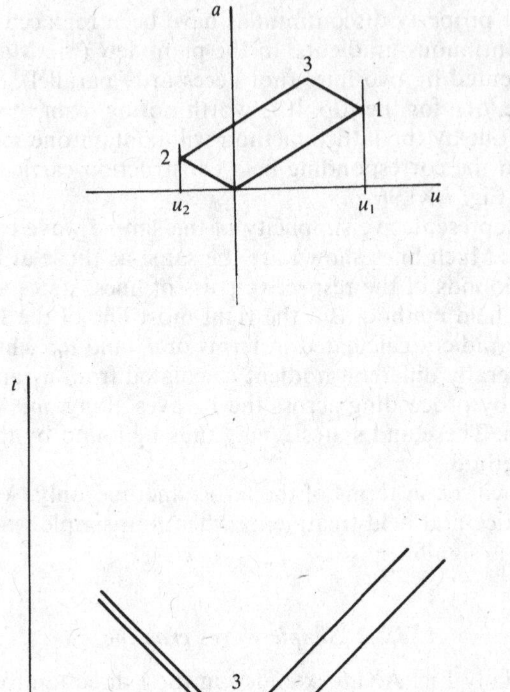

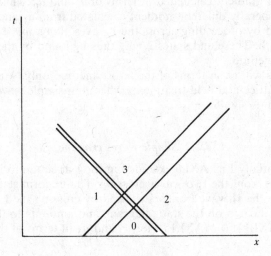

Fig. AXI.10 Simple waves crossing.

or out of an open end. The Method of Characteristics has a standard approach to these cases. Subsonic flow is assumed.

There is an important difference between outflow and inflow. In Fig AXI.12 the condition for outflow is that $u > 0$, (regardless of $a$). Conversely, that for inflow is $u < 0$. (If the opening is at the left-hand end of the tube, these conditions reverse.) Downstream of the exit in outflow conditions are highly three dimensional, and analysis is virtually impossible (Fig. AXI.12(*a*)). Use is therefore made of the empirical observation that pressure, $p$, just inside the duct is equal to that of the surrounding medium downstream, $p_\infty$, for all $t$.

For the inflow case, conditions just inside the mouth of the tube are taken to relate to conditions outside via the steady-flow energy equation (Fig. AXI.12(*b*)):

$$c_p T_\infty + 0^2/2 = c_p T + u^2/2$$

With $c_p = R\gamma/(\gamma - 1)$ and $a = \sqrt{(\gamma R T)}$ the steady-flow energy equation becomes:

$$a^2 + (\gamma - 1)u^2/2 = a_0{}^2 \qquad \text{(AXI.16)}$$

This is the equation of an ellipse – the so-called *steady-flow ellipse.* Drawn into the state diagram (Fig. AXI.12(*c*)) it defines $u$ in terms of $a$ at the open end *when there is inflow.*

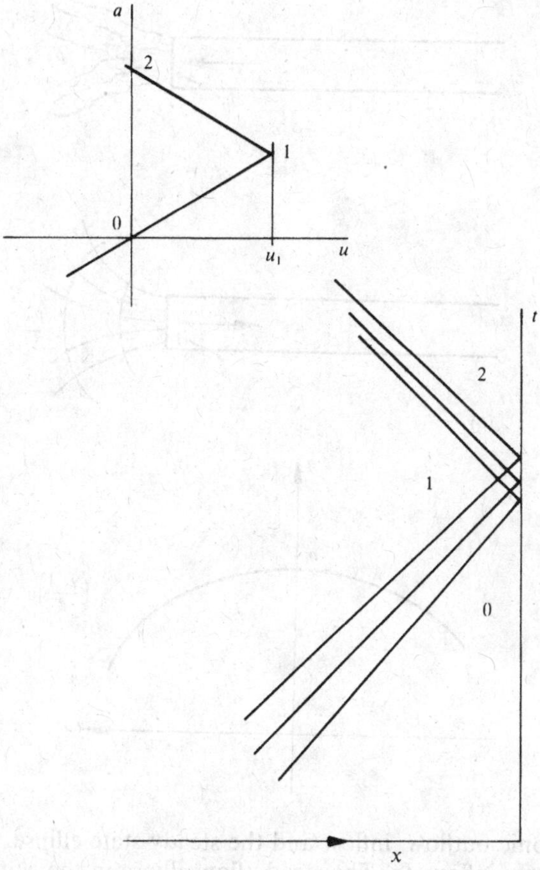


Fig. AXI.11 Reflection of simple compression wave at a wall.

Fig. AXI.13 shows a rarefaction wave of the **II** family arriving at the open, left-hand end of a duct, where there is inflow ($u > 0$). States 1 and 2 are known. State 3 must be determined in order to establish whether the reflected **I** wave is a compression or an expansion wave, and to find its direction in the $t$–$x$ plane.

State 3 is found from the simultaneous conditions (*a*) that it lies on the steady-flow ellipse through $a_0$ and (*b*) that it is reached by following a (zero strength) **II** wave out of state 2. The construction is shown on the state diagram (inset). $a_3 > a_2$ so the reflected wave is compressive.

Calculation involving the steady-flow ellipse in particular is facilitated by the use of variables normalized in the manner traditional to the Method of Characteristics. Dividing Eqn (AXI.16) by $a_0$ gives:

$$(a/a_0)^2 + (\gamma - 1)(u/a_0)^2/2 = 1 \qquad \text{(AXI.16a)}$$

On the physical, or Mach, plane, distance $x$ may be normalized by $L$, the length of the duct. Time, $t$, is normalized by the time for one wave traverse in the initial,

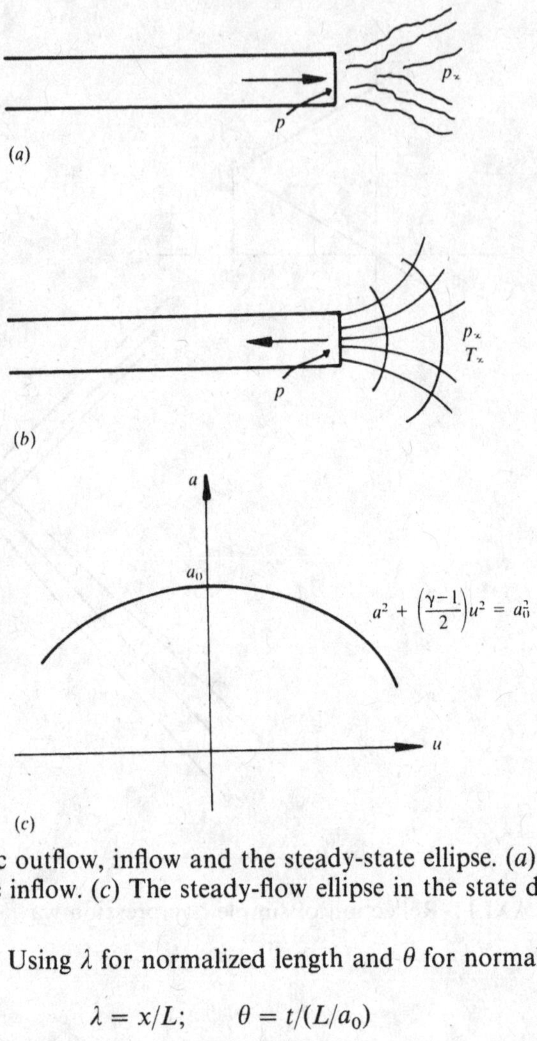


Fig. AXI.12 Subsonic outflow, inflow and the steady-state ellipse. (*a*) Subsonic outflow. (*b*) Subsonic inflow. (*c*) The steady-flow ellipse in the state diagram.

still gas, i.e., by $L/a_0$. Using $\lambda$ for normalized length and $\theta$ for normalized time:

$$\lambda = x/L; \qquad \theta = t/(L/a_0)$$

The gradient, $d\theta/d\lambda$, of a right- or left-running wave in fluid at $a_0$ and $u = 0$ is unity (i.e. $\pm 45°$) in normalized coordinates. The fact facilitates initial layout of the Mach plane when graphical techniques are being used, and makes it easy to set the gradient of individual Mach line segments by reference to the printed grid of graph paper.

The equations for change on the state diagram become

$$\Delta(u/a_0)_{\text{I}} = -[2/(\gamma - 1)]\Delta(a/a_0)_{\text{I}} \qquad \text{(AXI.17a)}$$

$$\Delta(u/a_0)_{\text{II}} = +[2/(\gamma - 1)]\Delta(a/a_0)_{\text{II}} \qquad \text{(AXI.17b)}$$

By choosing the axes of the state diagram to be $a/a_0$ and $[(\gamma - 1)/2]u/a_0$ all state change paths are at $\pm 45°$, resulting in a considerable reduction of work when a solution is being obtained graphically. From the homentropic assumption:

$$(a/a_0) = (p/p_0)^{(\gamma-1)/2\gamma} \qquad \text{(AXI.18)}$$

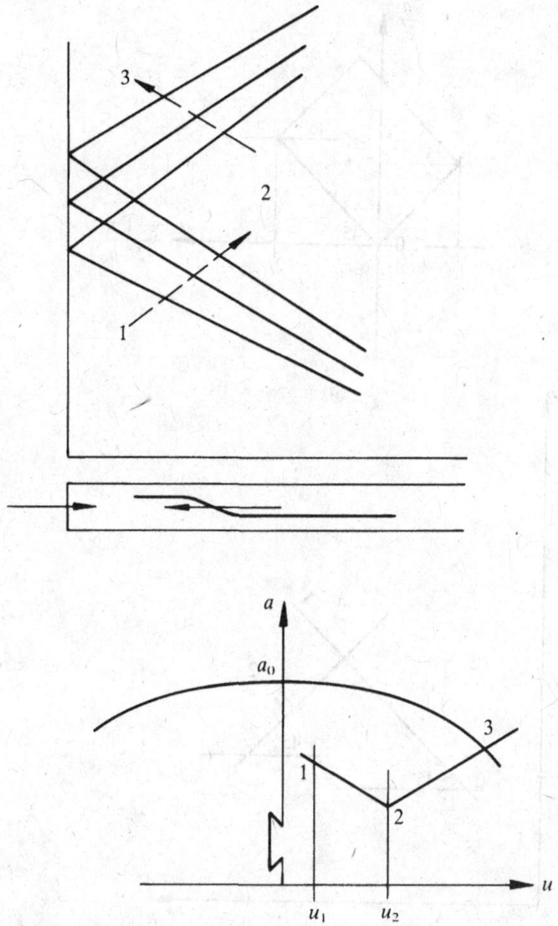


Fig. AXI.13 Left-travelling rarefaction wave reaching open end of duct during sustained inflow.

Fig. AXI.14 shows skeleton Mach and state planes in terms of normalized variables. This will be the pattern for all future Mach and state diagrams.

## AXI.8 The simple, finite wave in field and lattice interpretations

A simple wave is a continuous distribution of wavelets travelling in *one* direction – left to right. Fig. AXI.15 shows a right-running **I** wave and reviews the two interpretations – field and lattice – of the finite pressure distribution.

Crossing the wavelet system from right to left on a wave of the **II** family amounts to proceeding on the state diagram from the origin (point $u_0/a_0$, 1.0) to the state corresponding to conditions behind the wave. Depending upon choice of field as opposed to lattice method, the way in which the flow field is subdivided is slightly different. The continuous lines in the Mach plane represent individual waves in the lattice system. In this system the line on the state diagram is divided as indicated by the numerals. Uniformity of subdivisions is the recommended

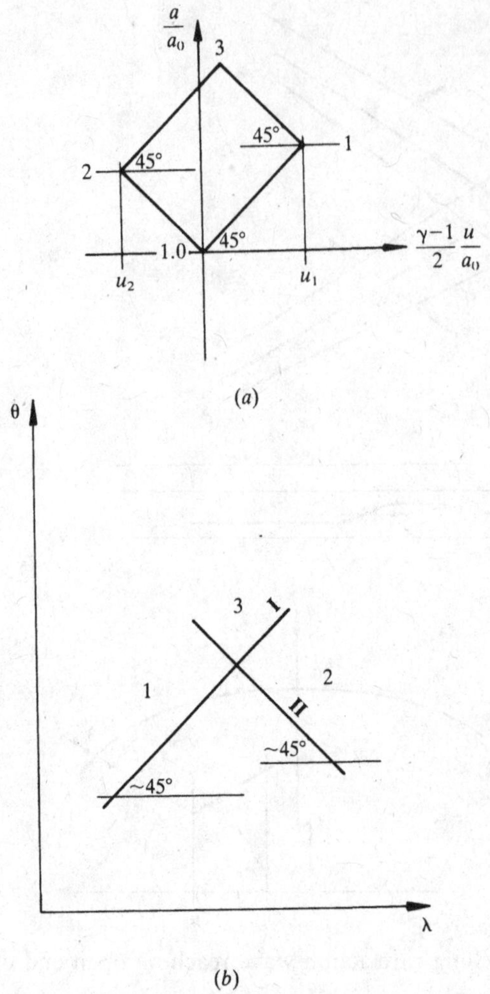


Fig. AXI.14 Skeletons of state and physical planes in terms of normalized variables. (*a*) State plane. (*b*) Physical plane.

option but is not essential. The individual Mach lines proceed with gradient, $d\theta/d\lambda$ computed on the basis of specific states, 1, 2, 3 etc. The profile of pressure vs $x$ at the selected time instant shows the outline of the wave to consist of straight-line segments having gradient discontinuity, but no discontinuity – infinitesimal or otherwise – in the property, $p$ or $a$ itself. The dashed characteristic directions correspond to the field interpretation, and thus represent a plan view of infinitesimal steps in the distribution of thermodynamic properties. In between the dashed characteristics are fields in which properties are uniform.

Individual uniform properties are those indicated by the respective alphabet characters. It will be seen that individual subdivisions of the state line in the field method bisect those for the lattice method. The characteristics in the field system have gradients, $d\theta/d\lambda$, which are calculated in terms of properties in the adjacent fields at either side, and which are therefore the mean of the gradients of the bounding lattice characteristics. It follows that a given problem subdivided as

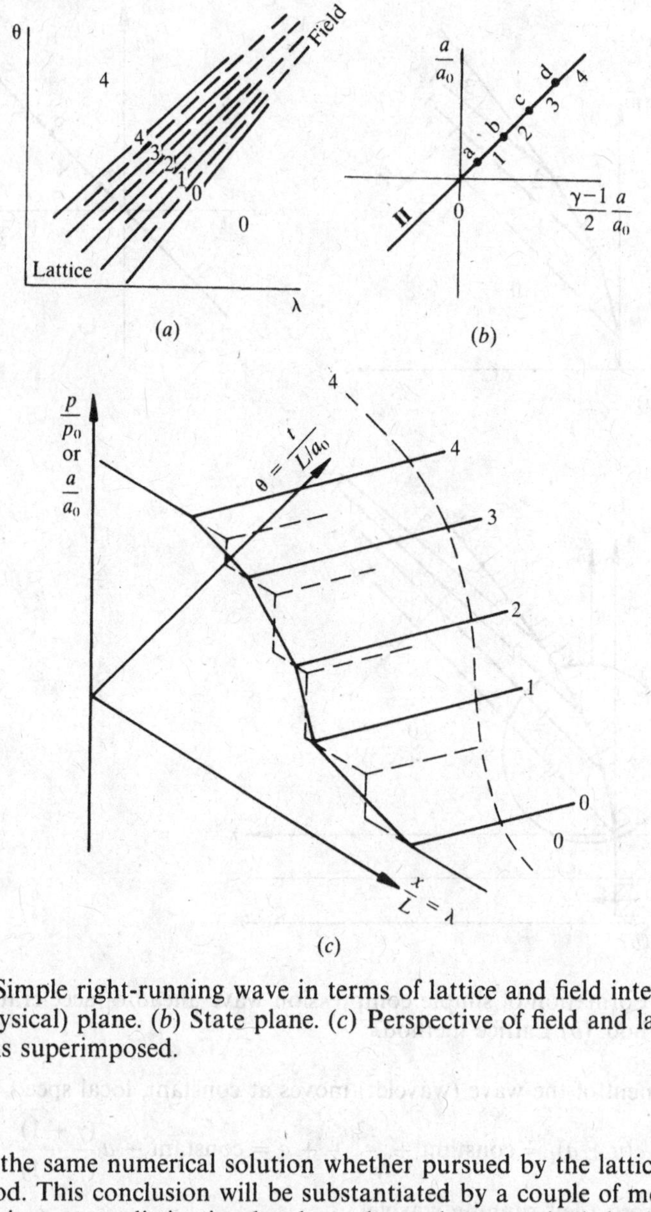


Fig. AXI.15 Simple right-running wave in terms of lattice and field interpretations. (*a*) Mach (physical) plane. (*b*) State plane. (*c*) Perspective of field and lattice interpretations superimposed.

shown yields the same numerical solution whether pursued by the lattice method or field method. This conclusion will be substantiated by a couple of more specific cases after an important distinction has been drawn between the behaviour of compression waves on the one hand and expansion waves on the other.

## AXI.9 Finite simple compression and expansion waves

A right-running (I) wave is considered. In either lattice or field interpretation:

$$u - \frac{2a}{\gamma - 1} = \text{constant across the wave}$$

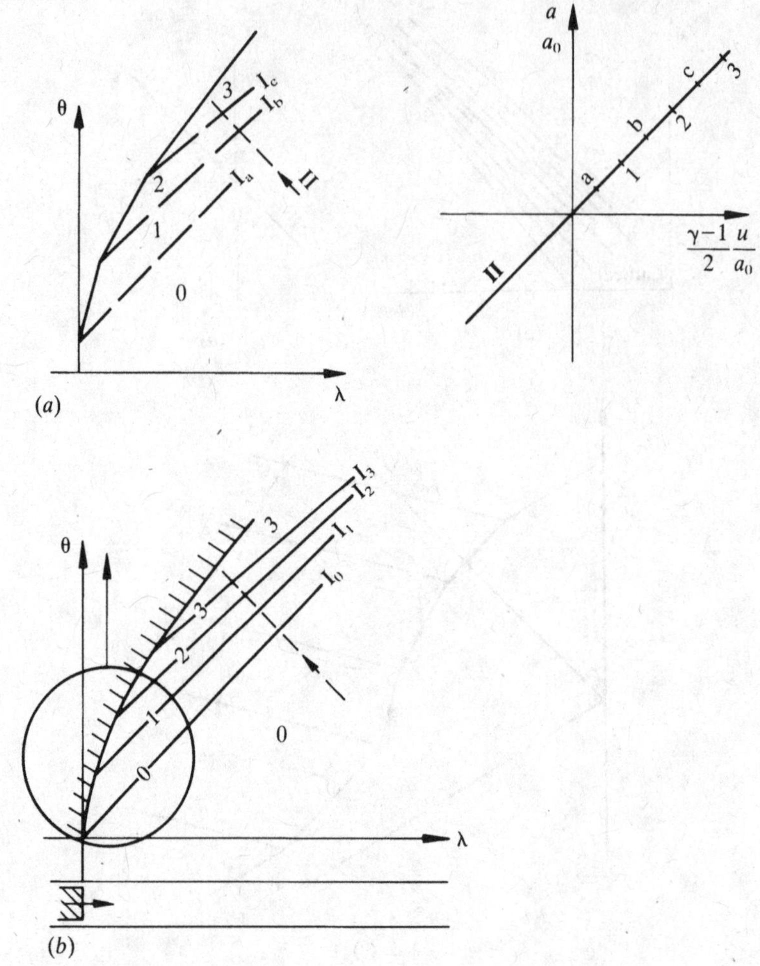


Fig. AXI.16 Formation of simple compression wave ahead of accelerating piston. (*a*) Field method. (*b*) Lattice method.

Each component of the wave (wavelet) moves at constant, local speed, $u + a$

$$(u + a)_{\mathrm{I}} = \text{constant} + \frac{2a}{\gamma - 1} + a = \text{constant} + a\frac{(\gamma + 1)}{(\gamma - 1)} \qquad \text{(AXI.19a)}$$

Similarly, across a left-running wavelet

$$u + \frac{2a}{\gamma - 1} = \text{constant}$$

Each wave moves at constant, local speed, $u - a$

$$(u - a)_{\mathrm{II}} = \text{constant} - \frac{2a}{\gamma - 1} - a = \text{constant} - a\frac{(\gamma + 1)}{(\gamma - 1)} \qquad \text{(AXI.19b)}$$

Across a compression wave sonic speed increases and *a compression wave steepens.* Sonic speed, $a$, decreases across an expansion, and *an expansion wave flattens out.*

## AXI.10 Generation of finite simple compression wave

Fig. AXI.16 shows the formation of a compression wave ahead of a piston accelerating from rest to a final velocity $u_f$. The total change in state across the resulting wave may be represented in the state diagram by the point on the (zero-strength) **II** characteristic at $u/a_0 = u_f/a_0$.

In the lattice method the piston curve is continuous, and labelled at selected values of $u$ up to $u_f$. If the increments in $u$ are uniform, the subdivisions of the state line are also uniform. Individual Mach lines (the continuous lines 1, 2, 3, . . .) in the physical plane travel at $d\theta/d\lambda$ calculated in terms of the states 1, 2, 3, . . . . Since both $u$ and $a$ are increasing the slopes of successive Mach lines decrease, so that, all things being equal, there is eventual convergence.

The corresponding construction in the field method sees piston motion as a succession of phases at uniform velocity. The closest approximation to the lattice case is achieved by arranging for the discontinuities in velocity to occur at the lettered points mid-way between the lattice labels (see both physical and state diagrams). The first field wave propagates at a velocity given by the mean of conditions either side – i.e. at the mean of states 1 and 2 of the lattice method. It thus follows a path mid-way between lattice lines 1 and 2, as shown in the figure.

## AXI.11 Finite simple expansion wave

Fig. AXI.17 is the expansion equivalent of the compression case just dealt with. All computational steps are comparable, except that, with piston velocity reaching increasingly negative values, the properties line in the state diagram shows continuously decreasing $a/a_0$. Successive Mach lines have increasing gradient in the $\theta - \lambda$ plane, and accordingly diverge. As noted in an earlier section, the field construction involving the same number of subdivisions of state across the wave leads to one fewer Mach lines in the physical plane than does the lattice method.

A type of expansion wave for which there is no compressive equivalent is the *expansion fan.* An impulsive expansion (Fig. AXI.18) results when a piston is put impulsively into motion with a velocity which brings about expansion. The total change in the fluid states may be indicated on the state diagram as before, simply by locating the terminal value of velocity, $u_f$. These changes of state occur at a singularity at the origin in the Mach plane, but the first wavelet to propagate from the singularity into the still gas must (in the lattice method) do so at a gradient given in terms of $u_0$ and $a_0$. Similarly, a wavelet must propagate at conditions $u_f$ and corresponding $a_f$. The neatest way of constructing intermediate wavelets is by a uniform subdivision of the state line, as shown in Fig. AXI.18. The field and lattice methods give comparable results if subdivisions of states for the former lie mid-way between those of the latter, as indicated.

Figs AXI.19($a$) and ($b$) are perspective sketches of the expansion fan in the physical plane in the lattice and field representations respectively. Fig. AXI.19($c$) is an attempt to superimpose the two pictures and to show why numerical results derived from both are the same.

An unsteady flow problem involving waves of one family only is scarcely a problem, since the solution may be defined in a few graphical or numerical steps – as suggested by the foregoing treatment of simple compression and expansion waves. The problems of real interest are those involving 'non-simple' waves – the repeated crossing of waves of both families.

A typical case is that arising out of the reflection of a simple wave at the closed

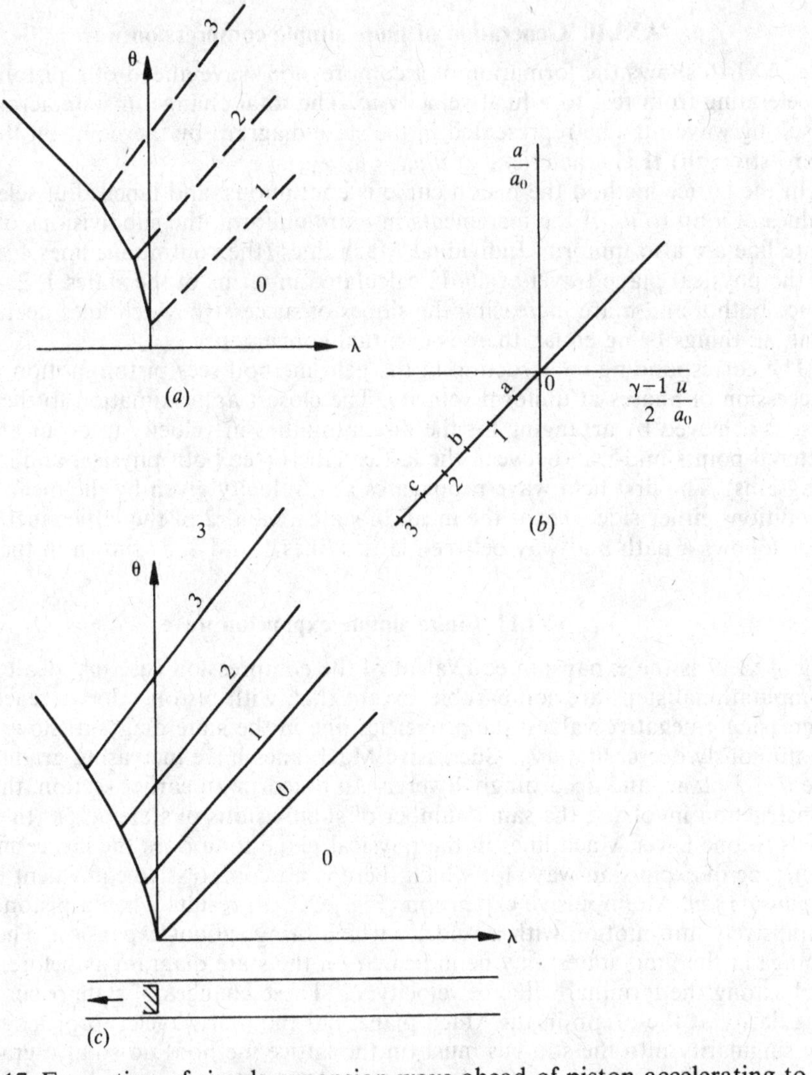


Fig. AXI.17 Formation of simple expansion wave ahead of piston accelerating to the left. (*a*) Physical plane (field method). (*b*) State diagram. (*c*) Physical plane (lattice method).

(or open) end of a duct. Fig. AXI.20 is an example, showing the arrival of an expansion fan at the closed end of a tube. Initial conditions permit the properties of all of the original fan elements to be specified up to the point where they meet the **II** wave reflected from the first **I** wave. Thereafter the characteristics net consists of quadrilaterals which can only be sketched approximately until the gradients of individual line elements have been determined. This is achieved by integrating systematically through the mesh to give the values of a previously unknown state in terms of two sets of states which are known. Each such integration step is known as a *unit process*. The unit process for the field method

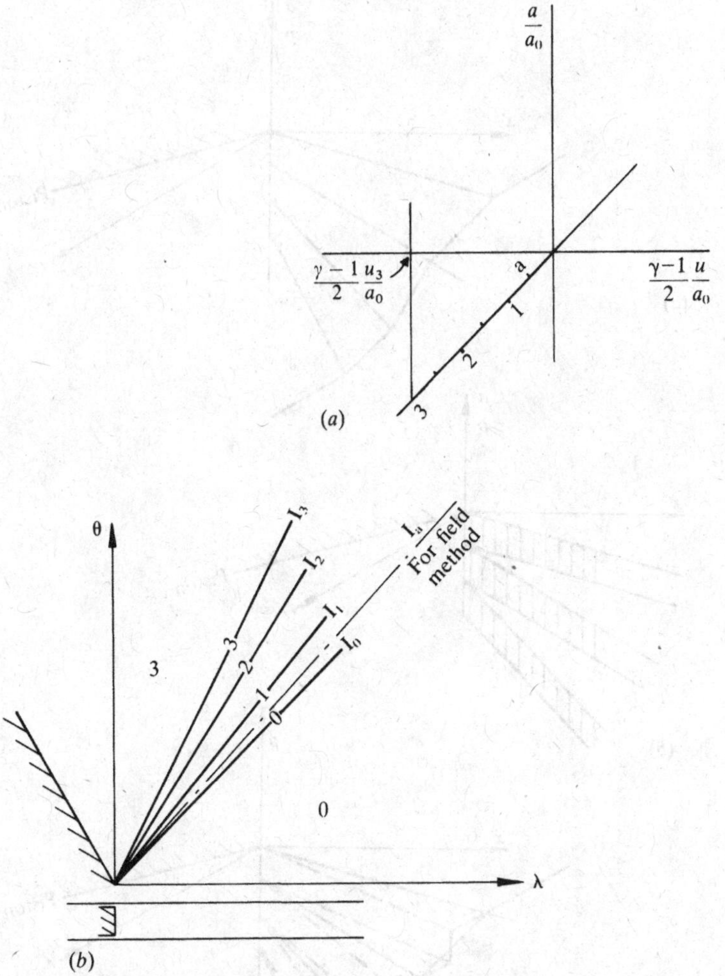


Fig. AXI.18 Generation of expansion fan due to impulsive start of piston motion. (*a*) State plane. (*b*) Physical plane (lattice method).

differs in detail, although not in principle, from that for the lattice method. The former will be illustrated first.

## AXI.12 The unit process

### *AXI.12.1 Lattice method*

Fig. AXI.21(*a*) is an enlarged version of that part of Fig. AXI.20 of current interest. Individual Mach line interactions are numbered. The states at points 1, 2, 3 and 4 are known. Those at 5, 6, 7, 8 and 9 are to be found.

It is known that at point 5 $u = 0$, so that proceeding along the **I** direction from point 2 gives $a_5$ (see state diagram, Fig. AXI.21(*b*)). The gradient of line

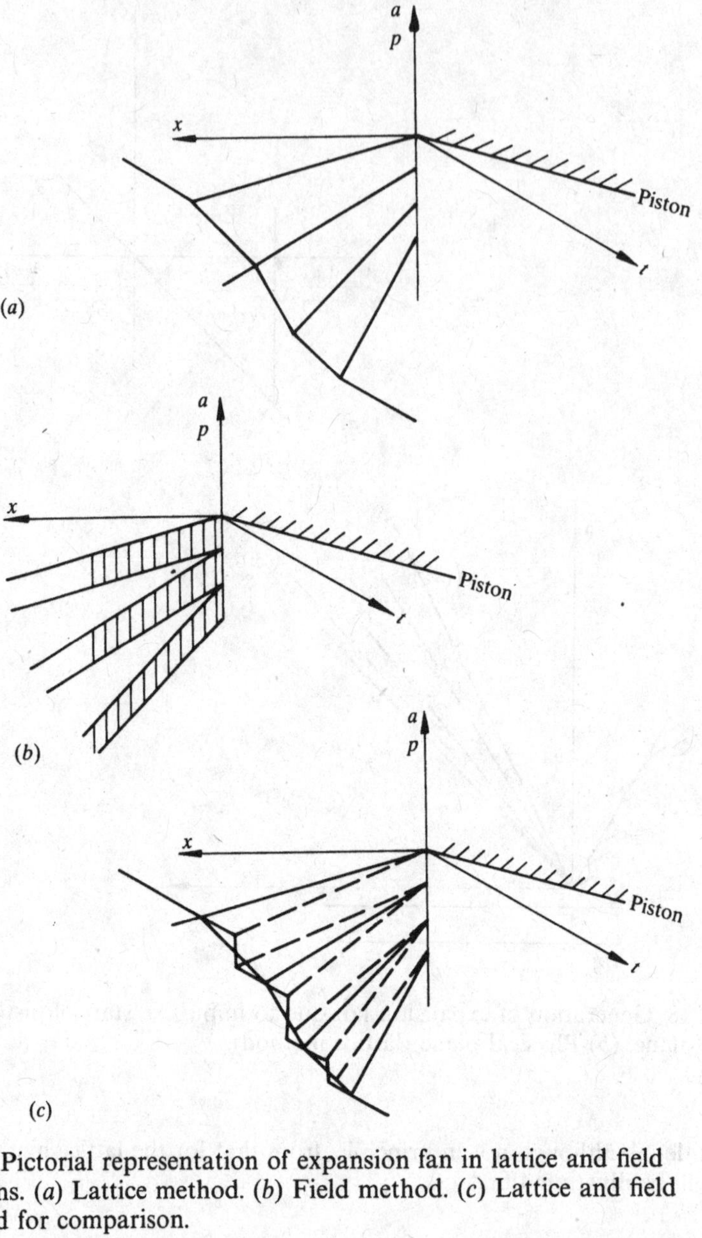


Fig. AXI.19 Pictorial representation of expansion fan in lattice and field interpretations. (*a*) Lattice method. (*b*) Field method. (*c*) Lattice and field superimposed for comparison.

element 2–5 may now be calculated from the mean of states 2 and 5 and laid in accurately. The calculation of state 6 from 3 and 5 typifies the unit process in the lattice method. State 6 is attained by travelling along the **I** characteristic from 3 *and* by travelling along the **II** characteristic from 5. The intersection of the two state lines (Fig. AXI.21(*b*)) thus determines state 6. In the physical plane the gradients of the Mach line elements 3–6 and 5–6 may now be computed from the means of states 3 and 6, and 5 and 6 respectively. With states 6 determined, the same unit process may now be used to find states 7 from 4 and 6.

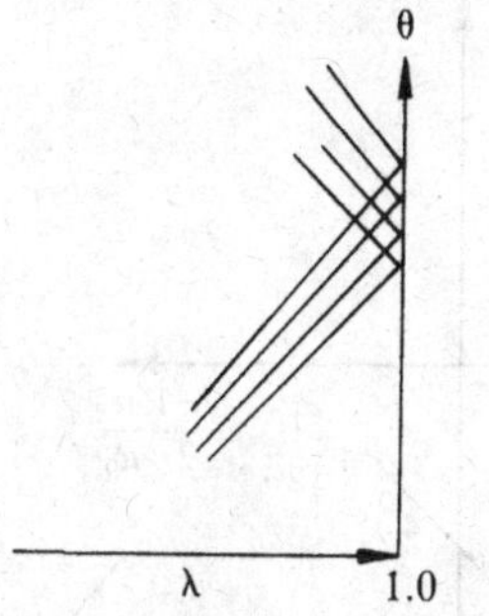


Fig. AXI.20 Expansion fan arriving at the closed end of a duct.

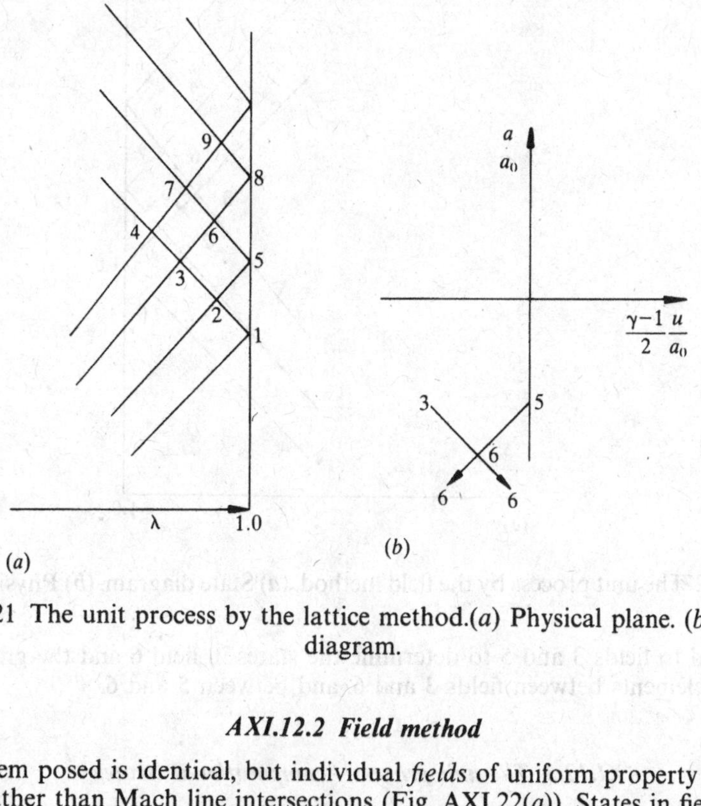


Fig. AXI.21 The unit process by the lattice method.(*a*) Physical plane. (*b*) State diagram.

### AXI.12.2 Field method

The problem posed is identical, but individual *fields* of uniform property are now labelled rather than Mach line intersections (Fig. AXI.22(*a*)). States in fields 0, 1, 2 and 3 are known. Those in 4, 5, 6, 7 and 8 are to be calculated.

By analogy with the lattice case, conditions in field 4 are fixed by the fact that, for this particular case of the closed tube, $u$ there is zero. $a_4$ thus follows immediately in the state diagram (Fig. AXI.22(*b*)). The gradient of the Mach line element between 1 and 4 is calculated in terms of the mean of $u$ and $a$ between the two fields and the line element marked.

States in field 5 are reached both by a **I** wave through field 2 and also by a **II** wave through field 4. The intersection of corresponding lines on the state diagram fixes state 5. The gradients of Mach line elements between fields 2 and 5 and fields 4 and 5 may now be calculated and the line elements ruled in. The unit process is

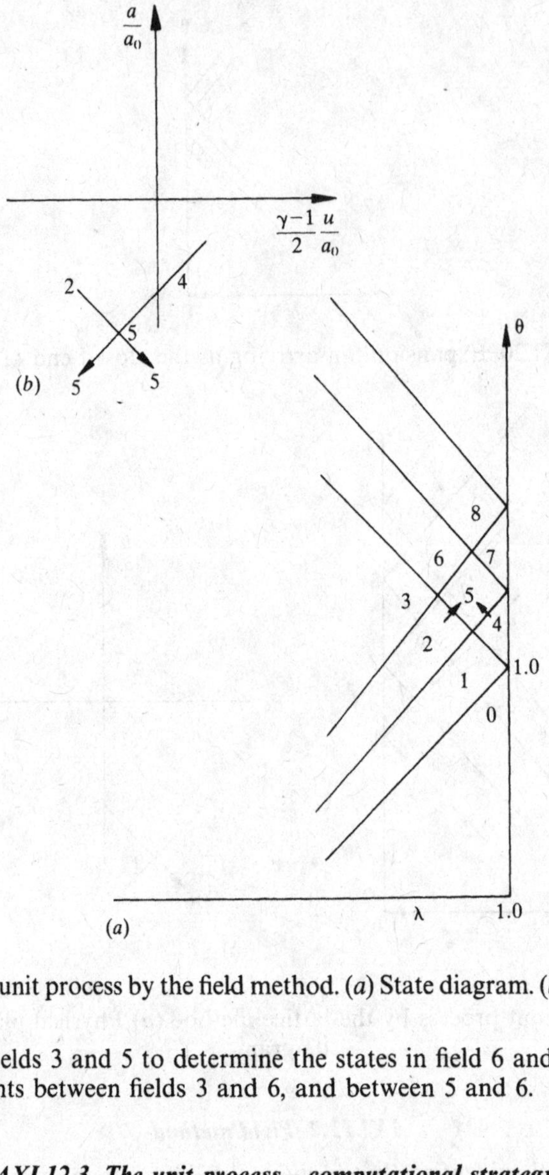


Fig. AXI.22 The unit process by the field method. (*a*) State diagram. (*b*) Physical plane.

now applied to fields 3 and 5 to determine the states in field 6 and the gradients of Mach line elements between fields 3 and 6, and between 5 and 6.

### AXI.12.3 The unit process – computational strategy

Lattice method:

(1) In the *Mach plane* Decide on a suitable number of Mach lines. Do a rough sketch showing all the lines. Label *intersections*.

(2) In the *state plane* Locate the state of a new point (say, 6) from those at known points, say 3 and 5.

(3) In the *Mach plane* *Accurately* construct from 3 to 6 at

$$\frac{d\theta}{d\lambda_1} = \frac{1}{\underline{u}_{36}/a_0 + \underline{a}_{36}/a_0}$$

and from 5 to 6 at

$$\frac{d\theta}{d\lambda_{II}} = \frac{1}{\underline{u}_{56}/a_0 - \underline{a}_{56}/a_0}$$

Field method:

| | |
|---|---|
| (1) In the *Mach plane* | Choose a suitable number of Mach lines, sketch and label *spaces.* |
| (2) In the *state plane* | Locate unknown state 5 from known states in adjacent fields 2, 4. |
| (3) In the *Mach plane* | Accurately determine the gradient of the Mach line element between fields 2 and 5. |

$$\frac{d\theta}{d\lambda_{II}} = \frac{1}{\underline{u}_{25}/a_0 - \underline{a}_{25}/a_0}$$

and between 4 and 5

$$\frac{d\theta}{d\lambda_{I}} = \frac{1}{\underline{u}_{45}/a_0 + \underline{a}_{45}/a_0}$$

Precisely the same logical and numerical sequence for the homentropic flow case may be followed in a version of the process programmed for computer. The computer equivalent of the unit process is dealt with in Chap. 11. It remains to deal with the friction and heat transfer terms which were ignored while the fundamentals of the Characteristics method were presented. It will be recalled that these appear in the state equations only (Eqns (AXI.9)). The expressions for the Mach line gradients remain as for the homentropic case, although there is an additional characteristic direction (the particle path) given by Eqn (AXI.8c), reflecting the possibility of property gradient discontinuities due to the effects of friction and temperature on individual fluid particles.

A method outlined by Shapiro[1] is presented first: Fig. AXI.23(*a*) is a schematic of the now familiar unit process for the left- and right-running waves, but with typical particle paths superimposed. (Comparable particle paths exist in the homentropic case, of course, and may be plotted into the Mach plane if desired, but their construction is not a necessary part of the solution process.) Fig. AXI.23(*c*) is the state plane. It is assumed that Eqns (AXI.8), (AXI.10) and that for $dp_{path}$ are being used – i.e., that calculation is essentially in terms of *a* and *u*. (It is possible to reformulate so that calculation proceeds in terms of *u* and *p*. Advantages are (*i*) that one is frequently more interested in pressure than in acoustic speed (or temperature), and (*ii*) that the final term in Eqns (AXI.10), i.e., the entropy gradient term, calls for inelegant interpolation. This is not necessary when computing in terms of *u* and *p*.)

Solution is assumed to have attained a stage where states at points 1 and 2 are known. $A_x$ and $dA_x/dx$ are known as if functions of *x*, and preliminary estimates of *q* and *F* are available. Integration of the path equation has proceeded along several path lines between 1 and 2 so that *p* and its derivative in the *x* direction are known locally. The sequence of steps described below substitutes the unit process of the homentropic case. Computation is ideally in terms of normalized variables, $\theta$, $\lambda$, $u/a_0$, $a/a_0$ etc. The use of absolute variables in the account which follows is for reasons of compactness of presentation only.

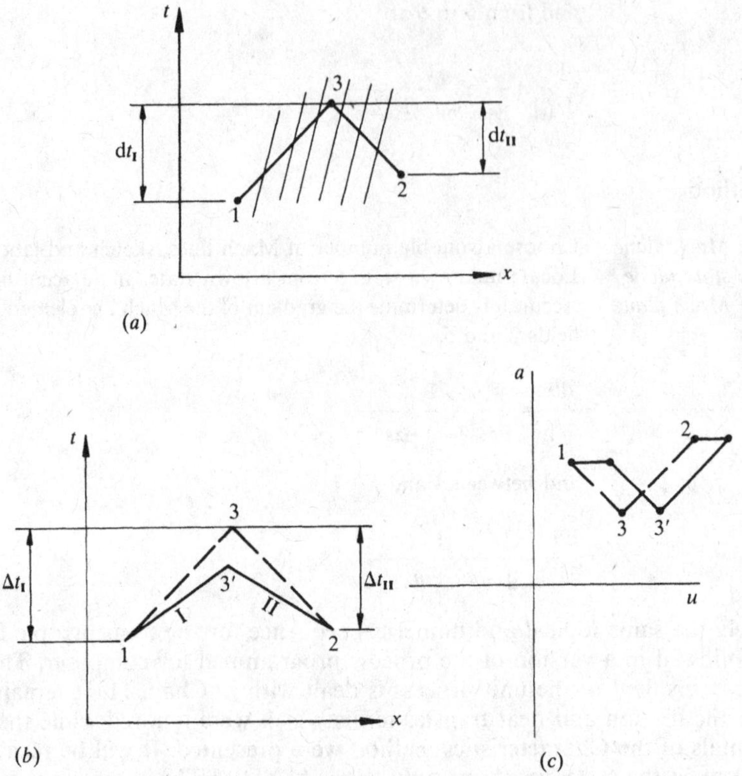


Fig. AXI.23 Shapiro's unit process[1] when friction and/or heat transfer are involved calls for iteration between physical and state planes. (*a*) Characteristic directions in the physical plane. (*b*) Approximate (dashed) and corrected (full) Mach lines. (*c*) Approximate (dashed) and corrected (full) state plane construction.

(1) A first approximation to point 3 in the state plane is obtained by provisionally ignoring all terms in Eqns (AXI.10) which depend on the d*t*s. The process is shown exaggerated by dashed lines in Fig. AXI.23(*c*).
(2) Using the provisional means $\underline{u}_{1-3}$, $\underline{a}_{1-3}$, $\underline{u}_{2-3}$ and $\underline{a}_{2-3}$, provisional **I** and **II** characteristics are constructed in the Mach plane, as shown by dashed lines in Fig. AXI.23(*b*). The vertical distances on the *t* axis between points 1 and 3 and between points 2 and 3 are first approximations to $dt_I$ and $dt_{II}$ respectively.
(3) The time dependent terms of Eqns (AXI.10) may now be tentatively calculated using the provisional $dt_I$ and $dt_{II}$. These terms add algebraically to *u* as indicated in the schematic state diagram. Using the improved mean values, $\underline{u}_{1-3}$, $\underline{a}_{1-3}$, $\underline{u}_{2-3}$ and $\underline{a}_{2-3}$ the construction in the physical plane may now be improved, leading to the more accurate point 3′.

The computational cycle repeats until convergence is satisfactory.

Shapiro's approach appears apt and comprehensive until one comes to implement it in detail. At that stage it is found that the method lacks specific means of feeding information carried along path lines into the Mach line construction process between successive Mach line intersections. The difficulty disappears if it is agreed to use a unit process *precisely* the same in concept as that for the homentropic flow *except that* the intersection point is between two Mach lines *or* between a Mach line and a path line as appropriate. Chap. 11 explains more fully, and demonstrates that such a unit process is readily programmed for computer.

Indeed, in this computer age, the rôle of the manual graphical procedure is merely to give an insight into, and practice with, the method. The solution of anything other than a trivial problem calls for digital processing. This has its own, special challenges which Chap. 11 deals with.

# Appendix XII
# Frequently-used data

## AXII.1 Perfect gases

Over a limited range of pressures and temperatures close to ambient, it may be assumed that the properties listed against the gases of Table AXII.1 are constant, and that pressure, $p$, density, $\rho$, and temperature, $T$, are related by $p = \rho RT$. $R = \underline{R}/\underline{M}$, where $\underline{R}$ is the universal gas constant with value 8.3143 kJ/kmol K.

Some of the gases listed are unlikely candidates for routine use in a Stirling cycle machine. On the other hand, their availability makes possible the experimental extension of the performance envelope to values of the dimensionless parameters which would be difficult to achieve by varying pressure, rpm etc. alone.

Table XII.1. *Properties of selected gases near ambient conditions*[1,2]

| Gas | Nom. molar mass $\underline{M}$ kg/kmol | Gas constant $R$ kJ/kg K | Specific heat $c_p$ kJ/kg K | Specific heat $c_v$ | Specific heat ratio $c_p/c_v = \gamma$ |
|---|---|---|---|---|---|
| $H_2$ | 2 | 4.12 | 14.20 | 10.08 | 1.41 |
| He | 4 | 2.08 | 5.19 | 3.11 | 1.67 |
| Ne | 20 | 0.415 | 1.03 | 0.62 | 1.66 |
| $N_2$ | 28 | 0.297 | 1.04 | 0.74 | 1.4 |
| CO | 28 | 0.297 | 1.04 | 0.75 | 1.4 |
| Air[a] | 29 | 0.287 | 1.01 | 0.72 | 1.4 |
| $O_2$ | 32 | 0.260 | 0.92 | 0.66 | 1.4 |
| Ar | 40 | 0.208 | 0.52 | 0.31 | 1.67 |
| $CO_2$ | 44 | 0.189 | 0.85 | 0.66 | 1.28 |

[a] Volumetric and molar composition: 21% $O_2$, 79% atmospheric nitrogen
Gravimetric composition: 23.2% $O_2$, 76.8% atmospheric nitrogen

## AXII.2 Conversion factors – non-SI to SI. (From various sources, including Ref. 3.)

An advantage of the dimensionless formulation adopted throughout the main text is that the dimensionless variables assume the same numerical values in *any* consistent set of units. The entries which follow cover the need for occasional conversion from one system to another.

*Base units of the International System (SI)*

| | | |
|---|---|---|
| length | metre | m |
| mass | kilogram | kg |
| time | second | s |
| thermodynamic temperature | kelvin | K |

*Length*

| | |
|---|---|
| 1 ft (foot) | = 0.3048 m |
| 1 in (inch) | = 0.0254 m |

*Area*

| | |
|---|---|
| 1 $ft^2$ | = 0.092903 $m^2$ |
| 1 $in^2$ | = $645.16 \times 10^{-6}$ $m^2$ |

*Volume*

| | |
|---|---|
| 1 $ft^3$ | = 0.028316 $m^3$ |
| 1 $in^3$ | = $16.387 \times 10^{-6}$ $m^3$ |

*Mass*

| | |
|---|---|
| 1 lb | = 0.45359 kg |
| 1 oz | = 0.02835 kg |

*Mass flux rate*

| | |
|---|---|
| 1 $lb_m$/hr $ft^2$ | = $1.355 \times 10^{-3}$ kg/$m^2$ s |

*Density*

| | |
|---|---|
| 1 $lb_m/ft^3$ | = 16.0185 kg/$m^3$ |
| 1 $lb_m/in^3$ | = $27.68 \times 10^3$ kg/$m^3$ |

*Force*

| | |
|---|---|
| 1 $lb_f$ | = 4.44822 N |

*Pressure (stress)*

| | |
|---|---|
| 1 Pa | = 1 N/$m^2$ |
| 1 $lb_f/ft^2$ | = 47.8803 N/$m^2$ |
| 1 $lb_f/in^2$ | = 6.89476 kN/$m^2$ |
| 1 bar | = $10^5$ N/$m^2$ |
| 1 mbar | = 100 N/$m^2$ |
| 1 atm | = 0.101325 MN/$m^2$ |
| 1 mm Hg | = 133.322 N/$m^2$ |
| 1 torr | = 133.322 N/$m^2$ |
| 1 $kg_f/cm^2$ | = 9.80665 kN/$m^2$ |
| 1 $kg_f/m^2$ | = 9.80665 N/$m^2$ |
| 1 mm $H_2O$ | = 9.80665 N/$m^2$ |

*Viscosity (dynamic)*

| | |
|---|---|
| 1 Pa s | = 1 N m s |
| 1 $lb_m$/ft s | = 1.4882 kg/m s |
| 1 $lb_f$ h/$ft^2$ | = 0.1724 MPa s |
| 1 $lb_f$ s/$ft^2$ | = 47.880 Pa s |
| 1 $lb_f$ s/$in^2$ (1 reyn) | = 6894 Pa s |
| 1 cP (centipoise) | = 0.001 Pa s |

*Viscosity (kinematic), diffusivity*

| | |
|---|---|
| 1 $ft^2$/hr | = $2.583 \times 10^{-5}$ $m^2$/s |
| 1 $ft^2$/s | = 0.0929 $m^2$/s |
| 1 centistoke | = $10^{-6}$ $m^2$/s |

*Work, energy, quantity of heat*

| | |
|---|---|
| 1 J | = 1 N m |
| 1 Btu | = 1.05506 kJ |
| 1 ft $lb_f$ | = 1.35582 J |

*Power, heat flow rate*

| | |
|---|---|
| 1 W | = 1 N m/s = 1 J/s |
| 1 Btu/s | = 1.05435 kW |
| 1 Btu/hr | = 0.293071 W |
| 1 horsepower hp | = 745.7 W |
| 1 ft $lb_f$/s | = 1.35582 W |

*Temperature* (R indicates degrees Rankine. 1 degree R temperature difference = 1 degree F (Fahrenheit). 1 degree K (Kelvin) temperature difference = 1 degree C)

| | |
|---|---|
| 1 R | = $\frac{5}{9}$ K |

*Specific heat capacity, specific entropy*

| | |
|---|---|
| 1 Btu/$lb_m$ R | = 4186.8 J/kg K |
| 1 ft $lb_f$/$lb_m$ R | = 5.38032 J/kg K |

*Coefficient of convective heat transfer*

| | |
|---|---|
| 1 Btu/$ft^2$ hr °F | = 5.67826 W/$m^2$ °C |

*Thermal conductivity*

| | |
|---|---|
| 1 Btu/hr ft °F | = 1.7307 W/m K |
| 1 Btu in/$ft^2$ hr °F | = 0.144228 W/m K |
| 1 Btu in/$ft^2$ s °F | = 519.220 W/m K |

## AXII.3 Fluid properties as a function of temperature

Particularly in the case of reversed-cycle machines (coolers) the range of temperatures spanned between expansion and compression ends is sufficiently great that properties such as the specific heats, coefficient of dynamic viscosity and thermal conductivity cannot be considered uniform. The most convenient way of dealing with the temperature dependence in automated computation is to express it in functional form with a reference value of the property as a simple multiplier. This reference value gets taken up into the defining dimensionless groups, leaving a dimensionless expression controlling the variation with (dimensionless) temperature. The temperature dependence of coefficient of dynamic viscosity provides an example:

### *AXII.3.1 Coefficient of dynamic viscosity, μ*

The variation of $\mu$ with absolute temperature, $T$, may be expressed in terms of a Sutherland temperature, $T_{s\mu}$, and a specific value of $\mu$, viz, $\mu_0$, at temperature $T_0$:

$$\mu = \mu_0 \frac{T_0 + T_{s\mu}}{T + T_{s\mu}} \left(\frac{T}{T_0}\right)^{3/2}$$

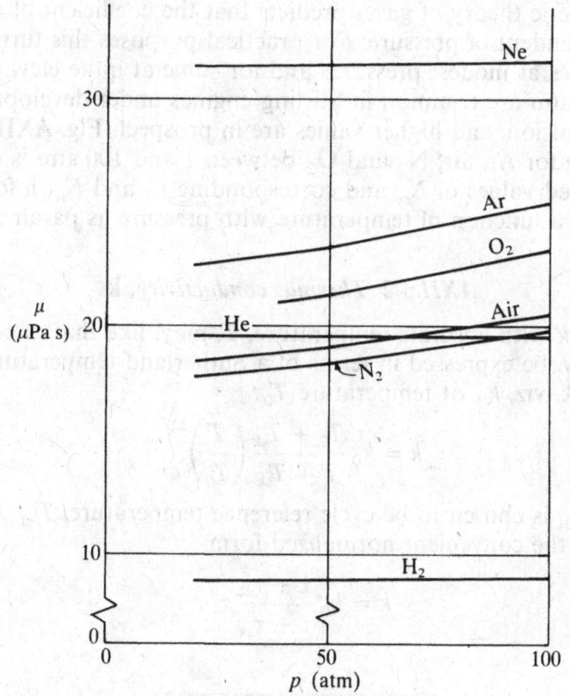


Fig. AXII.1 Variation of coefficient of dynamic viscosity with pressure for selected gases. (From Ref. 5.)

If temperature $T_0$, is chosen to be cycle reference temperature, $T_{ref}$, the foregoing expression reduces to the convenient normalized form:

$$\mu = \mu_0 \frac{1 + \tau_{s\mu}}{\tau + \tau_{s\mu}} \tau^{3/2}$$

Table AXII.2 contains values of $T_{s\mu}$ for some common gases.

Table AXII.2. *Sutherland temperatures, $T_{s\mu}$, for common gases*[4]

| Gas | $T_0$ K | $\mu_0$ Pa s | Sutherland temp, $T_{s\mu}$ K |
|---|---|---|---|
| $H_2$ | 300 | $0.0090 \times 10^{-3}$ | 84.4 |
| He | 300 | $0.0178 \times 10^{-3}$ | 80 |
| $N_2$ | 300 | $0.0179 \times 10^{-3}$ | 106 |
| CO | 300 | $0.0179 \times 10^{-3}$ | 136 |
| Air | 300 | $0.0184 \times 10^{-3}$ | 112 |
| $O_2$ | 300 | $0.0208 \times 10^{-3}$ | 139 |
| Ar | 300 | $0.0229 \times 10^{-3}$ | 144 |
| $CO_2$ | 300 | $0.0150 \times 10^{-3}$ | 222 |
| Ne | 300 | $0.0313 \times 10^{-3}$ | |

The simple kinetic theory of gases predicts that the coefficient of dynamic viscosity is independent of pressure. For practical purposes this turns out to be true for most gases at modest pressures and for some at quite elevated values. Pressures of 100 atm are common in Stirling engines under development for automotive propulsion, and higher values are in prospect. Fig. AXII.1 shows how the variation in $\mu$ for Ar, air, $N_2$ and $O_2$ between 1 and 100 atm is sufficient to influence computed values of $N_{re}$ and corresponding $C_f$ and $N_{st}$; $\mu$ for $N_2$ is shown in Fig. AXII.2 as a function of temperature with pressure as parameter.

### *AXII.3.2 Thermal conductivity,* k

The variation of $k$ with absolute temperature, $T$, may, like that of coefficient of dynamic viscosity, be expressed in terms of a Sutherland temperature, $T_{sk}$, and a specific value of $k$, viz, $k_0$, at temperature $T_0$:

$$k = k_0 \frac{T_0 + T_{sk}}{T + T_{sk}} \left(\frac{T}{T_0}\right)^{3/2}$$

If temperature $T_0$, is chosen to be cycle reference temperature, $T_{ref}$, the expression again reduces to the convenient normalized form:

$$k = k_0 \frac{1 + \tau_{sk}}{\tau + \tau_{sk}} \tau^{3/2}$$

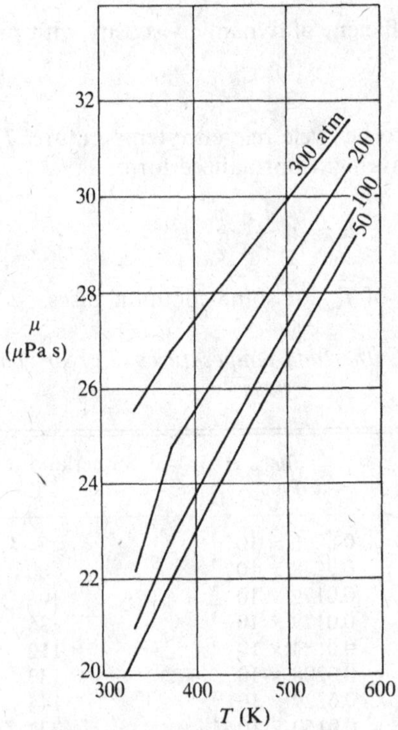


Fig. AXII.2 Coefficient of dynamic viscosity of nitrogen gas with pressure as parameter. (From Ref. 5.)

Table AXII.3 gives values of $k_0$ and corresponding $T_{sk}$ for some common gases.

Table AXII.3. *Sutherland temperatures, $T_{sk}$, for common gases*[4]

| Gas | $T_0$ K | $k_0$ W/m K | Sutherland temp, $T_{sk}$ K |
|---|---|---|---|
| $H_2$ | 300 | 0.163 | 166 |
| He | 300 | 0.151 | |
| $N_2$ | 300 | 0.024 | 166 |
| CO | 300 | 0.023 | 177 |
| Air | 300 | 0.024 | 194 |
| $O_2$ | 300 | 0.025 | 222 |
| Ar | 300 | 0.016 | 150 |
| $CO_2$ | 300 | 0.015 | 222 |
| Ne | 300 | 0.049 | |

## AXII.4 Van der Waals' equation

Van der Waals' equation of state proposes corrections $a$ and $b$ for molecular attraction and for finite molecular size respectively:

$$\left(p + \frac{a}{v^2}\right)(v - b) = RT \qquad \text{(AXII.1)}$$

Numerical values for $a$ and $b$ for a given gas are derived from the critical pressure, $p_{\text{crit}}$, and temperature, $T_{\text{crit}}$, for that gas:

$$a = \frac{27}{64}\frac{R^2 T_{\text{crit}}{}^2}{p_{\text{crit}}}$$

$$b = \frac{1}{8}\frac{RT_{\text{crit}}}{p_{\text{crit}}}$$

Table AXII.4 lists some critical temperatures and pressures.

Table AXII.4. *Critical temperatures and pressures for selected gases*[2]

| Gas | $T_{\text{crit}}$ K | $p_{\text{crit}}$ MPa |
|---|---|---|
| $H_2$ | 33.3 | 1.30 |
| He | 5.3 | 0.23 |
| Ne | 44.5 | 2.73 |
| $N_2$ | 126.2 | 3.39 |
| CO | 133.0 | 3.50 |
| Air | 133.2 | 3.77 |
| $O_2$ | 154.8 | 5.08 |
| Ar | 151.0 | 4.86 |
| $CO_2$ | 304.2 | 7.39 |

## AXII.5 Properties of selected constructional materials

### *AXII.5.1 Nominal (ambient-temperature) thermal properties of some metals and alloys having application to gas circuit design*

Table AXII.5 lists density, $\rho$, (kg/m$^3$), mean specific heat, $C$ (kJ/kg K), thermal conductivity, $k$ (W/m K) and thermal diffusivity (m$^2$/s) for candidate materials for heat exchanger, regenerator and working space design. Values for $\rho$ and $k$ are respective means between 0 and 100 °C.

Table AXII.5. *Thermal properties of selected metals and alloys between 0 and 100 °C*[6]

| Material | Density kg/m$^3$ | Specific heat kJ/kg K | Thermal conductivity W/m K | Thermal diffusivity $\times 10^6$ m$^2$/s |
|---|---|---|---|---|
| Copper (99.9%) | $8.913 \times 10^3$ | 0.385 | 362 | 105 |
| Brass (70/30) | $8.525 \times 10^3$ | 0.376 | 110 | 35 |
| Commercial bronze (Cu-90, Zn-10) | $8.800 \times 10^3$ | 0.377 | 188 | 56 |
| Aluminium bronze (Al-7, Fe-2.5, Cu-bal) | $7.89 \times 10^3$ | 0.376 | 67 | 23 |
| Aluminium alloy ASTM B211 | $2.79 \times 10^3$ | 0.96 | 171 | 64 |
| Titanium 35A | $4.50 \times 10^3$ | 0.51 | 15 | 7 |
| 80Ni-20Cr | $8.41 \times 10^3$ | 0.43 | 15 | 4 |
| Invar | $8.11 \times 10^3$ | 0.50 | 11 | 3 |
| Nickel 211 ASTM F290) | $8.73 \times 10^3$ | 0.46 | 44 | 11 |
| Monel alloy 400 | $8.84 \times 10^3$ | 0.42 | 22 | 6 |
| Incoloy alloy 800 | $8.01 \times 10^3$ | 0.50 | 12 | 3 |
| Inconel alloy 600 | $8.43 \times 10^3$ | 0.44 | 15 | 4 |
| Nimonic 80 | $8.19 \times 10^3$ | 0.46 | 11 | 3 |
| Steel SAE 1021 | $7.86 \times 10^3$ | 0.45 | 52 | 15 |
| Grey cast iron | $7.20 \times 10^3$ | | 45 | |
| Ni-Resist Type 1 ASTM A436 | $7.3 \times 10^3$ | 0.48 | 40 | 11 |
| Stainless steel (301) | $7.9 \times 10^3$ | 0.50 | 16 | 4 |

### *AXII.5.2 Thermal conductivity of selected stainless steels as a function of temperature*

Fig. AXII.3 is based on data converted from Ref. 7 showing that there may be a ten-fold variation in thermal conductivity, $k$, of certain stainless steels between cryogenic temperatures and the 800 °C or so of recent designs of heater head.

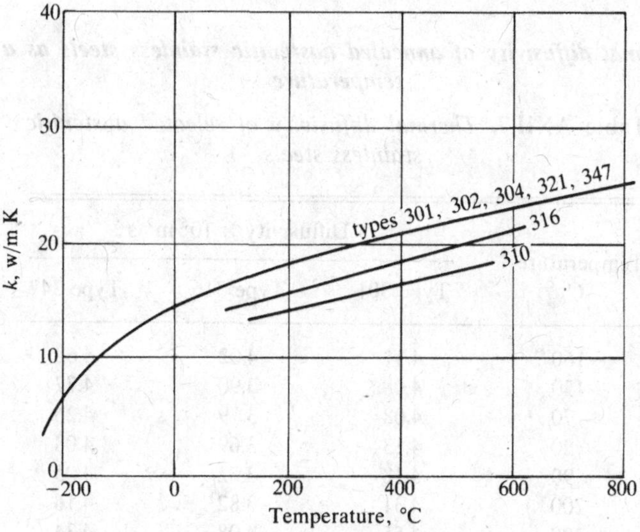


Fig. AXII.3 Thermal conductivity of various stainless steels in function of temperature. (From Ref. 7.)

### *AXII.5.3 Specific heats of annealed austenitic stainless steels as a function of temperature*

Table AXII.6. *Specific heat of selected austenitic stainless steels*[7]

| Temperature °C | Specific heat kJ/kg K | | |
|---|---|---|---|
| | Type 301 | Type 316 | Type 347 |
| −160 | 0.335 | 0.331 | 0.335 |
| −130 | 0.360 | 0.356 | 0.356 |
| −70 | 0.402 | 0.394 | 0.394 |
| 20 | 0.456 | 0.452 | 0.452 |
| 90 | 0.489 | 0.486 | 0.486 |
| 200 | 0.532 | 0.547 | 0.519 |
| 320 | 0.557 | 0.549 | 0.549 |
| 430 | 0.574 | 0.565 | 0.565 |
| 540 | 0.586 | 0.574 | 0.582 |
| 650 | 0.599 | 0.586 | 0.603 |
| 760 | 0.619 | 0.616 | 0.624 |
| 870 | 0.645 | 0.649 | 0.666 |

### *AXII.5.4 Thermal diffusivity of annealed austenitic stainless steels as a function of temperature*

Table AXII.7. *Thermal diffusivity of selected austenitic stainless steels*[7]

| Temperature °C | Diffusivity × $10^6$ $m^2/s$ | | |
|---|---|---|---|
| | Type 301 | Type 316 | Type 347 |
| −160 | 4.11 | 4.02 | 4.62 |
| −130 | 4.08 | 3.90 | 4.77 |
| −70 | 4.08 | 3.79 | 4.26 |
| 20 | 4.13 | 3.69 | 4.08 |
| 90 | 4.18 | 3.72 | 4.05 |
| 200 | 4.31 | 3.82 | 4.16 |
| 320 | 4.55 | 4.08 | 4.34 |
| 430 | 4.78 | 4.31 | 4.55 |
| 540 | 5.10 | 4.47 | 4.78 |
| 650 | 5.24 | 4.85 | 4.96 |
| 760 | 5.39 | 4.93 | 5.14 |
| 870 | 5.48 | 5.01 | 5.16 |

# References

## Chapter 1


1 Stirling R, Improvements for diminishing the consumption of fuel, and in particular an engine capable of being applied to the moving of machinery on a principle entirely new. Patent No. 4081, 1816.
2 Hargreaves C M, *The Philips Stirling Engine*, Elsevier, Amsterdam, 1991.
3 Finkelstein T, Air engines. *The Engineer*, 27 Mar., 492–7, 3 Apr. 522–7, 10 Apr. 568–71, 8 May, 720–3, 1959.
4 Anon, The centenary of the heat regenerator and the Stirling air engine. *The Engineer*, Dec. 1917, 516.
5 Finkelstein T, Regenerative thermal machines. *Battelle Technical Review*, May 1961.
6 Anon, The Philips air engine. *The Engineer*, 12 Dec. 549–50, 19 Dec., 572–4, 1947.
7 de Brey H, Rinia H, and van Weenen F L, Fundamentals for the development of the Philips air engine. *Philips Technical Review*, **9**(4) 1947, 97–124.
8 Schmidt G, Theorie der Lehmann'schen calorischen Maschine. *Z. des Ver. deutscher Ingenieure* **XV**(2) Jan., 1–12, Feb., 98–112, 1871.
9 Finkelstein T, Optimization of the phase angle and volume ratio for Stirling engines. Paper 118C, *Proc. SAE Winter Annual Meeting*, 1960.
10 Walker G, An optimization of the principal design parameters of Stirling cycle machines. *J. Mech. Engng. Sci.* **4**(3) 1962, 226–40.
11 Kirkley D W, Determination of the optimum configuration for a Stirling engine. *J. Mech. Engng. Sci.* **4**(3), 1962, 204–12.
12 Meijer R J, The Philips Stirling engine. Philips Research Laboratories, Eindhoven, May 1967.
13 Feurer B, Degrees of freedom in the layout of Stirling engines. *Proc. 53 Lecture Series*, von Karman Institute, Brussels, 12–16 Feb., 1973.
14 Urieli I A, computer simulation of Stirling cycle machines. PhD dissertation, University of the Witwatersrand, Johannesburg, 1977.
15 Creswick F, A thermal design of Stirling cycle machines. Paper 949C, *SAE Automotive Engineering Congress*, Detroit, Michigan, 11–15 Jan., 1965.
16 Iwabuch M and Kanzaka M, Experimental investigation into heat transfer under the periodically-reversing flow condition in heated tube. Paper C24/82, *I. Mech. E. Conference Stirling engines – progress towards reality*, University of Reading 25–6 March 1982.
17 Kubo M, Matsue J and Terada F, Up-to-date information on the NS30S Stirling

engine. Paper 889352, *Proc. 23 Intersociety Energy Conv. Engineering Conf., Denver, Colorado*, ASME, July 1988.
18 Yamashita I, Tanaka A, Azetsu A, Shinogama E, Endo N, Mizuhara K, Watanabe M, Yamada Y, Tanaka M, Chisaka F and Takahashi S, Fundamental studies of Stirling engine systems and components. *Bull. of Mech. Engng. Lab.* No. 47 Mechanical Engineering Laboratory, Namiki 1-2, Sakura-mua, Ibaraki-ken 305, Japan, 1987.
19 Sales catalogue, Alpha-N Design and Office Services Inc., 1823 Hummingbird Court, West Richland, Washington 99352, USA.
20 Tew R C Jr, Status of several Stirling loss characterization efforts and their significance for Stirling space power development. Paper 889280, *Proc. 23 Intersociety Energy Conv. Engineering Conf., Denver, Colorado*, ASME, July 1988.

## Chapter 2

1 Walker G, *Stirling Engines.* Clarendon, Oxford, 1980.
2 Urieli I, A computer simulation of Stirling cycle machines. PhD dissertation, University of the Witwatersrand, Feb. 1977.
3 Urieli I, A current review of Stirling cycle machine analysis methods. Paper 839113 *Proc. 18 Intersociety Energy Conv. Engineering Conf.* 1983, pp. 702–7.
4 Martini W R, Design manual for Stirling engines. Report submitted under US DoE contract, Grant 3152, Ed. 1 (undated – approx 1977).
5 Chen N C J and Griffin F P, A review of Stirling engine mathematical models. Draft report ORNL/TM, Oak Ridge National Laboratory under US DoE contract W-7405-eng-26.
6 West C D, *Principles and Applications of Stirling Engines.* van Nostrand, New York, 1986.
7 Finkelstein T, Generalized thermodynamic analysis of Stirling engines. Paper 118B, *SAE Winter Annual Meeting* 1960.
8 Finkelstein T, Cyclic processes in closed regenerative gas machines analysed by digital computer simulating a differential analyser. Paper 61-SA-21 *ASME Summer Annual Meeting*, Los Angeles, 11–14 June 1961.
9 Finkelstein T, Simulation of a regenerative reciprocating machine on an analogue computer. Paper 949F, *SAE Int. Congr. on Automotive Engineering*, Detroit, 11–15 Jan. 1965.
10 Finkelstein T, Walker G, and Joshi T, Design optimization of Stirling cycle cryogenic cooling engines by digital simulation. Paper K4, *Cryogenic Engineering Conf.* June 1970.
11 Kirkley D W, A thermodynamic analysis of the Stirling cycle and a comparison with experiment. Paper 949B, *SAE Int. Congr. on Automotive Engineering*, Detroit, 11–15 Jan. 1965.
12 Schock A, Stirling engine nodal analysis program. *J. Energy*, Nov–Dec 1978 **2**(6), 354–62.
13 Heames T J, Uherka D J, Zabel J C and Daley J G, Stirling engine thermodynamic analysis – a user's guide to SEAM1. Report ANL-82-59 Argonne National Laboratory, Argonne Illinois, Sept. 1982.
14 Tew R, Jeffries K and Miao D A, A Stirling engine computer model for performance calculations. US DoE/NASA report TM-78884, July 1978.
15 Organ A J, Gas dynamics of Stirling cycle machines. Paper C25/82, *I Mech E Conf. Stirling Engines, Progress Towards Reality*, Univ. of Reading, 25–6 Mar. 1982.

16 Organ A J, Gas dynamics of the temperature-determined Stirling cycle. *J. Mech. Engng. Sci.*, **23**(4), 1982, 207–15.
17 Sirett E J, The gas dynamics of the Stirling machine. First-year report on post-graduate study, Engineering Dept., Univ. of Cambridge, 29 Apr. 1981.
18 Rispoli F, The $\lambda$-scheme method applied to Stirling engines. Paper 859325, *Proc. 20. Intersociety Energy Conv. Engineering Conf.*, pp. 3.301–3.306, 1985.
19 Lowry H V and Haydn H A, *Advanced Mathematics for Technical Students.* Longmans, London Ed. 2 1957.
20 Shapiro A H, *Dynamics and Thermodynamics of Compressible Fluid Flow.* Ronald, New York, 1954, **II**.
21 Fokker H and van Eckelen J A M, Typical phenomena of the Stirling cycle as encountered in a numerical approach. Paper 789113, *Proc. Intersociety Energy Conv. Engineering Conf.* 1978, pp. 1746–52.
22 Berchowitz D M, Urieli I and Rallis C J, A numerical model for Stirling cycle machines. *Proc. Israel Joint Congress on Gas Turbines*, Haifa, 9–11 July 1979.
23 Ames W, *Non-linear Partial Differential Equations in Engineering.* Academic Press, New York 1965.
24 Rinia H and du Pré F K, Air engines. *Philips Tech. Rev.* **8**(5), 1946, 129–60.
25 Finkelstein T, Optimization of the phase angle and volume ratio of Stirling engines. Paper 118C, *SAE Winter Annual Meeting* 1960.
26 Organ A J, Dimensional analysis of pumping losses in a Stirling cycle machine. Paper 829280 *Proc. 17 Intersociety Energy Conv. Engineering Conf.*, 1982, pp. 1694–8.
27 Organ A J, Thermodynamic design of Stirling cycle machines. *Proc. Inst. Mech. Engrs.* Pt. C **201**(C2) 1987, 107–16.
28 Hutchinson R A, Assessment of regenerator heat exchanger experimentation and measurement and analysis requirements. Draft internal report Battelle Pacific Northwest Laboratories, 1 Dec. 1986.
29 Annand W J D, Heat transfer in the cylinders of reciprocating internal combustion engines. *Proc. Inst. Mech. Engrs.*, 1963, **177**(36), 973–7.
30 Organ A J, Perturbation analysis of the Stirling cycle. Paper 829281 *Proc. 17 Intersociety Energy Conv. Engineering Conf.* 1982, pp. 1699–704.
31 Rix D H, An enquiry into gas process asymmetry in Stirling cycle machines. PhD dissertation, University of Cambridge, 1984.
32 Rios P A and Smith J L Jr, An analytical and experimental evaluation of the pressure losses in the Stirling cycle. *Trans ASME Jnl. Engng for Power*, Apr. 1970, 182–7.
33 Rauch J S, Harmonic analysis of Stirling cycle performance – a comparison with test data. Paper 849173, *Proc. 19 Intersociety Energy Conv. Engineering Conf.*, Aug. 1984, pp. 2015–19.
34 Chen N C J, Griffin F P and West C D, Linear harmonic analysis of Stirling machine and second law analysis of four important losses. Paper 849141, *Proc. 19 Intersociety Energy Conv. Engineering Conf.*, Aug. 1984, pp. 2015–20.
35 Feurer B, Degrees of freedom in the layout of Stirling engines. *Proc. Stirling Engine Short Course*, von Karman Institute for Fluid Dynamics, Brussels, 12–16 Feb., 1973.
36 Qvale E B and Smith J L, A mathematical model for steady operation of Stirling type engines. *Trans. ASME J. Engng. for Power*, Jan. 1968, 45–50.
37 Bejan A, *Entropy Generation through Heat and Fluid Flow.* Wiley Interscience, New York, 1982.

38 Gedeon D R, The optimization of Stirling cycle machines. Paper 78913, *Proc. Intersociety Energy Conv. Engineering Conf.* 1978, pp. 1784–90.
39 Organ A J, Fluid particle trajectories in Stirling cycle machines. *J. Mech. Engng. Sci.*, 1978, **20**(1), 1–10.
40 Schmidt G, Theorie der Lehmann'schen calorischen Maschine. *Z. des Ver. deutscher Ing.* 1871, **15**(2), 1–12 (Jan), 98–112 (Feb).
41 White F M, *Fluid Mechanics.* McGraw-Hill, Tokyo, 1979.
42 Finkelstein T, Specific performance of Stirling engines. *Proc. 3rd. Conf. on Perf. of High Temperature Systems* (Ed. G S Bahn) Pasadena, Calif., Dec. 1964, **II**, Gordon and Breach, New York, 1969.
43 Walker G, Operations cycle of the Stirling engine with particular reference to the function of the regenerator. *J. Mech. Engng. Sci.*, 1961, **3**(4), 394–408.
44 Creswick F A, Thermal design of Stirling cycle machines. Paper 949C, *SAE Int. Cong. on Automotive Engineering,* Detroit, Michigan, 11–15 Jan. 1965.
45 deBrey H, Rinia H and van Weenen F L, Fundamentals for the development of the Philips air engine. *Philips Tech. Rev.* 1947, **9**(4), 1947, 97–124.
46 Cuttil W E, Malik M J, and Decker V L, *Investigation of a 3 kW Stirling cycle solar power system.* Vol. IX, *Power system control analysis.* Tech. Rep. WADD.TR-61-122 by Flight Accessories Lab., Aeronautical Systems Div., Air Force Systems Command, Wright Patterson Air Base, Ohio, Mar. 1962.
47 Urieli I and Berchowitz D M, *Stirling Cycle Engine Analysis.* Adam Hilger, Bristol 1984.
48 Finkelstein T, Analysis of practical reversible thermodynamic cycles. Paper 64HT37, *AIChE/ASME Conf. on Heat Transfer*, Cleveland, Ohio, 9–12 Aug. 1964.
49 Organ A J, The concept of 'critical length ratio' in heat exchangers for Stirling cycle machines. Paper 759151 *Proc. 10th Intersociety Energy Conv. Engineering Conf.*, Newark, Delaware 18–22 Aug. 1975, pp. 1012–19.
50 Gedeon D R, Numerical advection errors in Stirling cycle nodal analysis. Paper 849101 *Proc. 19 Intersociety Energy Conv. Engineering Conf.* 1984 pp. 1898–904.
51 Schock A, Nodal analysis of Stirling cycle devices. Paper 789191, *Proc. 13 Intersociety Energy Conv. Engineering Conf.*, San Diego, Aug. 1978, pp. 1771–9.
52 Gedeon D R, *A Globally-Implicit Stirling Cycle Simulation.* Gedeon Associates Athens, Ohio 45701, 1986.
53 Leach C E and Fryer B C, A 7.3 kW(e) radio-isotope energized under-sea Stirling engine. *Proc. Intersociety Energy Conv. Engineering Conf.*, 1968, pp. 330–44.
54 Hartree D R, Some practical methods of using characteristics in the calculation of non-steady compressible flow. Report AECU-2713, Technical Information Service, US Atomic Energy Commission, Oak Ridge, Tennessee, Sept. 1953.
55 Larsson V H, Characteristic dynamic energy equations for Stirling cycle analysis. Paper 819798, *Proc. 16 Intersociety Energy Conv. Engineering Conf.*, 1981, pp. 1942–7.
56 Taylor D R, The method of characteristics applied to Stirling engines. Paper 849177, *Proc. 19 Intersociety Energy Conv. Engineering Conf.* 1984 pp. 2037–42.
57 Anon FORTRAN mini-manual Mk. 11. Numerical Algorithms Group (NAG) Ltd., Oxford, Nov. 1983.
58 Walker G, An optimization of the principal design parameters of Stirling cycle machines. *J. Mech. Engng. Sci.*, 1962, **4**(3) 226–40.
59 Kirkley D W, Determination of the optimum configuration for a Stirling Engine. *J. Mech. Engng. Sci.*, 1962, **4**(3) 204–12.

60 Organ A J, Analytical optimization of the specific performance of a Stirling cycle machine. Research report, Instituto Technológico de Aeronautica (ITA), São José do Campos, 12200 SP Brazil 1971.
61 Takase C N, Otimização do desempenho especifico de motores Stirling. MSc. dissertation, ITA, São José dos Campos, 12200 SP Brazil 1972.
62 Walker G and Khan I, Theoretical performance of Stirling cycle engines. Paper 949A, *SAE Int. Congr. on Automotive Engineering*, Detroit, Michigan, 11–15 Jan. 1965.
63 Heames T J and Daley J G, SEAMOPT – Stirling engine optimization code. Paper 849103 *Proc. 19 Intersociety Energy Conv. Engineering Conf.* 1984, pp. 1905–12.

## Chapter 3

1 Rogers G F C and Mayhew Y R, *Engineering Thermodynamics, Work and Heat Transfer*. Longmans, London 1957.
2 Reader G T and Hooper C, *Stirling Engines*. Spon, London 1983.
3 Walker G, *Stirling Engines*. Oxford University Press, Oxford, 1980.
4 Holtz R E and Uherka K L, Reliability study of Stirling engines for solar dish/heat engine systems. Paper 879414, *Proc. 22 Intersociety Energy Conv. Engineering Conf.*, Aug. 1987, pp. 1885–1890.
5 Bejan A, *Advanced Engineering Thermodynamics*. Wiley, New York, 1988.

## Chapter 4

1 Creswick F A, Thermal design of Stirling cycle machines Paper 949C, *Proc. SAE Automotive Engineering Cong*, Detroit, Michigan, Jan. 1965.
2 Gedeon D, Scaling rules for Stirling engines. Paper 819796 *Proc. 16 Intersociety Energy Conv. Engineering Conf.*, Atlanta 1981, pp. 1929–35.
3 Organ A J, Dimensional analysis of pumping losses in a Stirling cycle machine. Paper 829280, *Proc. 17 Intersociety Energy Conv.. Engineering Conf.* 1982, pp. 1694–8.
4 Organ A J, Thermodynamic design of Stirling cycle machines. *Proc. Inst. Mech. Engrs*, 1987, **201**(C2), 107–16.
5 Organ A J, Thermodynamic analysis of the Stirling cycle machine – a review of the literature. *Proc. Inst. Mech. Engrs.* 1987, **201**(C6), 381–402.
6 Rix D H, An enquiry into gas process asymmetry in Stirling cycle machines. PhD diss., University of Cambridge 1984.
7 White F M, *Fluid Mechanics*. McGraw-Hill Kogakusha, Tokyo 1979 International student edition.
8 Shapiro A H, *Dynamics and Thermodynamics of Compressible Fluid Flow* **II**. Ronald, New York, 1954.
9 Kays W M and London A L, *Compact Heat Exchangers*. McGraw-Hill, New York, 1964.
10 Kays W M, *Convection Heat Transfer*. McGraw-Hill, New York, 1966.
11 Miyabe H, Takahashi S and Hamaguchi K, An approach to the design of Stirling engine regenerator matrix using packs of wire gauzes. Paper 829306, *Proc. 17 Intersociety Energy Conv. Engineering Conf.*, 1982, pp. 1839–44.

## Chapter 5


1 Lee K, Krepchin I P and Toscano W M, Thermodynamic description of an adiabatic, second-order analysis for Stirling engines. *Proc. 16 Intersociety Energy Conv. Engineering Conf.*, 1981, pp. 1919–24.

2 Lee K and Smith J L, Performance loss due to transient heat transfer in the cylinders of Stirling engines. *Proc. 15 Intersociety Energy Conv. Engineering Conf.*, 1980, pp. 1706–9.

3 Rix D H. An inquiry into gas process asymmetry in Stirling cycle machines. PhD. diss., University of Cambridge 1984.

4 Kays W M and London A L, *Compact Heat Exchangers.* McGraw-Hill, New York, 1964.

5 Squiers J C, Fluid flow resistance models for wire weaves. *Filtration and Separation*, Sept/Oct. 1984, 327–330.

6 Miyabe H, Takahashi S and Hamaguchi K, An approach to the design of Stirling engine regenerator matrix using packs of wire gauzes. Paper 829306, *Proc. 17 Intersociety Energy Conv. Engineering Conf.*, 1982, pp. 1839–44.

7 Pinker R A and Herbert M J, Pressure loss associated with compressible flow through square-mesh wire gauzes. *J. Mech. Engng. Sci.*, 1962, **9**(1), 11–23.

8 Masha B A, Beavers G S and Sparrow E M, Experiments on the resistance law for non-Darcy compressible gas flows in porous media. *J. Fluids Eng. Trans. ASME*, Dec 1974, 353–7.

9 Green L and Duwez P, Fluid flow through porous metals. *J. Appl. Mech.* 1951, **18**, 39–45.

10 White F M, *Fluid Mechanics.* McGraw-Hill Kogakusha, Tokyo 1979.

11 Su C C, An enquiry into the mechanism of pressure drop in the regenerator of the Stirling cycle machine. PhD. diss., University of Cambridge, 1986.

12 Forschheimer P, Wasserbewegung durch Boden. *Z. des Vereins Deutscher Ing.* 1901, **45**, 1782–8.


### *Further reading*

*Incompressible regime*


Chen N and Griffin F P, Effects of pressure drop correlations on Stirling engine predicted performance. Paper 839114, *Proc. 18 Intersociety Energy Conv. Engineering Conf.*, 1983, pp. 708–13.

Finegold J G and Sterrett R H, Stirling engine regenerators literature review. Report (numbered 5030-230) prepared for NASA Lewis Research Center by the Jet Propulsion Laboratory, 15 July 1978.

Grootenhuis P, A correlation of the resistance to air flow of wire gauzes. *Proc. Inst. Mech. Engrs.* 1954, **168**(34), 837–43.

Jonas P, The changes produced in an air stream by a wire gauze. *The Engineers' Digest*, May 1957, **18** No. 5, 191–3. (From *Czechoslovak J. Phys.* 1957, **7**(2), 202–12.)

Kim J C, An analytical and experimental study of flow friction characteristics for periodically reversing flow. Paper 73-WA/FE-13, ASME Nov. 11–15, 1973.

Wieghardt K E G, On the resistance of screens. *Aeronautical Quarterly* **IV** Feb. 1953, 186–92.

Rios P A and Smith J L Jr, An analytical and experimental evaluation of the pressure-drop losses in the Stirling cycle. *J. Engng. for Power, Trans ASME*, Apr. 1970, 182–9.

Robinson M and Franklin H, The pressure drop of a fibrous filter at reduced ambient pressure. *Aerosol Science* **3**, Pergamon Press, 1972. 413–27.

Thomson T A, The flow through a gauze screen in a straight duct. Aero. Tech. Note. 7801, Dept. Aeronautical Engineering, Univ. of Sidney, Jan. 1976.

de Vahl Davis G, The flow of air through wire screens. *Proc. 1st. Australasian Conf. on Hydraulics and Fluid Mechanics*, Univ. of Western Australia, 6–11 Dec. 1962, Pergamon 1964, pp. 191–212.

Walker G, Pressure drop across the regenerator of a Stirling cycle machine. *The Engineer*, 27 Dec. 1963, 1063–5.

*Compressible (or incompressible plus compressible) regime*

Adler A A, Variation with Mach number of static and total pressures through various screens. CB No. L5F28, National Advisory Committee for Aeronautics, Feb. 1946.

Anon, Pressure drop in ducts across round-wire gauzes normal to the flow. Engineering Sciences Data Unit No. 72009 ESDU, June 1972 (12 pp.).

Benson R S and Baruah P C, Non-steady flow through a gauze in a duct. *J. Mech. Engng. Sci.* 1965, **7**(4), 449–59.

Organ A J, An enquiry into the mechanism of pressure drop in the regenerator of the Stirling cycle machine. Paper 849006, *Proc. 19 Intersociety Energy Conv. Engineering Conf.*, San Francisco. Calif., 1984, pp. 1776–81.

## Chapter 6

1 Finkelstein T, Optimization of the phase angle and volume ratio for Stirling engines. Paper 118C *Proc SAE Winter Annual Meeting* 1960.

2 Urieli I and Berchowitz D M, *Stirling Cycle Engine Analysis.* Adam Hilger, Bristol, 1984.

## Chapter 7

1 Organ A J, Thermodynamic design of Stirling cycle machines. *Proc. Inst. Mech. Engrs.* 1987, **201**(C2), 107–16.

2 Bejan A, *Entropy Generation through Heat and Fluid Flow.* Wiley, New York, 1982.

3 Kays W M and London A L, *Compact Heat Exchangers.* McGraw-Hill, New York, 1964.

4 Su C-C, An enquiry into the mechanism of pressure drop in the regenerator of the Stirling cycle machine. PhD dissertation, University of Cambridge, 1986.

5 Malone J F J, A new prime mover. *J. Roy. Soc. Arts* 1930, **79**, 679–709.

6 Malone J F J, A new prime mover. *The Engineer* 1931, 97–101.

7 Urieli I and Berchowitz D M, *Stirling Cycle Engine Analysis.* Adam Hilger, Bristol, 1984.

## Chapter 8

1 Jakob M, *Heat Transfer* (in 2 volumes). Wiley, New York, 1949 and 1957.

2 Kim J C and Qvale E B, Analytical and experimental studies of compact wire-screen heat exchangers. *Adv. Cryogenic Eng.* 1971, **16**, 302–11.

3 Miyabe H, Takahashi S and Hamaguchi K, An approach to the design of Stirling engine regenerator matrix using packs of wire gauzes. Paper 829306, *Proc. 17 Intersociety Energy Conv. Engineering Conf.*, 1982, pp. 1839–44.
4 Schneider P J, *Conduction Heat Transfer*. Addison-Wesley, Reading, Mass., 1955.
5 Arpaci V S, *Conduction Heat Transfer*. Addison-Wesley, Reading, Mass., 1966.

## Chapter 9

1 Organ A J, Mechanical efficiency of the Stirling cycle machine with rhombic drive. *Proc. Intersociety Energy Conv. Engineering Conf.*, 1978, pp. 1841–52.
2 Creswick F A, Thermal design of Stirling cycle machines. Paper 949c, *Proc. SAE Automotive Engineering Cong.*, Detroit, Michigan, 11–15 Jan., 1965.

## Chapter 10

1 Finkelstein T, Generalized thermodynamic analysis of Stirling engines. Paper 118B, *SAE winter annual meeting* 1960.
2 Annand W J D, Heat transfer in the cylinders of reciprocating internal combustion engines. *Proc. Inst. Mech. Engrs.* 1963, **177**(36), 973–90.
3 Finkelstein T, Cyclic processes in closed, regenerative gas machines analyzed by a digital computer simulating a differential analyzer. Paper 61-SA-21 *J. Engng. for Industry*, Trans ASME 1961.

## Chapter 11

1 Hartree D R, Some practical methods of using Characteristics in the calculation of non-steady compressible flow. Report AECU-2713, Technical Information Service, US Atomic Energy Commission, Oak Ridge, Tennessee, Sept. 1953.
2 Rispoli F, The λ-scheme method applied to Stirling engines. Paper 859325, *Proc. 20 Intersociety Energy Conv. Engineering Conf.*, 1985, pp. 3.301–6.
3 Organ A J, The concept of 'critical length ratio' in the heat exchangers for Stirling cycle machines. Paper 759151, *Proc. 10 Intersociety Energy Conv. Engineering Conf.*, Am. Inst. Electrical and Electronics Engineers, Newark, Delaware, 18–22 Aug., 1975, pp. 1012–19.
4 Shapiro A H, *The Dynamics and Thermodynamics of Compressible Fluid Flow* (in 2 volumes) Ronald, New York, 1954.
5 Gedeon D R, Numerical advection errors in Stirling cycle nodal analysis. Paper 849101, *Proc. Intersociety Energy Conv. Engineering Conf.*, 1984, pp. 1898–1904.
6 Fokker H and van Eckelen J, Typical phenomena of the Stirling cycle as encountered in a numerical approach. Paper 789113 *Proc. Intersociety Energy Conv. Engineering Conf.*, 1978, pp. 1746–52.
7 Rix D H, An enquiry into gas process asymmetry in Stirling cycle machines. PhD dissertation, University of Cambridge 1984.
8 Organ A J, *An Introduction to Unsteady Gas Dynamics*. Cambridge University Engineering Department, 1990.
9 Organ A J, Gas dynamics of the temperature-determined Stirling cycle. *J. Mech. Engng. Sci.*, 1982, **23**(4), 207–16.
10 Sirett E J, The gas dynamics of the Stirling machine. First year report on research, University of Cambridge, April 1981.

## Chapter 12


1 Dowling A P and Ffowcs-Williams J E, *Sound and Sources of Sound.* Ellis Horwood, Chichester 1983.
2 Rix D H, An enquiry into gas process asymmetry in Stirling cycle machines. PhD diss. University of Cambridge 1984.
3 Lighthill J, *Mathematical Biofluid Dynamics.* Soc. for Industrial and Applied Mathematics, Philadelphia, 1975.
4 Finkelstein T, Optimization of the phase angle and volume ratio for Stirling engines. Paper 118C, *Proc. SAE Winter Annual Meeting*, 1960.


## Chapter 13


1 Bauwens L, Consistency, stability, convergence of Stirling engine models. Paper 900533, *Proc. 25 Intersociety Energy Conv. Engineering Conf.*, Am. Inst. Ch. Engrs., 1990, **5**, pp. 352–8.
2 Berchowitz D, Urieli I and Rallis C J, A numerical model for Stirling cycle machines. Paper 79-GT-Isr-16, *Proc. 1979 Israel Joint Gas Turbine Congress*, Haifa, ASME, July 9–11, 1979.
3 Tew R C, Correspondence with the author concerning the 3rd. Stirling Thermodynamic Loss Workshop, NASA-Lewis, Oct. 16–17, 1990.
4 Organ A J, The concept of 'critical length ratio' in heat exchangers for Stirling cycle machines. *Proc. 10 Intersociety Energy Conv. Engineering Conf.*, Newark, Delaware, 18–22 Aug., 1975, pp. 1012–19.
5 Roache P J, *Computational Fluid Dynamics.* Hermosa, Albuquerque, 1976.
6 Shapiro A H, *The Dynamics and Thermodynamics of Compressible Fluid Flow.* Ronald, New York, 1954.
7 Rix D H, A thermodynamic design simulation for Stirling cycle machines using a Lagrangian formulation. *Proc. Inst. Mech. Engrs.* 1988, **202**(C2), 85–93.
8 Crowley B K, PUFL, an 'almost-Lagrangian' gas dynamic calculation for pipe flows with mass entrainment. *J. Computational Phys.* 1967, **2**, 61–86.
9 Tew R, Jefferies K and Miao D, A Stirling engine computer model for performance calculations. Report No. DOE/NASA/1011-78/24 prepared for the US Department of Energy, NASA Lewis Research Centre, July 1978.


## Chapter 14


1 Organ A J, Mechanical efficiency of the rhombic-drive Stirling cycle machine. *Proc. Intersociety Energy Conv. Engineering Conf.*, 1978, pp. 1841–52.
2 Finkelstein T, Optimization of the phase angle and volume ratio for Stirling engines. Paper 118C, *Proc. SAE Winter Annual Meeting*, 1960.
3 Walker G, An optimization of the principal design parameters of Stirling cycle machines. *J. Mech. Eng. Sci.*, 1962, **4**(3), 226–40.
4 Kirkley D W, Determination of the optimum configuration of a Stirling engine. *J. Mech. Eng. Sci.*, 1962, **4**(3), 204–12.
5 Organ A J, Analytical optimization of the specific performance of a Stirling cycle machine. Research Report, Instituto Tecnológico de Aeronáutica, São José dos Campos, 12200 São Paulo, Brazil, 1973.
6 Takase C N, Otimização do desempenho específico de motores Stirling. MSc dissertation, Instituto Technológico de Aeronautica, São José dos Campos, 12200 São Paulo, Brazil, 1972.

7 Schmidt G, Theorie der Lehmann'schen calorischen Maschine. *Zeit. des. Vereins deutscher Ing.* **XV** Heft 2, Jan 1871, 1–12, Feb. 1981, 98–112.
8 Walker G and Khan I, Theoretical performance of Stirling cycle engines. Paper 949A, *SAE Int. Aut. Eng. Cong.*, Detroit, Michigan, 11–15 Jan., 1965.
9 Walker G and Agbi B, Optimum design configuration for Stirling engine with two-phase, two-component working fluids. Paper 73-WA/DGP-1, *Winter Annual Meeting of Diesel and Gas Engine Power Division of the ASME*, Detroit, Michigan, 11–15 Nov., 1973.
10 Wilde D J, *Optimum Seeking Methods.* Prentice-Hall, Englewood Cliffs, New Jersey 1964.
11 Finkelstein T, Generalized thermodynamic analysis of Stirling engines. Paper 118B, *SAE Winter Annual Meeting*, 1960.

**Chapter 15**

1 Finkelstein T, Air Engines. *The Engineer* 27 Mar., 492–7; 3 Apr., 522–7; 10 Apr., 568–71; 8 May, 720–3, 1959.
2 Sier R, *A History of Hot-Air and Caloric Engines.* Argus, London, 1987.
3 Ross A, Stirling cycle engines. Solar Engines, Phoenix, 1977.
4 Anon, Owner's manual No. 1, Stirling cycle engine, Solar Engines, Phoenix (undated).
5 Organ A J, Two-dimensional thermodynamic and flow analysis of the simple hot-air engine. *Proc. Inst. Mech. Engrs.* 1988 **202**(C1), 31–8.
6 Bird R B, Stewart W E and Lightfoot E N, *Transport Phenomena.* Wiley, New York, 1960.
7 Organ A J, The concept of 'critical length ratio' in heat exchangers for Stirling cycle machines. Paper 759151, *Proc. 10 Intersociety Energy Conv. Engineering Conf.*, Newark, Delaware, 1975, pp. 1012–19.
8 Kays W M and London A L, *Compact Heat Exchangers.* McGraw-Hill, New York, 1964, Ed. 2.

**Chapter 16**

1 Finkelstein T, Specific performance of Stirling engines. Paper 21 in Vol. II of *Proc. 3rd. Conf. on Performance of High Temperature Systems*, Pasadena, California, Dec. 1964. Ed. G S Bahn, Gordon and Breach, New York, 1969.
2 Wurm J, Kinast J A, Roose T R and Staats W R, *Stirling and Vuilleumier Heat Pumps – Design and Applications.* McGraw-Hill, New York, 1990.
3 Tew R C, Summary of 3rd. Stirling Thermodynamic Loss Workshop, NASA Lewis Research Centre, Oct. 16–17, 1990.
4 Daneshyar H, Private communication with the author.
5 Organ A J, Vasciaveo L and Long P J G, Back-to-back test for determining the pumping losses in a Stirling cycle machine. *Proc. 17 Intersociety Energy Conv. Engineering Conf.*, 1982, pp. 1694–8.
6 Martini W R, Design manual for Stirling engines. Report prepared for US Dept. of Energy under Grant No. 3152 administered by NASA Lewis Research Center (undated).
7 Organ A J, *Stirling Engine Thermodynamic Design – Without the Computer. mRT*, Cambridge, 1991.
8 Hamming R W, *Numerical Methods for Scientists and Engineers.* McGraw-Hill, New York, 1962.
9 Roache P J *Computational Fluid Dynamics.* Hermosa, Albuquerque, 1972.

### Appendix I

1 Finkelstein T, Air engines. *The Engineer*, 27 Mar., 492–7, 3 Apr., 522–7, 10 Apr., 568–71, 8 May, 720–3, 1959.

### Appendix II

1 Finkelstein T, Optimization of phase angle and volume ratio for Stirling engines. Paper 118c, *Proc. SAE Winter Annual Meeting* 1960.
2 Gradshteyn I S and Ryzhik I M, *Table of Integrals, Series and Products.* Academic Press, New York Ed. 4, 1965.

### Appendix V

1 Urieli I and Berchowitz D M, *Stirling Cycle Engine Analysis.* Adam Hilger, Bristol 1984.
2 Rix D H, An enquiry into gas process asymmetry in Stirling cycle machines. PhD dissertation, University of Cambridge, 1984.

### Appendix VI

1 Urieli I and Berchowitz D M, *Stirling Cycle Engine Analysis.* Adam Hilger, Bristol, 1984.
2 Rix D H, An enquiry into gas process asymmetry in Stirling cycle machines. PhD dissertation, University of Cambridge, 1984.
3 Cairelli J E, Thieme L G and Walter R J, Initial test results with a single-cylinder rhombic-drive Stirling engine. Report NASA TM-780919 by NASA Lewis Research Centre under Interagency agreement No. EC-77-A-31-1040, July 1978.
4 Thieme L G, Low-power baseline test results for the GPU-3 Stirling engine. Report NASA TM-79103 by NASA Lewis Research Center under Interagency agreement No. EC-77-A-31-1040, April 1979.

### Appendix VII

1 Bejan A, *Entropy Generation Through Heat and Fluid Flow.* Wiley. New York, 1982.
2 Organ A J, Thermodynamic design of Stirling cycle machines. *Proc Inst. Mech. Engrs.* **201**(C2) 1987, 107–16.

### Appendix VIII

1 Finkelstein T, Generalized thermodynamic analysis of Stirling engines. Paper 118B *SAE Winter Annual Meeting* 1960.
2 Organ A J, Thermodynamic design of Stirling cycle machines. *Proc. Inst. Mech. Engrs.* **201**(C2), 1987, 107–16.
3 Halpern V and Shtrikman S, Idealized adiabatic operation of regenerative cycle refrigerators. Copy acquired without details of source – possibly from *Advances in Cryogenic Engineering* – and numbered paper A1/2–26 undated, 1–6.
4 Rule T T and Qvale E B, Steady-state operation of the idealized VM refrigerator. *Advances in Cryogenic Engineering* **14** 1968, 343–52.

## Appendix IX


1 Organ A J, Thermodynamic design of Stirling cycle machines. *Proc. Inst. Mech. Engrs.* 1987, **201**(C2), 107–16.


## Appendix XI


1 Shapiro A H, *The Thermodynamics and Gas Dynamics of Compressible Fluid flow.* **II** Ronald, New York 1954.
2 Abbott M B, *An Introduction to the Method of Characteristics.* Thames and Hudson London, 1966.
3 Organ A J, *An Introduction to Unsteady Gas Dynamics.* Cambridge University Engineering Department 1990.


## *Further sources on the Method of Characteristics*


de Haller P, The application of a graphic method to some dynamic problems in gases. *Sulzer Technical Review* Pt. 1, 1945, 6–24.

Hartree D R, Some practical methods of using Characteristics in the calculation of non-steady compressible flow. US Atomic Energy Commission Report AECU-2713, Sept. 1953. Supplied by Technical Information Service, Oak Ridge, Tennessee, USA.

Issa R I and Spalding D B, Unsteady one-dimensional compressible frictional flow with heat transfer. *J. Mech. Eng. Sci.*, 1972, **14**(6), 365–9.

Jenny E, Berechnungen und Modellversuche ueber Druckwellen grosser Amplituden in Auspuff-Leitung. Doctoral diss.ETH 161, Technical High School, Zurich, 1949.

Jenny E, Unidimensional transient flow with consideration of friction, heat transfer and change of section. *Brown Boverie Review* 1950, **37** Pt. 11, 447–61.

Kahane A and Lees L, Unsteady, one-dimensional flows with heat addition or entropy gradients. *J Aeronautical Sci*, Nov. 1948, 665–70.

Lakshminarayanan P A, A finite difference scheme for unsteady pipe flows. *Int. J. Mech. Sci.* 1979, **21**, 557–66.

Lister M, The numerical solution of hyperbolic differential equations by the Method of Characteristics. Chap. 15 of *Mathematical Methods for Digital Computer* Edited by Ralston and Wilf, Wiley, London 1960.

MacLaren J F T, Tramschek A B Sanjines A and Pastrana O F, A comparison of numerical solutions of unsteady flow using solutions of the unsteady flow equations applied to reciprocating compressor systems. *J. Mech. Eng. Sci.*, 1975, **17**(5), 271–9.

Payri F, Corboran J M and Boada F, Modification to the method of characteristics for the analysis of the gas exchange process in internal combustion engines. *Proc. Inst. Mech. Engrs.* 1986, **200**(D4), 259–66.

Spalding D B A, procedure for calculating the unsteady, one-dimensional flow of a compressible fluid, with allowance for the effects of heat transfer and friction. Report UF/TN/D/2, Imperial College of Science and Technology, Exhibition Rd., London, Nov. 1969.

**Appendix XII**

1 Haywood R W, *Thermodynamic Tables in SI (metric) Units.* Cambridge University Press 1976 (Ed. 2).
2 van Wyken G J and Sonntag R E, *Fundamentals of Classical Thermodynamics – SI Version.* Wiley, New York, 1978.
3 Anderton P and Bigg P H, *Changing to the Metric System.* National Physical Laboratory, HMSO London 1972.
4 White F M, *Viscous Fluid Flow.* McGraw-Hill, New York, 1974.
5 Kaye G W C and Laby T H, *Tables of Physical and Chemical Constants.* Longman, London 1973 (1978 reprint), Ed. 14.
6 Anon, *Properties of Some Metals and Alloys.* International Nickel Co., New York, 1968, Ed 3.
7 Anon, *Mechanical and Physical Properties of the Austenitic Chromium-Nickel Stainless Steels at Elevated Temperatures* International Nickel Ltd., Millbank, London 1966.

# Index of proper names

# Subject index